AHMANSON · MURPHY
FINE ARTS IMPRINT

THE AHMANSON FOUNDATION
has endowed this imprint
to honor the memory of
FRANKLIN D. MURPHY
who for half a century
served arts and letters,
beauty and learning, in
equal measure by shaping
with a brilliant devotion
those institutions upon
which they rely.

The publisher gratefully acknowledges the generous support and contributions of the following benefactors:

The Art Endowment Fund of the University of California Press Foundation, which was established by a major gift from the Ahmanson Foundation

Duke University

The Graham Foundation for Advanced Studies in the Fine Arts

IMAGING DISASTER

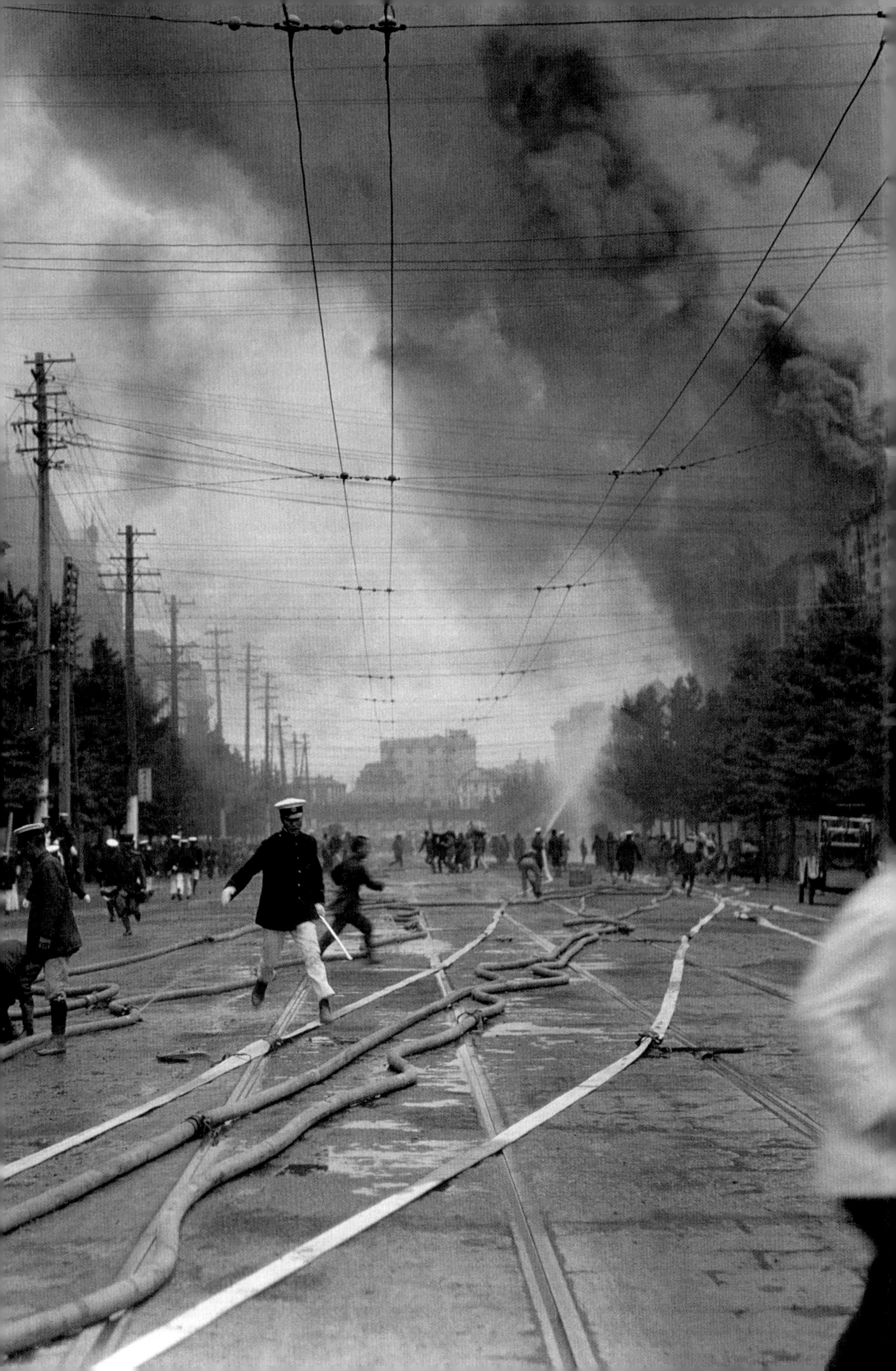

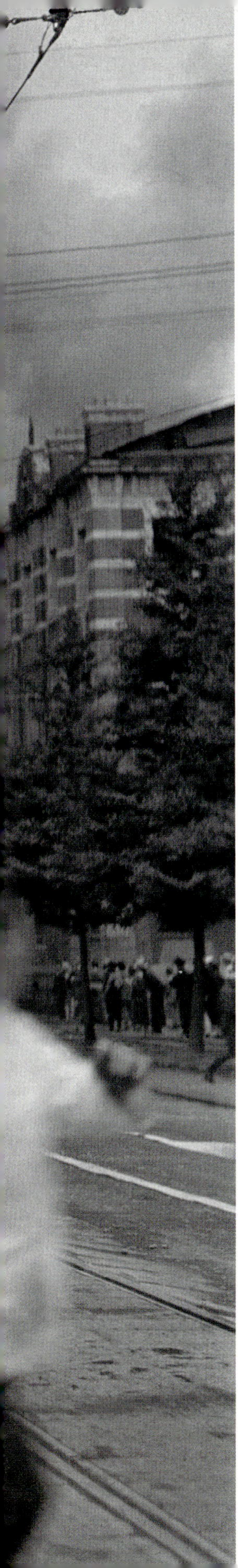

IMAGING DISASTER

Tokyo and the Visual Culture of Japan's Great Earthquake of 1923

GENNIFER WEISENFELD

UNIVERSITY OF CALIFORNIA PRESS BERKELEY LOS ANGELES LONDON

University of California Press, one of the most distinguished university presses in the United States, enriches lives around the world by advancing scholarship in the humanities, social sciences, and natural sciences. Its activities are supported by the UC Press Foundation and by philanthropic contributions from individuals and institutions. For more information, visit www.ucpress.edu.

University of California Press
Berkeley and Los Angeles, California

University of California Press, Ltd.
London, England

Title page and chapter opening details are as follows: title page, *Marunouchi* (fig. 3.28, p. 120); chap. 1, *The Kashima Deity Napping* (fig. 1.6, detail, p. 28); chap. 2, Map of Tokyo-Area Fires and Victim Gathering Sites (fig. 2.3, p. 42); chap. 3, *Large Whirlwind of Fire on Nakanomachi Street* (fig. 3.15, p. 99); chap. 4, Kawasaki Shōko, *Former Clothing Depot and the Yasuda Mansion after the Great Earthquake* (fig. 4.10, p. 141); chap. 5, Women washing in muddy groundwater (fig. 5.27, p. 197); chap. 6, *Quick Guide to the Land Boundary Readjustment* (fig. 6.15, p. 237); chap. 7, *The Earthquake Memorial Hall, Honjo* (fig. 7.15, p. 275); chap. 8, *Incendiary Bomb Threat* (fig. 8.3, p. 299). Full credit information appears in the captions for the listed images.

Library of Congress Cataloging-in-Publication Data

Weisenfeld, Gennifer S. (Gennifer Stacy), 1966–
Imaging disaster : Tokyo and the visual culture of Japan's Great Earthquake of 1923 / Gennifer Weisenfeld. — 1st [edition].
pages cm. — (Asia: local studies/global themes ; 22)
Includes bibliographical references and index.
ISBN 978-0-520-27195-1 (cloth : alk. paper)
1. Arts, Japanese—20th century—Themes, motives. 2. Kanto Earthquake, Japan, 1923. 3. Earthquakes in art. 4. Arts and society—Japan—History—20th century. 5. Earthquakes—Social aspects—Japan. I. Title.
NX584.A1W45 2012
704.9'49952032—dc23 2012003245

Manufactured in China

21 20 19 18 17 16 15 14 13 12
10 9 8 7 6 5 4 3 2 1

The paper used in this publication meets the minimum requirements of ANSI/NISO Z39.48–1992 (R 2002) *(Permanence of Paper).*

For Derek and Luci

CONTENTS

PREFACE

Anyone who has lived or worked in areas along the Pacific Rim "ring of fire" knows that earthquakes and related natural disasters are always a possibility. But having just completed years of research on the devastating 7.9-magnitude earthquake that flattened Tokyo and surrounding areas in 1923, perhaps I am more fixated on this topic than most. I vividly remember a friendly yet chilling chat with an elderly neighbor in the downtown Kachidoki district of Chūō ward in Tokyo, where I lived in a charming, if slightly rundown, old wooden *nagaya* (row house) on one of the small back streets of this working-class neighborhood. I had been told that my neighbor lived through the 1923 quake as a young child, and while hesitant to ask about his experiences, I eagerly sought his advice about evacuation plans and sites if another temblor hit. His answer was sobering: "Don't even bother. This place will burn to the ground." No matter how much we plan, nature offers no guarantees.

The 1923 quake is unparalleled in Japan's modern history in its combined seismic scale and extent of damage; in fact, it was one of the world's worst natural disasters of the early twentieth century. In terms of loss of life and material damage, it is still Japan's worst national disaster. But when the unprecedented 9.0-magnitude earthquake of 11 March 2011 hit off the coast of Tōhoku, the northeastern region of Japan, followed by hundreds of strong aftershocks, people throughout the world were reminded of modern humanity's fragility in the face of nature. Not only did the quake produce another massive tsunami that washed away the coastline, but also like its 1923 predecessor, it generated a torrent of images that saturated the visual field. Given that I was already deep into writing this book and thinking about the ways in which visual images mediate our vision and understanding of disasters, this new event offered fascinating echoes of the earlier earthquake. I had only to view a sampling of the work of contemporary photojournalists to see the recurrence of iconic imagery from previous catastrophes: spectacular scenes of destruction, panoramic and aerial vistas of devastated landscapes, grieving and contemplative figures in a sea of ruins, wounded objects of daily life, desperate message boards to locate missing family and friends, and rows of anonymous bodies (now sanitized by more stringent press taboos).[1] The compelling visual lexicon of disaster is enduring and undeniable.

How do these thousands of images of ruined landscapes and graphic representations of death and destruction act on us to evoke emotions or to provoke action? The producers of the images, consciously or unconsciously, stake a claim to a particular meaning for disasters, and they communicate this meaning to us through both form and content. My study of the visual responses in 1923 suggests that disasters provide opportunities for individuals, societies, and nations to articulate their diverse agendas for urban, social, political, or moral reform. After a disaster, everyone is a critic, even when the criticism is couched in the collectivist rhetoric of national resilience. The striking similarity between criticisms in 1923 and 2011—with accusations of political ineptitude and malfeasance and comments about the moral turpitude of modern society or the pitfalls of technological hubris—reveal powerful parallels between the past and the present.

Despite claims of a shared national trauma when disaster strikes, inequitable social structures and economic development policies expose certain members of the populace to greater risk and suffering than others. Recent newspaper editorials, such as Oguma Eiji's trenchant commentary in the *Asahi shinbun* on the Tōhoku region's particular structural vulnerabilities, have underscored this point. Oguma, a historical sociologist at Keio University in Tokyo who has written about the postwar origins of Japan's myth of a homogeneous nation, points out that the poorer rural regions of the northeast initially provided cheap labor and food for the country's burgeoning metropolises to the west and then, when faced with international competition and depopulation, had little choice but to allow hazardous nuclear energy development, which left them particularly vulnerable to the 2011 disaster.[2] Like the unfortunate victims of the 1923 temblor, who were disproportionately from the poorer neighborhoods of Tokyo's "low city" (*shitamachi*), Tōhoku residents are suffering more now because of historical class and regional inequities.

In 1923, the fundamental power of the Japanese nation-state defined the quake experience as a national triumph against adversity that produced the gleaming new Tokyo metropolis. The more recent disaster is still unfolding as this book goes to press, so we do not yet know what the master narrative of the event will be, particularly with the added horrific element of nuclear radiation contamination, and whether the Japanese state is strong enough to direct it. But it is already clear that in the national and international mass media, the nuclear disaster story has superseded the natural one, reinforcing scholars' common assertion about disasters: whatever their triggering event—natural or human—they all have aspects that are man-made.

In the past century, the twenty-four-hour news cycle has intensified the commercial demand for news around the globe.[3] New visual technologies have provided ever-more-sophisticated and spectacular images of death and destruction, whether via time-delay digital topographical maps that record the fireworks-like barrage of shocks and aftershocks or via scroll-over digital coastline maps that allow the viewer to recapitulate at will the devastation of the tsunami along the full northeastern edge of the archipelago.[4] Perhaps one of the greatest changes in visual culture is the ability of individuals to contribute to the dialogue through handheld personal devices, ubiquitous mobile phones, and miniature video recorders, which both supplement and bypass mass media through websites such as You Tube and a host of new digital social media. Like their roles in the so-called Twitter or Facebook revolutions in the Middle East and North Africa, these new media have forged innovative direct pathways to viewers around the world. One of the most compelling examples of this is the simply shot video posted on You Tube on 26 March 2011 by the despondent mayor of Minami Soma city in Fukushima prefecture, Sakurai Katsunobu, in which he sits in his office and explains the desperate plight of local residents and implores volunteers to bring critical supplies to the afflicted area.[5] Expressing a sense of abandonment by the central government and the energy company that runs the Fukushima Daiichi nuclear power plant, Tepco (Tokyo Electric Power Company), Sakurai's forlorn, solitary plea is one of the most powerfully enduring images of this catastrophe. Three months into the disaster, the video had received over four hundred thousand hits and had ignited a mass volunteer movement to help people in the region, transforming the unassuming Sakurai into a kind of folk hero. Still, such individual contributions are easily co-opted by the mainstream commercial media and have certainly not supplanted images produced by professional photographers and cameramen. Rather, they are often an evocative supplement that reinforces the mass media framework, providing a personal touch and a local particularity to a sadly universal story line.

After the Tōhoku quake, Japanese-born journalist and photographer Iwabu Takaaki, a twenty-year resident of the United States who lives in Raleigh, North Carolina (just down the road from where I teach at Duke University), published a personal narrative in the *News & Observer* online describing his experiences of the earthquake and explaining why he felt compelled to return to Japan to witness the aftermath himself. Despite being an expatriate and a global citizen, Iwabu pondered his attachment to Japan and the power of images to elicit his sense of national belonging: "Personally, I have always tried to avoid looking at the world through a lens of country or ethnicity—my little stand against shallow generalizations and nationalism. People are people, wherever we are born. Then why did my heart race as I was glued to the TV screen on March 11 and the following week? If I were a citizen of the planet, as I always try to view myself, why did I feel more pain watching Japan suffer than when I heard the news about disasters in other countries? Despite my long absence, I must admit that I still feel a strong sense of belonging in Japan. These are my roots." Impelled by the media images of the destruction in his homeland, Iwabu flew to Japan ten days after the tsunami hit and later reported his impressions: "Pundits say a natural disaster exposes the weakness of the society it strikes. They are right. What we have seen in Japan so far is a painful lack of leadership and organization." Yet, inspired by the strength of character, humility, and compassion of his countrymen, Iwabu lamented Japan's slow decline over the past twenty years, fearing that this catastrophe might be "the country's final shove toward the sidelines." Reiterating the long-held Japanese belief in the socially transformative possibilities of disasters, Iwabu imagined the moment as one of possible renewal for the Japanese people. "I would like to believe, however, that this is also giving Japan a chance to begin something new. . . . A tragedy of this magnitude must change a society. My hope is that it creates a sense of urgency and fosters a new sense of country among all. Somehow we have to go forward."[6]

With similar great admiration for the resilient spirit of my many dear friends and colleagues in Japan, and a deep awareness of the trials and opportunities posed by such tragic events, I offer my thoughts in the following pages about how we might think about disaster through images.

ACKNOWLEDGMENTS

The completion of this project would not have been possible without the support of many colleagues, friends, and family. The debts incurred have been great indeed, and simply mentioning them here hardly seems sufficient to express the deep and abiding gratitude I feel. I hope that the people who have supported me throughout this process already know how much I appreciate them. This book is a testament to their individual and collective contributions. But more importantly, they have all added to my life immeasurably, making my work a rich and rewarding experience.

As with any long-term project, this book has gone through numerous iterations, and I have benefited tremendously from constructive feedback along the way. A trusted reader and dear friend for many years, Amy Ogata provided invaluable comments on the first draft that helped crystallize my thinking. Jordan Sand and Sandy Isenstadt generously read later drafts, providing crucial insights that helped me structure the argument. I also owe thanks to the three anonymous reviewers for the University of California Press for their helpful comments. I hope they will all see the fruits of their labors here. Any shortfalls, of course, are my responsibility.

Excavating research materials has at times been an arduous process requiring the expertise of many professionals. No one has been more helpful than Kris Troost, head of International Area Studies and Japanese Studies librarian at Duke University. At all hours of the day or night, she responded to my barrage of queries, locating rare and seemingly unattainable sources. I stand in awe of her knowledge and resourcefulness, not to mention her kindness. Others have also provided invaluable library support for my project, including Yoriko Dixon, Maureen Donovan, Lee Sorenson, Carson Holloway, Reiko Yoshimura, Sachie Noguchi, and Hyejeong Choi. They are joined by a crack team of research assistants led by the incomparable Yukiko Kato, who has helped me in too many ways to mention. Suffice it to say that she has been a lifeline. In locating sources and navigating the complex process of image permissions, I have also relied on a resourceful group of young Japanese scholars: Kita Takaomi, Yamamoto Masako, Kōgo Eriko, Kito Sakiko, and Maezaki Shinya.

While in-laws usually offer primarily emotional support, my father-in-law, Stanleigh Jones, has been a constant language resource, helping me work through difficult translations and generally offering cheerful linguistic guidance. Anyone who knows the frustrations of translating Japanese into English can understand the enormity of his contribution. I thank him from the bottom of my heart for his good humor and willingness to help. Any remaining infelicities of translation or language, however, are mine.

Over the years, I have been extremely fortunate to have wonderful mentors in Japan who have deeply influenced me as a scholar. I cannot imagine what I would have done without the sage advice and critical assistance of Tano Yasunori, Mizusawa Tsutomu, Omuka Toshiharu, Kaneko Ryūichi, and Takizawa Kyōji. Their scholarship and kindness are inspiring. In addition, many friends and colleagues in Japan have supported my work. To them I express my deep appreciation: Shimura Shoko, Kawata Akihisa, Mashino Keiko, Ōtani Shōgo, Niita Tarō, Eriguchi Yūko, Shimizu Isao, Itakura Fumiaki, Tochigi Akira, Fujiki Hideaki, Hosokawa Shuhei, Takano Hiroyasu, Kogure Nobuo, Katō Hiroko, and Richard Catalano. I also extend heartfelt thanks to the Shibusawa Eiichi Memorial Foundation for its continued support of my work, particularly Kimura Masato, Koide Izumi, and Inoue Jun. In the United States and Europe, a number of people have kindly shared works in their collections or at museums: Helen Kulas, Eric van den Ing, Sawako Chang, Frederic Sharf, and the Rabitz family.

I also have the great fortune of being surrounded by an incredible group of colleagues in art history, history, and Asian studies in North Carolina's Research Triangle who have helped me in innumerable ways: Hans Van Miegroet, Sheila Dillon, Kristine Stiles, Patricia Leighten, Mark Antliff, Stan Abe, Neil McWilliam, Tom Rankin, Rick Powell, Annabel Wharton, Sara Galletti, Mark Olson, John Taormina, Bill Broom, Marion Monson, Ed Balleisen, Simon Partner, Sumathi Ramaswamy, Leo Ching, Ralph Litzinger, Hae-Young Kim, Richard Jaffe, David Ambaras, Morgan Pitelka, John Mertz, Aimee Kwon, Dominic Sachsenmaier, Claire Conceison, and Mark Driscoll. I would like to extend a special thanks to Jack Edinger, imaging specialist in the Department of Art, Art History, and Visual Studies at Duke, whose expert skills in photography, scanning, and Photoshop helped provide high-quality images for the book's illustrations. Friends, students, teachers, and colleagues around the world have also provided great support: Yoshiaki Shimizu, Jonathan Reynolds, Ken Oshima, Barbara Sato, Ming Tiampo, Peter Lurie, Mycah Braxton, Jenny Anger, David Odo, Cary Karacas, Alex Bates, Greg Smits, Janet Borland, Dora Ching, Gregory Levine, Melanie Trede, Kari Shepherdson-Scott, Ignacio Adriasola, Nancy Hamilton, Kimberly Slentz-Kesler, Ann Gleason, Catherine Szuch, and Holly Rogers. I am particularly grateful to my friend Andy Watsky, who always gives great advice and never fails to make me laugh. I am grateful to Kärin Wigen, Jeff Wasserstrom, and Hue-Tam Ho Tai, editors of the University of California Press series Asia: Local Studies/Global Themes, for including my book. I also thank everyone at the press who has helped shepherd this book to publication: Deborah Kirshman, Kari Dahlgren, Kim Hogeland, Eric Schmidt, Jacqueline Volin, and Adrienne Harris.

The writing of this book was supported by a Duke Endowment fellowship at the National Humanities Center, a scholarly utopia to which I hope to return soon. I also thank everyone at the Sainsbury Institute for the Study of Japanese Arts and Cultures for their ongoing support. And I thankfully acknowledge the important financial assistance of the Triangle Center for Japanese Studies and the Japan Foundation.

The Graham Foundation for Advanced Studies in the Fine Arts provided critical publication support for this book.

Duke University has unfailingly supported my research over the years. I am deeply appreciative of the extraordinary financial contributions that I have received from different parts of the university that have allowed me to include the many rich images in this book. I gratefully acknowledge the support of the Office of the Provost, Duke University and Provost Peter Lange, Trinity College of Arts and Sciences and Dean Srinivas Aravamunden, the Globalization and the Artist Project of the Duke University Center for International Studies and Executive Director Rob Sikorski, the Visual Studies Initiative at Duke University, and the Duke University Asian/Pacific Studies Institute and Director Simon Partner and Associate Director Yan Li.

Nothing is more important to me than my family. My loving and supportive parents, Susan and Jeffrey, have always been my bedrock. I dedicate this book to my wonderful, unflappable husband, Derek, and my dearest daughter, Luci, who have heard more than their share about earthquakes and disasters yet continue to fill my life with joy.

BURNT AREAS OF THE WORLD GREAT FIRES

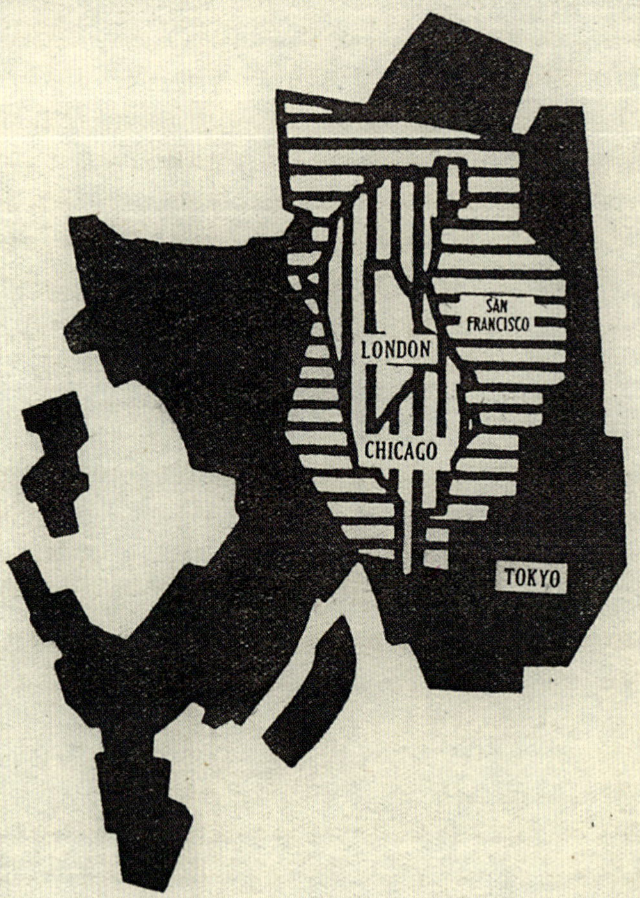

Cities	Year	Burnt Area	Loss of Wealth
		(sq.m.)	yen
London	1666 Sept. 2–6	1,768,603	107,300,000
Chicago	1871 Oct. 8–9	8,595,880	330,000,000
San Francisco	1906 April 18–21	12,165,344	750,000,000
Tokyo	1923 Sept. 1–3	33,477,836	5,506,386,034

I-1. Burnt Areas of the World['s] Great Fires, map and chart. Tokyo Municipal Office, *The Reconstruction of Tokyo* (Tokyo: Tokyo Municipal Office, 1933), 97.

INTRODUCTION

The Indian Ocean tsunami, Hurricane Katrina, the Sichuan and Haiti earthquakes—the experience of disaster is both universal and particular. Most of us understand these horrific events through a complex matrix of media, most of them visual, that attempt to record and ascribe meaning to destruction, chaos, and tragedy. Images mediate our experiences. How the visual functions in relation to disaster, however, requires close critical examination. The visual culture of disaster can produce an exploitative aesthetics of horror and spectacle that transforms viewers into unwitting voyeurs. At the same time, people can use images to reclaim disaster sites for social and political purposes. Images can impart emotional value to an event by humanizing, heroizing, and monumentalizing it. They can also erase or justify aspects of it. Images act on us, or as W. J. T. Mitchell has written, pictures "want" things from us and for themselves.[1]

Focusing on one landmark catastrophic event in the history of an emerging modern nation—the Great Kantō Earthquake that devastated Japan's imperial capital and its surrounding areas in 1923—I explore how different media produce modes of seeing, understanding, and, eventually, remembering. Major disasters focus the social energy of diverse media on one critical event. Such moments create a density and intensity of visuality that make them ideal to explore the influence of imaging practices.[2] But media do not exist in isolation. As I demonstrate through the excavation of tropes and motifs that circulated widely, visual culture constitutes an intermedia dialogue to which each medium brings a distinct yet critical inflection. In 1923, viewers encountered photography and film, and visualizations of scientific information, alongside hand-rendered prints, paintings, sculpture, and cartoons. They experienced a public visual sphere in which objective and subjective values were inextricably fused. Images played a central role in constructing the earthquake as a national event, rather than simply a local tragedy, that demanded solidarity from all Japanese people. Mass media and new scientific technologies globalized Japan's tragedy, inviting empathy from the world. Images of disaster erected a framework of visual authority that legitimated the state's responses to this national catastrophe. Yet the disaster was not the sole preserve of the state. A range of public and private entities, from the imperial household to the leftist avant-garde, used the earthquake to advocate their own visions for the future.

Although proceeding from natural causes, the 1923 earthquake became a metaphor for the relentless destruction of tradition by modernity. For many Japanese, the earthquake recapitulated and accelerated the ruptures and dislocations of modernization, which had been subjects of intense debate long before the temblor hit. The disaster not only crystallized these debates but also demonstrated the fragility of modern society: the greater the technological achievement, the more spectacular its destruction. The shock of the quake became the shock of the modern.

Disaster images do not emerge ex nihilo, however; they are part of a cumulative history of visual production. Images draw from local genealogies and pictorial traditions that include secular and sacred representations. As time goes on, they become imbricated in a global network of image making. My interest in this process, and in the images, spaces, and discourses that it produces, has led me to write this book.

Similarly, disaster is not static. It is an unfolding temporal landscape within which visual production must be historicized. In exploring diverse reactions to the Kantō earthquake over time, I identify the multiplicity and changing nature of this traumatic experience for the Japanese. Only by analyzing contending visual responses within disaster communities and how they are codified into collective memory to form a national narrative can we ultimately understand how major events like the Great Kantō Earthquake become history.[3]

1 September 1923, 11:58 A.M.

"At noon today a great earthquake struck, followed by great fires which have transformed almost all of the city into a sea of flame. Casualties are in the tens of thousands, transportation is still impossible, and there is no food or water."[4] So read the alarming telegraph sent to Osaka by the Kanagawa prefecture police chief at 9:01 P.M. on 1 September 1923, relayed by wireless from the steamship *Korea-maru* moored in

Yokohama Harbor. This was the first news of the disaster to reach outside the Kantō region. Earlier that day, at exactly 11:58 A.M., a massive earthquake registering 7.9 magnitude hit the Kantō region surrounding Tokyo, crushing or incinerating more than one hundred thousand people. Soon dubbed "the Great Kantō Earthquake" (Kantō Daishinsai), it wreaked unprecedented damage and razed nearly 44 percent of the land area of metropolitan Tokyo, the capital of the Japanese empire.[5] The quake similarly decimated the major commercial port city of Yokohama in nearby Kanagawa prefecture along with large portions of five other surrounding prefectures: Saitama, Shizuoka, Yamanashi, Chiba, and Ibaraki. Caused by slippage along the fault that transects Sagami Bay from northwest to southeast near Oshima Island, the earthquake produced a massive tsunami, reaching over forty feet high along the coastal cities ringing Tokyo and Sagami Bay.[6] The total monetary damage was estimated ten years later at 5,506,386,034 yen (roughly $1.65 billion in 1933 dollars and $26 billion in 2011 dollars), a sum close to a third of Japan's gross national product in 1930.[7]

No one knows exactly how many people perished in the earthquake and its subsequent conflagrations, for statistical figures are wildly divergent, and official reports surely underrepresented minority casualties. Some totals of the dead and missing top one hundred forty thousand. The death and destruction were clearly not evenly distributed throughout the capital. Areas in the easternmost part of the city, historically known as the "low city" (*shitamachi*), were hit much harder than the residential areas of the "high city" (*yamanote*), which were built on solid bedrock hills. The low city, largely built on land reclaimed from Tokyo Bay during the Tokugawa or Edo period (1603–1868), when Edo (modern-day Tokyo) was the capital of the shogunate, consisted of unconsolidated alluvial sediments and landfill that quickly liquefied with the seismic vibrations. The unrestricted industrial development in these poorer wards, which placed highly flammable material adjacent to residential areas, worsened destruction. Of the fifteen wards in the city, six from the low city were almost entirely destroyed: Nihonbashi, Asakusa, Honjo, Kanda, Kyōbashi, and Fukagawa. In addition, the post-quake fire damage was greatly exacerbated by a typhoon that whipped up tremendous firestorms and, combined with the heat from the conflagration, produced intense updrafts and high-velocity winds. The fires burned for three days.

No one in Japan, even those who were physically unharmed, was untouched by the 1923 calamity. A vivid outpouring of grief, sympathy, curiosity, nostalgia, anger, and Good Samaritanism flowed from the Japanese community. People registered these emotions in a vast body of visual and literary work that testifies to the potency of this traumatic event in the collective imagination.[8]

Although disastrous consequences of natural events can seem random, scholarly studies have demonstrated that preexisting social inequities often translate into disparities of suffering and victimization. In other words, we are not all equal in the face of catastrophe, despite rhetorical claims to the contrary. Poor people and the disenfranchised suffer more than other parts of the population. But while the experiences of disaster are diverse, nations have a deep-seated investment in expressing the collective nature of trauma to mobilize social action and to downplay dissension. This is certainly clear in the visualization of the Kantō earthquake. The response to the earthquake was both history in the making and the silencing of that history through active suppression of visible social fractures along the fault lines of class, ethnicity, and politics.

Modes of Imaging, Modes of Seeing

Visual representations of disaster markedly increased in Japan's early modern period from the early seventeenth century onward because of an expansion of the public sphere and innovations in woodblock printing. This visual production accelerated further with the advent of lithography, photography, and film, and because of the transnational/international nature of these new media, it also became connected to the rest of the world. While scholars have studied the remarkable surge in visual production in the first half of the nineteenth century, they have paid less attention to the period after the Ansei-era Edo earthquake of 1855.[9] Assessments of disasters in this later period consistently fail to integrate visual media substantively into historical analysis and therefore devote little attention to how modern disaster is communicated.[10] By analyzing the operation of these various media and their imaging practices, I excavate how the public visual lexicon of disaster was formed during and after the 1923 event. This visual lexicon played a critical mediating role in the production of the historical narrative of the quake and the solidification of its legacy. For seven years following the Kantō earthquake, when the city of Tokyo officially celebrated the "completion" of its reconstruction effort, thousands of images, objects, and spaces were produced to represent, reenact, and remember the experience of the disaster. The quake triggered cultural responses that ran from voyeuristic and macabre to sublime, spectacular to sacred. Mournful commemoration coexisted with emancipatory euphoria, and national expressions of solidarity could not suppress racist vigilantism and sociopolitical critique. The tenor of the work is both somberly respectful and comically irreverent. It is by both well-known and anonymous Japanese artists, photographers, architects, and designers who worked in all types of media: public and private, singular and reproducible. In their images, the high and the low commingle promiscuously; fine artists drew unabashedly from the popular press, and photojournalists informed documentary images with the artistic conventions of the romanticized ruin and the classical nude. Collectively, their work transformed a tragic event into a historic landmark, contributing to the production of a world culture of public disasters.

This study considers the central role of mass media, photojournalism, and documentary photography in the communication of "visible evidence" of the Kantō earthquake to a national and international audience. It is no exaggeration to say that images of the earthquake reached every Japanese citizen, as well as wide audiences in Japan's expanding colonial empire, other parts of Asia, and across Europe and the Americas. The thousands of registered newspapers (dailies that also published special-issue pictorials) and a growing number of popular magazines had wide distribution among the Japanese public, reaching not only the better-educated upper and middle classes but also the increasing working class. Almost half the households in Japan subscribed to a daily paper by 1924.[11] Some well-respected news presses, such as the *Ōsaka mainichi shinbun,* reached nearly seven hundred thousand readers by the time of the quake. The new respectability of the modern press stemmed from its close connections with prominent politicians who had backgrounds in journalism and were known to support the democratic cause and universal manhood suffrage.[12] Documentary press photographs had many afterlives as "soft news" in the profitable postcard business and in the myriad souvenir photo collections issued for years after the quake. These visual mementos of the event, particularly press- and government-issued commemorative volumes illustrating the disaster and the nation's reconstruction project, are found in collections across the globe as keepsakes of the historical moment and as official symbols of gratitude for international aid. They merge the documentary and the souvenir.

Of all the modes of representing 1923, photography was most valued for its purported truth-telling abilities. The mass media capitalized on the public's faith in the notion that photography is objective. Despite their documentary value, however, disaster images, above all photographs, cannot simply be read as transparent representations, nor are they sinister simulacra attempting to stand in for an elusive notion of "reality." Rather, visual production constitutes a component of the historical event itself, embodying specific social relations and interests.

Photographs of the 1923 disaster, while seeming to offer the immediacy of the quake experience, also demonstrate its irretrievability. As purportedly documentary records of historical events, photographic images express temporality in contrasting ways. They express immediacy and ephemerality as well as longevity and immortality, simultaneously capturing fleeting events at the moment they occur and preserving them for posterity. Recognizing their temporal dimension, we must ask what disaster photographs are trying to do. They record, inform, express, evoke, titillate, mobilize, preserve, and remember. They function in a variety of ways both at the time of production and in different contexts of presentation and consumption. Photographs carry traces of the past, present, and future. By underscoring the present temporal position of the viewer, who embodies the potential to act in the future, photographs communicate a sense of imminence while reinforcing a sense of distance. This paradox brings to mind the ruminations of Walter Benjamin on the photograph as historical index, in which he notes that it is in this moment of the "interruption and explosion of historical presentation that we engage the conditions of 'authentic' historical understanding, an understanding which, offering us the truth of time, tells us that history is something to which we can never be present."[13]

Photographic images of the Kantō earthquake quickly went from being indexical to iconic through their repeated reproduction and circulation in the mass media, inflecting visual responses in other media such as prints and painting. More than just instruments that mediated between an event and a viewership, visual imagery acquired a degree of cultural agency when tropes crystallized and, appearing to take on a life of their own, eclipsed competing portrayals of the disaster. In this regard, the photographic eye was profoundly influential in shaping general perceptions of the event, having an impact beyond its mimetic capabilities. Optical technology could serve as a weapon of control, whether through the commanding strategic vision of aerial photography used in disaster management or the dominating gaze of the camera that imposed order in the face of violent chaos. These visions also communicated the immense and historic scale of the catastrophe that accorded Japan the dubious distinction of premier status in the global theater of earthquake nations. The photographic eye legitimated its own visual authority.

Photographs are not merely representations; they are material objects that form what theorist Ariella Azoulay has termed a "civil contract." The civil contract of photography articulates how images form relationships—even if they are unequal power relationships—between people, between individuals and larger entities such as the government and the media, and, as in the Kantō quake, between nations. As Azoulay has argued, "Photography is an apparatus of power that cannot be reduced to any of its components: a camera, a photographer, a photographed, a photographed environment, object, person, or spectator. 'Photography' is a term that designates an ensemble of diverse actions that contain the production, distribution, exchange, and consumption of the photographic image. Each of these actions involved in the photographic event makes use of direct and indirect force."[14] In the civil contract, the viewer and even those represented in a photograph are as much active agents in the act of photography as the person pointing the camera. Therefore, in my analysis, rather than render the represented and viewing subjects as objectified and powerless within the photographic encounter, I try to elucidate the complex forms of ethical—and unethical—practice embodied in images of disaster.

Just as the cultural responses to 1923 were manifold, so were the modes of spectatorship they elicited. While the notion of spectatorship entails the act of beholding, I understand it as an embodied and multisensory act. I consider the significant synesthetic, somatic experience of disaster, particularly the associations among the visual, auditory, olfactory, and haptic senses. Moreover, I understand spectators as active participants in the formation of visual culture. As media circulate, the act of taking them in, of consuming them, inflects their meaning—and not necessarily in a neutral or harmless way. Through the act of seeing, spectators of the images of the Kantō earthquake could commit ocular violence. They could also assert representations that countervailed the state's predominant visual message of collective tragedy, national resilience, and communal altruism. Viewers often confounded this official narrative by absorbing images in unanticipated ways, even consuming them for touristic and erotic pleasure.

Photography and documentary film present what Elizabeth Cowie has termed "the spectacle of actuality," producing viewing pleasure simply through their purported representation of the real.[15] Film expanded the photographic eye during the 1923 quake by offering the enhanced technological thrills of the moving image. Over forty documentary films, short and feature length (mostly silent and a few later talkies), were originally made of the event. However, only ten are known to have survived. Some were made during or immediately after the disaster, and others were produced several years later when the process of rebuilding was already well under way. The latter films often combined earlier, original film footage with scenes of rebuilding, presenting a "before and after" narrative of national recovery. A diverse array of organizations—prominently including government entities such as the Ministry of Education (Monbushō), the national [Earthquake] Reconstruction Agency (Fukkō Kyoku), and Tokyo municipal government—produced the films. By the mid-1920s, film was already an important mass medium in Japan, with over 153.7 million tickets sold per year. The Home Ministry (Naimushō) was solely responsible for regulation of the film industry, which, unlike other mass media, did not enjoy constitutional protection of free speech.[16] The state privileged the cinematic apparatus for narrating the nation, and film was closely identified with the representation of national projects, which included narratives of national trauma.[17] Along with exhibits of photographs and other visual media, film screenings were presented throughout the country as part of fund-raising campaigns for rebuilding the national capital. Still, at this time, films were largely seen as entertainment, so many of the cameramen who shot the original footage of the quake had worked for commercial entertainment film studios such as Nikkatsu and Shōchiku. The films display an impressive range of amusing cinematic techniques, from moving ground and aerial shots taken from trains, automobiles, and airplanes to modernist-style artistic montages with dramatic fades and cuts, demonstrating the oft-quoted axiom of pioneering documentarian John Grierson that "documentary is the creative treatment of actuality."[18]

Many visual forms that circulated in the mass media constructed an edifice of knowledge and expertise through their purportedly objective presentation of visible evidence. Scientific visualization technologies, specifically, made strong claims for their universal validity. Maps charting everything from regional and national damage to casualties by neighborhood, or the dramatic seismograms recording the violent terrestrial movement caused by the quake, sought to persuade audiences of their utility and visual authority, particularly in the comparative world culture of disaster. At the same time, each medium represented not only the disaster of 1923 but also itself: while claiming visual authority, each revealed its partiality and limitations.

Maps are visual propositions that give form to social and political interests.[19] Despite their seeming neutrality, they are anything but neutral or objective. A wealth of scholarship on maps has shown how they naturalize knowledge and, by extension, power by producing the bounded entities they purport to show. By authorizing knowledge, they exert force and impel social action such as amelioration or reconstruction in the case of disaster. Their temporal and spatial frameworks make implicit connections by harkening back to earlier earthquakes in the region and well-known historical narratives of tragedy. They also build up a scientific record of seismic activity of potential utility in remediating the current situation or coping with future quakes. Mapping binds together discrete entities. In 1923, it consolidated distinct townships and prefectures into the amalgamated Kantō region, a major geographic area whose national importance implicitly exceeded that of any one municipality—a strategic move that historian Narita Ryūichi has also identified in the renaming of the quake as the Great Kantō Earthquake. Maps establish visual authority over the territory they depict. Yet they are by nature selective and partial. And as Denis Wood has rightly noted, "The map's authority is the social manifestation of its factuality," which "is established by the social assent given to the proposition it embodies."[20]

On the international stage, maps' macroview can delineate the geobody of the nation in crisis—or in triumph. On one simple schematic map in the Tokyo municipal government's 1933 commemorative reconstruction volume (see fig. I.1), the excessive blackness of the burnt Tokyo landmass tragically subsumes the hatched areas of its disaster counterparts: London, Chicago, and San Francisco. With this comparison, the nation's resurrection of the capital after such a massive catastrophe is that much more impressive. The authority of the map is buttressed by the statistical chart below it that quantifies the damage in terms of the burned area of the city and total cost. The map's visual language provides a definitive historical interpretation of the event: Japan's recovery from the Great Kantō Earthquake, by far the largest urban disaster in the history of the world, dwarfing all previous great fires, was an unparalleled national triumph against adversity that bears witness to the unique resilience of the Japanese people and the superlative leadership of their government.

The mass circulation of seismograms similarly took visible evidence to the people. The development of seismology in the late nineteenth century established the seismograph as an international instrument in a worldwide network and thus enabled comparability. In being recorded by seismographs around the world, Japan's earthquake became a global event. The potential predictive ability of seismograms and detailed maps of seismic activity throughout the affected region offered a scientific salve for the intense collective anxiety, implying that the knowledge embodied in these visualizations might mitigate the risks of natural disaster, although they had no such ability.

Such types of information design, which Edward Tufte has called "the cognitive paradise of explanation," do not merely lay out data but rather convey specific meanings through the visual disposition of the data.[21] So when state bureaucrats responsible for assessing the earthquake damage, such as Home Minister Gotō Shinpei (1857–1929) and his cadre of technocratic urban planners, presented their plans for postquake reconstruction in a vast array of statistical diagrams—pie charts, bar charts, graphs, and maps with overlaid statistical information—they sought to present them as the result of a technical, depersonalized process of rationalization, in order to simplify the complexity of the city into something legible and manageable. The bright, cheery colors of these diagrams conveyed a strong, authoritative sense of positivistic (scientific) optimism. Their goal was to facilitate state interventions into the social life of the city to make it a more productive, hygienic, and, ostensibly, safer metropolis on a par with other capitals around the world.

In contrast, artists resoundingly asserted the subjective, emotional nature of the disaster experience. They produced a plethora of prints, sketches, paintings (in ink and oil), and cartoons, not to mention a number of significant sculptural and architectural monuments, in response to the 1923 earthquake. By the 1920s, the modern Japanese art world was a large and diverse community working in a range of traditional, Western-style (*yōga*), and neotraditional hybrid forms. Each medium conveyed different visual qualities and associations that amplified the cultural resonance of the event. Woodblock prints, for example, which had a long artistic tradition in Japan and were also the primary news medium until the late nineteenth century, structured the spectacular destruction and ruins of the quake within the popular conventional taxonomy of picturesque famous sites (*meisho*). Familiar modes of visual consumption imparted layers of meaning, producing iconic disaster sites in the symbolic landscape of the modern metropolis.

The 1920s saw an unprecedented interpenetration of mass culture and high art around the world, typified by the imaging of the 1923 earthquake.[22] While works in the fine arts appealed primarily to middle- and upper-class viewers, the groups most likely to attend art exhibitions and visit galleries, the circulation of reproductions of these works in newspapers, magazines, and popular books, as well as through monthly subscriptions to woodblock print services, expanded their viewership.[23] It was also common for print publications to commission artists for cover designs or interior illustrations. Even Japanese department stores sponsored art exhibitions for the general public. Just as the tremendous interest in the quake brought the more recondite aspects of seismology into the public sphere, so did art and the heroic figure of the artist become popular conduits through which people could experience the subjective and affective nature of disaster.

Alongside the virtual space of the mass media, the Japanese produced real spaces to exhibit the images and objects of the catastrophe. These venues were also used to present propositions for rebuilding the nation's capital. Public and private organizations, most prominently the Tokyo metropolitan government, sponsored major exhibitions, and eventually a permanent memorial hall and museum, to narrate the story of disaster and reconstruction. The exhibitionary logic of these spaces yoked the agonistic somatic associations of the venerable debris of the quake to the visual rhetoric of progressive urban planning, thereby enfolding mourning and memory into reconstruction and modernization. Yet the deformed personal objects on display continued to serve as powerful material witnesses to the tragedy, metonymically communicating the presence of the dead and eliciting an intensity of communal bereavement that could not be eased simply by the progressive ideology of the new modern metropolis.

Amid the widespread visual lamentations over the tragedy, not all responses to the earthquake disaster were melancholic. Visual satire and humor have long been an important element in responses to catastrophe in Japan. Laughing in the face of calamity was not only a means of reclaiming normalcy through stress relief and psychological catharsis; it was also a method of conveying powerful social and moral criticism. Cartoonists (*mangaka*) and caricaturists (*fūshigaka*) working for the popular press took on the many ironies of the postquake moment and the slippery politics of reconstruction, specifically the conflicted perceptions of its main architect, Home Minister Gotō.

Visual satire actively shaped popular perceptions of the quake by giving voice to the central debates about the connection between disaster and modernity.

As humor scholar Marguerite Wells has noted, satire (*fūshi*) is humor "directed at the faults, vices, or follies of individuals or institutions, always with the intention of exposing and correcting them."[24] With its strong moralistic imperative, satire often has an uncomfortably aggressive style that grows out of the producer's sense of his or her role as a guardian of ideals or an upholder of the social good. Satire can also display a kind of moral absolutism in which the satirist sets forth standards against which to measure society and finds it wanting. Thus, visual satire aims to challenge social norms and often critiques the perceived baser qualities of human nature.[25] Disasters provide ideal opportunities to articulate these social criticisms in the interest of facilitating change.

Disaster itself is a paradox, destructive and constructive, both horrific and sublime in nature. While wreaking physical devastation and psychological trauma, it also creates space for reflection and renewal. Without a doubt, the social upheaval in the wake of disaster provides fertile ground for creativity. And the physical marks of catastrophe brutally etched into the land, the city, and onto the human body can be simultaneously repulsive and alluring.

Observing the cultural responses to catastrophic events over the course of Japanese history, we can see that they expose the ambiguous distinctions between natural and man-made disaster, a particularly timely issue now, when a recent spate of natural catastrophes around the world has brought questions of social and national culpability to the fore. Only through the intercession of man do natural phenomena become "disasters" for humankind, so, in that sense, there are no purely natural disasters.

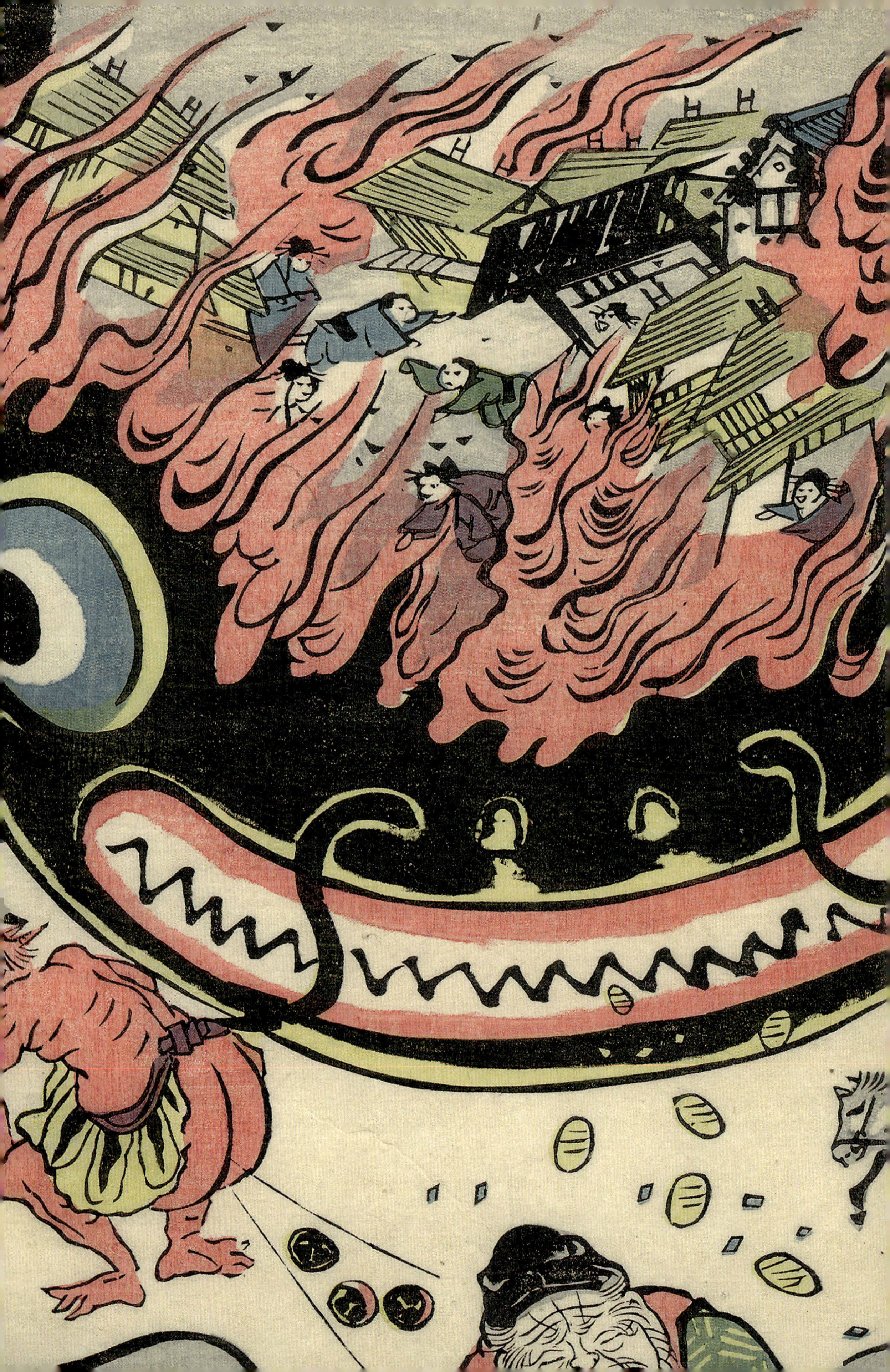

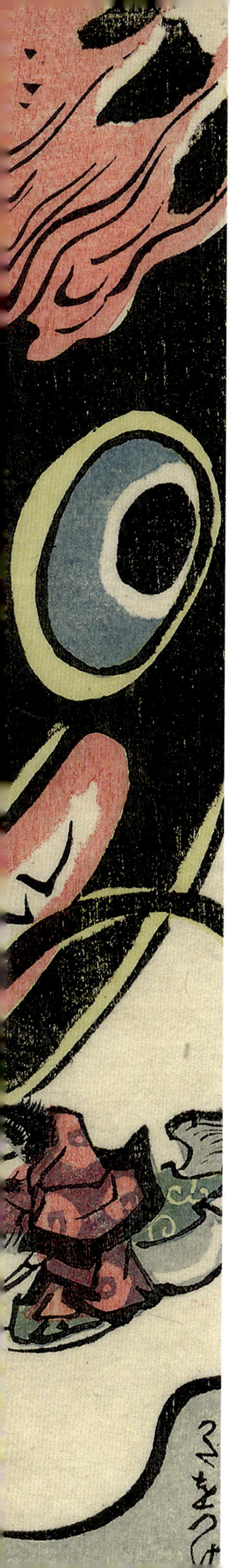

1

EARTHQUAKES IN JAPAN

A Brief Prehistory

As an archipelago created by the highly active tectonic zone where the Pacific and Philippine plates collide with the Eurasian plate, Japan is regularly rocked by volcanic eruptions, typhoons, earthquakes, and tsunamis. That has stimulated a particularly rich Japanese tradition—reaching all the way back to the beginning of its recorded history—of visual responses to calamitous events. One might even say that disaster has been a generative force in Japanese culture. It certainly has been a catalyst for change.

Disaster, commonly *saigai* or *saika* in Japanese, usually suggests tragedy, ruin, catastrophe, and calamity. It implies misfortune and adversity. But natural events are not inherently disasters; disasters are made. As Haruno Ogasawara poignantly notes, the designation of disaster is predicated on a moral and sociological interpretation of an event that perceives it as disrupting society with negative repercussions.[1] Disruptions, of course, can be bad or good, depending on the interpreter's status or position. In Japan, earthquakes historically have been considered transformative, even numinous events associated with contemporary social and political circumstances. They could be both devastating and renewing. And as Kitahara Itoko, one of the foremost scholars of disaster in Japan, has emphasized, the study of disaster should not be just a historical chronicling of damage and loss; it should be an interdisciplinary exploration of the dialectical relationship between destruction and reconstruction in the context of social formations.[2] Disaster is a defining feature of Japan's cultural landscape, and, consequently, the country's general belief system has integrated the cyclicality of destruction and renewal. Since Kitahara and others have devoted a lifetime to elucidating the extraordinary history of disaster in Japan, my intention here is not to repeat their efforts but to undertake the more modest project of illuminating a select set of premodern religious and philosophical beliefs, as well as imaging practices, that contributed to the genealogy of Japan's modern culture of disaster, particularly as manifested during the Great Kantō Earthquake of 1923.

In this chapter, I focus on the deep-seated belief in the moral connections between human action and disaster (natural or man-made) and the role of spiritual activities, particularly a range of imaging practices, in helping to prevent or ameliorate such circumstances. These imaging practices reveal the humorous and playful amalgamation of horror and parody and highlight the amusing fusion of the moralistic and the macabre that produced spectacular forms of visual entertainment.

Interpreting Disaster

The common jocular aphorism for the four most powerful forces of nature, "earthquake, thunder, fire, and father" (*jishin, kaminari, kaji, oyaji*) suggests that earthquakes are one of the most feared natural phenomena in Japan—comparable to the wrath of the stern Japanese patriarch.[3] The earliest texts in classical Japanese, such as the *Nihon shoki* (Chronicles of Japan), dating to the sixth century, use the term *nai* to describe a tremor of the earth (*na* refers to the land, and *i* indicates the verb *to be*).[4] The common current term for earthquake, *jishin* (literally, trembling of the earth), upon which the name of the modern scientific study of earthquakes—seismology (*jishingaku*)—is based, seems to first appear in documents from the early part of the eighth century (734) in the Shōsōin temple repository of Tōdaiji Temple in Nara. It was in common use by the beginning of the seventeenth century as indicated by its transliteration as *gixin* in the comprehensive 1603 Jesuit dictionary *Nippo jisho* (Vocabulario da lingoa de Iapam com adeclaração em Portugues) that translated Japanese words and their pronunciations into Portuguese.[5] The other common current term, *shinsai* (quake disaster), did not appear until the year after the Ansei earthquake of 1855, but by 1892 it was being used officially for the establishment of the national Earthquake Disaster Prevention Research Council (Shinsai Yōbō Chōsakai).[6]

Most Japanese people from the premodern period (before 1600) up through the nineteenth century believed that natural disasters were the result of imbalances in the five elements of nature caused by social impurities directly linked to human behavior; therefore, they thought that appropriate actions could be taken to ameliorate the situation. On the occasion of major disasters, rulers took extreme measures such as moving the capital and changing the imperial reign name of the period to dissociate themselves from previous rulership. This reflected the Confucian belief in the mandate of heaven that linked equilibrium in the cosmos with proper governance, a set of beliefs that had a pervasive influence in Japan along with Buddhism and Shinto. Throughout the medieval period in Japan, earthquakes figured prominently in didactic historical texts to articulate causal links between natural disasters and the moral turpitude of the current rulership or the immoral behavior of the general populace, notions that continued to have resonance up through the modern period.

Two medieval tales exemplify this understanding of earthquakes as expressions of the terrible state of moral affairs, particularly during the period known as the latter day of the Buddha's law (*mappō*): the *Hōjōki* (An Account of My Hut, 1212) by the exiled Buddhist monk Kamo no Chōmei and the military epic *Heike monogatari* (The Tale of the Heike, 1242).[7] The *Hōjōki* is a moralizing tale about the period after the Genpei civil wars (1180–85) and the profligate behavior in the age of the decline of the Buddha's law. The tale expounds on the transience of life (*mujō*), describing five disasters that befell the residents of the imperial capital, Kyoto, including a large-scale earthquake in July 1185. Like many interchangeable Japanese tales of natural and man-made disaster, the *Hōjōki* describes tragic human suffering in vivid detail as a warning to readers about the punishments that lay in wait for their actions.

A story of the internecine Genpei warfare, the *Heike monogatari* similarly features the 1185 earthquake, among others. The tale evinces an overwhelming pessimism as it chronicles the downfall of the Heike (Taira) clan. Reading the rumblings of one quake for the emperor, the chief of the Board of Divination ominously predicts terrible destruction in the future because of moral lapses: "The charter of the divination indicates that the earthquake this time is a warning for far more than minor self-control. When I consulted the Konko-kyō, one of the three texts of divination, the article said: 'Within a year, or within a month, or within a day, a great disaster will come.' This is an emergency." Responding to another quake, people exclaim, "The god of the earth has been angered."[8] The final great earthquake that hit Kyoto, and the ensuing terror, is once again described (and later illustrated) in all its horrific detail and is seen as the retribution of the vengeful spirits of the Heike and the murdered child emperor Antoku.

Traditionally, a range of Buddhist sutras enumerated various misfortunes brought on by improper adherence to the Buddhist law. The thirteenth-century cleric Nichiren was the first to bring them all together in his 1260 treatise *On Establishing the Correct Teaching for the Peace of the Land* (Risshō ankokuron), written soon after the devastating earthquake that hit the shogunal capital Kamakura in 1257. In it, he coined the term "three calamities and seven misfortunes" (*sansai shichinan*) to describe the result when both rulers and their people turned against the true teaching of the Buddhist law. The three calamities had greater and lesser versions. The greater calamities that would destroy the world were fire, wind, and water; the lesser that would cause human society to perish were high grain prices or inflation due to famine, warfare, and pestilence. The additional seven misfortunes were pestilence and epidemics; foreign invasion and aggression; internal strife; extraordinary changes in the heavens, such as those signaled by the appearance of comets and meteors; solar, lunar, and stellar irregularities; abnormal weather, such as unseasonable storms; and abnormal climatic conditions, such as prolonged droughts.[9] This Buddhist worldview did not distinguish natural and man-made disasters from each other, as they all were linked to human moral rectitude.

The Buddhist notion of ritual impurity, *kegare,* which was merged with similar Shinto beliefs, views things that are unclean as an offense to the gods and redolent of human guilt and sin and is associated with death, disaster, and disease. A major function of religious practitioners was to enact rituals to cleanse this pollution, and they developed a host of purification and apotropaic rituals that they performed regularly to divest communities of perceived contamination and prevent the recurrence of disasters. The *tsuina* ceremony (or, demon exorcism), for example, is still conducted all over Japan at the end or beginning of every year to expel demons from the community and thereby avert misfortune.

For most of Japanese history, people have believed that the dead and the living are connected. As historians of religion Jacqueline Stone and Mariko Walter have amply demonstrated, "Buddhism was the pre-eminent spiritual technology for consoling and pacifying the dead," and the Buddhist doctrine of an ethicized afterlife—where deeds were rewarded or punished by pleasant and painful circumstances and in which worldly bonds persisted beyond death—enjoined the living to dispose of bodies properly.[10] Religious ceremonies were essential for coping with the spiritual predicament of allaying the vengeance of the souls of the dead, who posed a real and present danger to the future prospects of the living if they were not properly conveyed back into the samsaric cycle of rebirth or given salvific rebirth in the Pure Land (*ōjō*).

This was particularly important after the untimely deaths caused by disasters, when the "feeding of hungry ghosts," the important ritual of *segaki,* had to be performed. This was a joint memorial service conducted for the souls of disaster victims in which offerings were made to appease the tormented souls of unattended wandering spirits (*muen botoke*) and hungry ghosts (*gaki*) residing in a liminal purgatory; the *segaki* guarded the spirits of the dead from the malevolent hungry ghosts and protected them from entering into this purgatory.

Visual Spectacle and the Macabre

While the imaging of disaster in Japan, if you include warfare, dates back to Japan's earliest pictorial traditions, two visual genealogies coalesced in the mid-nineteenth century to form the backbone of earthquake imagery that was transmitted into the modern period. One emanates from the work of the well-known eighteenth-century painter Maruyama Ōkyo, who produced his now famous three-fascicle *Handscroll of Misfortunes and Fortunes* (Shichinan shichifuku zukan, 1768), commissioned by Abbot Yūjō of Enmanin Temple in Ōtsu in 1765, which illustrated the misfortunes of natural and human disasters in the first two scrolls (fig. 1.1 shows the first scroll), with a final scroll detailing the leisurely, luxurious lives of the aristocracy. The scroll dialectically ties together misfortune and fortune within the Buddhist cyclical notion of reincarnation and provides a powerful visual polemic for righteous behavior in accordance with Buddhist principles. Ōkyo took his cue from Abbot Yūjō's eclectic vision as articulated in the prelate's preface to the scroll and preparatory drawings (still extant), for which Yūjō drew from a range of ecclesiastical sources.[11] But perhaps most important, the abbot incorporated local elements that he felt were most relevant to his parishioners, such as earthquakes, which were particularly prevalent in the Kansai region near Ōtsu and the imperial capital, Kyoto, but were never specified in any sutra texts. Yūjō focused on the contemporary, real-life experiences of the people around him to make his message more potent.[12] His mission was to proselytize righteous behavior by instilling fear of realistic divine punishment and the incentives of earthly rewards.

Ōkyo's representations of disaster drew heavily from historical models for compositions and motifs, particularly classical illustrated handscrolls, but they also resonated with contemporary Edo trends in naturalism and realism, for which the painter was known as a pioneer. At Yūjō's request, Ōkyo combined real-life elements with some aspects of historical Buddhist images of the six realms of transmigration or rebirth (*rokudō-e*), which include portrayals of those condemned to hell and the purgatory-bound hungry ghosts at the lower echelons of the karmic order, along with elements from traditional stand-alone pictorial depictions of hell (*jigoku-e*). These expressive representations of suffering beings in infernal landscapes became stylistically codified over time, although they never lost their evocative potential to instill fear in viewers. Ōkyo and Yūjō were able to update this genre and make it feel distinctly contemporary. And

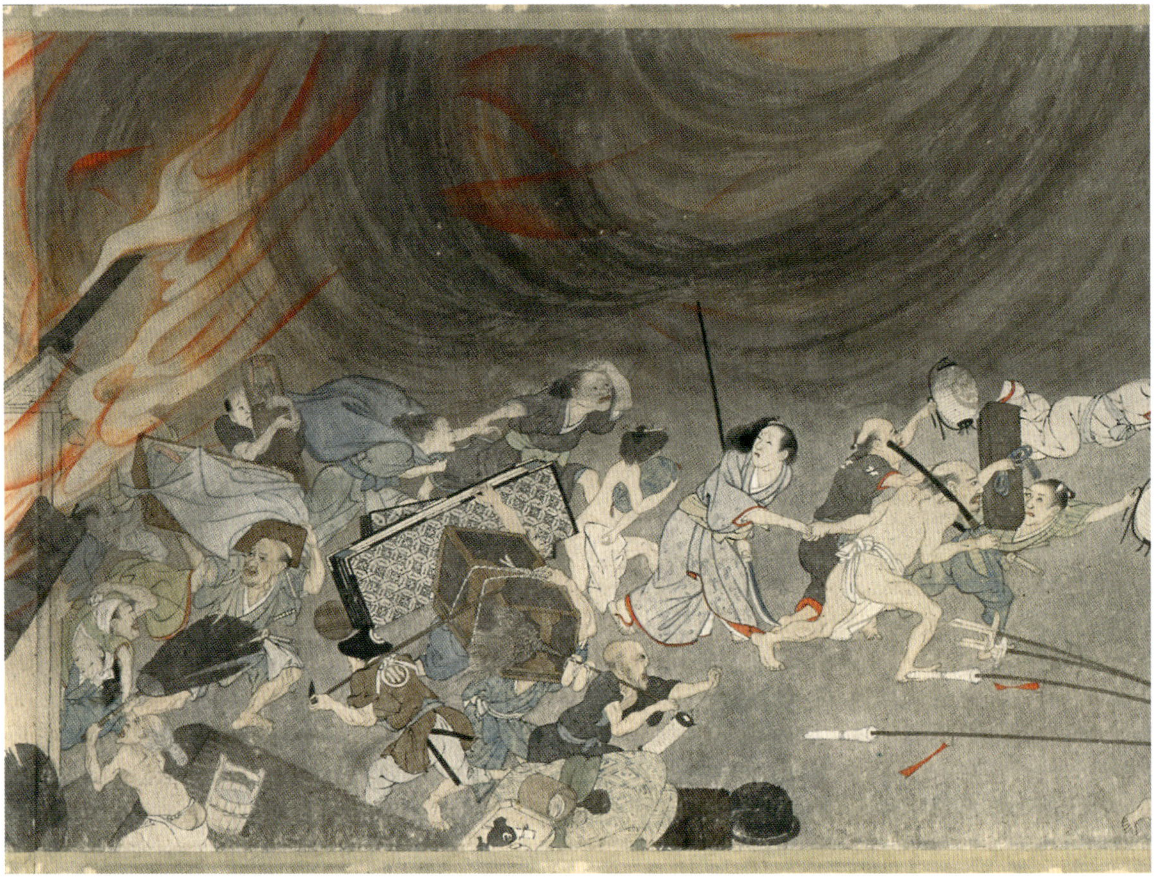

1.1 Maruyama Ōkyo (1733–95), *Handscroll of Misfortunes and Fortunes* (*Shichinan shichifuku zukan*), first of

Ōkyo's handscrolls continued to have a significant impact on pictorial representations of disaster through extensive copying by followers of the popular Maruyama-Shijō school of painting, who practiced widely into the modern period.

One surviving late nineteenth-century anonymous underdrawing (*shita-e*) copy of some of Ōkyo's misfortune sections shows this continued spectacular visualization of earthquakes, with infernal cyclonic firestorms bearing down on fleeing refugees as they shriek in terror, dragging their few salvaged worldly belongings.[13] The cyclonic fires are rendered with a combination of black ink (*sumi*) and vermilion red mineral pigment to express the dynamism and searing intensity of the blaze. Burning, bloody bodies writhe in agony in the whirling conflagration, only inches from fleeing, terrified mothers clasping infants to their breasts. The violent nature of the tragic events

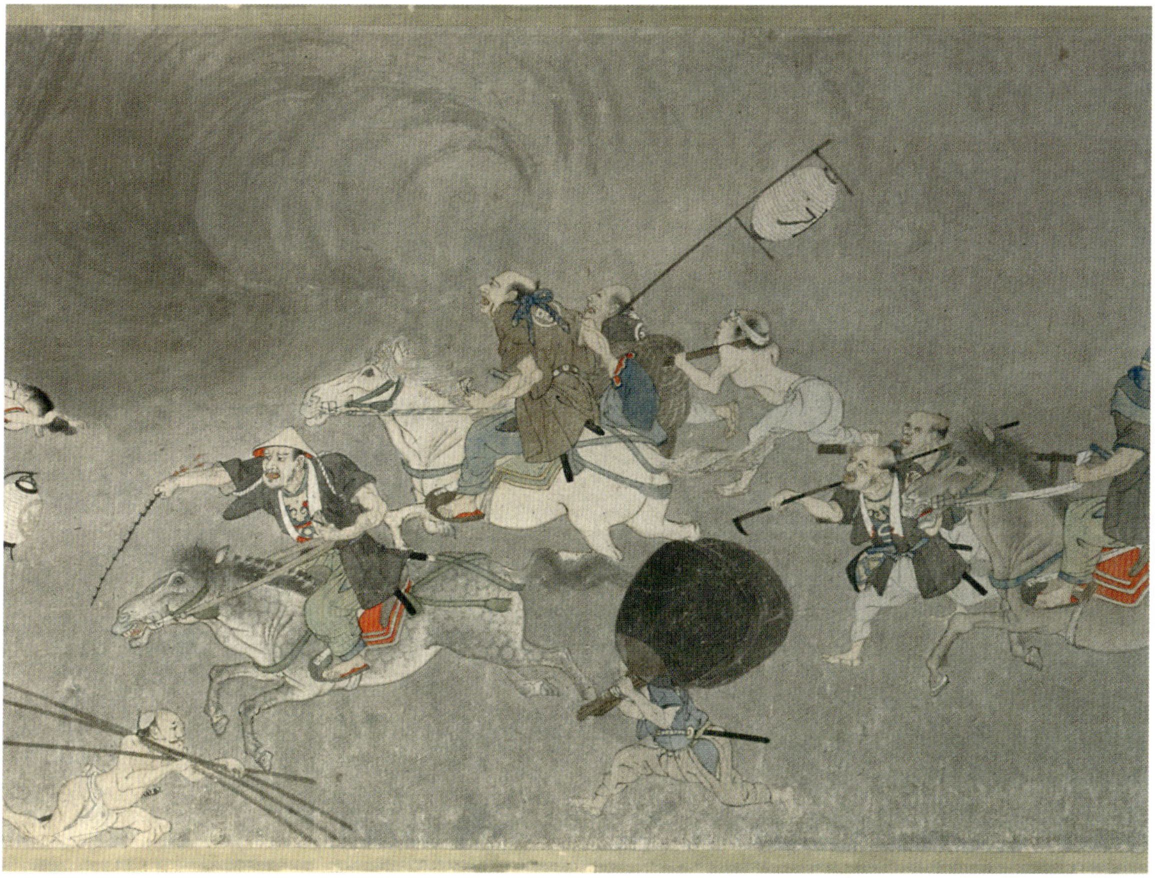

three handscrolls, ink and color on paper, 1768. Collection of Jōtenkaku Museum, Shōkokuji.

is conveyed in exquisite detail. All of the misfortune images—which include torrential floods and rainstorms, dramatically zigzagging lightning bolts striking people (fig. 1.2), and mythic demonic chimera such as *tengu* (monstrous birdlike creatures) and giant serpents menacing towns, carrying off children, and gnawing on bloody human limbs—capture the truly spectacular quality of disaster, simultaneously evoking fear and excitement. This spectacularized and macabre mode of visualizing the cautionary tales of disaster continued unabated into the modern period.

Such imagery is evident in the important visual chronicle of firsthand accounts of the devastating Ansei earthquake, *Ansei kenmonshi* (Ansei-Era Observations, March 1856), which appeared soon after the quake hit on 11 November 1855. Published anonymously because of the concern about censor punishment, the text (together with some

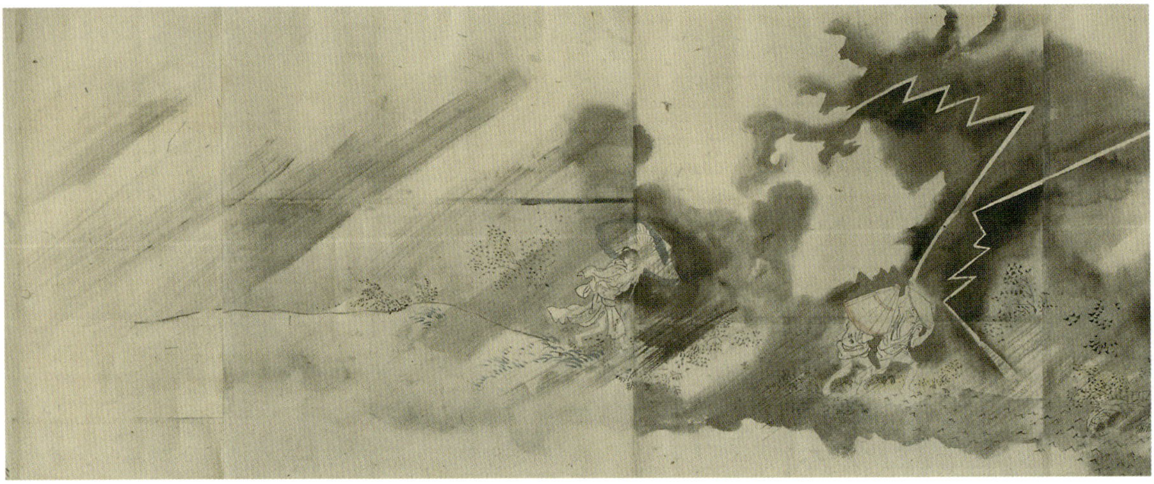

1.2 Maruyama-Shijō school, scene of lightning storm, underdrawing, handscroll, ink and color on paper,

illustrations) is widely believed to have been authored by Ippitsuan Eiju II, although it has also been credited to the well-known writer Kanagaki Rōbun. Some illustrations in the volume are by well-known *ukiyo-e* print designers Utagawa Kuniyoshi (known as Ichiyūsai Kuniyoshi; 1798–1861) and Utagawa Yoshitsuna (active 1848–68), as well as at least two other artists.[14]

The images in *Ansei-Era Observations,* combined with the vividly descriptive text, clearly provided visual entertainment as well as moral lessons. Stylistically, they drew heavily from the explicitly rendered gruesome scenes of the supernatural and macabre that were enormously popular in late-Edo visual culture. Renowned print designers such as Katsushika Hokusai and Utagawa Kuniyoshi produced extensive corpora of these images. This predilection for the macabre was also abundantly evident in the increasing focus on gruesome stories of criminality in the news.[15] There was no shortage of blood and gore in Edo popular imagery, a feature that was carried through to the subsequent Meiji period (1868–1912) by skilled print designers such as Tsukioka Yoshitoshi and Kawanabe Kyōsai. Disaster imagery could not be separated from visual entertainment, even when it conveyed strong moralistic messages.

The merging of the moral and the macabre is apparent in the depiction of the *segaki* purification ritual in a double-page illustration in the first volume of the *Ansei-Era Observations;* it shows a mob of disfigured blue corpses clustering before a group of Buddhist priests who are behind an altar performing the rituals (fig. 1.3). The text reads, "The many horrid deaths of people in the recent earthquake, though it was a natural

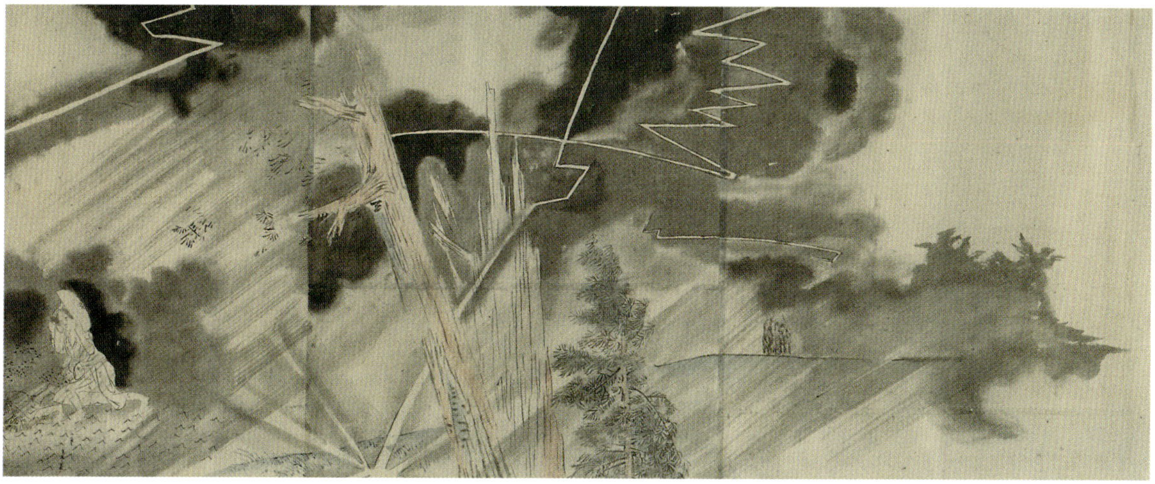

c. 19th century. Collection of the author.

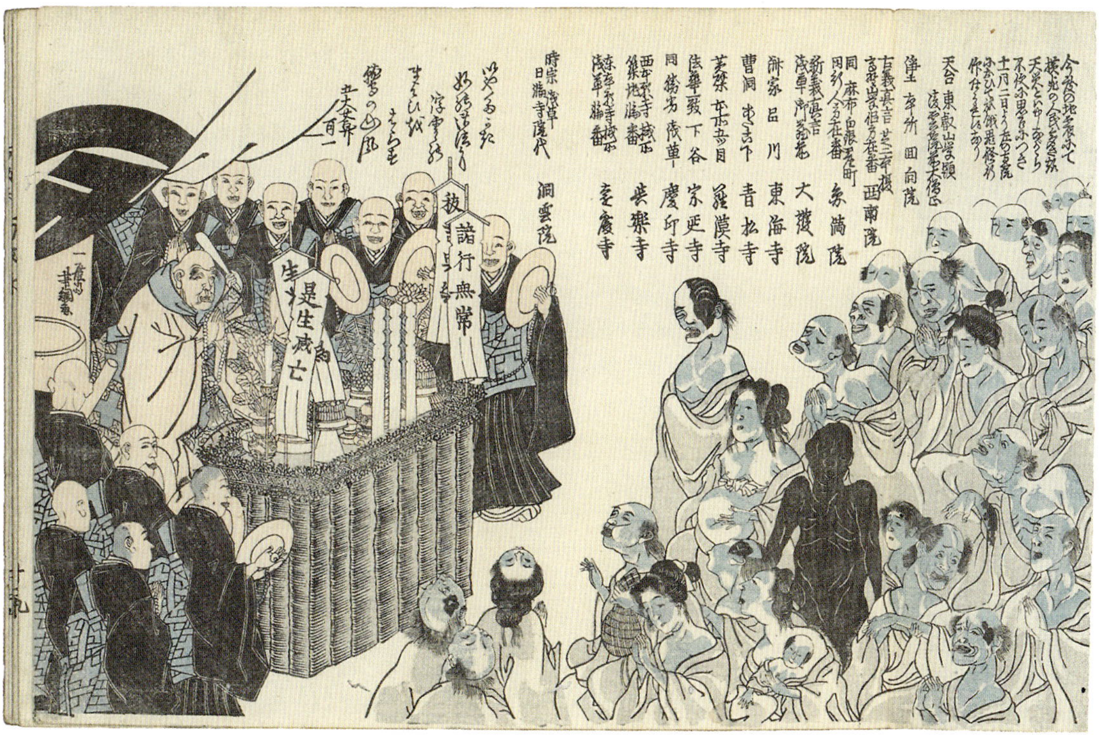

1.3 Utagawa Yoshitsuna, *segaki* purification ceremony for the spirits of Ansei earthquake victims, woodblock print. *Ansei-Era Observations* (*Ansei kenmonshi*), vol. 1 (Edo, March 1856). Collection of Nihon Shakai Jigyō Daigaku Toshokan.

disaster [therefore they deserve their own suffering], were so piteous that Segaki services were held on 2 December at the following temples . . . [to console their spirits]." A list of temple names follows.[16] The grisly features of individual figures rendered by the artist Utagawa Yoshitsuna include one with half his skin missing to reveal the skull beneath, one completely charred from fire, a pregnant woman with her distended blue belly, and another woman gripping her dead infant. These sufferers all look beseechingly toward the clergy for their salvation. While the service protected both the dead and the living, the spectacularly ghastly scenes frightened and titillated the public.

Apotropaic and Allegorical Catfish

One unique and enduring motif in Japanese disaster imagery is the catfish. The catfish image emerged out of talismanic maps, known as *Dai Nihonkoku jishin no zu* (Great Japan Earthquake Maps), which were designed to prevent earthquakes and foretell their future consequences. These maps originated in the fourteenth century but were popularized in the mid-seventeenth century on the covers of yearly almanacs and were sometimes used in *tsuina* demon-expelling ceremonies. Known as Gyōki-type maps after the Nara-period Buddhist monk who is credited with creating them, some of these maps depict a dragonlike serpentine creature surrounding the archipelago of Japan, with the internal domains delineated. Unlike specific regional representations, these images pictured the entire country in potential peril and unified in its fate. Early Buddhist texts identified four causes of earthquakes: fire (*kashin*), dragons (*ryūjin*), birds (connected to the Indian god Garuda, *Karura* in Japanese), and the Indian god Indra (*Taishaku*). These maps specifically evoke the dragon imagery—with the creature seemingly about to constrict and choke off the lifeblood of the country.

Over two centuries of reproduction, the dragon became conflated with a large cosmic fish—later specifically a catfish. (Catfish often act strangely before earthquakes, perhaps because they can sense the first small tremors as they swim in the mud close to the ground). The titles of the maps, *Jisoko namazu no zu* or *Chitei namazu no zu* (Pictures of Catfish beneath the Land), in calendar divination books (*ōzassho sanzeshō*) reveal the conflation of the dragon and the fish, and the images show the visual transformation of the dragon into an increasingly rounded figure of a catfish (*namazu*).[17] These map images also mark the first use of the protective keystone (pivot stone) motif, or *kaname ishi,* next to the dragon/fish's head. This keystone, which marks the place where the Kashima deity (*Kashima daimyōjin)* descended from heaven, was associated with the Kashima Shrine in modern-day Ibaraki prefecture, where it protrudes from the ground. From as early as the beginning of the fourteenth century, people believed that the stone was buried deep in the earth, resting on the head of the dragon/fish beast at the earth's core. Through the force of the Kashima deity, the stone (or in some cases the deity's sword) subdues this beast and prevents it from moving and causing earthquakes. This connection established Kashima as a national protector, and the shrine promoted this identity to bolster its importance and social status.

The bound versions of the *Ise koyomi* (Ise Almanac) later printed in Edo city include stylized versions of these talismanic maps, which were issued yearly to ward off earthquakes and to predict month by month when and where earthquakes might occur. They continued to have currency into the nineteenth century. An *Ise koyomi* image was included in the 1830 publication *Jishinkō* (Treatise on Earthquakes) by the private academy instructor Kojima Tōzan (1761–1831), whose lecture was transcribed by his student Kojima Tōrōan (fig. 1.4). The project was prompted by a massive earthquake in the Kyoto area that year. This image shows a dragon/fish creature surrounding an amorphous image of Japan that appears more like viscera than a map. The four directions are indicated, and on the upper portion to the right of the east cartouche, the familiar thirty-one-syllable *tanka* poem reads, "As long as the god of Kashima is majestically restraining it, even if there might be earthquakes, the Kaname stone will never come out of the earth" (yurugu tomo / yomoya nukeji no / kaname ishi / Kashima no kami no / aran kagiri wa). The months are segmented around the figure under the dorsal scales, going counterclockwise from the two o'clock position, indicating the predictions for each period.

This was the age of the mass-produced printed broadsides (*kawaraban*) that served as Edo newspapers. The new *kawaraban* media merged with the older traditions of talismanic dragon/fish maps after the massive Ansei earthquake of 11 November 1855 to generate one of the most interesting and unique genres of Japanese disaster representation—the catfish print (*namazu-e*)—which would continue to have resonance up through the twentieth century as an allegorical harbinger of divine intervention to rectify an ailing society. The Ansei quake has a particularly important place in Japanese history, for it damaged large portions of the Tokugawa capital, Edo, which was a major center of commercial and cultural activity in the mid-nineteenth century, not to mention the seat of government rule. It was also seen as the culmination of a series of three devastating earthquakes, starting with the 1847 Shinano quake and followed by the 1854 Tōkaidō earthquake. The survival of certain structures, such as the central hall of the Asakusa Sensōji Temple dedicated to Kannon, despite the total destruction of their surroundings, was seen as evidence of divine intervention. Correspondingly, the collapse of the stone watchtower on the perimeter of Edo Castle was deemed to presage the imminent collapse of Tokugawa authority.[18]

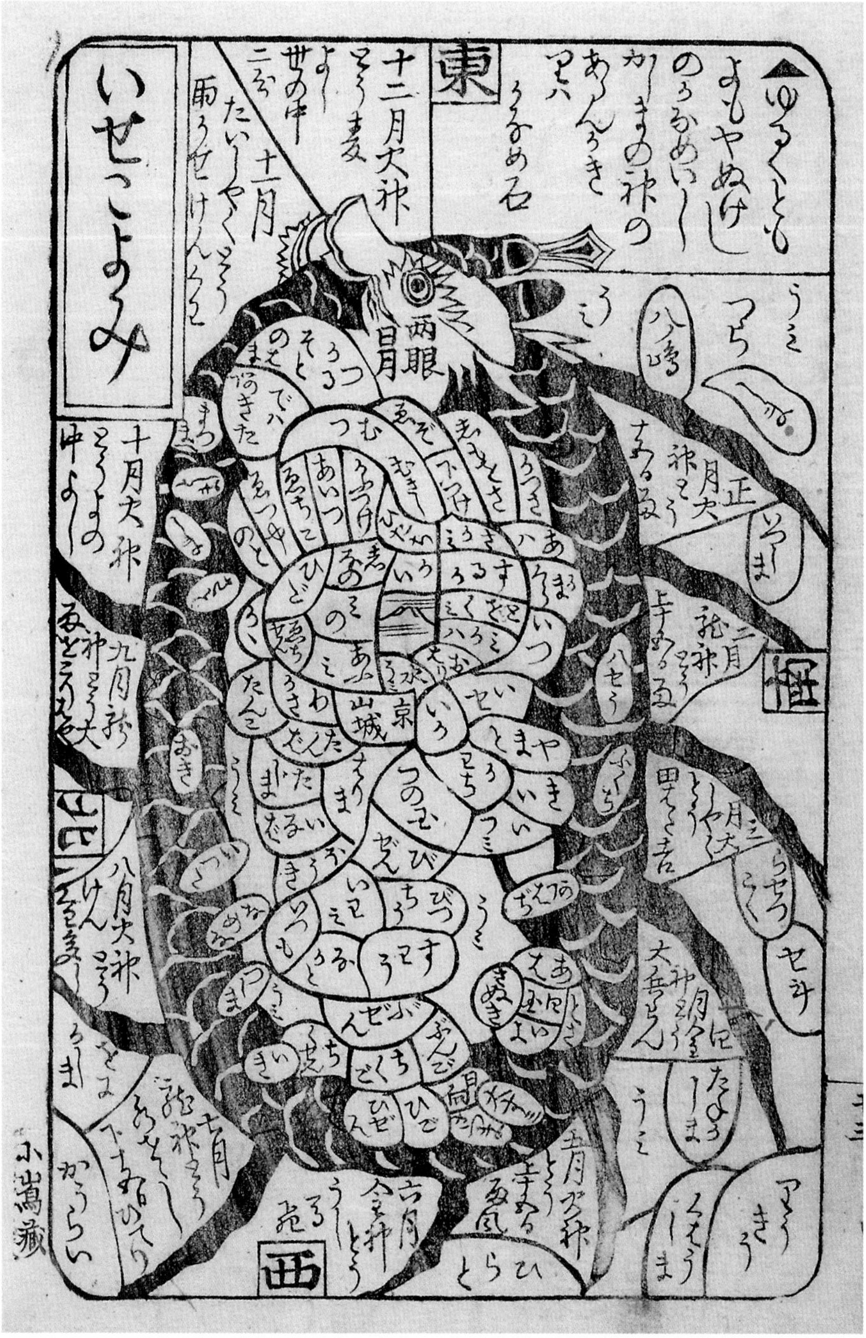

1.4 Artist unknown, *Ise Almanac* (*Ise koyomi*), showing a Gyōki-style map of a dragon/fish creature beneath the earth, woodblock print. Kojima Tōzan, *Jishinkō* (Treatise on Earthquakes) (Kyoto, 1830). Collection of Nihon Shakai Jigyō Daigaku Toshokan.

The Ansei quake inspired a tremendous surge in literary and artistic production, much of which was issued anonymously by commercial (*gesaku*) artists, a subject that has been considered at length by Andrew Markus. Within one month after the quake, single-sheet, standard *ōban*-size (10 × 15 inch) apotropaic and allegorical prints featuring catfish started to appear in the popular press in great numbers. They quickly became subject to regulation and censorship, but their popularity ensured their continued production illegally. Over four hundred varieties of catfish prints are known. Throughout the Edo period, people believed that at moments of moral or social crisis, particularly when "spiritual reconstruction" (*seishin fukkō*) was needed, the catfish would get loose, causing earthquake tremors by shaking the country on its back.[19] The destructive catfish was both deliverer of moral reckoning and facilitator of world renewal or world rectification (*yonaoshi* or *yonaori*).[20] Therefore, the Ansei catfish prints presented earthquakes as having restorative effects as well as destructive repercussions.

A look at some of the thematic concerns of Ansei catfish prints as visual satire can help us understand the continuities and ruptures enacted in the visual production of 1923. Kitahara has identified two distinct yet intermingled messages in catfish imagery: fortune and misfortune. The people venerate the "large" catfish as a messenger from heaven because it reveals deleterious social ills in an effort to bring about the eventual betterment of society—a view that was generally held by the upper classes, particularly the samurai class. In contrast, the "small" catfish, as a symbol of the myriad misfortunes of disaster, is reviled and attacked by the people—a view generally held by the working classes.[21] This dichotomy is complicated, however, by the fact that many *namazu-e* express sympathy for the plight of the working classes, presenting the earthquake as a positive event that will precipitate a much-needed redistribution of wealth. The theme of class inequity undergirds many of the works. In one print, euphoric figures gaze rapturously at a catfish spouting coins like a whale—perhaps referencing a well-known scene of whaling off the Gotō islands from Hokusai's *Oceans of Wisdom* series (1833). In this *namazu-e,* people happily look on as they imagine the forging of a new, more equitable society. The print can also be read as a critique of the enormous amount of profiteering that took place in reconstruction-related business, a commonly recurring subject. Issues of class equity and social abuse are also addressed in a number of prints that depict the catfish delivering divine punishment (*tenken* or *tenbatsu*) to the rich or ruling class for their moral lapses, materialistic excesses, and inept leadership.

According to Cornells Ouwehand, through the anthropomorphization of the catfish and the personification of earthquakes, the *namazu-e* could represent human players in disasters from all angles, as destroyers, renovators, enemies, and heroes. The catfish

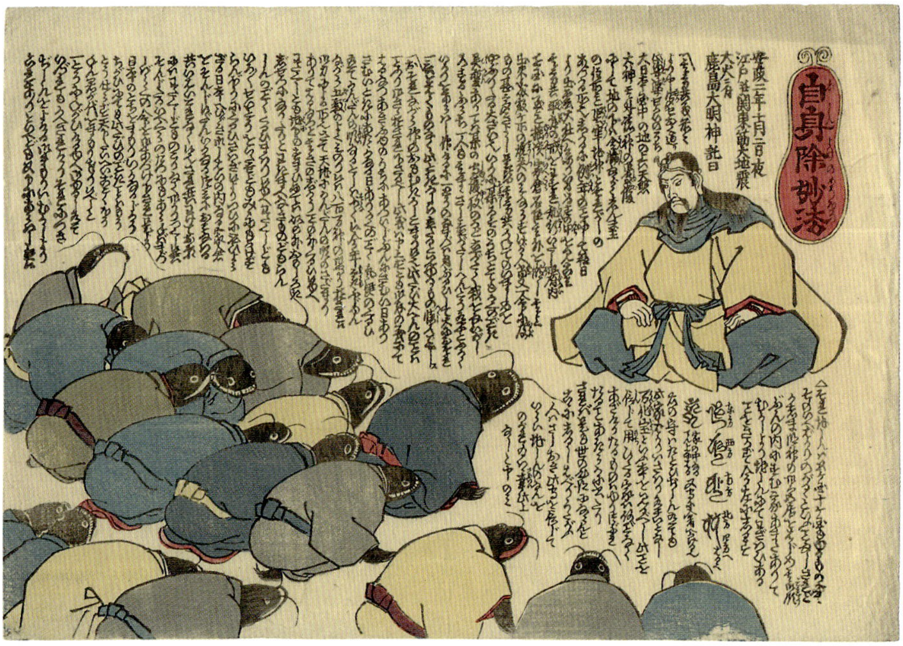

1.5 Artist unknown, *Magic Spell for Protection from Earthquakes (Jishin yoke no myōhō)*. Japanese, Edo period, c. 1855 (Ansei 2). Woodblock print (nishiki-e); ink and color on paper. Horizontal ōban. Photograph © 2012 Museum of Fine Arts, Boston, William Sturgis Bigelow Collection, 11.21985.

mirror human society in all its ambiguity.[22] As such, they epistemologically transpose earthquake disasters into something understandable. *Namazu-e* indicate a more active notion of personal will in relation to earthquakes, as evidenced in the earnest images of mobs of townsmen kneeling in prayer to the brightly labeled protective keystone at Kashima Shrine—*Safety Protective Keystone* (Anshin kaname ishi) is one example—which demonstrate their attempts to control natural phenomena through pious action. At the same time, images sharply satirized the connection between human action and disaster. An amusing pendant image to the townsmen shows a group of clothed catfish-men, grinning with teeth bared from ear to ear, similarly bowing in contrition to the Kashima deity for their transgressions in causing the Ansei quake along with others throughout the country (fig. 1.5). They agree to sign a formal declaration of their good intentions. The title, *Magic Spell for Protection from Earthquakes* (*Jishin yoke no myōhō*), uses the homonym *jishin* (oneself) instead of the characters for *jishin* (earthquake), so that it can also be read "magic spell for protection from oneself," ironically indicating the problematic "self" of the pictured catfish that caused the earthquakes, or more philosophically, pointing to the causal relationship between human action and disasters.[23] The title cartouche in the image is shaped like a tapered gourd container (*hyōtan*), a common apotropaic image associated with subduing troublesome catfish that was sometimes placed on the fish's head instead of the stone or the sword.

In recent work, Gregory Smits has focused on Ansei catfish prints in the context of Japanese urban society at a time of uncertainty and has related them thematically to the weakening of the social and political order. Among other things, he reads the prints as expressing popular anxiety over the political pressure American Admiral Matthew Perry and his Black Ships exerted on the weak Tokugawa shogunate.[24] Anxiety about foreign incursions onto Japanese soil and the international situation in general was indirectly responsible for the magnitude of the quake's damage in that large quantities of gunpowder stored throughout the city for protective purposes exploded in the ensuing conflagration.[25]

Along with the sophisticated multilayered wordplay common throughout Japanese print culture, particularly *ukiyo-e* (images of the floating world of the pleasure quarters), a tinge of parody and a comedic veneer often infused the underlying moral message of earthquake representations. As people gain distance from traumatic events and their lives begin to return to normalcy, they often use humor to highlight the absurdities of the event and relieve the stresses of the disaster. But perhaps most important in Japan was the ability to use parody as a powerful form of veiled social critique, at a time when open public criticism was strictly forbidden.

In one well-known, satirical catfish print (fig. 1.6), Ebisu, one of the gods of good fortune, who has taken over watch for the Kashima deity while he is at a meeting of the gods at Izumo Shrine, falls asleep on the protective keystone, thus letting the catfish loose. As the hapless Kashima races back on the right, the catfish wreaks destruction upon Edo. Gold and silver coins are shaken loose from the city, indicting the wealthy elites and heralding the redistribution of wealth. The squatting scatological figure on the left shows the thunder deity engaged in the popular Edo pastime of competitive farting, also known as "thunder farting." The drums excreted by the god evoke the thunderous sound of the earthquake, one of its most fearful aspects.[26] Quasi-comic and decidedly scatological in tenor, this image reminds viewers that even the gods fall down on the job; hence, they themselves must be even more vigilant.

Anger and revenge against the destructive catfish (as scapegoat) was a common theme. One print, for example, depicts a frightened, beseeching catfish and its children being set upon by a vision of the vengeful Kashima deity and an angry mob of armed citizens seeking retribution. In other whimsical examples, the catfish and his cohort engage in more endearing human activities. One such print (fig. 1.7) displays a catfish, dressed in a summer kimono (*yukata*), playing the "earthquake game" (*jishin ken*) with

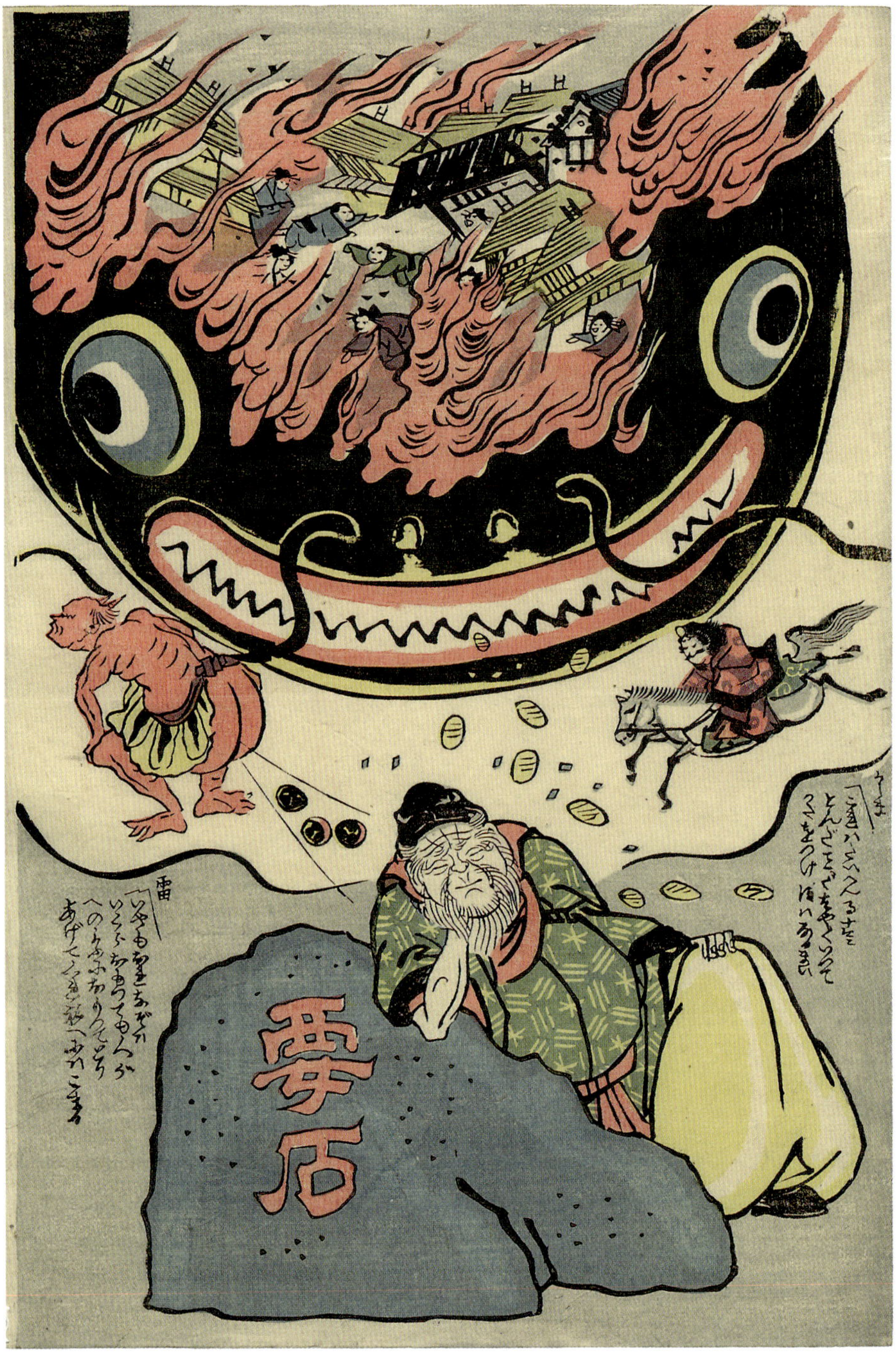

1.6 Artist unknown, *The Kashima Deity Napping* (untitled in Japanese; catfish and the *kaname-ishi*). Japanese, Edo period, c. 1855 (Ansei 2). Woodblock print (nishiki-e); ink and color on paper. Vertical ōban. Photograph © 2012 Museum of Fine Arts, Boston, William Sturgis Bigelow Collection, 11.38628.

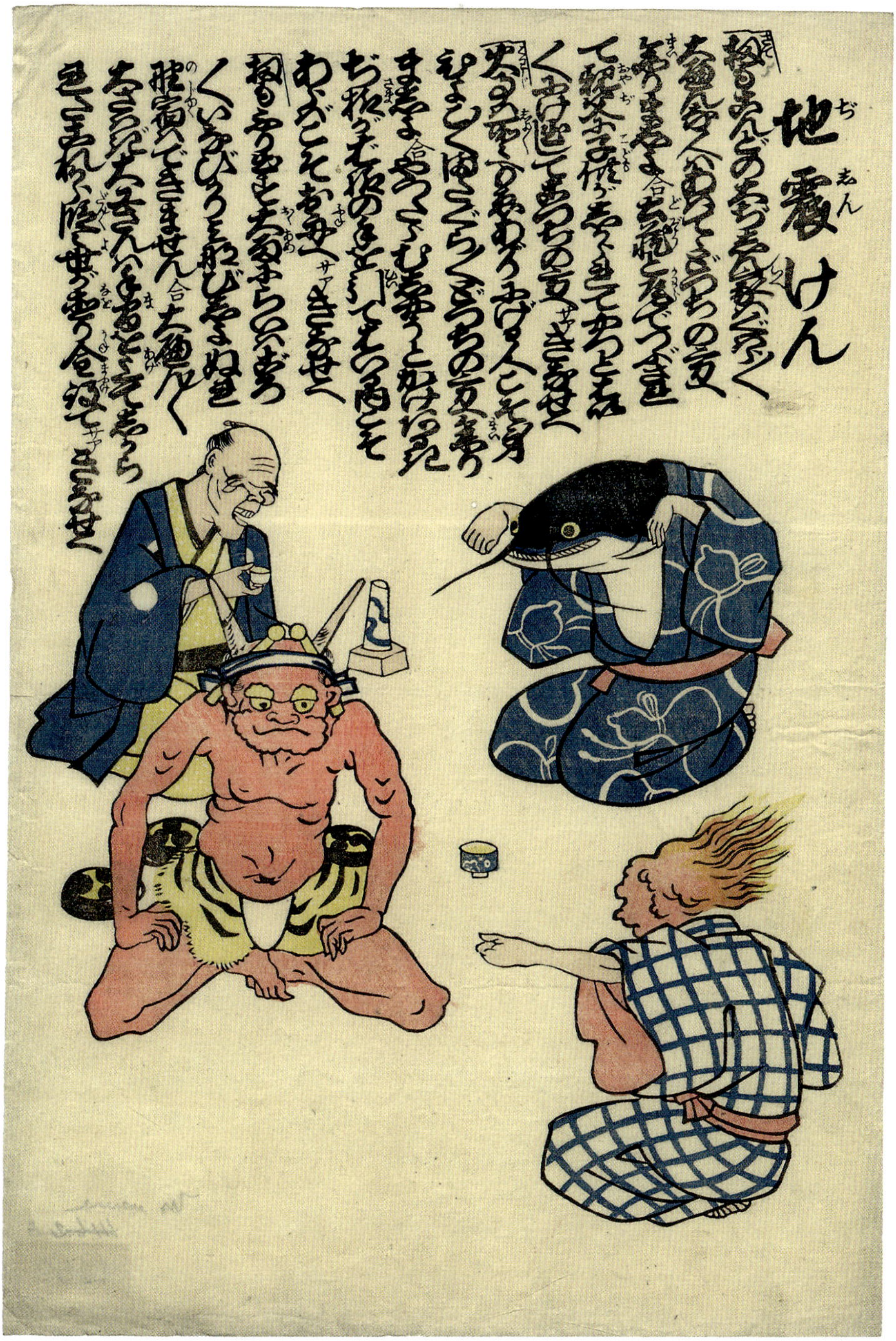

1.7 Artist unknown, *Earthquake Ken Game* (*Jishin ken*). Japanese, Edo period, c. 1855 (Ansei 2). Woodblock print (nishiki-e); ink and color on paper. Vertical ōban. Photograph © 2012 Museum of Fine Arts, Boston, William Sturgis Bigelow Collection, 11.21960.

the gods of thunder and fire, who were often paired with him. The humor in this image lies in the comical representation of these fantastic characters playing the well-known children's game *kitsune ken,* or "fox fists," involving hand gestures (a game akin to rock-paper-scissors), and the allegorical pictorialization of the common aphorism about the four most fearful things: "earthquake, thunder, fire, father." The print shows the catfish (earthquake) playing with the red thunder god and the fire god, whose back is turned to the viewer, as the stern father looks on from behind. In the end, the gods are busy playing games while the fate of humanity is left to chance. The *namazu-e* demonstrate the centrality of visual satire in responses to disaster.[27]

Disaster Miscellany

Commercial publishing greatly expanded in the Edo period, fueling a highly profitable market for news and nonfiction books about subjects like disasters. A new amalgamated genre that emerged during the Edo period, the disaster miscellany (*tensai zassan*), brought together historical and personal anecdotal reportage, scientific documentation, and dramatic literary storytelling.[28] A prime example was the illustrated three-volume *Ansei-Era Observations,* which combined traditional disaster representation showing elevated bird's-eye and close-up views of the cityscape and its inhabitants under siege by firestorms with more totalizing modes of visualization, such as the fold-out panoramic view of the burning city (fig. 1.8).[29] Significantly, this compilation also included scientific images, such as a very early illustration of the construction of a seismograph and a systematic chart of circles representing the quake's series of shocks and aftershocks with their relative magnitude indicated by size.[30]

This textual and visual collage draws attention to the multiplicity and self-reflexive nature of the disaster experience by combining eyewitness accounts, personal anecdotes, stories of supernatural miracles, moralistic tales of virtuous deeds, and purportedly objective historical and scientific reports.[31] This function is highlighted by the image of the catfish in the frontispiece of the first volume—a male figure draped in a hooded cloak that is widely believed to be a self-portrait by Kuniyoshi, suggested by the poem at the upper left by the artist's teacher and patron Umeya Kakuju (fig. 1.9). The catfish's eyes peer out above the man's face at the front of the hood, and the gray skin envelops him like a monk's robe. This feature is a clear reference to the first patriarch of Zen Buddhism, Bodhidharma (Daruma in Japanese), who often wore a hooded red robe. The figure peers through a cracked hole in a wall, and the text on the facing page notes that the author recorded what he "saw and heard" of the earthquake. The viewer is made aware of the act of witnessing through the screen of destruction.

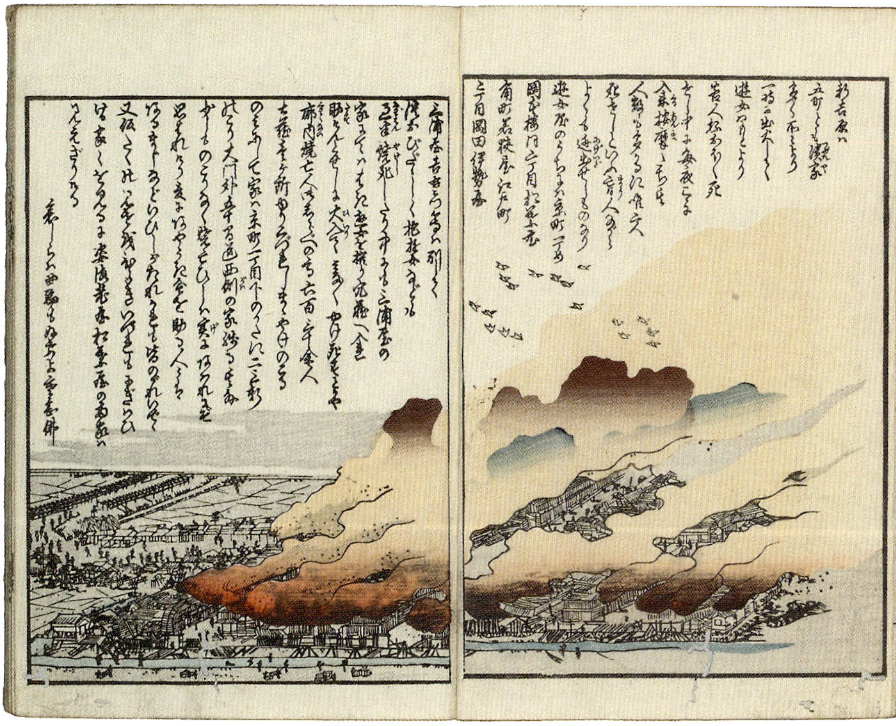

1.8 Artist unknown, Bird's-eye view of Edo in flames, woodblock print. In *Ansei-Era Observations* (*Ansei kenmonshi*), vol. 2 (Edo, March 1856). Collection of Nihon Shakai Jigyō Daigaku Toshokan.

Cloaked in the catfish, the figure sees the event from the creature's perspective. This personification puts man in the catfish's position, enabling him to witness his own act of destruction. The image mirrors a popular *namazu-e* print design that similarly features a portrait of Bodhidharma peering through a destroyed wall in which the figure's eyes and other facial features are made out of coins—a satirical take on earthquake profiteering.[32] *Ansei-Era Observations* cleverly points to the creative fusion of the objective and the subjective in disaster reportage, not to mention disaster's myriad commercial profit opportunities for carpenters, plasterers, and publishers.

Looking toward the Modern

Woodblock print media dominated the representation of disaster in Japan all the way up to the late nineteenth century. As Japan moved into the world theater of modern nations in the latter part of the nineteenth century, the country's history of disaster joined a global history of calamitous events, which was further melded through science

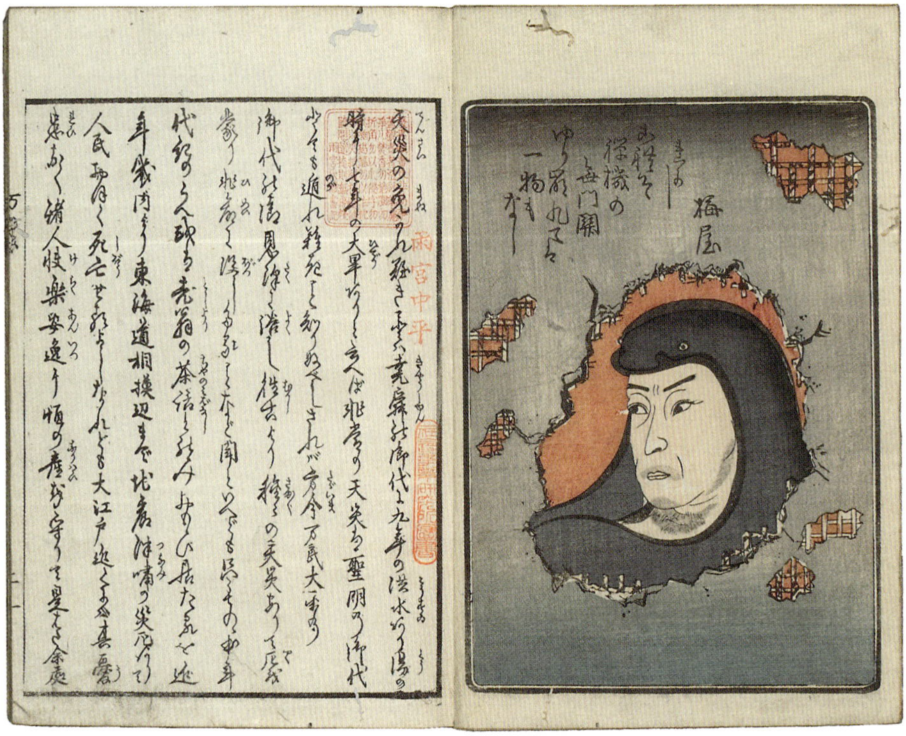

1.9 Attributed to Utagawa Kuniyoshi, figure with catfish hood, believed to be a self-portrait, woodblock print. *Ansei-Era Observations* (*Ansei kenmonshi*), vol. 1 (Edo, March 1856). Collection of Nihon Shakai Jigyō Daigaku Toshokan.

and technology exchange and international relief aid. Destructive earthquakes and volcanic eruptions have occurred all over the world, but the eruptions of Krakatoa in 1883 and Vesuvius in 1906 and the devastating San Francisco quake of 1906 loomed particularly large in the Euro-American public imagination of disaster in the early twentieth century. In fact, Krakatoa's global news coverage revealed the high level of interest in disasters among the general public around the world, which the Japanese news media quickly embraced. One of the earliest photographs published in a Japanese newspaper, appearing in the *Yomiuri shinbun,* was of the eruption of the Mount Bandai volcano northwest of Tokyo in July 1888, an event that was also covered in the world press.[33]

Kitahara has spotlighted the early impact of photography on disaster reportage in the mid-Meiji period, particularly in the representation of the Shōnai earthquake of 1894. By the mid-1890s, a decade of disasters in Japan, mass-published regional newspapers using photography had largely displaced woodblock print media such as *kawaraban* (broadsides) and *nishiki-e* (brocade pictures; multicolor, single-sheet woodblock prints) as the primary means of communicating news information about disasters and other current events. Kitahara even argues that the putative verism of photography supplanted personal literary accounts of disasters, as survivors no longer felt the need to record their experiences for posterity. While this may have occurred in the short term, it certainly was not the case in 1923, when multitudes of literary and visual representations of the Great Kantō Earthquake appeared in every conceivable medium.[34] Japan's premodern imaging of disaster continued to exert a strong influence well into the modern period, underscoring the moral connections between the natural, the human, and the divine; finding humorous and playful ways to meld horror and parody in visual satire; and capitalizing on the spectacular entertainment value of the macabre.

I argue here for a more expansive view of the intermedia nature of the modern visual culture of disaster. While producers of images relied on photography (and to a lesser extent film) as a touchstone of "the real" to provide visible evidence, they also continued to embrace a broad array of media with deep historical connections to disasters of the past, recognizing that these traditional tools lent their own distinct subjectivity, aesthetics, and historical resonances to the tangible and intangible aspects of disaster.

Earthquakes are unequivocally visual as well as physical experiences, and the abundance of images of disaster in the twentieth century attests to the importance of visuality in the modern communication of major historical events. However, the imaging of disaster is not merely a by-product or representation of history—it is history. In its time, the Great Kantō Earthquake was one of the largest and most devastating disasters in the world. What follows is an exploration of Japan's heterogeneous attempts to record and give meaning to the traumatic events of 1923—its first major national disaster on the international stage—in visual terms.

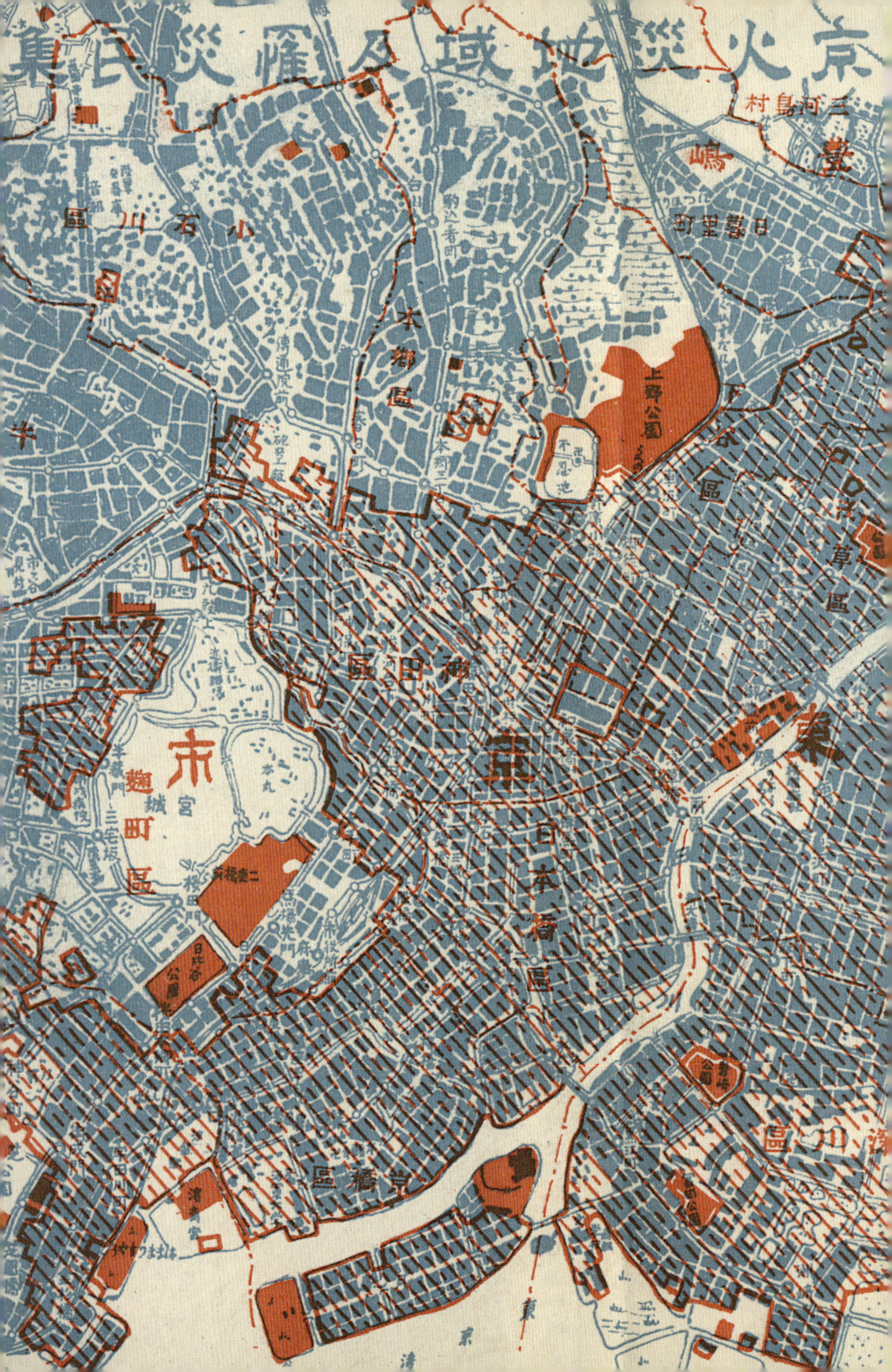

京火災地域及罹災民集
三河島村
日暮里町
小石川區
本郷區
上野公園
不忍池
下谷區
淺草區
神田區
東京市
麴町區
宮城
本丸
二重橋前
日比谷公園
日本橋區
京橋區
深川區

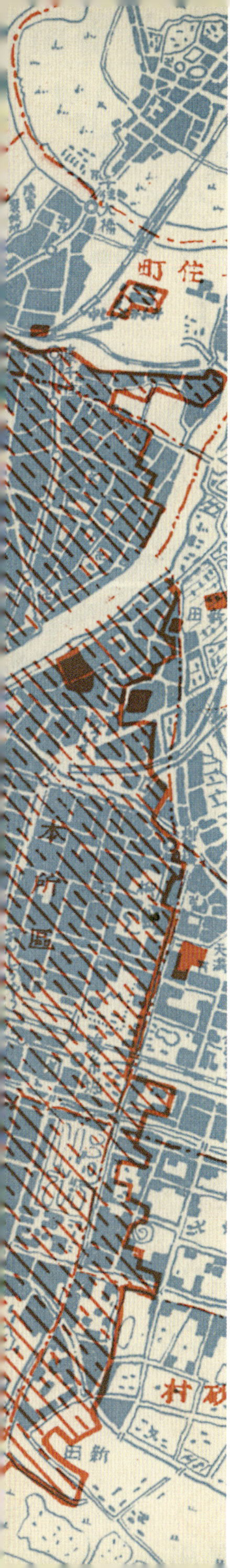

2

THE MEDIA SCALE OF CATASTROPHE

September 1, 1923, has been stamped indelibly on the mind of the Japanese nation, for on that day, at noon, Tokyo and six other prefectures in Kanto were visited by a terrific earthquake, the fire following which devoured hundreds of thousands of houses, destroyed all means of communications, ran up an incalculable death list, created a million and a half refugees, destitute of homes, food, or clothing, and shattered family ties.

Tokyo and Yokohama, the two greatest cities in the districts affected, of whose great buildings and cultural equipment the whole nation has been proud, changed into masses of fire-swept debris, the only remains of the seat of civilization of the Far East.

Japan, however, will not succumb before nature's ravages; such a day of national calamity marks the beginning of a test of the nation's stamina.

The Osaka Mainichi has undertaken to publish photographs, taken at great risk and sacrifice by members of the photographic corps of the paper, to enable the readers to realize the damage wrought by the calamity, and supply the public with a costly souvenir of the greatest calamity in the world—a souvenir that can be passed down through the generations.

FOREWORD TO ŌSAKA MAINICHI SHINBUNSHA, *KANTŌ SHINSAI GAHŌ* (KANTŌ EARTHQUAKE PICTORIAL), 15 SEPTEMBER 1923

When the Ōsaka Mainichi Shinbunsha published its bilingual three-volume photographic pictorial of the Great Kantō Earthquake just two weeks after the event, the calamity had already been captured in thousands of images that circulated on a national and international media highway. Commercial photographers and photojournalists produced the most abundant and immediate images of the quake, which were transmitted in newspapers, special-issue newspaper pictorials, commemorative photography collections, illustrated survivors' accounts, and sets of commemorative postcards. These photographic images functioned as both news and souvenirs, rendering their consumers/viewers, inside and outside the devastated locale, into both witnesses and voyeurs. Images in the news media and those issued by respected publishing houses carried the visual authority of supposed facticity. As such they both produced and became the historical record of the event. These photographs also heroized the vision of their producers, who pointed out that they had taken them "at great risk and sacrifice" to assist the world in comprehending the event. Many photodocumentary images quickly went from being indexical to iconic through their repeated reproduction and circulation, even setting the grammar for visual responses in other media such as prints and painting. An intermedia dialogue emerged based on the purported verisimilitude of the photodocumentary as a touchstone for narrating the event, but it revealed the inadequacy of any one medium, including the "realistic" mode of photography, to express the totality of the disaster.

In this chapter, I examine how the extensive photodocumentary coverage in the mass media allowed people throughout Japan (and throughout the world) to share vicariously in the plight of Tokyo residents living out a life-and-death drama on the streets of the capital, a supreme example of what Paul Virilio has called "the *media scale* of catastrophes and cataclysms that dress the world in mourning."[1] I argue that these visual images helped collectively produce the event known as the Great Kantō Earthquake by codifying the tropes and motifs that would form an enduring visual lexicon of disaster. By asserting visual authority, the diverse perspectives of the photographic eye and its technologically mediated vision had an enormous impact on people's perception of the event. Together with the camera, other scientific and technologized modes of visualization, including cartography and seismography, produced visible evidence to buttress this visual authority. In its social impact, disaster photography embodied a civil contract that forged multilayered power relationships. It simultaneously inscribed and pointed to the limits of visual authority by revealing the heterogeneity of viewing experiences. At a psychological level, the imagery sought to instill national resilience and empathic mourning yet still unavoidably indulged the eye in the perverse pleasures of disaster viewing.

The sublimated iconoclasm of poststructuralism has generated an overwhelming distrust of images and vision in theoretical discourse in the past few decades. And popular cultural theorists such as Susan Sontag have maintained a critical skepticism while acknowledging the power of images to evoke empathy for those touched by calamity.

According to Sontag, "Nonstop imagery (television, streaming video, movies) is our surround, but when it comes to remembering, the photograph has the deeper bite. Memory freeze-frames; its basic unit is the single image. In an era of information overload, the photograph provides a quick way of apprehending something and a compact form for memorizing it. The photograph is like a quotation, or a maxim or proverb. . . . Conscripted as part of journalism, images were expected to arrest attention, startle, surprise. . . . The image as shock and the image as cliché are two aspects of the same presence."[2] Yet the image fatigue of the late twentieth-century visual world of which Sontag spoke grew out of a mass culture that was radically different from that of the 1920s, when photography still retained its spectacular aura for mimetic reproduction and the mass media had been circulating photographic images of disaster globally for only a little over three decades. In 1923, the imaging of disaster ultimately attested to the Japanese nation-state's survival and its tenacity in the face of adversity. This was highly significant because of the unprecedented, "historic" nature of the catastrophe. In the first few days after the quake, there were rumors throughout Japan and abroad that Tokyo and Yokohama had been completely obliterated. Visible evidence proved otherwise. The cyclical nature of destruction in Japan precluded an apocalyptic vision, even when the nation was seemingly on the brink of total annihilation (this would change in the postwar period after the introduction of the atomic bomb).

The photographic eye is not singular and the print media are not univocal. In fact, the imaging of the Kantō disaster displays a cacophonous world of widely diverging experiences and objectives, perhaps not surprising in a heterogeneous society, although it is not always immediately evident from the many rhetorical efforts to galvanize the populace through a moralistic, unifying discourse of resilience and optimism. Therefore, one must tease out the strands of meaning in these visual representations and ascertain how the visual became imbued with social significance.

Technologies of Seeing and Visible Evidence

Before the advent of popular radio, print news media were the primary means of disseminating information throughout Japan. For the most part, the Kantō quake destroyed all telecommunications like the wireless and the telephone.[3] While the publishing industry and the news media were devastated, they recovered quickly.[4] At least twelve newspapers were being published in Tokyo at the time, and two of the five major papers, *Hōchi shinbun* and *Tōkyō nichinichi shinbun,* were able to resume printing by the fourth day after the quake. Others, such as the *Tōkyō asahi shinbun* and *Yomiuri shinbun,* were up and running again by 12 and 15 September, respectively.

However, this brief break did not stop the dissemination of information. Although still predominantly a regional business, newspapers from throughout the nation published information about the quake—some fact, some fiction—that streamed out of the capital. There were also numerous public official statements (*kanpō*) providing updates about the areas of destruction (including regions surrounding the capital), locations of refugee facilities, state of telecommunications and transportation, status of foreign aid, and activities of international dignitaries, such as the American ambassador. Even in the first month after the earthquake, news publications collectively began to parse the visual lexicon of Japan's massive natural disaster and to codify the tropes that would define it.

Some of the earliest press images of the disaster were from the areas surrounding the capital and ran in Osaka-based news outlets. The coverage indicated how difficult it was for journalists to get close to the center of the devastated region. It marked the sudden stream of attention inward toward the capital and the flow of refugees outward away from the devastation. The *Ōsaka mainichi shinbun* was one of the first newspapers to publish photographs of the disaster in a special edition. With its reporters unable to reach Tokyo, the newspaper featured scenes of the city of Numazu in Shizuoka prefecture. On the day after the event, the front page pictured large fissures in the ground, a house roof dramatically in ruins, and a close-up of displaced refugees waiting at a train station. The second page featured Osaka residents preparing supplies to send to the stricken region and a pilot boarding his airplane to fly to Tokyo to provide assistance. Subsequent issues had aerial shots of billowing clouds of smoke from the smoldering landmass of the city and ground shots of all aspects of the destroyed cityscape.[5] Some formal portrait lozenges were scattered throughout the images of destruction, spotlighting well-known individuals who had perished or been injured in the disaster.

The photographic eye directly affected perception. Aerial photographs of the damaged areas dwarfed the people on the ground, emphasizing the expanse of devastation. Such images inherently express scale and magnitude and speak more of civilizational and urbanistic destruction than of individual lives lost. Photographs of burning land seen at a distance have their roots in the print images of Edo. For example, images in *Ansei-Era Observations* show Edo from a bird's-eye panoramic view, highlighting large swaths of billowing smoke rising from the burning city. The new heights achieved in aeronautic technology amplified the vastness of the photographic gaze and further depersonalized the connection between viewer and devastation. They also added a layer of technological mediation, interposing the aircraft as well as the camera. Aerial photography was already an important strategic technology in war and disaster reconnaissance imaging abroad. The "logistics of military perception" developed during World

War I, according to Virilio, had as much tactical importance as the timely supply of ammunition. As a "watching machine," the aerial camera turned the photographic eye into a weapon.[6] Moreover, when catastrophe struck, aerial photography became not merely a reproductive technology but also a critical method of disaster management: relief operations benefit enormously from immediate information about the extent of damage and the unfolding destruction, and are better able to deploy emergency fire and rescue teams efficiently. Aerial photography is still a critical means of identifying priorities in postearthquake relief operations, providing information for logistical planning and, critically, for determining access routes.[7]

While drawing attention to the devastating extent of the damage, aerial photography's commanding view also communicated a controlling gaze in images of the 1923 quake: if the disaster could be presented in its totality, it could somehow be controlled. However, media efforts to establish a commanding vision also articulated the limits of such control as critical sites still in jeopardy remained concealed from view. On the cover of *Asahi Graph*'s (*Asahi gurafu*) special pictorial issue "Full Record of the Great Earthquake" (*Daishinsai zenki*) (fig. 2.1), which proclaimed itself "the most well-organized record and pictorial report," a single large official aerial photograph of the burning capital taken by a naval aircraft shows white smoke gusting off the page, obscuring two-thirds of the ground.[8] The ordered grid of the city and the pronounced roadways, the only identifiable details from this height, seem about to be consumed by the unstoppable conflagration. The same photograph, reproduced in the *Ōsaka asahi shinbun* with the caption "Raging fires consuming the imperial capital," injects the scene with more ominous portent by adding the critical descriptive detail that the white smoke is shrouding the Imperial Palace—the home of the empire's divine ruler—and thereby calling into question the survival of the nation's leadership.[9]

Some aerial photographs, such as the iconic photograph of the Marunouchi district that ran in the *Hōchi shinbun* (fig. 2.2), emphasized the technological mediation of the photographic eye by incorporating the wing of the airplane as a framing device. Such views reinforced both the tremendous expansion of modernity's vision, literally and figuratively, and its ever more disastrous consequences. Nevertheless, the state again tried to instrumentalize the totalizing vision of aerial film in 1926, when it sought to demonstrate the success of large-scale urban reconstruction in the fifteen-minute film *Reconstruction of the Imperial Capital Seen from an Airplane* (*Kokusen nite fukkō no teito e*). (The film was shot by well-known cameraman Shirai Shigeru and produced by the Ministry of Education [Monbushō].)[10] Through movement, film increased the temporal and spatial expansiveness and impact of the aerial vision. Yet while the semblance of totality was useful for the Japanese government, it was deceptive, because

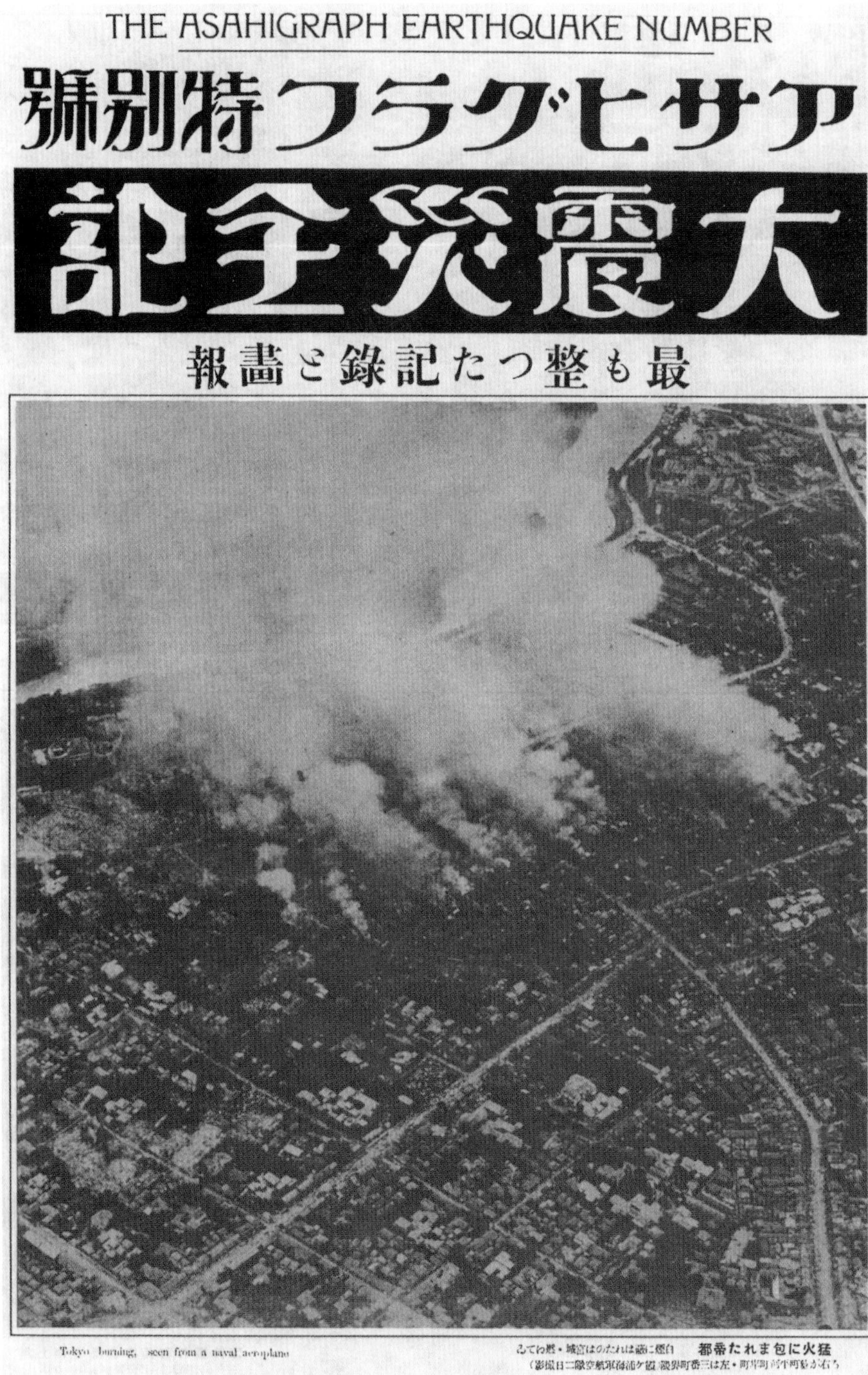

2.1 Aerial photograph of the burning capital, front page. "Full Record of the Great Earthquake" (*Daishinsai zenki*), special issue, *Asahi Graph,* October 1923.

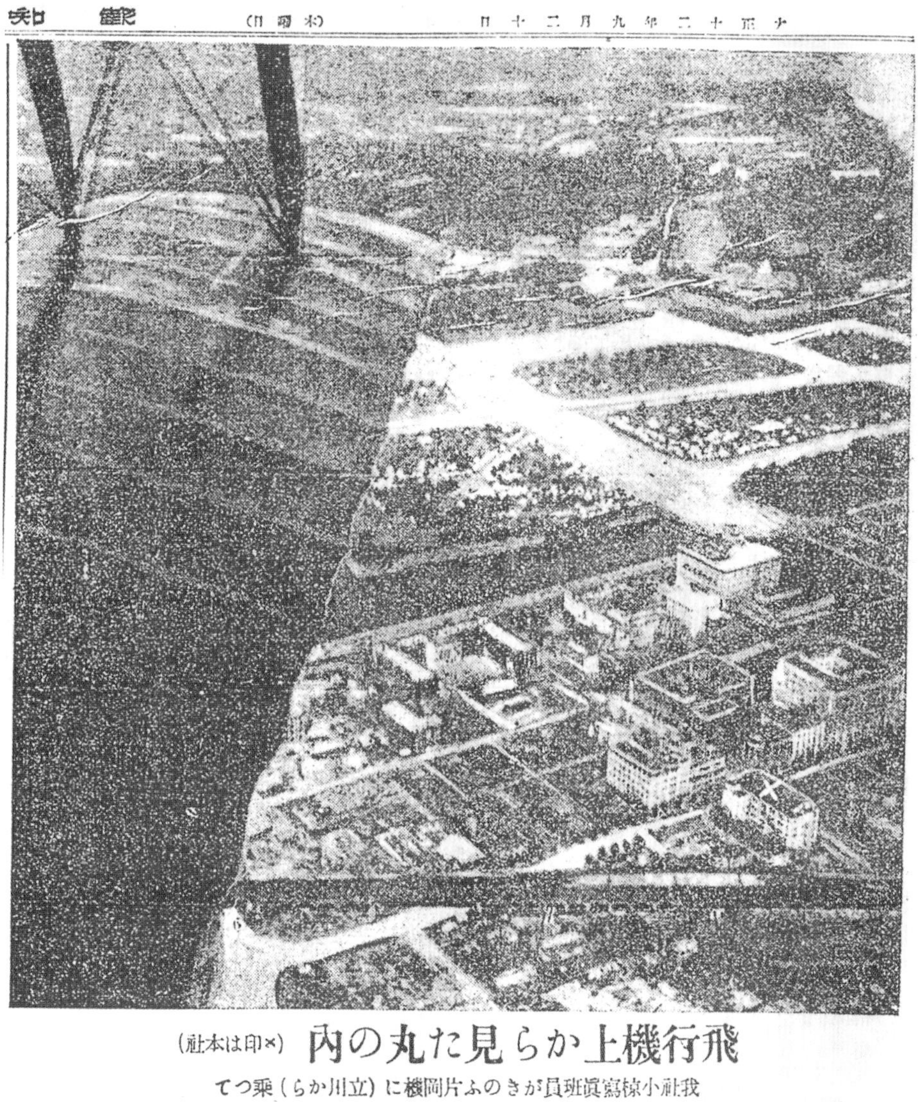

2.2 *Marunouchi Seen from an Airplane* (*Hikōkiue kara mita Marunouchi*), photograph. *Hōchi shinbun*, 20 September 1923, A.M. ed., 7. By permission of Yomiuri Shinbun.

no view could encompass the full implications of the event or the ramifications of reconstruction.

Schematic maps of surviving transportation networks, refugee facilities, and the deployment of military personnel throughout the region affected by the quake also visually supported the authority of the state by implying control over a vast area of devastation. Like aerial photography, such maps were a powerful logistical weapon. The

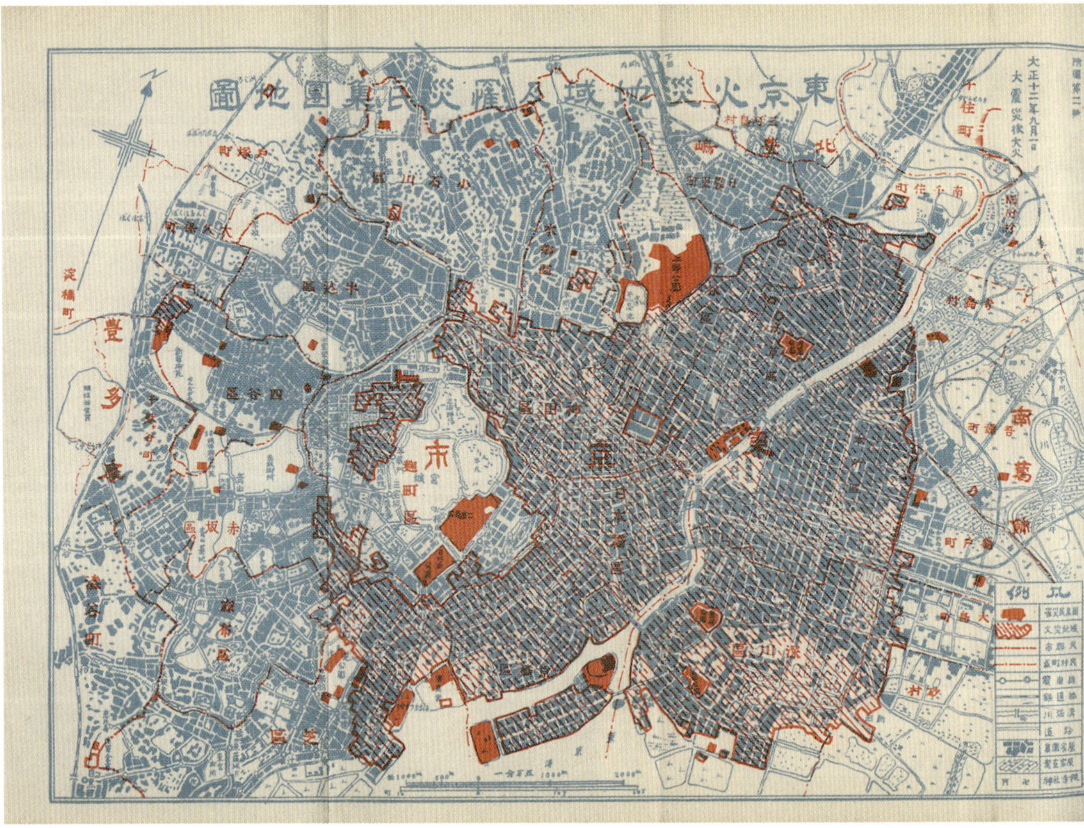

2.3 Map of Tokyo-Area Fires and Victim Gathering Sites (Tōkyō kasai chiiki oyobi risaimin shūdan chizu), with hatches indicating areas of destruction. Dai Nihon Yūbenkai Kōdansha, *Taishō Daishinsai daikasai* (The Great Taishō Earthquake and Conflagration) (Tokyo: Dai Nihon Yūbenkai Kōdansha, 1923).

detachment conveyed by the scale and magnitude of aerial photography was echoed in the many chillingly clinical geographical mappings of the disaster, which covered the region with red hatches indicating the complete destruction of vast, blighted sections (fig. 2.3). Many Taishō maps, building on their increasingly detailed late-Edo precursors, were scientifically cartographic, delineating the exact layout of the metropolitan area, and the graphic red hatching overlaid on the schema of the city is utterly dispassionate in marking the decimation of nearly half of its land area, lending visual authority through apparent facticity.[11] The technique of blotting out large areas of the city with red pigment to indicate fire damage was common in late-Edo mapping. For example, it appears in the news *kawaraban* (broadside) maps of fire-blighted areas of Kyoto destroyed during the Kinmon or Hamaguri Rebellion (Kinmon no Hen or Hamaguri

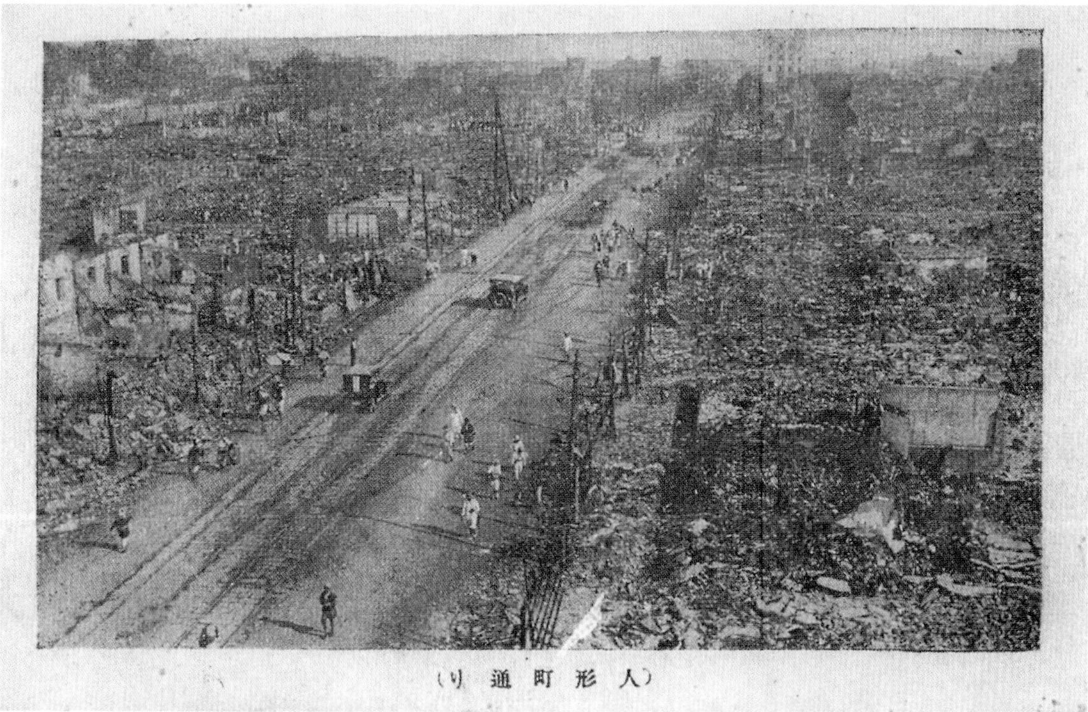

2.4 *Ningyōchō Street* (*Ningyōchō dōri*), postcard, 1923. Ishii Toshio Collection. By permission of Japan Publishing Copyright Association.

Gomon no Hen) of 1864, when the Tokugawa shogunal forces quashed an uprising by antiforeign imperial loyalists.[12] The 1923 maps refined this practice. Yet they are not entirely lifeless, for the small patches of red indicating safe havens for refugees look eerily like blood hemorrhaging across the image.

In photographs, the subtle differences in perspective among elevated and panoramic scenes of the devastated cityscape give hints to the scope of the disaster. Some images visually extended the horizon line of the urban ruins, giving Tokyo the look of a never-ending postapocalyptic wasteland (fig. 2.4). These open, decimated vistas expose recently inhabited land to the observer and show new views born out of the destruction; where a dense metropolitan cityscape previously blocked the gaze and directed it along the axes of major boulevards, now one's gaze can extend for miles unobstructed. This new visibility is reinforced through the picturing of areas commonly known for their urban density and large buildings: Nihonbashi, Ginza, Marunouchi, Kanda, and Kyōbashi, for example. In areas where the burned-out carcasses of monumental Western-style edifices remained, the viewer can see through their empty, charred shells

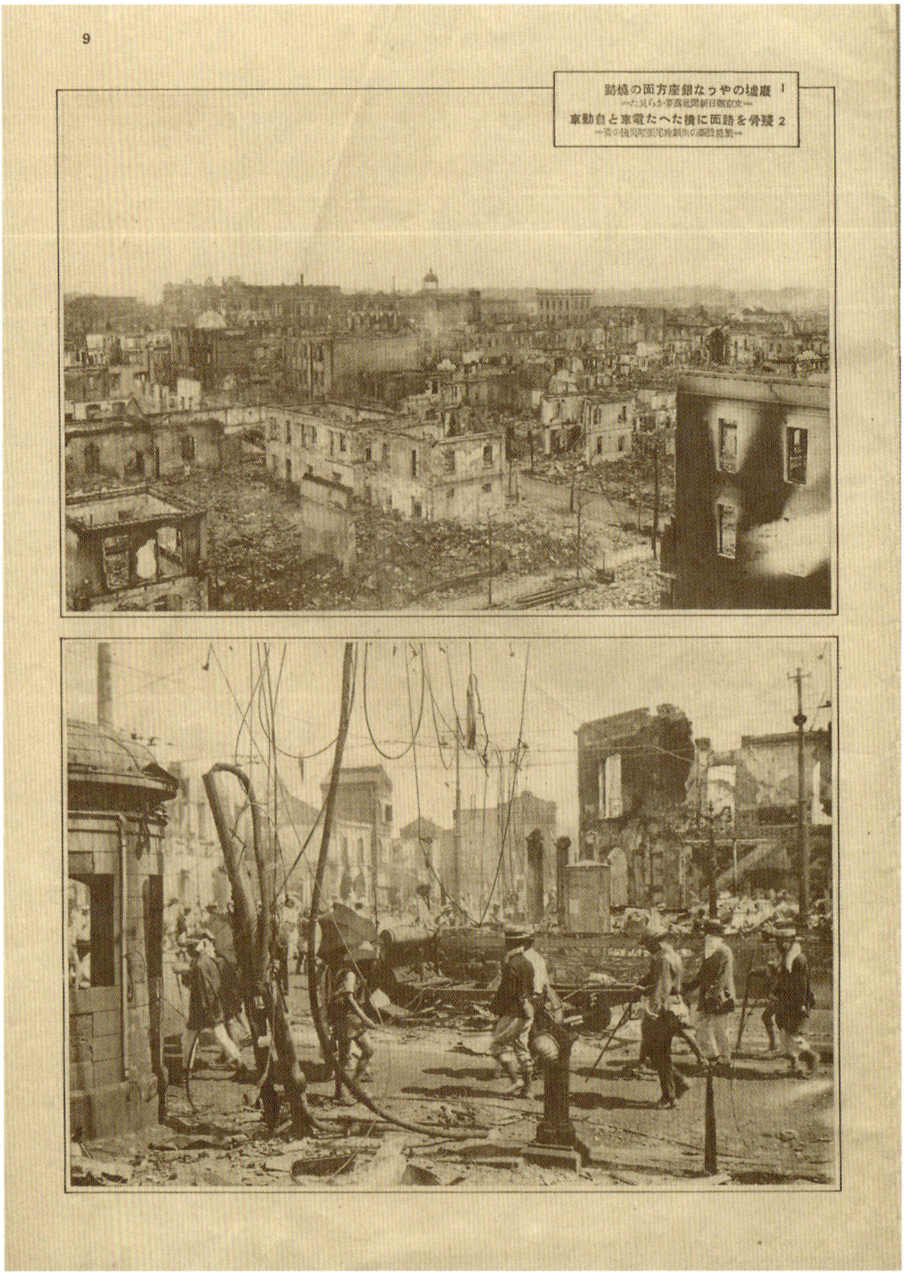

2.5 *The Burned Expanse toward Ginza That Looks like Ruins* (*Haikyo no yōna Ginza hōmen no yakeato*) *(aerial view, top)*; *Burned Skeletal Remains of Trams and Cars Lying across the Road* (*Zankotsu o romen ni yokotaeta densha to jidōsha*) *(ground view, bottom)*, photographs. Ōsaka Asahi Shinbunsha, *Daishinsai shashin gahō* (Photographic Pictorial of the Great Earthquake), vol. 2 (Osaka: Ōsaka Asahi Shinbunsha, 25 September 1923), 9.

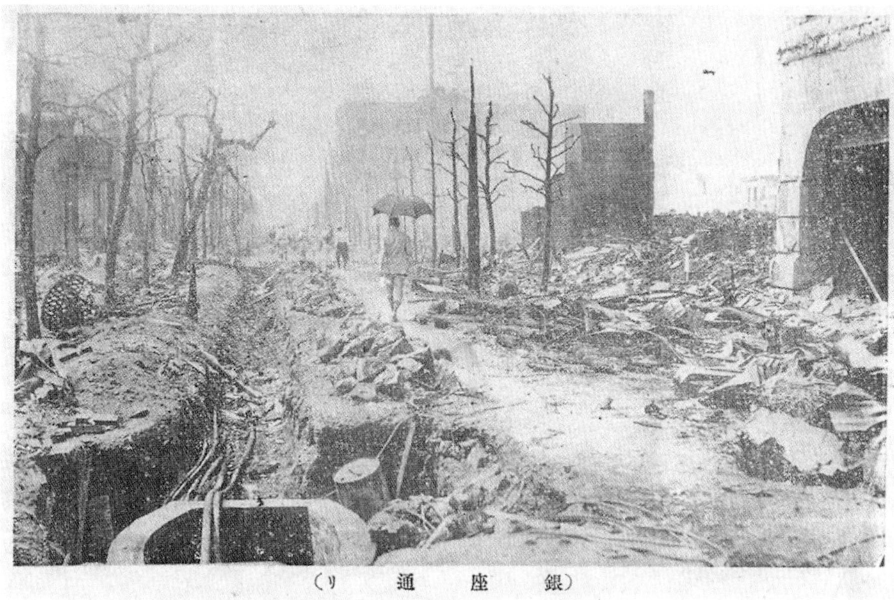

2.6 *Ginza Street* (*Ginza dōri*), showing a lonely figure on the once-fashionable boulevard, postcard, 1923. Ishii Toshio Collection. By permission of Japan Publishing Copyright Association.

into the distance (fig. 2.5 top). The known scale of these buildings established a visual benchmark for perceiving the magnitude and enormity of destruction, and the transformation of the solid, opaque structures into open screens reiterates the evisceration of the city itself.

In contrast to the aerial images, photographs taken from a ground perspective invited the viewer to identify with figures making their way through the treacherous rubble (fig. 2.5 bottom). These images reduce the city to human scale, thereby acquiring emotional resonance and allowing the viewer to identify with the plight of the survivors. In one photograph (fig. 2.6), a single male figure holding an umbrella and walking away into the murky distance amid the ruins of the formerly fashionable Ginza district imbues the scene with pathos. Such images of the city carry a tragic irony in their contrast with the myriad familiar images of well-dressed ladies and gentleman sauntering down the elegant commercial boulevards of the Ginza, Tokyo's equivalent of the Champs-Élysées in Paris or Fifth Avenue in New York. The popular leisure activity of "strolling in the Ginza" (a practice known as *ginbura*), past its fashionable shops and department stores, is profoundly transformed in these photographs; clearly there was no window-shopping to be done here. Postquake *ginbura* was a promenade through ruins.

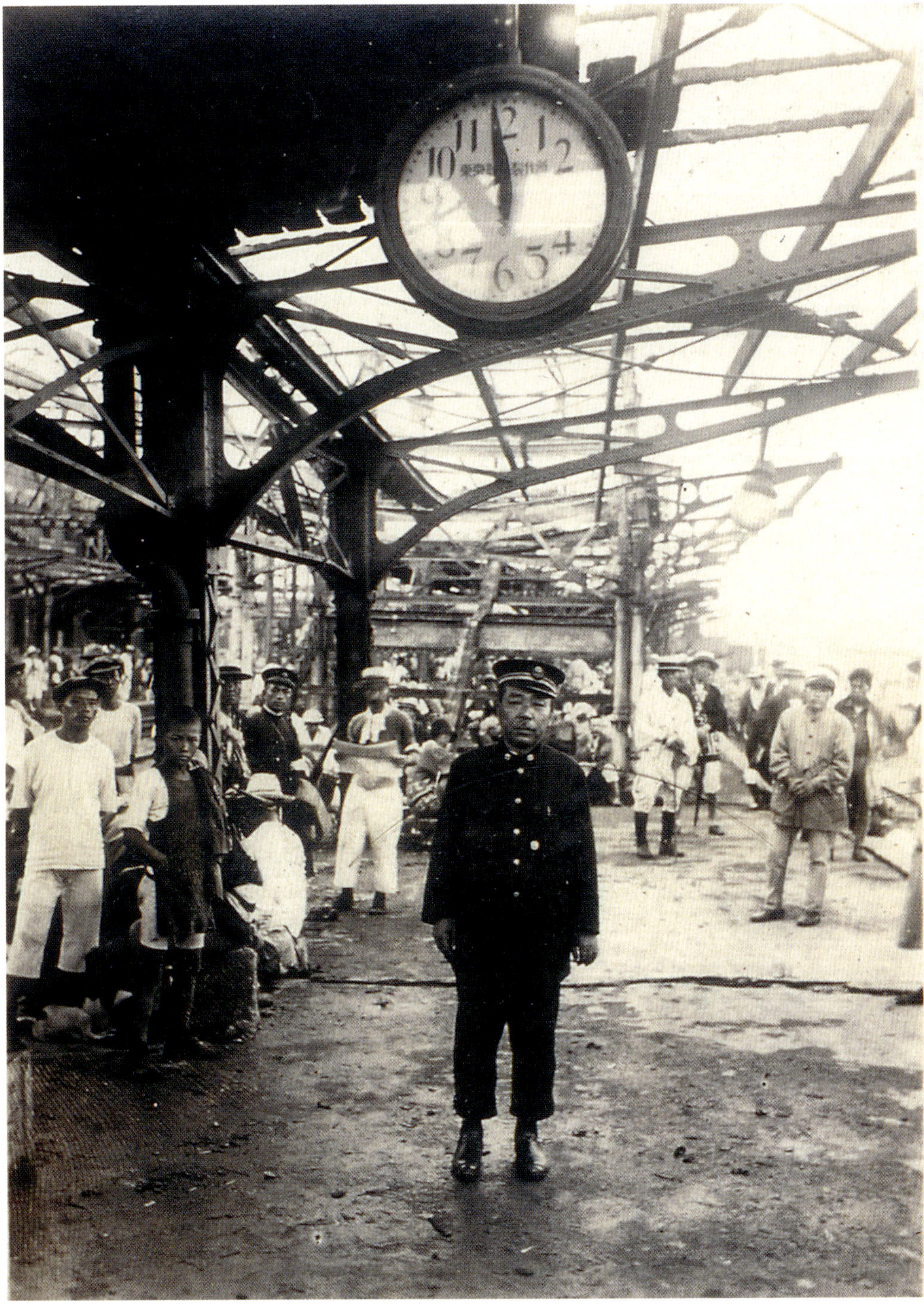

2.7 Artist unknown, *Yokohama Station*, photograph, 1923. Collection of Yokohama Kaikō Shiryōkan.

Eduardo Cadava has noted photography's ability to mediate vision and experience in relation to trauma, particularly in its imaging of ruins: "The image of ruin tells us what is true of every image: that it bears witness to the enigmatic relation between death and survival, loss and life, destruction and preservation, mourning and memory." An "image of ruin often speaks of the death and impossibility of the image," its "inability" to tell a story. The image, then, communicates "a kind of silence in the face of loss and catastrophe" but also the "capacity to bear the traces of what it cannot show, to go on, in the face of this loss and ruin, to suggest and gesture toward its potential for speaking."[13] The viewer's identification with the lone figure in the ruins signals these sentiments of loss and survival. In its quietude, the photograph speaks eloquently about the sense of isolation one feels even in the midst of Japan's greatest collective tragedy. It also reiterates the temporal distance from the event, a divide that is produced the moment the photograph is created, which, in Benjamin's terms, transforms all photographic images into the ruins of history.[14]

Time seemed to stand still when the quake hit, an emotion literalized by the many iconic images of clocks as they were arrested by the temblor at exactly 11:58 A.M., the moment when people's lives were irreparably changed. In one image (fig. 2.7), a conductor stands on the train platform in Yokohama, his arms haplessly drooping by his sides, his pants almost boyishly too short for his frame, and gazes stunned at the camera while the frozen bystanders around him create a visual arc that leads the eye to the clock looming above that hauntingly reads 11:58. This luminous image is almost surreal. The clock serves as a metonym for the traumatic disruption of the quake and recapitulates photography's own interruption of the flow of time.

While capitalizing on photography's interruptive capacity to produce the visible evidence of history, media organizations were also acutely aware of its epitaphic character, which prompted widespread attempts to reanimate news images through creative editorial layouts and photo collage of pictures in different sizes and shapes. For example, the cover of *Ōsaka asahi shinbun*'s third special issue on 4 September, bearing the bold headline "The Great Chaos of Tokyo" (*Daikonran no Tōkyō*) (fig. 2.8), set pictures of varied sizes at skewed angles on the page and overlapped images for effect. This editorial style had just recently come into common use in print media with the inauguration of graphic pictorial magazines like *Asahi Graph,* which began publishing in late January 1923. The approach allowed viewers to apprehend the disaster from multiple perspectives, viewing it simultaneously from the air, on the ground, at a distance as part of a vast expanse, or from close up as an intimate scene of death and destruction. It sutured all of the views of the photographic eye into one animated visual experience.

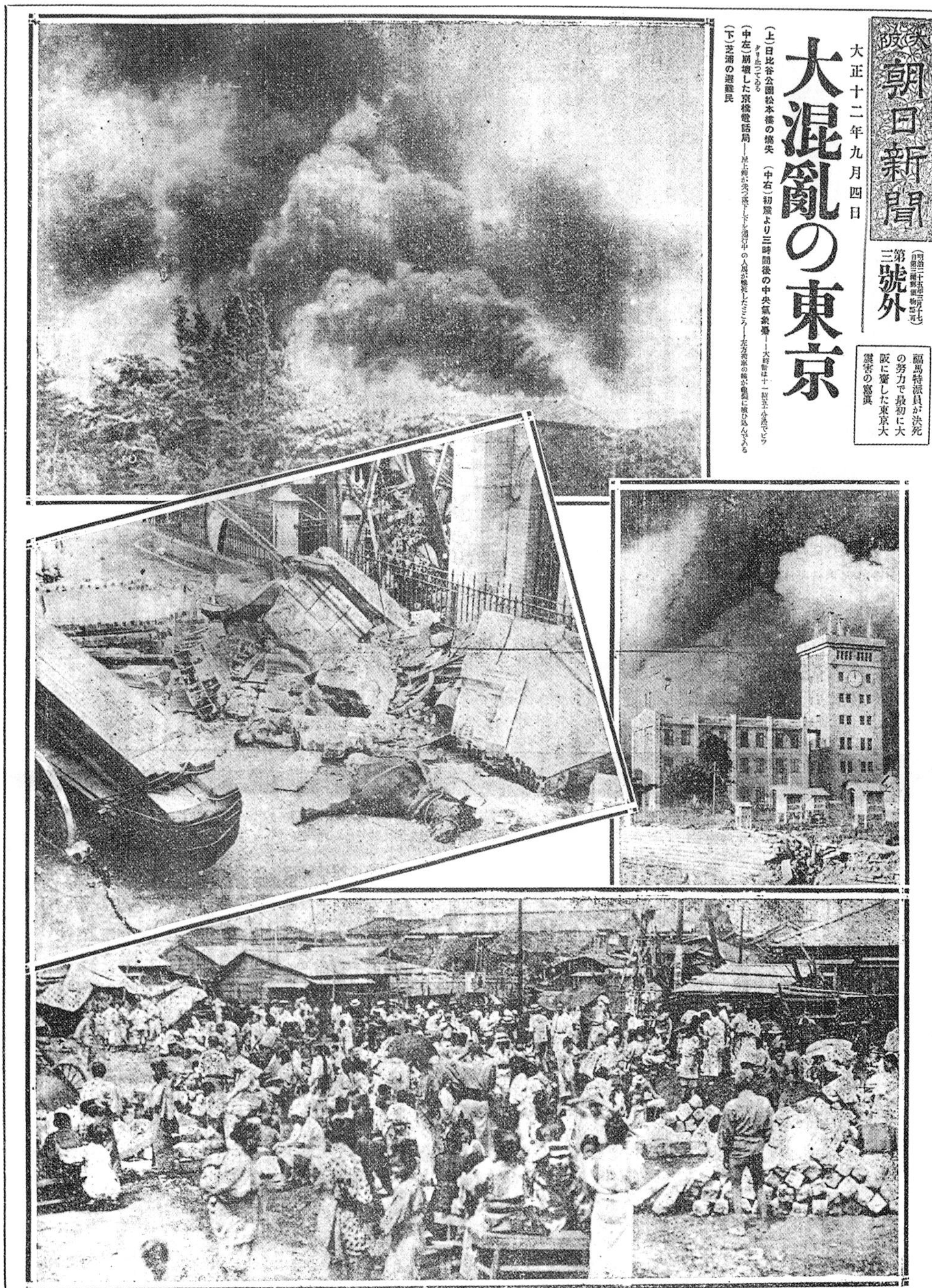
大阪朝日新聞

第三號外

大正十二年九月四日

大混亂の東京

福馬特派員が決死の努力で最初に大阪に齎した東京大震害の寫眞

(上)日比谷公園松本樓の燒失

(中右)初震より三時間後の中央氣象臺

(中左)崩壞した京橋電話局

(下)芝浦の避難民

2.8 "The Great Chaos of Tokyo" (*Daikonran no Tōkyō*), photograph, front page. *Ōsaka asahi shinbun,* special edition, 4 September 1923.

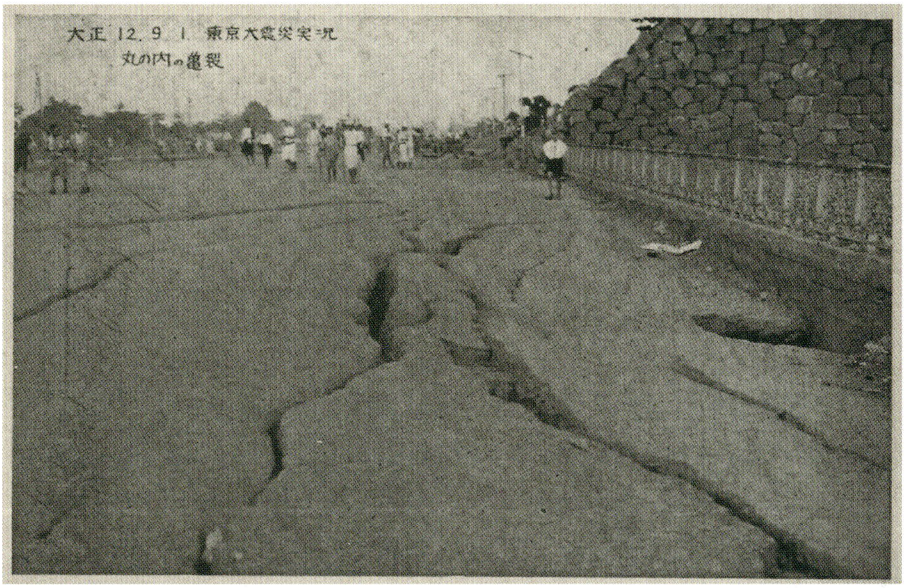

2.9 *Large Fissures in the Ground in Marunouchi* (*Marunouchi daijiware*), postcard, 1923. Collection of the author.

Setting the images off kilter was a way to simulate the dynamism of the news being pictured and, in this case, to express the off-kilter nature of the situation on the ground, particularly the dramatically collapsed structures being photographed around the city. The many special-issue pictorials and souvenir historical photo albums issued in the months and years after the quake continued to use this photo-collage style in an effort to simulate the modern ability to see the world in multiple modes simultaneously.

Manipulation of the photographic eye could also reanimate images to amplify the quake's physical disfigurement of the earth's surface as well as the man-made creations that sat upon it. Images captured the massive fissures (*jiware*) in the ground created by the temblor's violent activity—giant gaping wounds that came to symbolize the physical and psychic scars left by the experience. They visually expressed the magnitude of the earth's displacement. Dozens of images shot from dramatic camera angles further exaggerated the size of the fissures. One frequently reproduced image, taken in Marunouchi near the Imperial Palace (fig. 2.9), uses the photographic eye to enlarge the fissures while diminishing the scale of the figures in the distance, metaphorically juxtaposing the power of nature with the powerlessness of man or imperial rule itself. It also self-consciously heightens the viewing terror by making the fault lines appear to run directly under the viewer, as if the earth were literally cracking underfoot. Such images of violent fracture literalize the movement of the earth and the intensity of the shocks that rent the ground asunder. They effectively animate the forces of nature, giving the ground tentacles that seem to reach out from the earth's core.

Seismography's Scientific Vision

With the development of the seismograph, the seismic force that produced these fissures in the ground could be translated into a scientific vision of magnitude. Seismograms are graphic records of ground motions at a particular location over time. Seismographs consist of a seismometer (a sensitive instrument, such as a pendulum, to measure ground motion), a precise timing element, and a recording device; in the 1923 quake, the seismograms were recorded on a piece of paper attached to a rotating drum. Japanese scientists, such as Ōmori Fusakichi (1868–1923), were major contributors to the science of seismology and seismography, particularly improving the understanding of the relationship between the duration of tremors and the location of the hypocenter of seismic activity. The expression of magnitude in a world-historical context became more important with the inauguration of modern seismology. Gregory Clancey has explicated Japan's pioneering role in the development of seismology and pointed out that "in objectifying seismicity and seeking to inscribe it onto paper with machines, nineteenth-century science did its best to make the sudden and sublime into the gradual, regular, and, above all, understandable."[15] Yet Clancey does not consider the impact of the seismograph's ability to visualize the disaster in a new way, which provided an abstract index for posterity of the spasmodic shocks and aftershocks of the earth's movement. Whereas the width and depth of the fissures gave one indication of the violence of terrestrial movement, the frenetic vertical strokes of the seismograph provided an alternate visualization of the quake's intensity. It also recorded the duration of individual shocks, not their cumulative results. The 1923 quake was the first earthquake in Japan to leave profuse seismometrical data of the main shock and aftershocks. For the first time, the Japanese public could see its experience abstracted into pure magnitude and duration. This also made the quake more easily comparable to other seismic events.

The only full scientific recording in Japan of the 1923 temblor (consisting of seventeen hundred separate quakes over three days) was the seismogram recorded at Tokyo Imperial University (fig. 2.10).[16] With seismology's newly recognized scientific authority, it became powerful visible evidence of the relative magnitude and gravity of the Kantō quake in world history. The 1923 seismogram appeared repeatedly in popular news media and commemorative pictorials. It is even featured on the open-

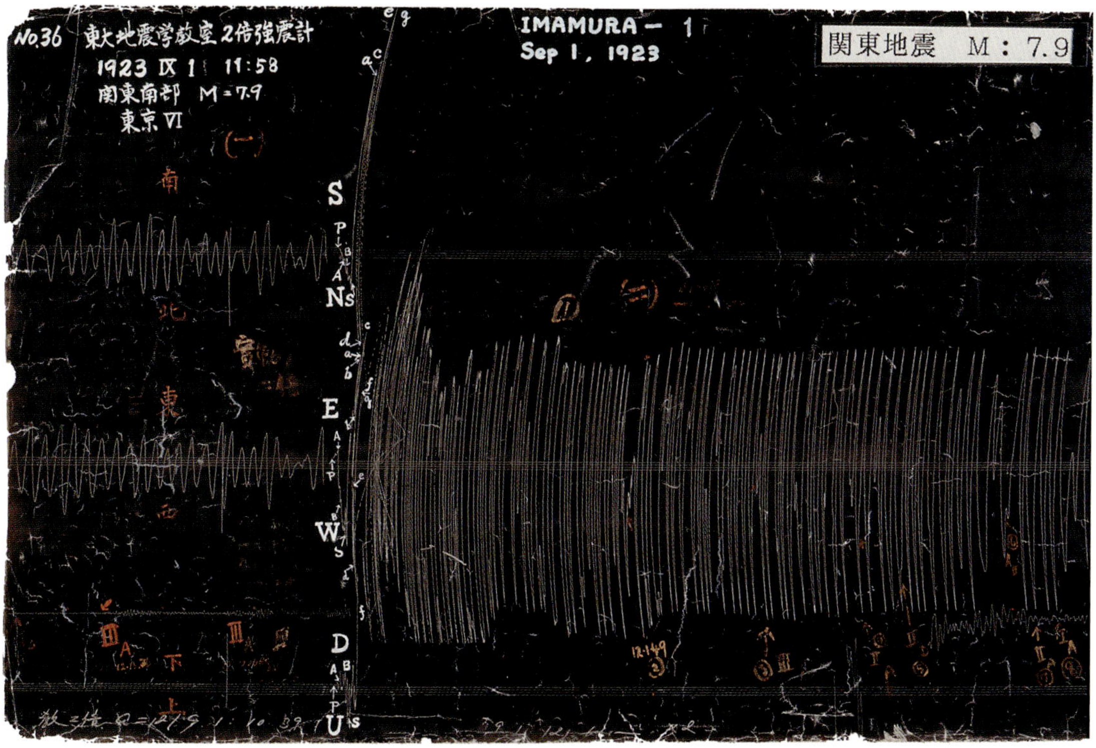

2.10 Seismographic reading of the Great Kantō Earthquake recorded at the Earthquake Research Center at Tokyo Imperial University, Hongō, 1 September 1923. Courtesy of Earthquake Research Institute, University of Tokyo.

ing pages of the Home Ministry's official commemorative pictorial volume issued on the third anniversary of the quake in 1926, where it is paired with the iconic photograph of the clock tower on the Central Meteorological Observatory that stopped at exactly 11:58 A.M. (fig. 2.11)—suggesting that the violence of the recorded shocks literally arrested time. The pairing of images visually translates disaster's shocking somatic and perceptual experiences into the captured temporality of history. The media also used seismograms to convey the expansive reach of the quake as well as its severity—for example, by featuring seismographic readings in the Kansai region hundreds of miles away from the capital. The *New York Times* even reported on 2 September 1923 that seismographic observatories around the world, from Hawaii to London, recorded the "five hour" Japanese quake.

The scientification of the seismic force of the earthquake made it objectively comparable to other events, which established the Kantō quake's preeminent stature as the

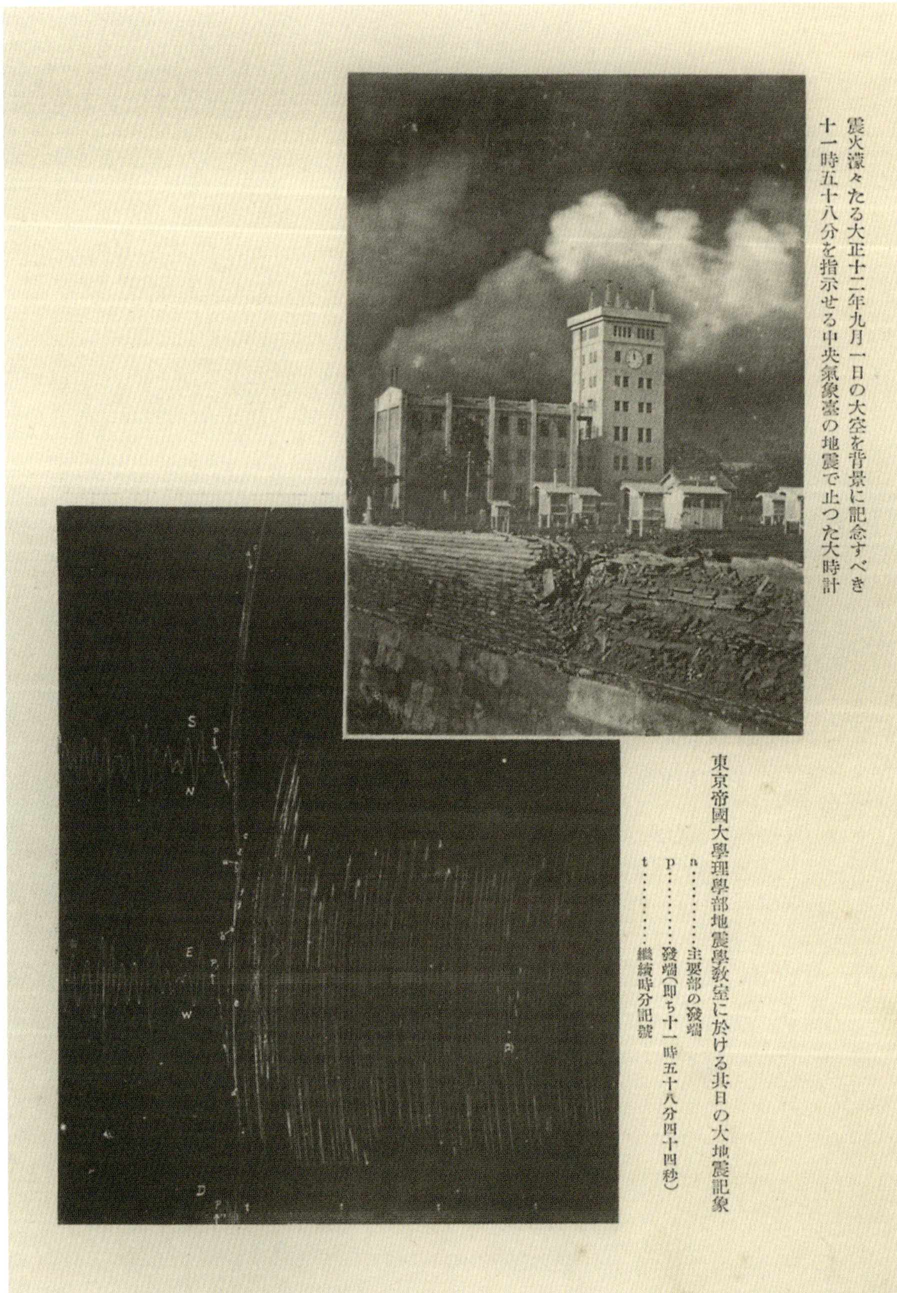

2.11 Clock tower on the Central Meteorological Observatory, stopped at exactly 11:58 A.M., 1923, photograph *(top)*; seismographic reading of the Great Kantō Earthquake at the Earthquake Research Center at Tokyo Imperial University, photograph *(bottom)*, 1923. Fujisawa Morihiko, ed., *Taishō Shinsaishi shashinchō* (Photograph Volume of the History of the Taishō Earthquake) (Tokyo: Naimushō Shakaikyoku, 1926).

largest and most powerful of any recent earthquake disasters in Japan or around the world. It gave the event relative *historical* gravitas. The repeated representation of the seismogram gave the actual physical object recorded at Tokyo Imperial University's Engineering Department Seismology Division the aura of a singular historical artifact. The official seismographic readings were framed together and prominently exhibited in the first section of the earthquake memorial museum later erected in Honjo. Through the collective visuality of the disaster (the images produced, the visual authority they conveyed through new visualization technologies, and the public's faith in the facticity of this body of visible evidence), the museumification of such artifacts solidified the master narrative of the event and conveyed its historical significance.

The Limits of Visual Authority

As with all disasters, the Great Kantō Earthquake awakened a deep-seated fear about the possible dissolution of social order into chaos. The day after the quake, the newly installed prime minister, Yamamoto Gonbei (Gonnohyōe, 1852–1933), a former admiral in the Japanese navy who had served as prime minister once before (1913–14), declared martial law for Tokyo and Yokohama, which continued until 16 November.[17] By 8 September, twenty-one infantry regiments, six cavalry regiments, eighteen engineer battalions, two railway regiments, two telegraph regiments, as well as flying corps and medical corps, had been mobilized in the area. Of the fifty thousand troops, thirty-five thousand army personnel were posted around Tokyo alone. A week after the declaration of martial law, on 7 September, the government spelled out in greater detail the "extraordinary measures" necessary to maintain public order. Emphasizing the precedence of national community, the edict gave the police broad authority to "suppress such meetings, newspapers, magazines, or advertisements as are calculated to be detrimental to the maintenance of public order." It also gave the commander of martial law in the Kantō districts (*Kantō kaigen shireibu*), Fukuda Masatarō (1866–1932), the power to examine mail and telegrams, close land and sea routes, and conduct house searches.[18]

The moment of extreme chaos generated powerful visual assertions of authority. Officials of all varieties appear throughout the media coverage of the event: policemen, army personnel, the fire brigade, medical staff, and government bureaucrats (fig. 2.12). Photographs show them actively engaged in critical disaster activities: shepherding hordes of refugees to safety, extinguishing fires, reconnecting vital telecommunications, tending to the injured, and valiantly standing guard against the threat of imminent chaos. The authorities approached the masses as both sympathetic vic-

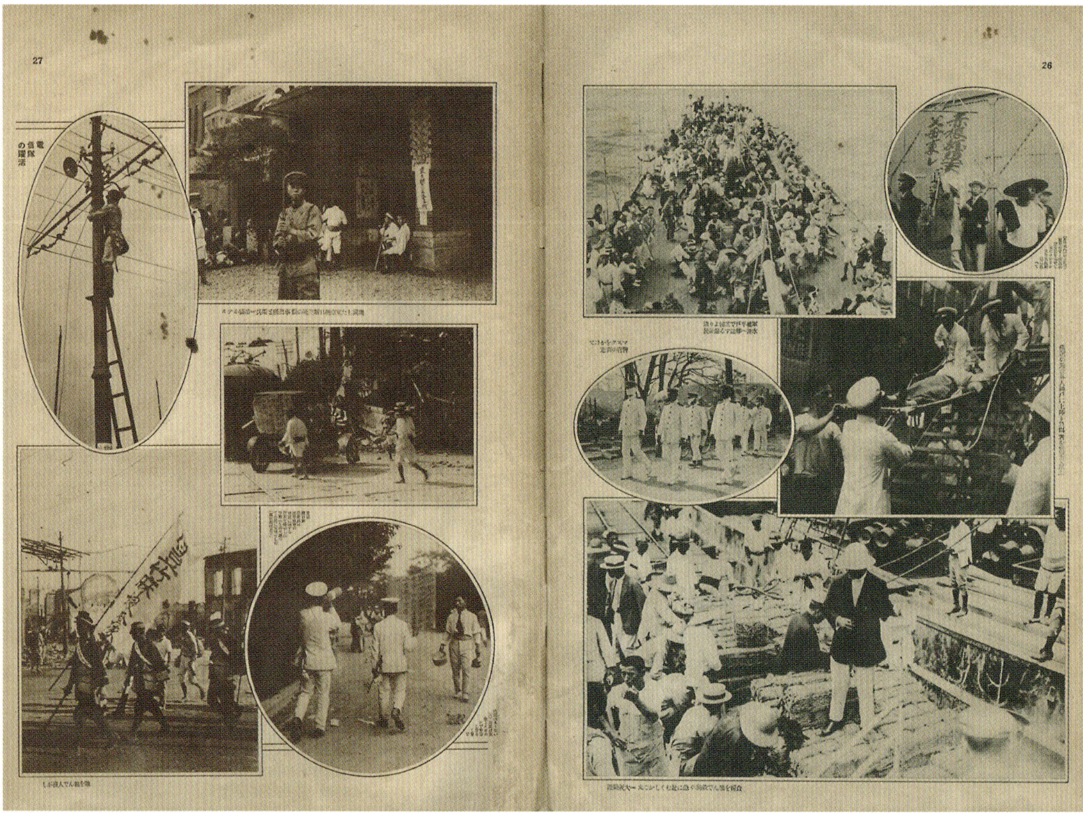

2.12 Authorities in Tokyo after the earthquake, photographs. Ōsaka Asahi Shinbunsha, *Daishinsai shashin gahō* (Photographic Pictorial of the Great Earthquake) (Osaka: Ōsaka Asahi Shinbunsha, 15 September 1923), 1:26–27.

tims and potentially harmful assailants who could at any moment degenerate into violent mobs. Widely circulated maps charted the deployment of troops throughout the Kantō region both to assuage the fears of the people by conveying a sense of security and to forewarn any crowds massing in refugee centers that unruliness would not be tolerated. Presentation of this visible evidence of purported control over the area conveyed authority even when eyewitness accounts indicated that the authorities were not in complete control.

To direct the flow of news, immediately after the quake hit on 1 September the Japanese government issued the press a "Notification of Request for Cooperation" (*Kyōryoku irai tsūchō*) that effectively turned all public media into quasi-official organs of the state. Intended to preserve peace and prevent disorder, this edict prohibited the publication of certain kinds of information and inflammatory images of dead bodies.

The state's control of the press and heavy censorship were in effect for over two months after the earthquake, until the end of November.[19] Under the auspices of the government, the mass media immediately began a moral campaign to create social solidarity under the purportedly august and enlightened rulership of the imperial family, with the emperor as the pater familias of the national family-state. This campaign was effectively a form of social control that erased or vilified behavior deemed incompatible with the interests of the national collective.

The two types of imperial earthquake imagery that bolstered the nation's authority were benevolence and steadfast leadership. Early images, such as one that ran in the 4 September special issue of *Shin Aichi,* the Nagoya-based newspaper, show a distant photograph of the Imperial Palace with the caption "The Imperial Palace that has been opened to refugees," expressing the imperial household's profound benevolence toward its people.

Numerous pictures of the prince regent Hirohito (who later became the Shōwa emperor) circulated in the press, showing him in military uniform meeting with the home minister and military officials or surveying the damage on horseback. On 17 September, Hirohito conducted a horseback inspection at Ueno Park with General Fukuda (fig. 2.13).[20] Equestrian images projected strength and command. The elevated position of the prince regent, high on his horse, underscored his nobility, reminiscent of the age-old Japanese practice of elevating the ruler above his subjects during shogunal processions in Tokugawa times. The figure with his back to the viewer saluting the crown prince, perhaps Fukuda, in a sense stands in for the imperial subject and implicitly calls the populace to attention. Such images also reinforced the notion that the command structure was intact; the rule of law was functioning. This message was supported by the circulation of portraits of Yamamoto's rapidly selected cabinet ministers in the press from the day after the quake, which signaled the continuation of the government's venerable bureaucracy. Photographs of the active home minister, Gotō Shinpei, former mayor of Tokyo and one of the main architects of the capital's reconstruction plan, also appeared frequently in the news media to give visible form to the government, temporarily shorn of its monumental public edifices.

One image, immortalized across the media, served as a visual metaphor of both the state's authority and national solidarity under divine rulership: the surviving statue of Joan of Arc that remained heroically standing in Kanda in front of the Franco-Anglo-Japanese Women's Higher School (Futsueiwa Kōtō Jogakkō) as the building collapsed around her (fig. 2.14). In film footage shot soon after the quake, which the Ministry of Education later assembled into a feature-length film about the city's disaster and its reconstruction, two Catholic nuns carefully descend an exposed internal staircase left

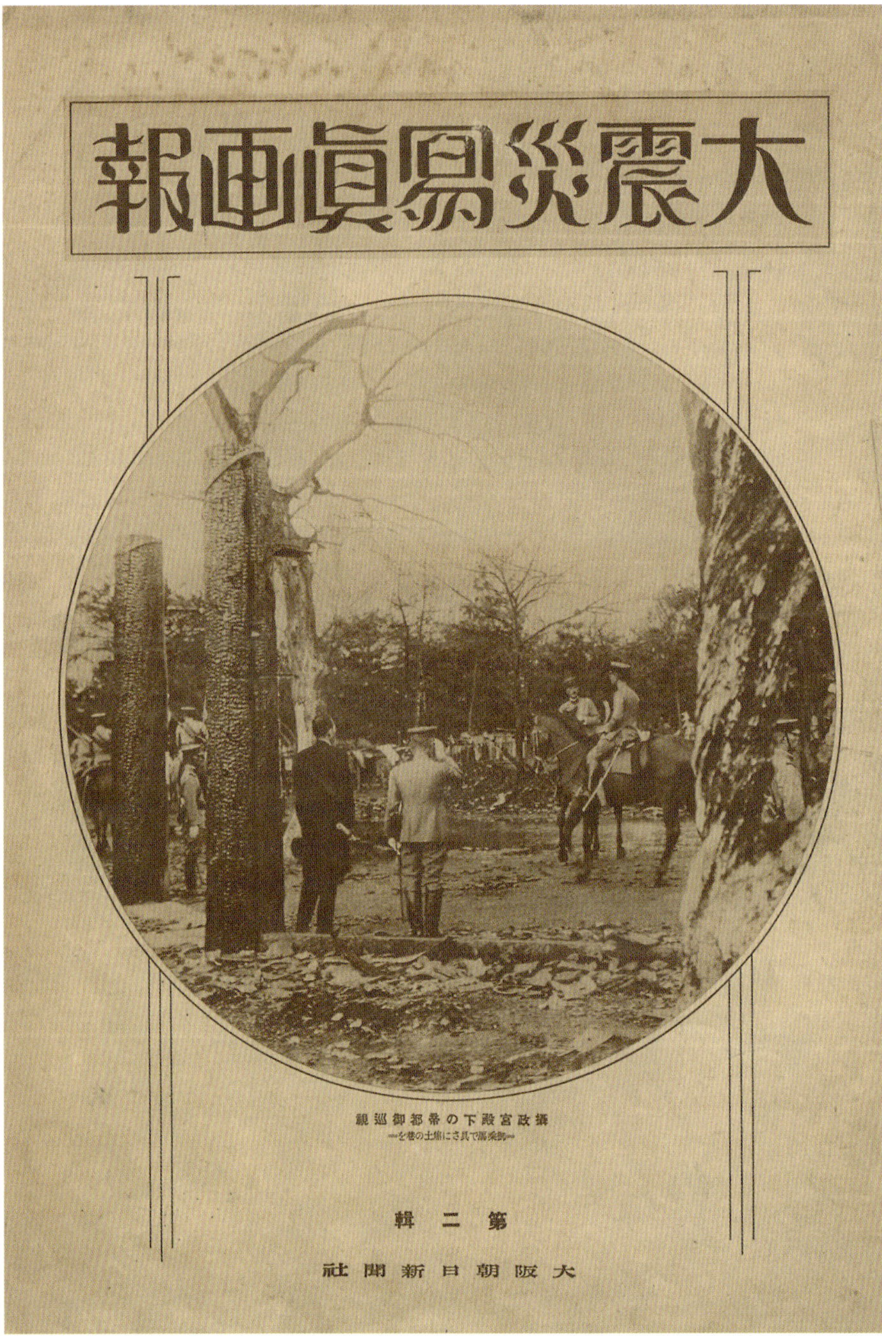

2.13 *The Prince Regent's Inspection of the Imperial Capital* (*Sesshō kyūdenka no teito go-junshi*), photograph, cover. *Daishinsai shashin gahō* (Photographic Pictorial of the Great Earthquake), vol. 2 (Osaka: Ōsaka Asahi Shinbunsha, 25 September 1923).

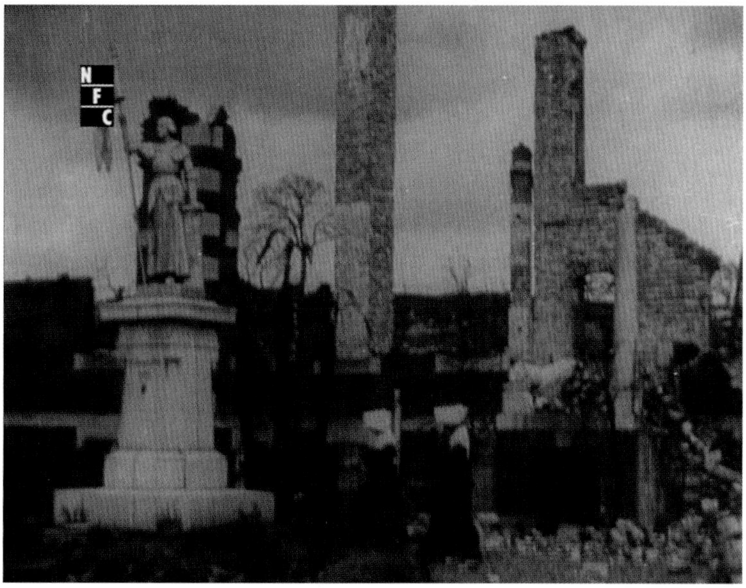

2.14 Statue of Joan of Arc in Kanda, still standing after the quake, screen shot. *Actual Conditions of the Great Kantō Earthquake and Fire* (*Kantō taishin taika jikkyō*), 1923. Produced by Tōkyō Shinema Shōkai and the Ministry of Education. By permission of Tōkyō Kokuritsu Kindai Bijutsukan Film Center

bare by the collapse of the school building that had once stood there and then walk over to the valiantly erect statue of Joan of Arc that adorned the school exterior. The French icon—a pious maiden queen symbolizing martial, religious, and national fortitude (who was burned at the stake)—became an ironic emblem in postquake Japan, unscathed and brandishing her standard while the building she guarded lay in rubble around her. The image of Joan of Arc epitomized the underlying message of visual images of authority: resilience, piety, and unity in the face of adversity. The "pathos of fact" in the official filmic narratives of national tragedy was designed to instill patriotism and evoke a sympathetic response among the audiences who were gathered in temple and shrine grounds and schools throughout the country for fund-raising events for the reconstruction of the capital.[21]

The limits of such visual authority, however, were in evidence everywhere. Despite the official prohibition against representations of dead bodies, there was no central authority in place to impose the ban, and ghastly images of corpses in every conceivable disposition abounded, catering to the morbid curiosity of consumers. The world news spotlighted the horrific body count under headlines exclaiming, "Dead Are in

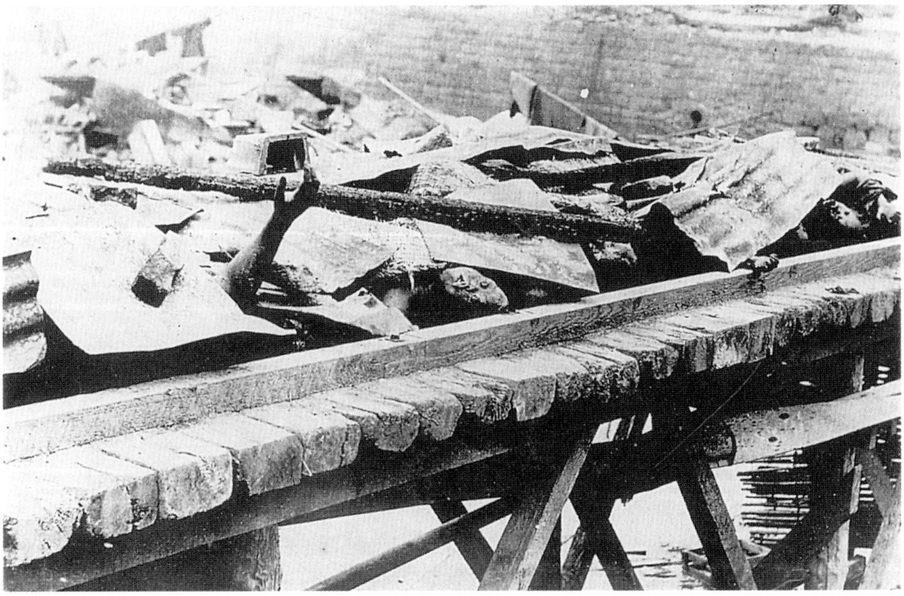

2.15 Naruse Keijirō, *Burned Corpse on the Shinbashi Earth Bridge* (*Shinbashi Dobashi no ue no yakishitai*), photograph, 1923. Ishii Toshio Collection. By permission of Japan Publishing Copyright Association.

Heaps in Tokio Streets"; and graphic photographs documented the gruesome scenes of mass incineration, while also singling out individual victims in the most gory and voyeuristic detail.[22] Without regard for the press ban, *Hōchi shinbun* and other papers featured large photographs of heaped corpses.[23] Photographic representations of mass carnage were repeated throughout the locales of the city, particularly in postcards, emphasizing the widespread loss of life, capitalizing on the tragedy, and inflaming public sentiment.

Images attempted to evoke empathy by personalizing the tragedy through the presentation of the dead as individuals. Stiff hands reach out from a pile of debris to beseech the viewer. However, the distorted faces and charred bodies in many images make the individuals unrecognizable, and they no longer seem human. These fossilized remains lie frozen in a rigid Pompeiian stasis, which, seen retrospectively through the lens of the atomic bomb, evokes images of Hiroshima and Nagasaki (fig. 2.15). Despite the immense discomfort the viewer experiences looking at these kinds of graphic images, like all photographs, the images are devoid of other critical sensory qualities such as olfactory sensations. Even so, such images are discomfiting because they implicate the viewer—then and now—as a voyeur. This exploitative objectification of the dead is underscored when the pictorial compositions show survivors gazing

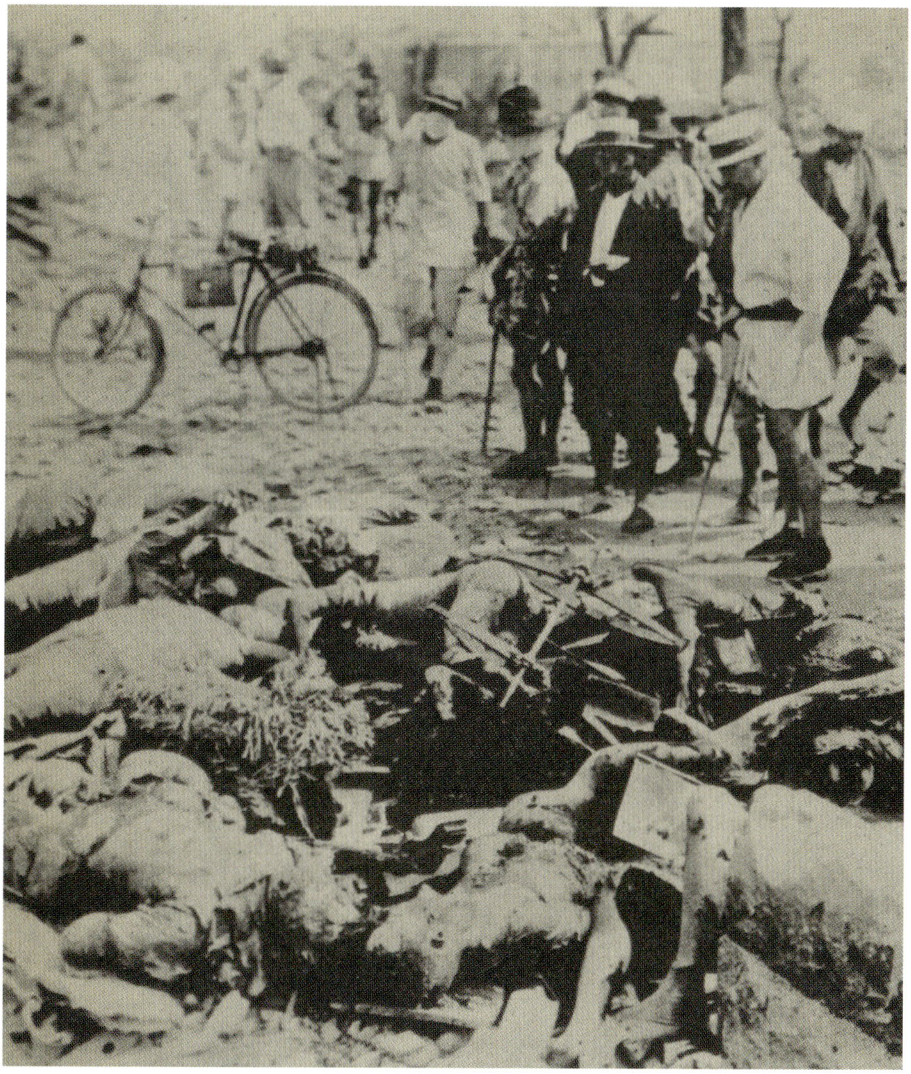

2.16 *People Searching for Missing Family in front of the Yasuda Gate,* postcard, 1923. Private collection.

intently upon the exposed bodies (fig. 2.16). In such cases, looking is a form of ocular aggression as well as historical witnessing.

The voyeuristic relationship between the viewer and the image, the living and the dead, is thrown into greater relief in a widely circulated news and postcard image of the pond in the Shin Yoshiwara licensed prostitution quarter (fig. 2.17). Thousands of women who worked in the brothels perished, unable to get out of the gated district. Like many victims throughout the city who sought refuge from the fires in the exten-

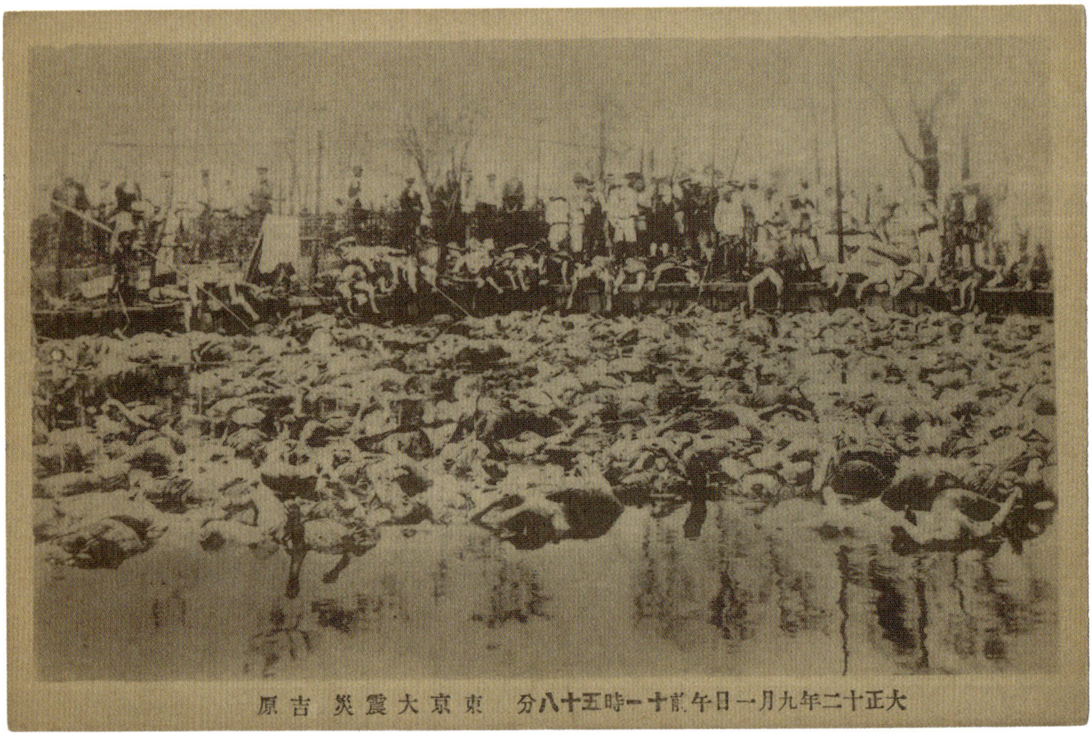

2.17 *Taishō 12, 1 September, 11:58* A.M. *Great Tokyo Earthquake Yoshiwara* (*Taishō 12-nen 9-gatsu tsuitachi gozen 11:58 Tōkyō Daishinsai Yoshiwara*), postcard, 1923. Collection of the author.

sive waterways of Tokyo, numerous women of Yoshiwara drowned or were boiled alive when they fled into the district's large pond, Benten Ike. This image clearly puts the dead in relation to the living, who seem to be gawking at the scene as they help pull the corpses onto dry land—an impression exacerbated by the presence of the photographer. The eerie image of the lifeless legs left dangling over the edge of the water reads almost like seated spectators among the standing live observers. And again there is a latent irony in this image displaying death, as these women also lived their lives regularly on display for customers in the latticed showrooms (*harimise*) that lined the main street of the prostitution district. Here a scopophilic economy has replaced a sexual one. Amid the death and destruction, the social hierarchy remains intact.[24]

Splayed bodies with charred genitals may on the surface present this libidinal voyeurism as incidental to viewing human remains, but the prominent foregrounding of the victims' anatomy, the frequency of such images in representations of the Yoshiwara pleasure quarters, the long history of oblique erotica in Japan, and the moralizing dis-

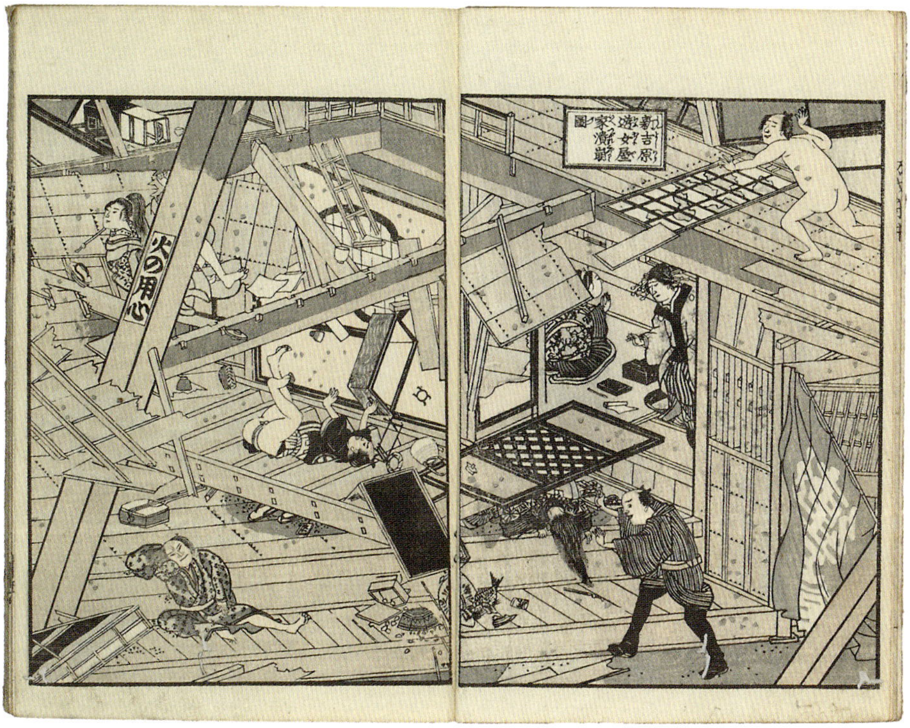

2.18 *Picture of the Destruction and Collapse of a Brothel in Shin Yoshiwara* (*Shin Yoshiwara yūjoya ie tsuburekuzushizu*), woodblock print. *Ansei-Era Observations* (*Ansei kenmonshi*), vol. 2 (Edo, March 1856). Collection of Nihon Shakai Jigyō Daigaku Toshokan.

course of divine punishment (*tenken* or *tenbatsu*) that prevailed after the earthquake support a more complex and unstable reading. While these images are clearly necrophilic and autoerotic, they are also cautionary, seeming to imply a kind of Sodom and Gomorrah divine punishment for the sybaritic inhabitants of the capital, exemplified by (but not limited to) the denizens of the pleasure quarters (here standing in for the modern national Japanese citizen). And lest we think that the scopically titillating display of bodies or the focus on Yoshiwara in disaster was novel, *Ansei-Era Observations* again offers a visual reminder of the Ansei quake's Edo precursors in an illustration that was widely reproduced after the 1923 quake. It illustrates the interior of a collapsed teahouse in the pleasure quarters (fig. 2.18), where people in various stages of undress are shown tossed to the ground.[25] A naked man stands on the roof; another is unsteadied by the quake and lands with his backside up. Courtesans lie crushed under fallen screens, and one sits in the back room with her legs thrown into the air, the opening of her kimono suggestively blocked by a fallen pillar with the popular warning sign

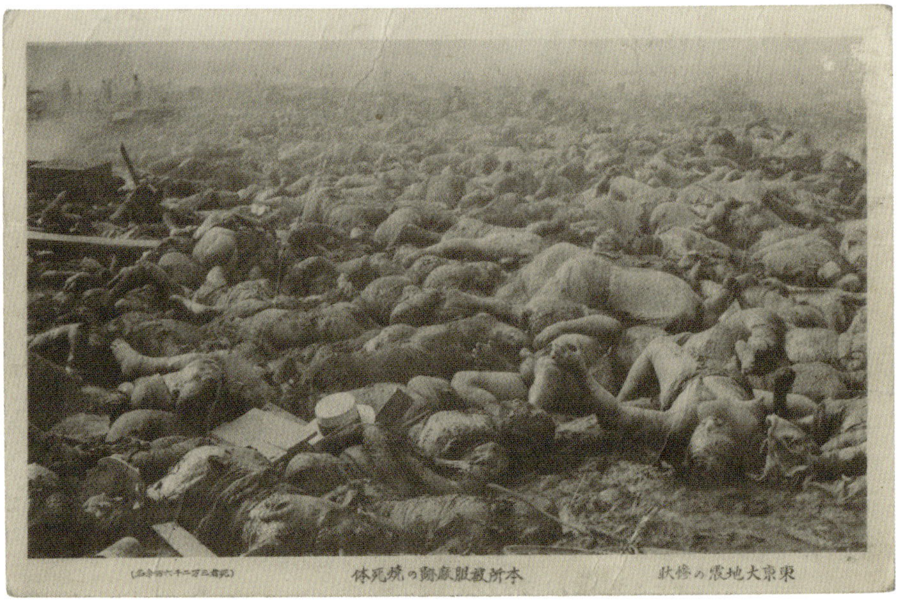

2.19 *The Tragedy of the Great Tokyo Earthquake: Burned Corpses at the Honjo Clothing Depot* (*Tōkyō Daijishin no sanjō: Honjo Hifukushō no yakeshintai*), postcard, 1923. Collection of the author.

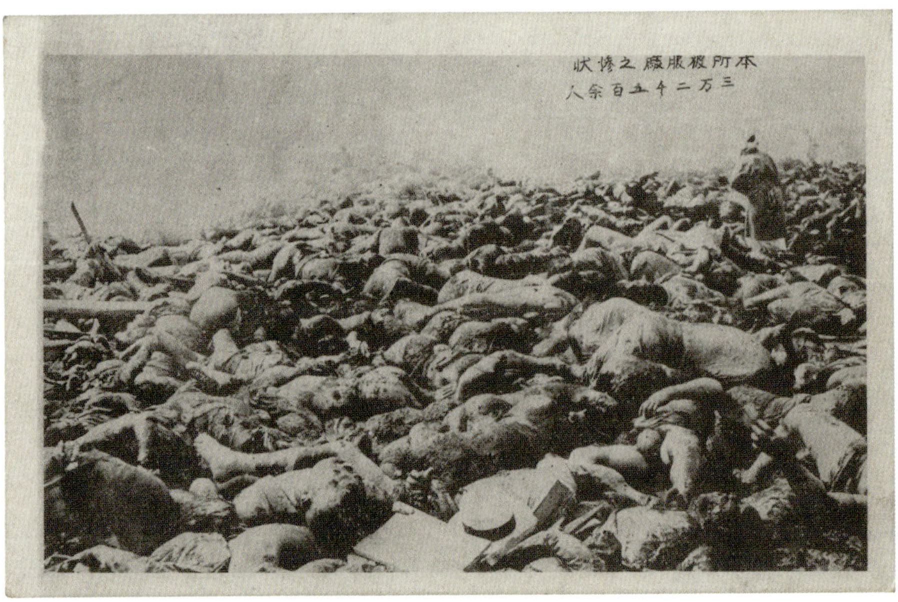

2.20 *The Tragedy of the Honjo Clothing Depot, More Than 32,500 People* (*Honjo Hifukushō no sanjō, sanman nisen gohyaku yojin*), postcard, 1923. Collection of the author.

"guard against fires" (*hi no yōjin*). Undoubtedly, the satirical placement of the cautionary sign to cover the woman's genitals would have been read in the libidinal terms appropriate to the pleasure quarters, which generated its own kind of heat, and it could also be read as a comment on the immoral acts that brought on the disaster in the first place. Ill-gotten gains have been turned to ash and cinders: the dead cannot take it with them. Images of the Yoshiwara district, whether in 1855 or 1923, were simultaneously titillating, tragic, and castigating.

While pictures showed the dead at multiple locations throughout the city, no site associated is more tragic than the large open space formerly occupied by the Army Clothing Depot (Rikugun Hifukushōato) at Honjo, which was one of the deadliest sites in the city.[26] More than thirty thousand people perished here. The site was one of the few open areas in the ward and a supposed sanctuary for fleeing refugees that became an inescapable tunnel of fire, causing indescribable misery that was indelibly etched into the collective memory of the residents of the nation's capital. Authorities later decided that the official earthquake memorial should be erected on the site of the depot, because for the survivors of the quake it represented the nadir of human suffering and loss.

Photographic images of the quasi-sacrosanct site were not impervious to objectification or the eroticized gaze. The composition and depth of field in one photographic postcard (fig. 2.19) bring the exposed body of an individual female figure into sharp focus in the right foreground, drawing the eye to her prominently visible breasts and rigid extended arm set against a seemingly interminable expanse of death in the background. Other images, however, showed charred bodies heaped into mounds as if melted into a single mass. The figures are undifferentiated; male or female, adult or child, they are merely a lump of intertwined human flesh. In one well-known image (fig. 2.20) that appeared in the news and as a commemorative postcard, the mass of charred bodies is pictured from a low angle, giving the impression of a rising mound that continues indefinitely. A small, almost imperceptible figure near the top makes her way through the mound of tangled corpses, clearly searching for relatives in the pile. Because the Honjo Clothing Depot was the deadliest site in the capital and the final resting place of the earthquake victims killed there, mass interred on-site, visualizations of it are retrospectively invested with intense significance.

The imaging of the catastrophic loss of life at the depot offers a case study of how disasters unfold over time and how important the temporal aspect is in viewing. It also demonstrates the instrumentality of disaster imagery and particular sites of trauma for evoking affective responses among the viewing public while pointing out their generic visual interchangeability. Images of refugees with their belongings at the clothing

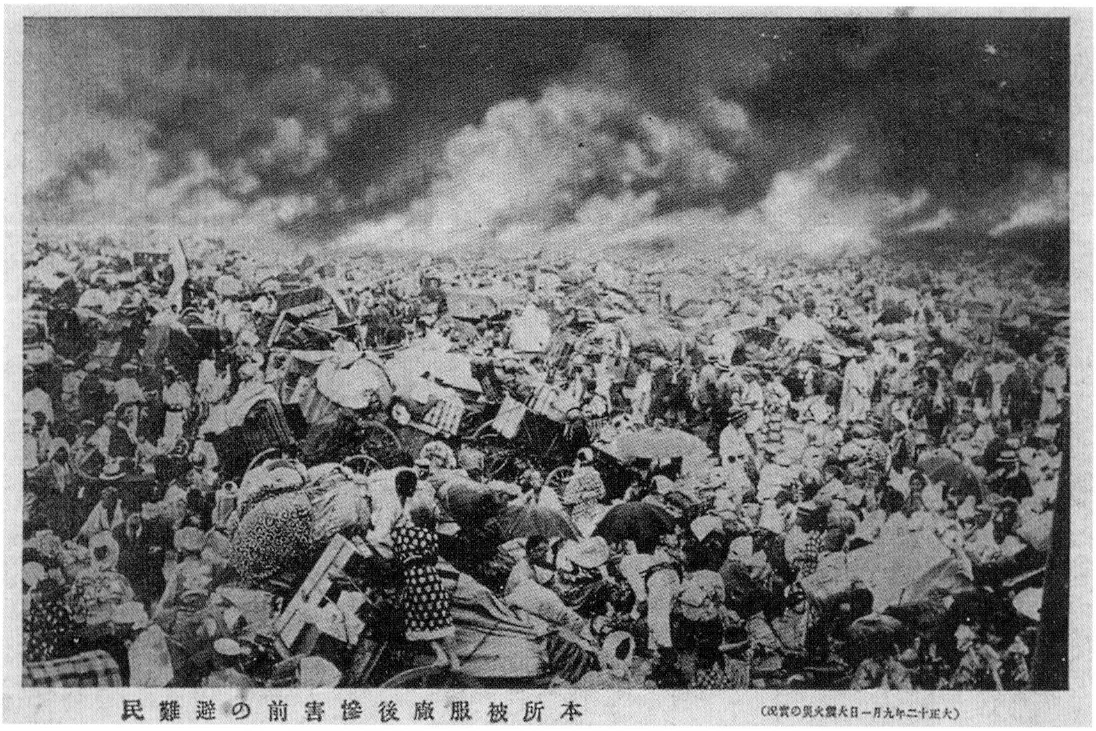

2.21 *Refugees before the Tragedy at the Former Honjo Clothing Depot* (*Honjo Hifukushōato sanjō mae no hinanmin*), postcard, 1923. Ishii Toshio Collection. By permission of Japan Publishing Copyright Association.

depot, gathered in what they believe to be a safe haven from the crumbling cityscape and watching the encroaching fire, carry a sense of impending doom with the retrospective knowledge of their imminent incineration (fig. 2.21). In fact, survivors recalled that the gathered people were in good spirits, even picnicking, at the depot as they waited out the aftershocks, unaware of the quickly changing winds that would produce an unimaginable vortex of fire at the site later in the day. The pictured belongings themselves offer a visual cue for the tragedy because viewers are keenly aware that these earthly possessions were highly incendiary and accelerated the spread of the fire. The news photograph, originally from *Yomiuri shinbun,* is hand-labeled at the top "Honjo clothing depot one moment before the horrific tragedy. 1 September, around 2:00 o'clock," anticipating the ensuing events.[27] The image recapitulates the temporal unfolding of the quake, underscoring both the participants' perilous unawareness on the ground and the viewers' painfully acute awareness of what is to come. Other postcards make this before-and-after temporal frame explicit by superimposing an inset

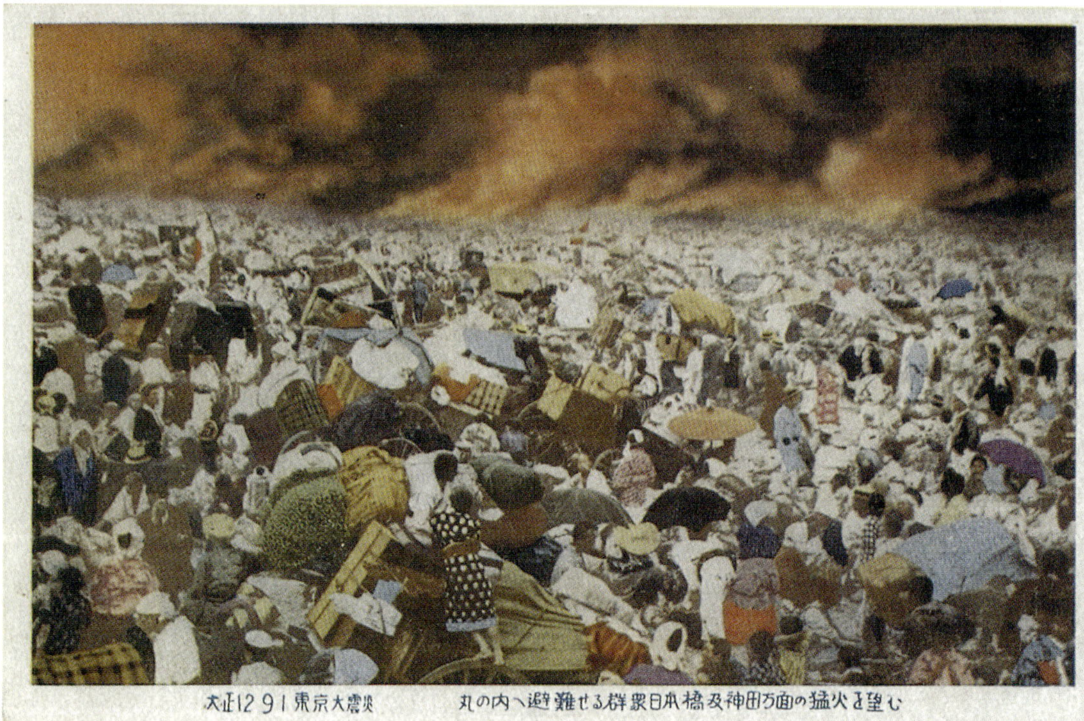

2.22 *Crowds Seeking Refuge in Marunouchi; Anticipating the Raging Fires from the Direction of Nihonbashi and Kanda* (*Marunouchi ni hinanseru gunshū Nihonbashi oyobi Kanda hōmen no mōka o nozomu*), a color version of the black-and-white scene in fig. 2.21 showing the difference in affective impact, postcard, 1923. Ishii Toshio Collection. By permission of Japan Publishing Copyright Association.

photograph of the piled bones of the dead on top of the living. Comparison of this black-and-white image with its color-enhanced postcard version (fig. 2.22) also subtly demonstrates the spectrum of affective impact.[28] The coloring of the victims' belongings spotlights this tragic motif while eerily contrasting the overall cheery tenor with the specter of imminent death. In retrospect, this combination produces an uncanny effect, with the aesthetics of fear revealing the anxiety aroused by exposing hidden threats within the familiar.

Then one notices that the color postcard is differently labeled "Crowds Seeking Refuge in Marunouchi," not Honjo, giving a disturbing sense of the interchangeability of death scenes and implicitly devalorizing the site of the clothing depot by undermining the authenticity of the visible evidence itself. The facticity of photography and other scientific modes of imaging upon which social trust is founded proves tenuous in this case.

Visibility/Invisibility

While the state-directed media campaign after the 1923 quake effectively bolstered national unity and rechanneled potentially disruptive social energies into altruistic and samaritan activities, it could not stanch the voyeuristic pleasures of disaster viewing, nor could it avert a surge of intense public anger toward marginalized individuals in Japanese society, mostly Korean colonial subjects residing in Japan. These groups became scapegoats for the calamity, along with some other foreigners and leftist political figures who were viewed as social subversives. There were several high-profile targeted assassinations of prominent members of the socialist and anarchist labor movement by the military police, who used the opportunity of the quake to strategically remove political opponents of the government. In the attacks now known as the Kameido Incident (3–5 September), ten activists associated with the Nankatsu Labor Union were killed, and in the Amakasu Incident (16 September), one of the best-known, charismatic leftist leaders, anarchosyndicalist Ōsugi Sakae, his common-law wife Itō Noe, and Ōsugi's young nephew were murdered in police custody by Special Higher Police Lieutenant Amakasu Masahiko and his officers.[29]

Ōsugi's murder was front-page news in the mass media. Portraits of him, Itō, and his nephew, along with their accused assailants, ran on the cover of a special edition of the *Ōsaka mainichi shinbun* on 8 October, still three weeks after the events (fig. 2.23). Ōsugi and Itō were portrayed as celebrities, with Itō sporting a fashionable jaunty hat and Ōsugi debonair in his stylish explorer headwear.[30] Even the seven-year-old nephew was given a prominent, full-length portrait. The charged assailants also received extensive media coverage. Higher-class and insider status afforded subjects more visibility, even when they were purported leftist subversives or indicted murderers. This coverage could not have been more different than the absence and anonymity of other victims of the postquake violence in the media. There was a proliferation of rumors and gossip (*ryūgen higo*) that accused "subversive" or "malcontent" resident Koreans (*futei senjin*) of sedition, indicating that they had set off explosions and poisoned well water, and these rumors incited widespread retributive violence, particularly among self-appointed vigilante squads (*jikeidan*) organized by local neighborhoods throughout the region.[31] Estimates suggest that well over six thousand of the twenty thousand resident Koreans in the Kantō region (in addition to many Chinese) were murdered in the aftermath of the quake. Scholars have argued that Koreans were not just the victims of bigoted xenophobia in chaotic times but were specifically targeted as threatening colonial subjects who had been voicing desires for national independence from Japanese rule and evoking intense fear in the Japanese imagination.[32] Despite the army's and police force's

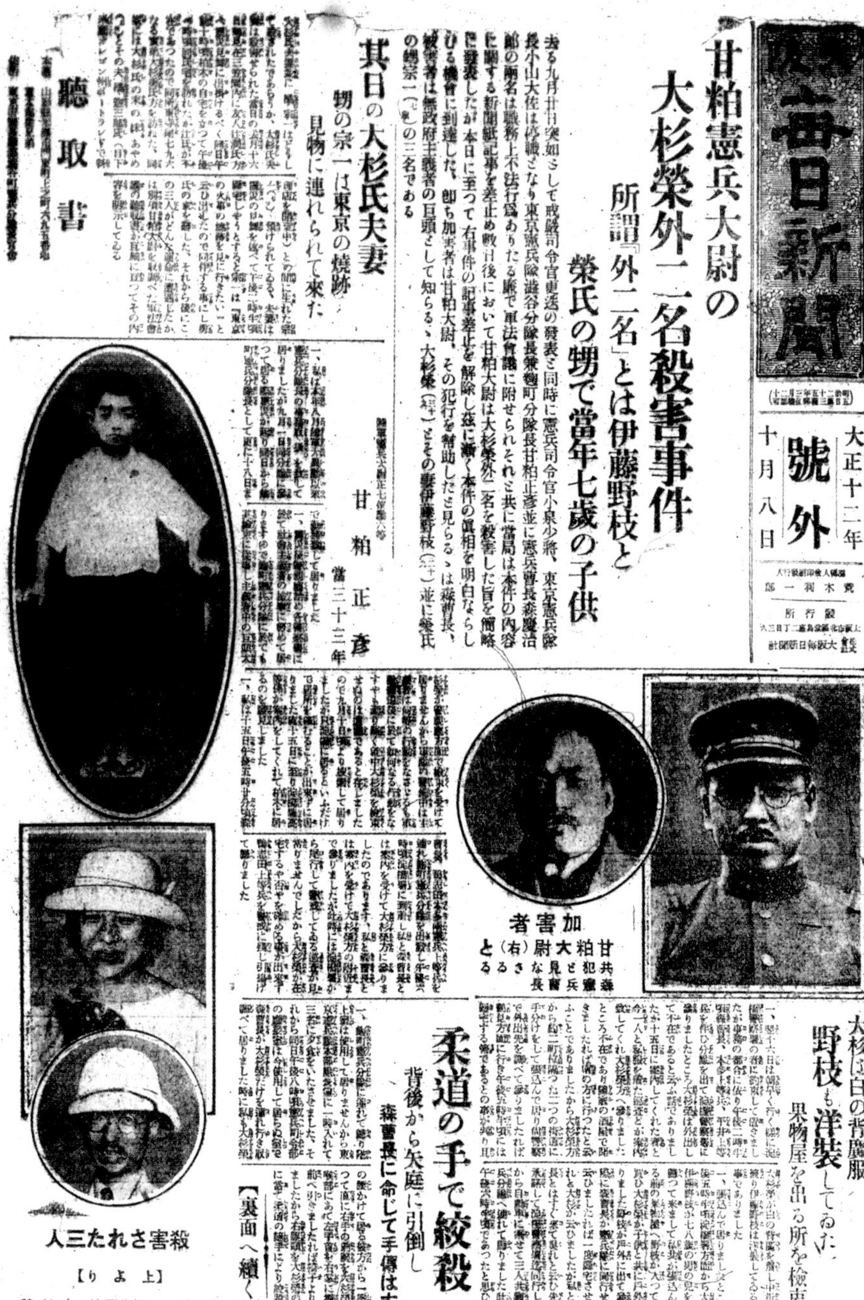

大阪毎日新聞

大正十二年十月八日 號外

甘粕憲兵大尉の大杉榮外二名殺害事件

所謂「外二名」とは伊藤野枝と榮氏の甥で當年七歳の子供

去る九月廿日突如として戒嚴司令官更迭の發表と同時に憲兵司令官小泉少將、東京憲兵隊長小山大佐は停職となり東京憲兵隊澁谷分隊長兼麴町分隊長甘粕正彦並に憲兵曹長森慶治郎の兩名は職務上不法行爲ありたる廉で軍法會議に附せられそれと共に當局は本件の內容に關する新聞紙記事を差止め數日後において甘粕大尉は大杉榮外二名を殺害した旨を簡略に發表したが本日に至つて右事件の記事差止を解除し茲に漸く本件の眞相を明白ならしむる機會に到達した、卽ち加害者は甘粕大尉、その犯行を幇助したと見らるゝは森曹長、被害者は無政府主義者の巨頭として知らるゝ大杉榮(三十九)とその妻伊藤野枝(二十九)並に榮氏の甥宗一(七歳)の三名である

其日の大杉氏夫妻

甥の宗一は東京の燒跡見物に連れられて來た

聽取書

甘粕正彦 當三十二年

加害者 甘粕大尉(右)と共犯と見らるゝ森憲兵曹長

大杉は白の背廣服 野枝も洋裝してゐた

果物屋を出る所を檢束

柔道の手で絞殺

背後から矢庭に引倒し 森曹長に命じて手傳はす

【裏面へ續く】

殺害された三人

【上より】

橘宗一 伊藤野枝 大杉榮

2.23 Front-page coverage of the murders of Ōsugi Sakae and Itō Noe. *Ōsaka mainichi shinbun,* special edition, 8 October 1923. By permission of Mainichi Shinbunsha.

claims to be protecting this vulnerable population, the press later revealed these authorities to be organizers, instigators, and cohorts of the vigilantes in the persecution. But their participation was not discovered until nearly two months after the quake when the media ban on covering the killings was lifted on 20 October 1923.[33]

Though the literature contains a number of in-depth analyses of these atrocities, my goal here is to expand the sphere of analysis to the visual. Strict censorship laws prohibited representations of dead bodies and other potentially disturbing subjects, including the massacre of Koreans (although, as we have seen, these measures could not stem the tide of contraband images available on the street that featured burned and bloated corpses in neighborhoods throughout the city). There was a total absence of direct coverage of the Korean massacre for two months after the quake due to the immediate media ban promulgated under the guise of benevolent "government protection." Colonial subjects were already subordinated within the imperial structure of assimilation policy, their labor exploited and their autonomy suppressed.[34] Expressing the deep-seated Japanese fear of reigniting uprisings in colonial Korea, visual culture recapitulated this subjugation through erasure. Still, the visuality of the Korean massacre was one of its most powerful aspects, repeatedly discussed in written memoirs, because the bodies, distinct from the burned corpses of the conflagration, showed clear evidence of intentionally inflicted torture and violence—their hands bound, their noses cut off, their eyes gouged out, their stomachs cut open, their limbs severed, and their skin marked by multiple lacerations. Not only did people, including many traumatized children, witness these brutal murders, but they also saw the bodies discarded along the sides of the roads.[35] Visual culture both reveals and conceals in its visualizations. Although the photodocumentary images of the earthquake appeared to be displaying the bodies' stories so transparently, they in fact conflated the widely divergent histories of the mangled corpses into one generic narrative. Hidden beneath the surface of the earthquake presentations, visible only to the knowing eye, is an alternative history.

Writing later in his memoirs, cameraman Shirai Shigeru, who shot the film for the full-length feature *Actual Conditions of the Great Kantō Earthquake and Fire* (*Kantō taishin taika jikkyō*), noted that he had filmed the lynching of Koreans near the Honjo Army Clothing Depot by survivors of the quake, but the authorities deemed this imagery too great a threat to the maintenance of public peace (*chian*) and confiscated the film he shot at the site.[36] The censorship of such inflammatory material that implicated "victims" as perpetrators assured the unblemished image of the depot as a site of tragic *innocent* suffering. The desire to convey innocence in victimhood to create a uniting symbol for the Japanese nation is evident in many discourses of the disaster and in a variety of earthquake motifs, such as the repeated use of mothers and children as well as the elevation of the dead as "heroic spirits" (*eirei*). Advocates of reconstruction continually invoked these victims' tragic sacrifices, using this theme to urge the citizenry to dedicate itself to rebuilding a new modern metropolis in the victims' memory. The discourse of innocent victimhood has parallels with other national narrations of historical tragedies, from the 1906 San Francisco quake to the terrorist attacks of 9/11.[37] The media's visual authority perpetuated notions of resilience, unity, and innocence despite overwhelming evidence to the contrary.

Mediating Dark Tourism

Most of the photographs I have discussed so far were produced within two weeks of the earthquake. In just a month's time, the print media had demonstrated its miraculous recovery by producing over seventy special earthquake issues in the month of October alone, further intensifying the media scale of the catastrophe. Also that month, the prominent publishing house Kōdansha, whose offices had been destroyed, published 500,000 copies of its collection *The Great Taishō Earthquake and Conflagration* (*Taishō Daishinsai daikasai*) under a special government dispensation. A collection of photographs, information, and narratives, this modern disaster miscellany was the first book on the earthquake to appear, and it was circulated throughout Japan via magazine distribution routes, resulting in a significant and enduring expansion of book distribution that eventually reinforced Tokyo's status as the center of print capitalism. It is estimated that between 120,000 and 400,000 copies of the publication were sold.[38]

The dramatic watercolor cover of *The Great Taishō Earthquake and Conflagration* (fig. 2.24) shows a city engulfed in crimson flames, with the composition unfolding from the front of the volume to the back. Red was a central hue in the earthquake palette, representing the color of the intense heat that tinted the sky and of the smoldering embers that were all that remained of beloved buildings, but it could not be expressed in black-and-white photography. Tinted postcards tried to compensate through the addition of color, but most documentary images were monochromatic. Ironically, the hand-rendered media could more evocatively express the colorful visual sensations of the experience. On the Kōdansha cover, the red articulates the swirling flames of the ravaging firestorms through modulated shading and subtly emblazons the burning structures of the cityscape amid the inferno, revealing the broken carcass of the iconic Asakusa skyscraper, the Ryōunkaku or Cloud-Surpassing Tower, popularly known as the Twelve Stories, as a deathly specter in the upper left, immediately adjacent to the title. The structure's spectacular demise will forever be associated with the Taishō disaster. On the back, a shrouded gray image of Sensōji Temple at Asakusa stands

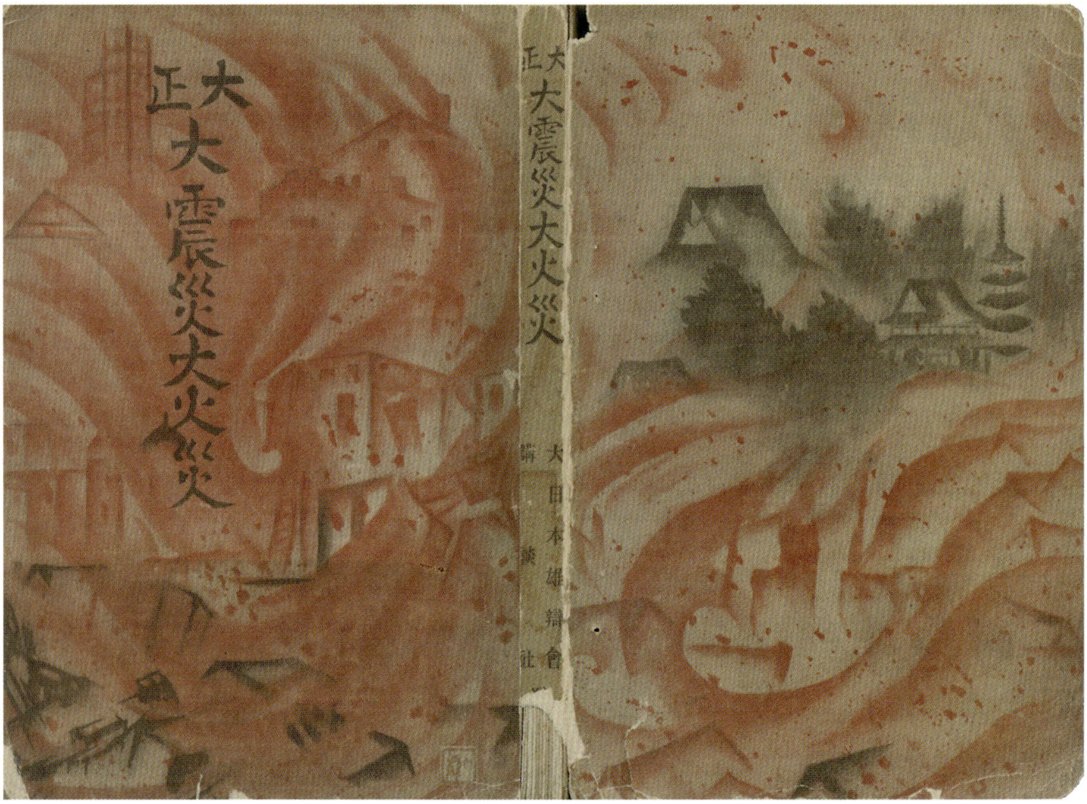

2.24 Front and back covers of the first book published about the earthquake, print reproduction of original watercolor. *Taishō Daishinsai daikasai* (The Great Taishō Earthquake and Conflagration) (Tokyo: Dai Nihon Yūbenkai Kōdansha, 1923).

defiantly above the sea of flames, saved from destruction at the last minute by a freak change in wind direction—or by the miraculous hand of the compassionate bodhisattva Kannon, as the legend went. This collection featured black-and-white photographs that had already appeared in the news media and demonstrates the rapid codification of tropes from scientific and technological expressions of visible evidence and visual authority to iconic sites around the city famous for their spectacular destruction and sublime state of ruin.

Drawing on the special-issue news compilations, the Kōdansha volume can be considered the first comprehensive souvenir album of personal accounts and collections of photographs, and it reveals the blurry boundaries between reportage, spectacle, tourism, and remembrance already in evidence in the pictorial news compilations and in the prolific industry of disaster postcards. News was indistinguishable from souvenirs, as the same photographs were widely circulated in the form of commemorative postcards, often sold in sets.[39] Postcards were introduced to Japan in 1873 with the new national postal system. They were a vivid symbol of Japan's modernity, marked by the inauguration of a new communications system in the Meiji period. Already a very popular medium for depicting scenes from the Russo-Japanese War (1904–5), postcards of natural disasters were commonplace in Japan by the time of the 1923 quake.[40] These commemoratives were predominantly treated as collectibles. Extant envelopes for sets indicate that Kantō earthquake postcards were produced far and wide, from Hakata on the southern island of Kyūshū to Osaka in the western Kansai region. Printed in collotype, offset, and gravure, many photographic postcards, along with original photographic prints, were commissioned by Tokyo publishers (*hanmoto*), which had been burned out of business but recognized the potential profit in earthquake images. In the capital, they approached the printing press Kōmura Insatsujō, one of the few Tokyo-based presses to survive the quake. Hundreds of varieties of postcards were issued. Publishers sold most of them on the street, and many of the images sold out quickly. The intimate, pocket-size format of the postcard lent itself to personal consumption. Since the vast majority of 1923 disaster postcards that survive have no writing on them, they were likely treated more as collectibles than as a form of postal communication. Many were put into albums, creating new ways to combine images and create visual cultures of disaster for home viewing. Accordion-style albums allowed for personalized, serial organization of images that produced unique, imagistic narratives of the event. The album pages were also two-sided and could be stretched out to view a series of images on recto and verso.

The success of the Kōdansha volume and the brisk trade in disaster souvenir postcards are indicative of a thriving business in "disaster tourism," part of a larger phenomenon of "dark tourism" (also known as thanatourism). The media scale of the Kantō earthquake disaster engendered both empathic mourning and a touristic desire to see the authentic sites of destruction. Consumers of the mass media were impelled to make their own pilgrimages. Long-standing fascinations with death-related visitor sites, attractions, and exhibitions throughout history have recently become a significant focus in tourism studies. Such tourist sites

> often trad[e] under the guise of remembrance, education and/or entertainment, which attract people eager to consume real and commodified death. Indeed, the act of touristic travel to sites of death, disaster and the macabre is becoming a pervasive cultural activity within contemporary society. From visiting Nazi death camps in eastern Europe as part of a wider holiday itinerary, to enjoying family picnics on battlefields of northern France, or

> purchasing souvenirs of genocide at Ground Zero, to allowing schoolchildren to gaze upon tools of torture from yesteryear at the London Dungeon, are all illustrations of the seemingly macabre.[41]

Yet the purported expression of respect for the dead and the educational benefits of disaster tourism often mask a clandestine, scopophilic pleasure at the core of the experience. The commercial market exploited this long-standing Japanese fascination with the macabre.

Vicarious tourist images and descriptions of the Great Kantō Earthquake fueled a surge in dark tourism in 1923. Ghastly newspaper accounts and grisly photographs of death and destruction quickly attracted hordes of sightseers. It was not enough to experience the devastation secondhand; people wanted to walk through the disaster zone themselves—in a sense inserting themselves into the history of the event as well as satisfying their morbid curiosity and desire to have an authentic experience. *Yakeato kenbutsu* or "fire ruin sightseeing" became a popular activity for people from all over the nation immediately after the quake, with close to two million people entering Tokyo in the two weeks following the tremor.[42] Like visitors in other times and places—who carried home a piece of the fallen Berlin Wall or mementos from Ground Zero at the World Trade Center, for example—many people in Tokyo wanted visual and material souvenirs of the disaster.

The number of sightseers in Tokyo was so extensive, in fact, that their presence became an impediment to reconstruction, and people were encouraged not to enter the city unless they had an absolute necessity to do so.[43] The leisure nature of this activity, however, foregrounded the difference between sightseers and survivors; the former position afforded safety from physical harm and a certain reflective distance—not to mention the fact that it highlighted the inherent class divisions between the majority of people who had suffered the most in the earthquake and those who came out relatively unscathed. Moreover, the sightseers' ability to move through the landscape in a time of martial law and restricted mobility was a marker of class.[44]

In the Kōdansha volume, a photograph titled *Fissures That Should Make You Tremble with Fear* (*Senritsu subeki jiware*) shows a man standing in one of the deep fissures up to his hips (fig. 2.25). His stylish walking attire and that of his companions, who wear pressed suits and panama hats, immediately mark the group as tourists engaged in disaster sightseeing, or in this case *jiware kenbutsu* (fissure sightseeing). Their relaxed attitude marks the disaster as an event of the past. Certainly, had there been any imminent danger of aftershocks, the man would not be standing in a potentially lethal location where he could be crushed at any moment.

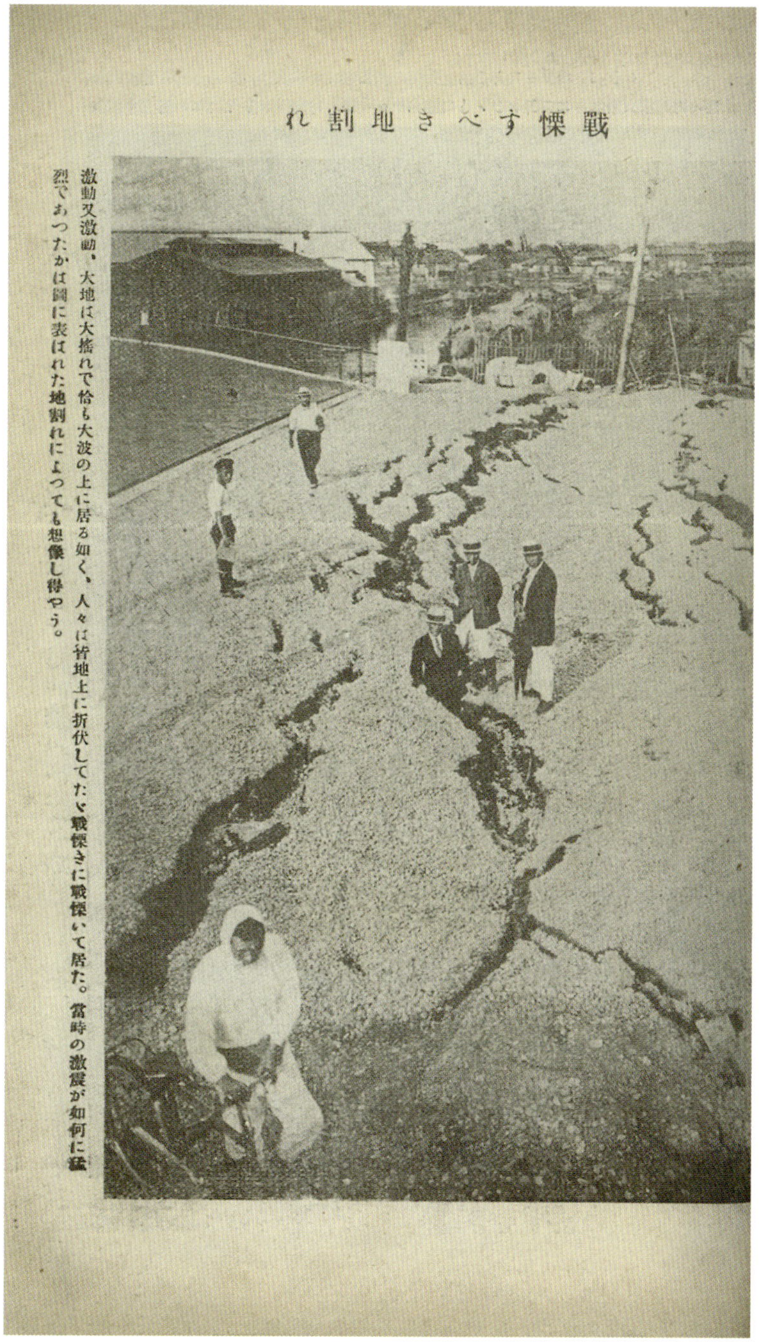

戦慄すべき地割れ

激動又激動、大地は大搖れで恰も大波の上に居る如く、人々は皆地上に折伏してたゞ戦慄きに戦慄いて居た。當時の激震が如何に猛烈であつたかは圖に表はれた地割れによつても想像し得やう。

2.25 *Fissures That Should Make You Tremble with Fear* (*Senritsu subeki jiware*), fissure sightseers, photograph. *Taishō Daishinsai daikasai* (The Great Taishō Earthquake and Conflagration) (Tokyo: Dai Nihon Yūbenkai Kōdansha, 1923).

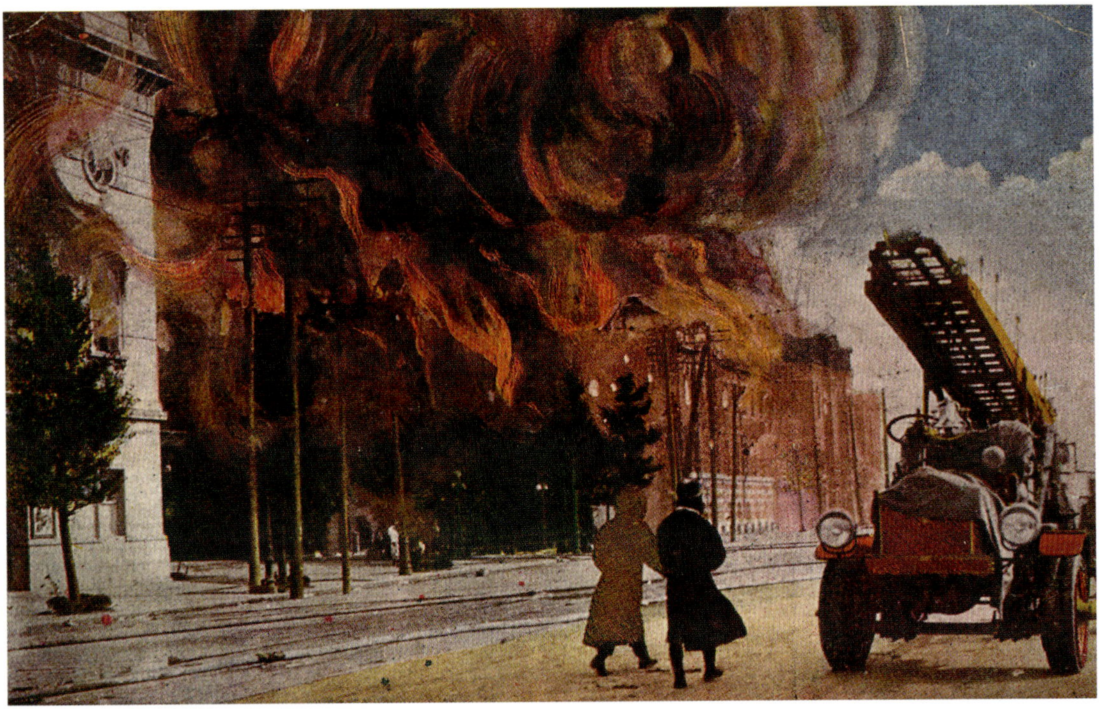

2.26 *Metropolitan Police Department and the Imperial Theater in the Ferocious Flames* (*Mōkachū no Keishichō to Teigeki*), postcard, 1923. Collection of the author.

The market for disaster sightseeing and the desire for souvenirs fueled a profitable postcard business, with thousands of examples produced in the early days after the quake. Mirroring the photographic eye in the pictorial volumes, photographic postcards surveyed the damage both from aerial perspectives and from the ground. However, they distinguished themselves through the widespread addition of color, which was thought to enhance the verisimilitude while aestheticizing the image.

Although still photographs could not capture the spectacle of the quake in real time, hand or machine application of color tinting to black-and-white images that were mass printed sought to provide an immediacy and layering to the otherwise largely static scenes. The dynamic "raging flames" (*mōka*) were added to photographic postcards to lend movement to the images, yet the stillness of the pictures speaks of an inability to reenact the event in a temporal or visceral way; rather, it reinforces the irretrievability and distance of history. Similarly, images of streets engulfed in flames (fig. 2.26), tinted to communicate imminent danger and terror, are perhaps more striking because of their inability to convey the true nature of the experience.

Empathic Mourning

It was clear from the beginning that despite its regional moniker, the Kantō Daishinsai was a *national* disaster that imperiled all of Japan and the larger empire. As the capital of the empire and the imperial seat of power, Tokyo was the jewel in the crown, and reconstruction of the capital was crucial to national prosperity. Japanese relief operations from the Kansai area in the west poured into Kantō, as did assistance from around the world.[45] Buttressing this empathic spirit of collective suffering and mourning were the purportedly true (and often sensationalized) stories (*jitsuwa*) about fellow Japanese citizens in peril and the noble deeds and heroic actions of their countrymen.[46]

As a fellow empire with interests in the Pacific region that had suffered a devastating earthquake in recent memory, the United States took intense interest in the plight of the Japanese. Americans were also concerned about the fate of their many countrymen serving in Japan as diplomats, businessmen, and Christian missionaries. The day after the quake, the *New York Times* bore the front-page headline "Great Earthquake and Fire Ravage Tokio and Yokohama; Many Perish, Buildings Collapse; Survivors Flee in Panic"; a map of the country showed the eastern coast of Japan darkened like a gangrenous limb (fig. 2.27).

On 3 September, most of the *New York Times* front page was still devoted to the earthquake under the massive headline "100,000 Dead in Japanese Earthquake; Tokio, Yokohama and Nagoya in Ruins; Millions Destitute; Fires Still Raging." Headlines below continued: "200,000 Buildings Are Razed in Tokio," "Martial Law Declared in Panic Following the Fire and Earthquake," "Dead Are in Heaps in Tokio Streets," "Survivors Are without Food and Water—10,000 Casualties in Yokohama, But the Foreign Section Escapes—Wires Still Out."

Such global media coverage of the disaster dramatized the event and sought both to titillate distant audiences and engender their empathic mourning, an emotion touchingly expressed in a newspaper illustration titled *The Stricken Sister* (*Uchihishigareta imo*) (fig. 2.28) that ran in the *San Francisco Chronicle,* whose home city had itself experienced a devastating earthquake less than two decades earlier and which had a sizable Japanese population.[47] With downcast eyes, the kimono-clad figure of Japan (her identity on her sleeve like a Red Cross armband) is compassionately embraced by the allegorical female figure of "World-Wide Sympathy." The two women sit amid the devastated, infernal Tokyo landscape with flames at their backs and the tsunami lapping at their feet.

U.S. President Calvin Coolidge had immediately sent a cable to the Taishō emperor expressing his sympathy for the nation and notifying the Japanese authorities that the

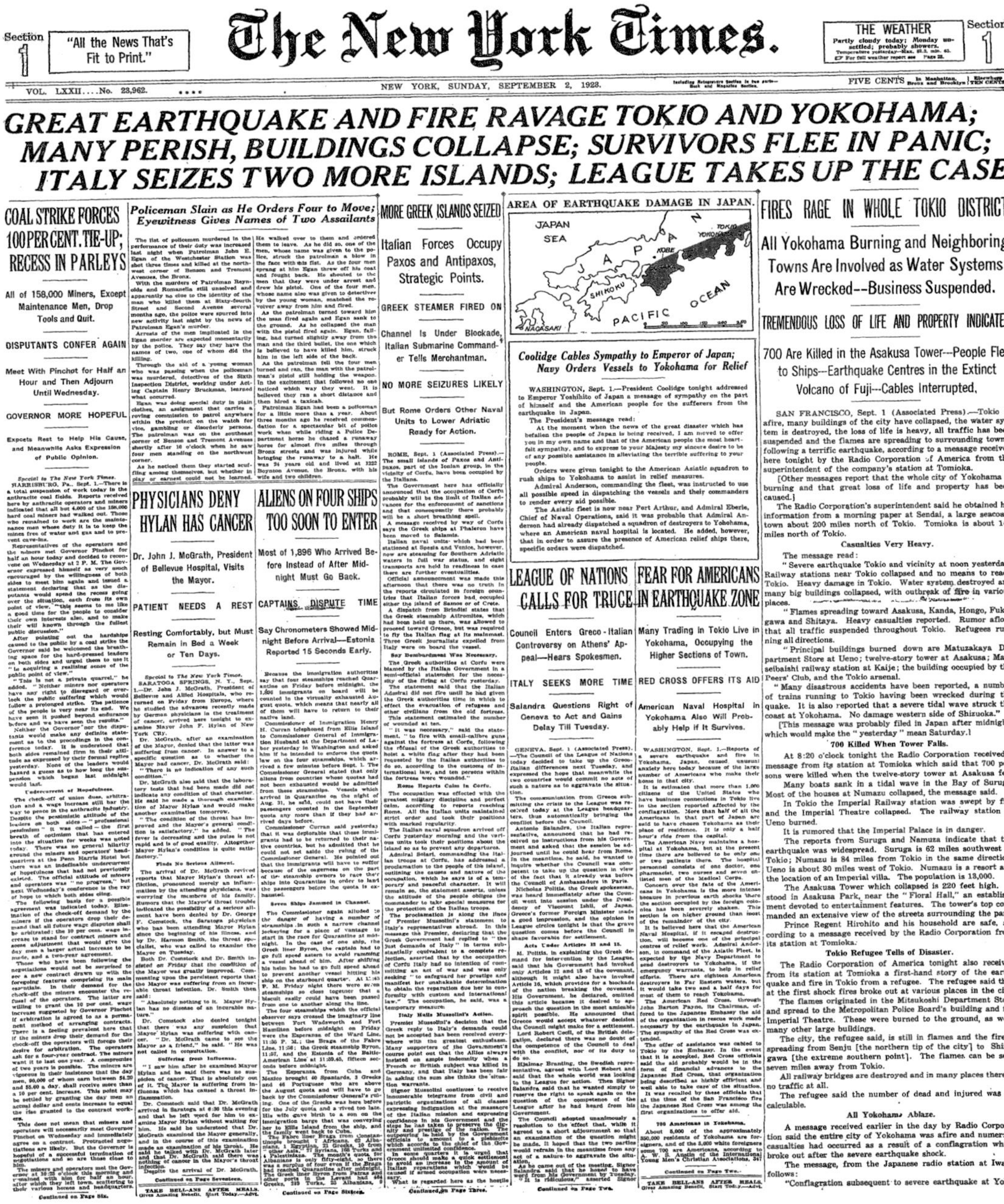
"All the News That's Fit to Print."

The New York Times.

VOL. LXXII....No. 23,962. NEW YORK, SUNDAY, SEPTEMBER 2, 1923. FIVE CENTS

GREAT EARTHQUAKE AND FIRE RAVAGE TOKIO AND YOKOHAMA; MANY PERISH, BUILDINGS COLLAPSE; SURVIVORS FLEE IN PANIC; ITALY SEIZES TWO MORE ISLANDS; LEAGUE TAKES UP THE CASE

COAL STRIKE FORCES 100 PER CENT. TIE-UP; RECESS IN PARLEYS

Policeman Slain as He Orders Four to Move; Eyewitness Gives Names of Two Assailants

MORE GREEK ISLANDS SEIZED

Italian Forces Occupy Paxos and Antipaxos, Strategic Points.

AREA OF EARTHQUAKE DAMAGE IN JAPAN.

FIRES RAGE IN WHOLE TOKIO DISTRICT

All Yokohama Burning and Neighboring Towns Are Involved as Water Systems Are Wrecked--Business Suspended.

TREMENDOUS LOSS OF LIFE AND PROPERTY INDICATED

700 Are Killed in the Asakusa Tower---People Flee to Ships---Earthquake Centres in the Extinct Volcano of Fuji---Cables Interrupted.

Coolidge Cables Sympathy to Emperor of Japan; Navy Orders Vessels to Yokohama for Relief

PHYSICIANS DENY HYLAN HAS CANCER

ALIENS ON FOUR SHIPS TOO SOON TO ENTER

LEAGUE OF NATIONS CALLS FOR TRUCE

FEAR FOR AMERICANS IN EARTHQUAKE ZONE

2.27 Front-page coverage of the earthquake in the *New York Times,* 2 September 1923.

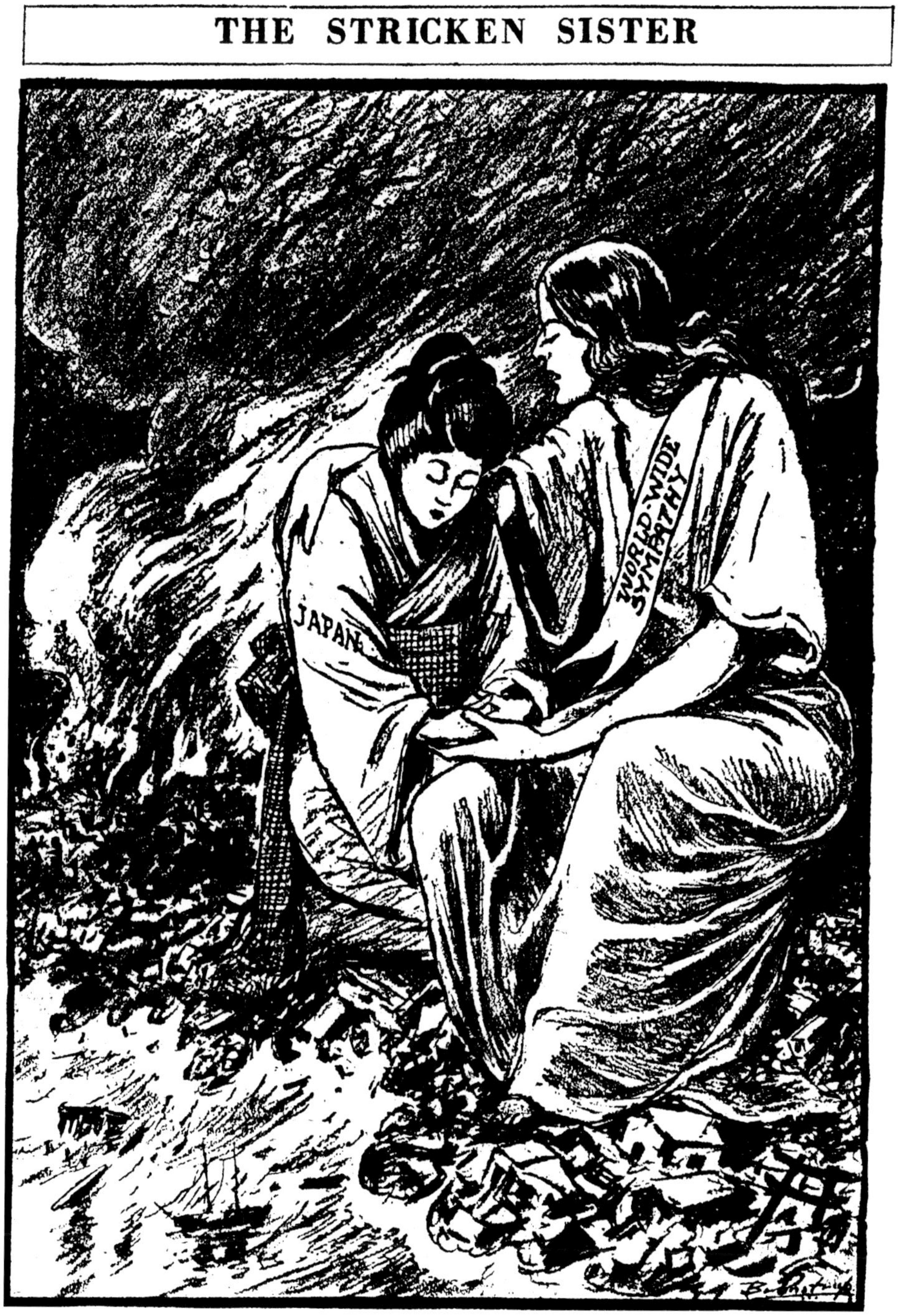

2.28 Artist unknown, *The Stricken Sister. San Francisco Chronicle,* 4 September 1923, 22.

American Asiatic fleet was being dispatched for rescue operations.[48] He wrote, "At the moment when the news of the great disaster which has befallen the people of Japan is being received, I am moved to offer you in my own name and that of the American people the most heartfelt sympathy, and to express to your Majesty my sincere desire to be of any possible assistance in alleviating the terrible suffering to your people."[49]

News of the disaster in Japan triggered a large-scale relief operation overseen by the American Red Cross, and just five days after the quake, on 6 September, a large advertisement sponsored by the Red Cross and the Japan Relief Committee ran in the *New York Times* imploring citizens of Greater New York to "Help Japan!"[50] Framed to one side of the ad boldface text in capital letters exclaims, "Minutes Mean Lives," followed by the desperate plea, "*Some* man, some woman, some little child will live if you answer this call *today.* Send as much as you can—*And at once.*"

Just as Japan had dedicated itself to assisting San Francisco after the earthquake, sending approximately $245,000 in aid, an amount far exceeding that of any other nation, the United States returned the assistance many times over in 1923. Over ten thousand Japanese living in California had been affected by the San Francisco quake, yet ironically they seem to have received almost none of Japan's aid because it was channeled elsewhere by relief organizations. Concerned about inciting incipient anti-Japanese racist sentiments, the Japanese-American community chose to form its own relief organizations rather than request official relief from the Red Cross.[51] Temporary amity between nations clearly did not solve the problems of racism on the ground—a message that took on even greater resonance a year after the Kantō quake, when the United States passed a widely supported anti-Japanese immigration law restricting the number of Japanese who could enter the country. For their part, the Japanese authorities, while clearly in desperate need of assistance immediately after the quake, still displayed great ambivalence about accepting aid for fear of losing sovereignty and international standing.[52]

In 1923, despite these incipient tensions, relief campaigns were mobilized throughout the United States to help the Japanese earthquake victims, who far outnumbered the estimated three thousand dead in San Francisco, and disaster-relief stamps pairing the sacred image of Mount Fuji with the Red Cross logo were even issued to raise funds.[53] New York City alone committed $1 million to Japan's recovery. The Red Cross was reported to have requested $5 million to underwrite its operations in Japan. The organization ultimately collected over twice that amount.

The Japanese-American community saw the Kantō disaster as extending to its diasporic community, and it was instrumental in supporting the relief effort. Twenty percent of the aid coming from the United States was from people of Japanese descent.[54]

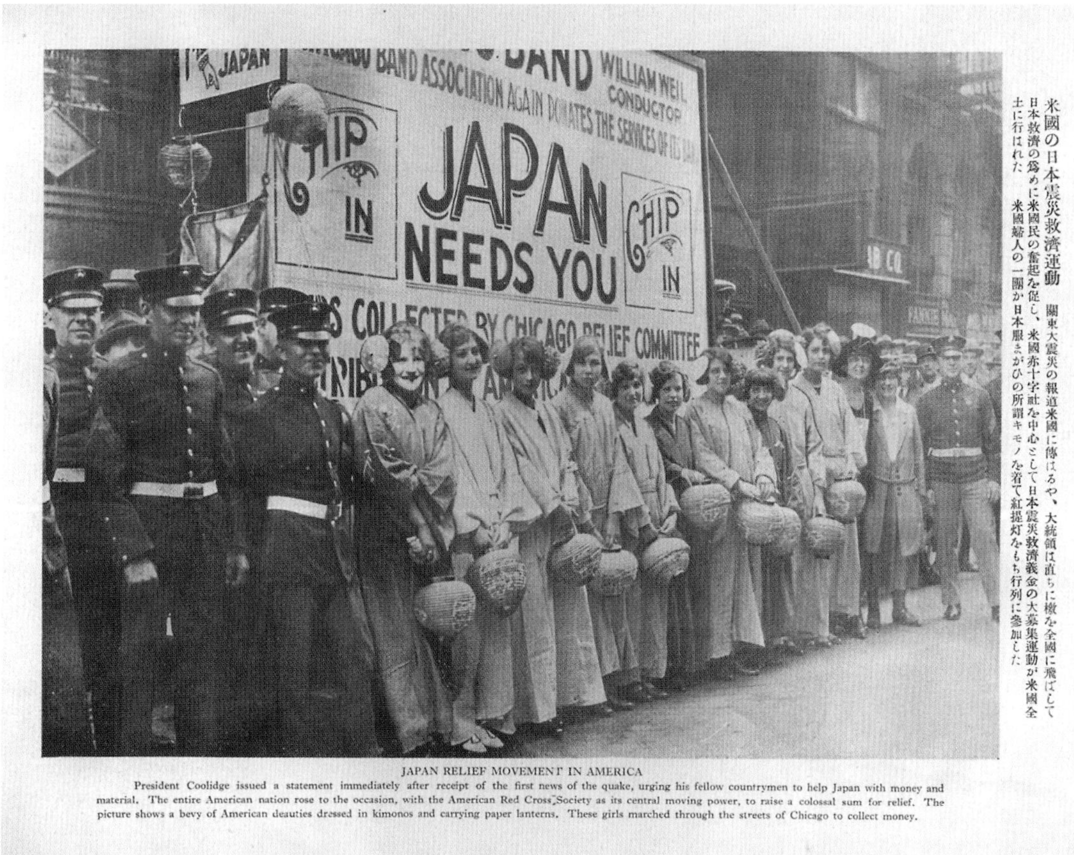

2.29 "Japan Needs You," relief fund-raising campaign, Chicago, photograph. *Tōkyō nichinichi shinbun,* 13–14 October 1923. Reproduced in Ōsaka Mainichi Shinbunsha, *Kantō Shinsai gahō,* vol. 3 (Osaka: Ōsaka Mainichi Shinbunsha/Hatsubaijo Tōkyō Nichinichi Shinbunsha, 28 October 1923). By permission of Mainichi Shinbunsha.

Japanese news organizations eagerly followed the coverage of Japan's disaster in the foreign press and reported on foreign-aid operations to gauge the extent of international empathy, which was seen to reflect Japan's standing in the theater of nations. For two days in mid-October, the front page of *Tōkyō nichinichi shinbun* featured a photograph of American women in Chicago dressed up in kimonos, wearing exotic face makeup, and holding paper lanterns under the sign "Japan Needs You: Chip In" (fig. 2 29).

The success of these fund-raising efforts is charted in the visually striking official diagrams of international aid compiled by the Tokyo Institute for Municipal Research in 1930, which record the staggeringly large amount of contributions from the United States, so large, in fact, that the U.S. bar on the chart would have extended well off the

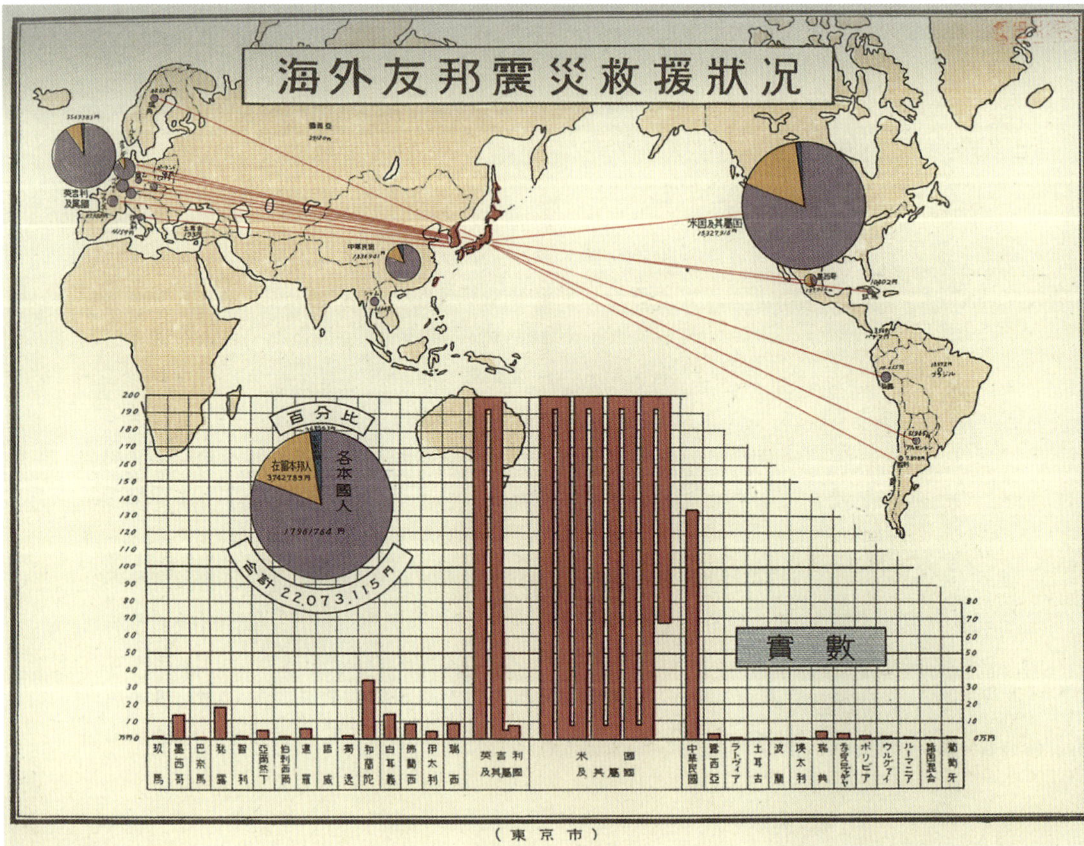

2.30 Overseas Allied Nations Earthquake Disaster Relief Situation, chart. Uchiyama Zenzaburō, ed., *Teito fukkō jigyō taikan* (General Survey of the Imperial Capital Reconstruction Project) (Tokyo: Nihon Tōkei Fukyūkai, 1930), 1:ch5, 17

page and had to be folded in a serpentine fashion to fit within the graph (fig. 2.30).[55] Superimposed on the world map, with Japan prominently in the center, international relief aid radiates from around the world into the island nation. The largest contributors, the United States, England, and China, are clearly marked with outsized pie charts that correlate with the bar chart below, visually reinforcing their stature in the spectrum of compassionate nations that aided Japan.

The national and international media scale of the Great Kantō Earthquake is indicative of the powerful and pervasive visuality of disaster in the early twentieth century. It also reveals how the media consciously used certain imagery to motivate people at home and abroad to participate in or contribute to the reconstruction effort. As Kevin Rozario has argued, despite the high moral values of charitable, humanitarian organizations like the American Red Cross, which viewed their activities as among the highest achievements of civilization, they achieved their phenomenal fund-raising success because of their active use of sensationalistic mass culture to manipulate people's emotions—in formats often indistinguishable from mass entertainment. The goal of the individual fund-raiser was "to inspire the noblest sentiments of the men and women assembled before him, describing incidents of violence and suffering to appall and shock them into acts of responsible compassion."[56] The use of mass media to inculcate empathy and generate aid continued to escalate throughout the twentieth century, expanding in global scope and scale into the new millennium. As a result, some individuals have begun to experience "compassion fatigue," prompting art critic Jeffrey Hughes to ask, "Does mass culture's encyclopedic indexing of catastrophe construct our sense of a present or immanent apocalyptic reality? Or do we use cognitive dissonance—a kind of not looking—as a method of coping?"[57] Undoubtedly, viewers dealing with the contemporary onslaught of media "clutter" are heirs to the potent visual legacy of the early twentieth century, when the lexicon of disaster was still emerging. Yet we must consider these two very different time periods in historical context. The avid acquisition of Kantō earthquake–related news and memorabilia by the general populace expressed a deep-seated desire to possess and personalize the experience rather than inure itself to the emotional impact of the event. And the pursuit of direct earthquake experience among collateral survivors in response to quake images spoke to a strong need for immediate, authentic eyewitness encounters to embody the somatic experience rather than to make it more palatable by keeping it at a distance. Image production and consumption were essential parts of the coping process after disaster, even as they generated a strong—and dark—touristic inclination.

Photographs, film, and other scientific modes of representation produced the media scale of the 1923 catastrophe. They presented visible evidence of its magnitude and devastating consequences, demanding national solidarity and international empathy. The factual nature of these modes of imaging constructed an armature of visual authority, but close examination shows the limits of that authority. For everything that was made visible in the imaging of the event, many things were made invisible as well, and such erasures are just as revealing.

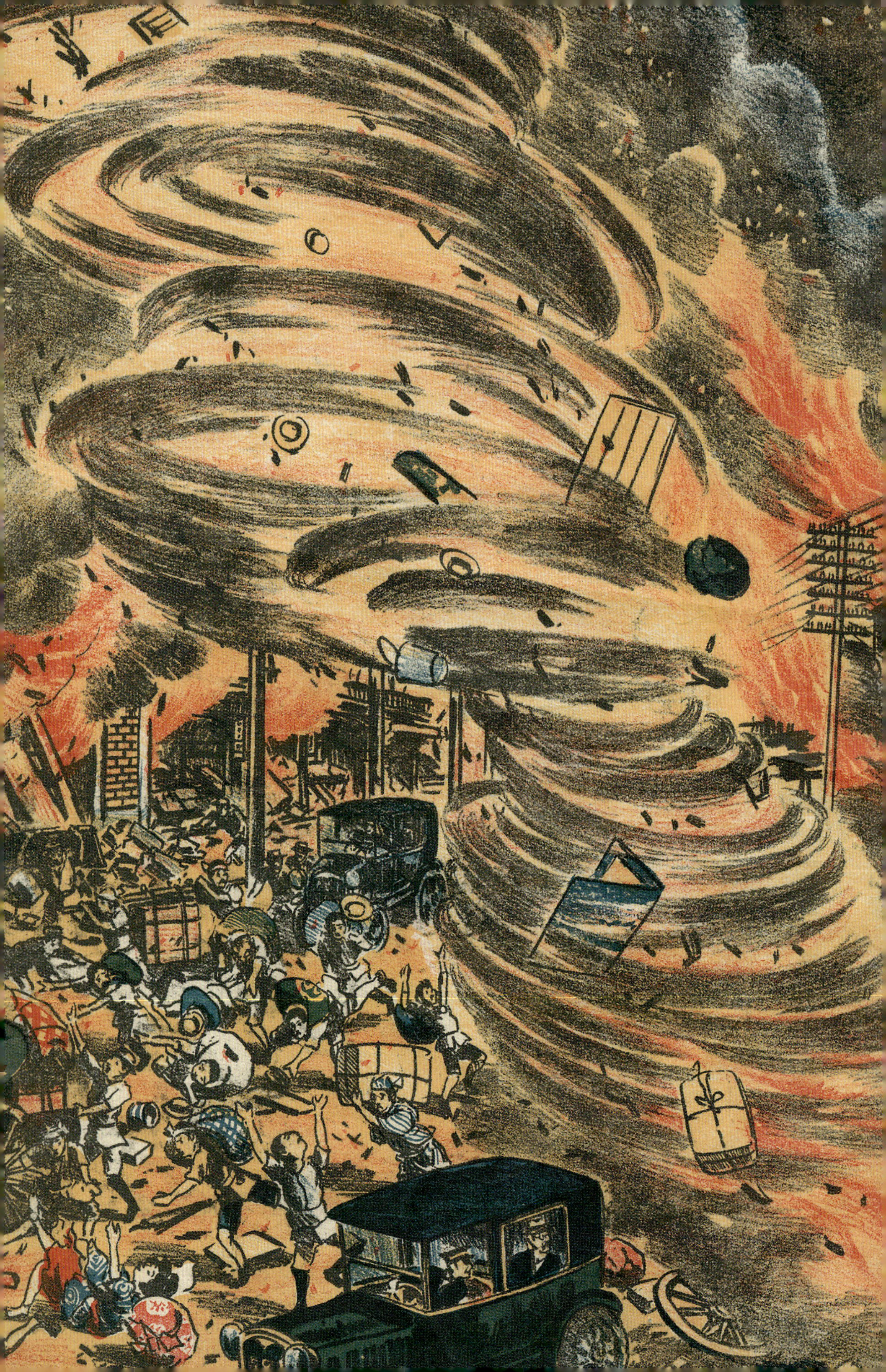

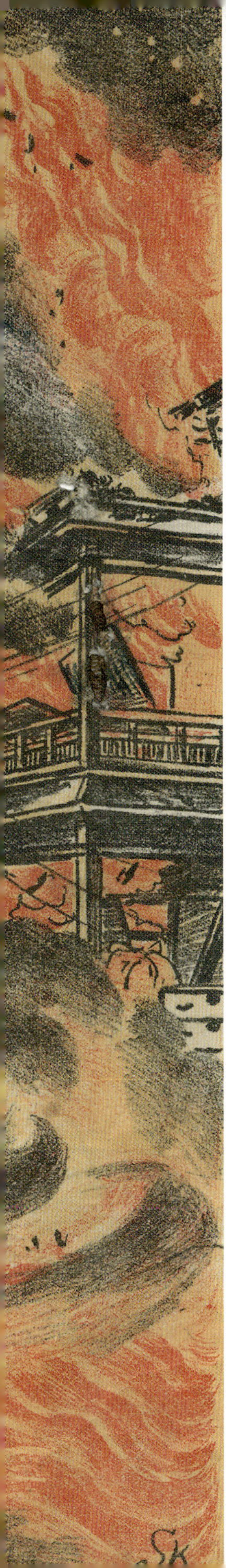

3

DISASTER AS SPECTACLE

A MAJESTIC CROWN OF WHITE BILLOWING SMOKE rises above the dome of a Western-style building in the photograph *Fire Seen in Okubo, 3 P.M.* (fig. 3.1).[1] In this moment frozen in time; the spotlighted clouds have a deific presence as they loom ominously over the darkly shadowed buildings and ground below. The composition of the picture is perfectly balanced, with the arc of the plume mirroring the slope of the landscape. This magnificently horrible plume was captured throughout the visual media, immortalized in ink, oil, and graphite, and on celluloid. Such representations are a vivid reminder of the constructed nature of disaster imagery, whose producers had a vested interest in spectacle and aesthetic impact while providing visible evidence of the event. The producer of spectacle must actively engage the public gaze to produce emotions such as marvel, horror, sorrow, and pity. Spectacle is a form of phatic image targeted to force viewers to look and to hold their attention.[2] With disaster, spectacle records agony in one's consciousness through the sense of sight and imbues the viewing experience with emotion. Spectacle produces spectators who are distinct from participants. But these spectators are not necessarily detached. Spectatorship can also be a participatory encounter that constitutes a form of social action. Hovering in the murky middle ground between sight and imagination, disaster spectacles assault the sensibilities while morally implicating the spectator in the event. While critics have cautioned that extensive viewing of such sensationalized images can blunt people's moral judgment and produce an analgesic effect, spectacle should always make us think about the power and politics of looking.

Spectacle is certainly about entertainment. It is also about com-

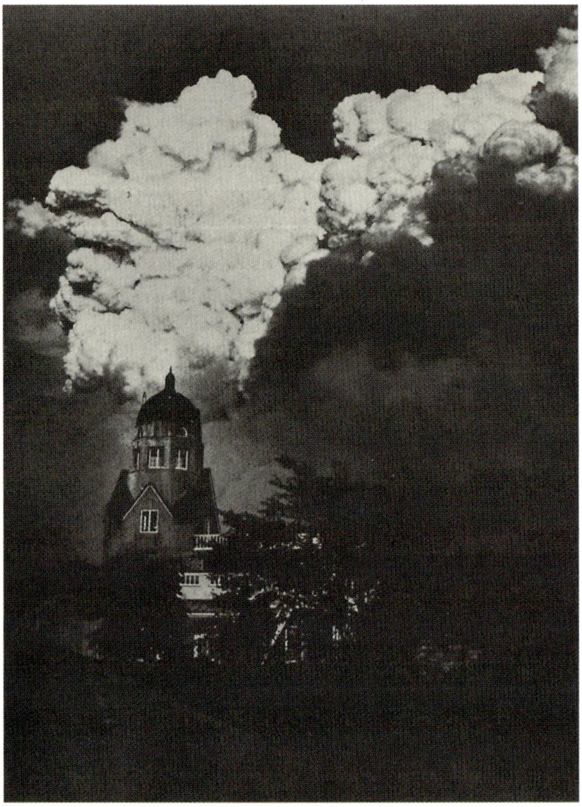

3.1 *Fire Seen in Okubo, 3 P.M.* ("The clouds of conflagration caused by the great earthquake fire on the 1st of September, 1923, as seen from Okubo, suburb of Tokyo, at 3 P.M. of the same day"), photograph, 1923. Tokyo Municipal Office, *The Reconstruction of Tokyo* (Tokyo: Tokyo Municipal Office, 1933), 1.

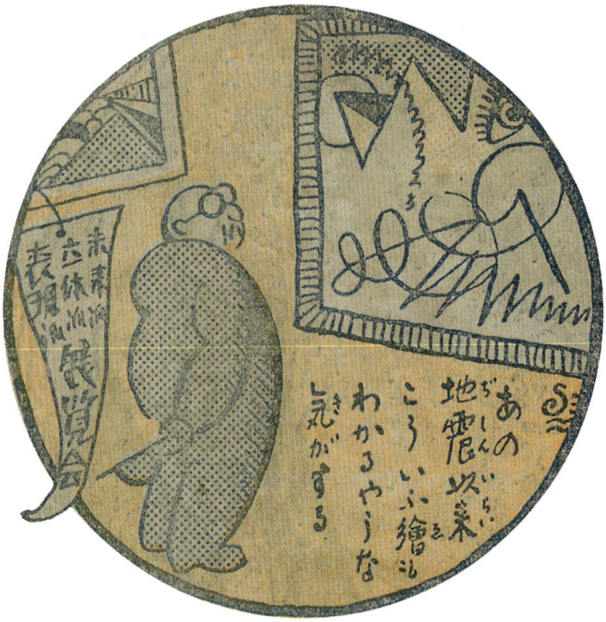

3.2 Arita Shigeru, "Somehow I get the feeling that after that earthquake, I understand these paintings better," cartoon. *Jiji manga* 132, 7 October 1923, 3. Collection of Ohio State University Billy Ireland Cartoon Library & Museum. By permission of Sangyō Keizai Shinbunsha.

modification, since sensations are packaged and consumed. Even as the news media serve a critical public function in emergency situations, historically they have still exploited the affective, sensational aspects of major disasters for profit. Guy Debord, Jean Baudrillard, and a host of theorists after them have critically argued that in the "society of the spectacle" that emerged with mass commercialism at the turn of the twentieth century, images have invariably been controlled by a dominant capitalist ideology and the state, which have dictated social relations and consumer desires, expunging individual subjectivity. Baudrillard has even called the postmodern age of the televisual a condition of hyperreality in which the viewer experiences and feels things only through the mediation of simulations. In the early twentieth century, with mass media's increased separation of spectators from spectacle—mediated through the modern technologies of photography and cinema—the wonder and visual pleasure of spectacle was clearly derived as much from the medium as from the object of the camera's gaze. Moreover, as Elizabeth Cowie has aptly noted, documentary images, photographic or cinematic, produce pleasure simply through their purported re-presentation of actuality. At the same time, they produce an interrelated "desire for reality as knowledge and as spectacle."[3] The aestheticized spectacle became amplified through other media, such as prints and paintings, which, unlike photography and film, were not tethered to the representational or the real, giving them artistic license to express the subjective visual sensations of catastrophe.[4] The mass media produced a frame of reference for the public imagination of disaster. Artists adorned that frame.

Modernity was itself an ongoing spectacle of new technology, from the weapons of war to the burgeoning metropolis, which, together with the dynamo of capitalism, produced its own kind of "creative destruction."[5] Interpreting modernity's perceived fracturing of vision and dissolution of the unified object, modernist artists aimed to create an explosive impact in their work and to portray radical ruptures similar to the earthquake. Modernism's experimentation with expression and abstraction opened a new realm of imaging that further explored disaster's affective and spectacular qualities. Modernist imagery offered something the press could not: a subjective window into the intense sensory experience of disaster spectatorship. Many critics noted this uncanny congruence between the shock of the quake and the shock of modernism. A small, pithy cartoon by Arita Shigeru (1890–?) in *Jiji shinpō*'s Sunday supplement *Jiji manga* (Cartoons of Current Affairs) (fig. 3.2) takes up this theme in a humorous scene of a portly, dignified man looking at a display of abstract modernist paintings next to a banner announcing "Futurism, Cubism, Expressionism: Exhibition." He muses, "Somehow I get the feeling that after that earthquake, I understand these paintings better."

In examining the spectacular elements of disaster pictures, it quickly becomes apparent that the representation of observable phenomena is always processed through the imagination, blurring the boundaries between the real and the fantastic. Fact, fiction, documentary, and melodrama commingle. Considering why people might desire spectacular imagery, we must reflect more generally on what these images say about the relationship between viewer and event. On the one hand, spectacular disaster images, in all their titillating and frightening aspects, seek to tap into the viewer's desire for emotional authenticity as a means of experiencing the sensory aspects of the disaster. They actively engage the sensorium to stimulate the embodied experience. On the other hand, various entities in addition to the mass media often seek to capitalize on the production of crisis and fear through spectacle, such as the Japanese insurance firms that purveyed commodified safety nets after the Kantō earthquake and the imperial authorities who sought to rechannel public panic into expansion of government authority and the security state.[6] Home minister and former mayor of Tokyo Gotō Shinpei even attempted to radically overhaul the capital's urban planning under the guise of urban renewal and disaster prevention, promoting projects on a scale akin to the invasive, deeply political incursions of nineteenth-century French bureaucrat Baron Haussmann in Paris. Still, recognizing the heterogeneity of responses to the Kantō catastrophe, I concur with Martin Jay's conclusion that a plurality of visions mitigates any "monocular regime of spectacle seduction," as viewers assemble their own responses to disaster from a diversity of experiences and sources.[7]

Disaster, War, and Mass Media Spectacle

While disaster has become a mainstay of contemporary popular entertainment, in the pretelevision era of the Kantō quake, a range of other mass media filled the public's need for ever more intense vicarious thrills through images of calamity. Already in the 1890s, popular illustrated magazines like the monthly *Fūzoku gahō* (Manners and Customs in Pictures), published in Tokyo by Tōyōdō (1889–1916) and issuing over 130,000 copies per year at its peak, frequently ran special issues devoted to the seemingly endless string of disasters that had recently afflicted the nation. With the onset of the Meiji period (1868–1912), coverage of contemporary events was no longer prohibited, although the press was still heavily regulated and censored. Tragic accounts of death and destruction, along with stories of heroic rescue, were catalogued by many sources in great detail, all copiously illustrated with vividly rendered lithographs—some in color and some inserted as folded pull-outs—showing spectacularly destroyed villages, families being violently swept away by flood waters, and even mutilated human remains in ghastly array. One horrific image pictured crows picking out the eyes of bloody corpses. Such graphic depictions of corpses in decay and composition (*kusōzu,* pictures of the nine stages of the decaying corpse) were, in fact, common in Buddhist imagery from the thirteenth through the nineteenth centuries, used as visual meditation aids for ardent religious devotees hoping to liberate themselves from sensual desires and affections.[8]

In *Fūzoku gahō,* deathly images merged with the sensationalism of Japan's budding modern media. The cover of the November 1891 special issue on earthquakes greeted readers with a colorful image of the destruction of towns on the Nōbi plain in central Japan when a massive earthquake, described as more intense than even the historic Ansei quake of midcentury, struck at the end of October. Continuing previous pictorial traditions and evoking the standard *namazu* imagery of the past, the issue number (35) appears on the left side of the cover in a cartouche shaped like a tapered gourd (*hyōtan*) and bearing the head of a catfish. A dramatic double-page spread of fire destruction follows (fig. 3.3), with flying bodies and limbs everywhere. The extended bloody hand of a man breaking through the lattice of a toppled shoji screen underscores the personal tragedy, drawing the reader's eye to details of individual suffering. And on the back cover, a special announcement from the emperor and the imperial household expresses condolences to the people of the Nōbi area. Combining dramatic stories of filial piety and heroism with expressions of imperial compassion, the journal legitimized its media sensationalism by offering overarching moral messages of social solidarity and state benevolence.

The heightened sensory experience of disaster is often equated with that of combat during wartime, an uneasy combination of exhilaration and horror. Japan has a clear pictorial genealogy linking war and disaster imagery that dates back to at least the twelfth century; such early images often elided the distinctions between human and natural violence.[9] In the Meiji period, popular Japanese woodblock prints chronicling the Sino- and Russo-Japanese wars as news reportage linked the motifs of spectacular death and destruction to national narratives of sacrifice. The prints were similarly prone to sensationalism, graphically depicting the bloody battles and grand heroics of Japan's modern military.[10] Set against the spectacular background of the battlefield with dramatic plumes of smoke from artillery fire and the bloody carnage of casualties strewn across the landscape, war imagery used dynamic compositions to spotlight narratives of individual heroism often tinged with the tragedy of selfless sacrifice (fig. 3.4). Japanese soldiers were rendered as powerful and dignified, sometimes even given Caucasian facial features and body types to emphasize Japan's modernity and civilizational progress in the world theater of imperialism. Enemy combatants were

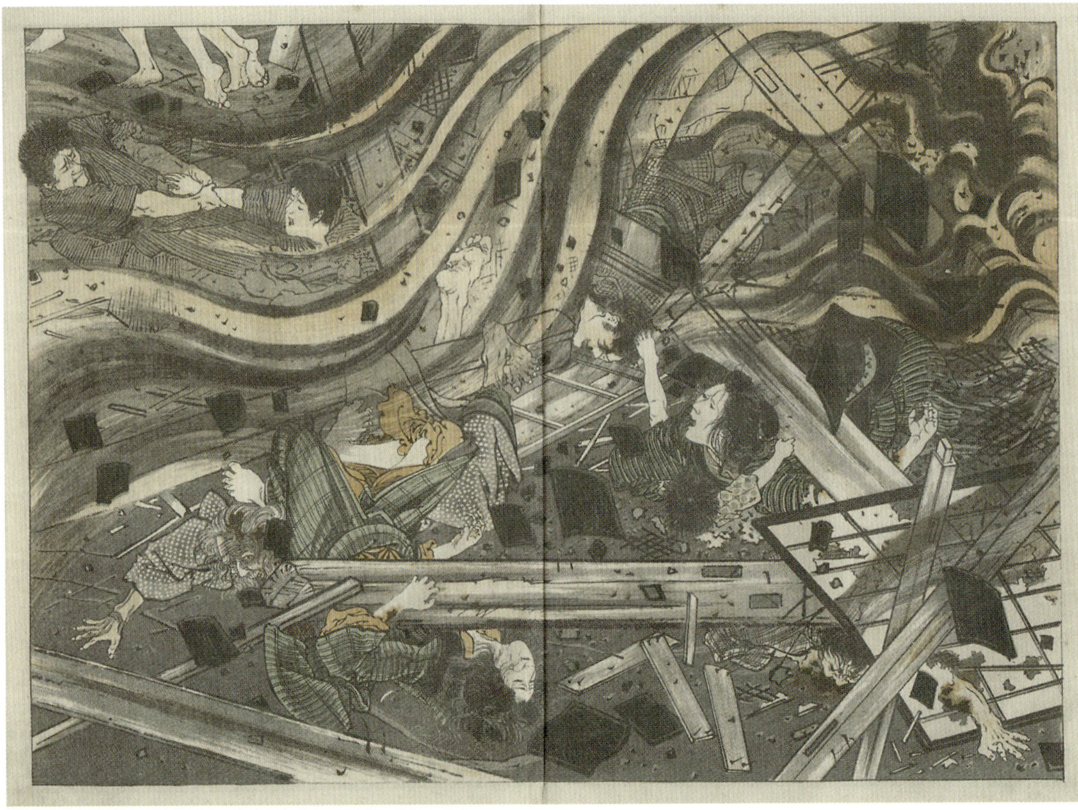

3.3 Artist unknown, scene of the Nōbi earthquake, *Fūzoku gahō* (Manners and Customs in Pictures), no. 35 (30 November 1891).

mostly shown in ignominious poses, except for select images of Russian officers who were admired as noble Western adversaries.

The Russo-Japanese War was a groundbreaking transnational media event, and many photographic and cinematic images of it were produced and distributed widely in Japan, Great Britain, and the United States. This pioneering "mediatization" of warfare in a global context—and serialization of death (to borrow Virilio's terms)—also fueled a national mass market for spectacular imagery that included the tremendously popular "soft news" visual mass media of postcards. Wartime reportage made visible the spectacular actuality of modern technological warfare, even when it did not depict battle scenes or the "body horror" of casualties, which were certainly included but perhaps in fewer numbers than one might expect. The imposing armaments and apparatuses of modern warfare, the unprecedented number of mobilized troops, and the titillating reality of Japan's ability to destabilize the West's seeming political monopoly on

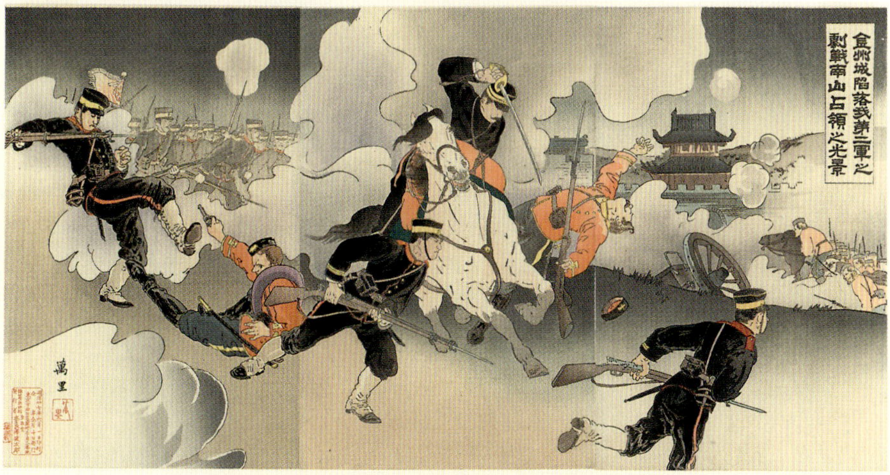

3.4 Banri (active 1904). Publisher: Narazawa Kenjirō. Block cutter: Umezawa horu. *Scene of Our Second Army Occupying Nanshan in a Fierce Battle at the Fall of Jinzhoucheng* (*Kinshūjō kanraku waga dainigun no gekisen Nanzan senryō no kōkei*). Japanese, Meiji era, 1904 (Meiji 37), printed June 1, published June 10. Woodblock print (nishiki-e); ink and color on paper. Vertical ōban triptych, 35.9 × 70.5 cm (14 1/8 × 27 3/4 in.). Photograph © 2012 Museum of Fine Arts, Boston, Jean S. and Frederic A. Sharf Collection, 2000.446a-c.

modernity by successfully waging war in the theater of Western nations all provided national spectacle.[11]

Some wartime images explicitly featured spectacular destruction, such as the postcard of the explosive demolition of a covered protective ditch (caponiere) at a Russian fort east of Kikwan Hill near Port Arthur (fig. 3.5). Without its caption, the image is indistinguishable from depictions of natural volcanic eruptions, such as the heavily reported eruption of Mount Bandai in Fukushima prefecture just a few years earlier, in 1888, or the later eruptions of Mount Usu in Hokkaidō in 1910, and the one on the island of Sakurajima off Kyūshū in 1914 (fig. 3.6).[12]

Modern weaponry, locomotives, and skyscrapers—modernity's parade of new technological offerings—provided an ongoing spectacle that included increasingly fantastic scenes of destruction that transformed the visuality of disaster. While modernity produced its own new cyclical order of "creative destruction," daily demonstrating its awesome power to obliterate the old in the "march of progress," any belief that its technological achievements subjugated the natural world was upended by their spectacular destruction in natural disasters, fueling a consuming interest in visualizations of catastrophe.[13] To borrow a term from Hariman and Lucaites, the scale of destruction in modern warfare and disasters constantly put in question "modernity's gamble"—

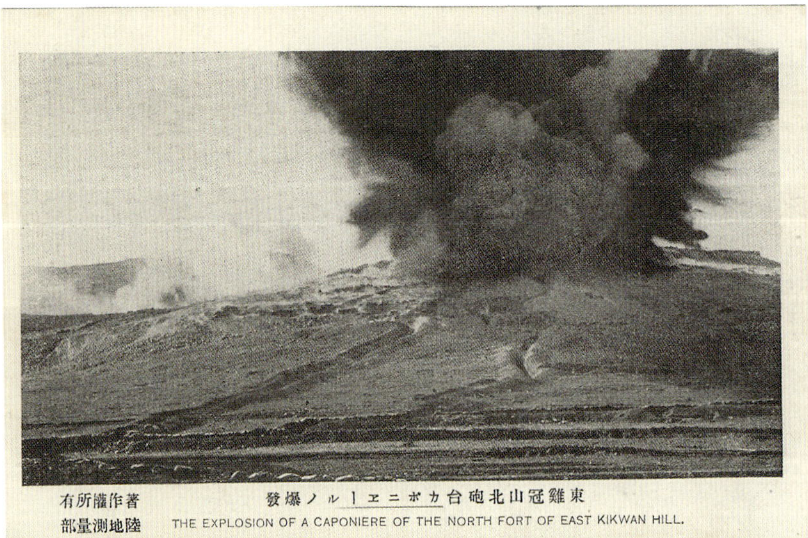

3.5 Artist unknown. Publisher: Ogawa Isshin, 1860–1929. *The Explosion of a Caponiere of the North Fort of East Kikwan Hill* (*Higashi Keikanzan kita hōdai no kaponiēru no bakuhatsu*). Japanese, late Meiji era, 1904–5. Place of creation: Japan. Relief halftone; ink on card stock (postcard). Overall, 8.8 × 13.8 cm (3 7/16 × 5 7/16 in.). Photograph © 2012 Museum of Fine Arts, Boston, Leonard A. Lauder Collection of Japanese Postcards, 2002.3261.

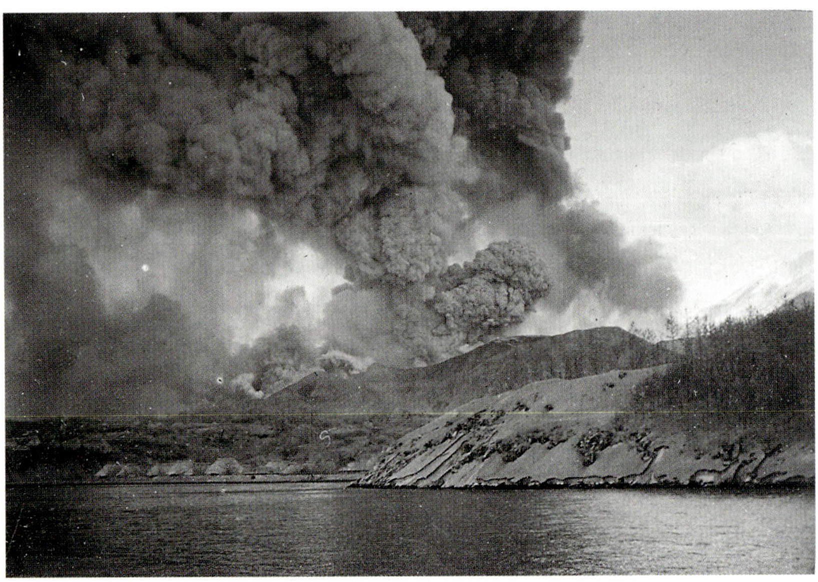

3.6 Volcanic eruption on Sakurajima, Kyūshū, photograph, 1914. Collection of Kokuritsu Kagaku Hakubutsukan.

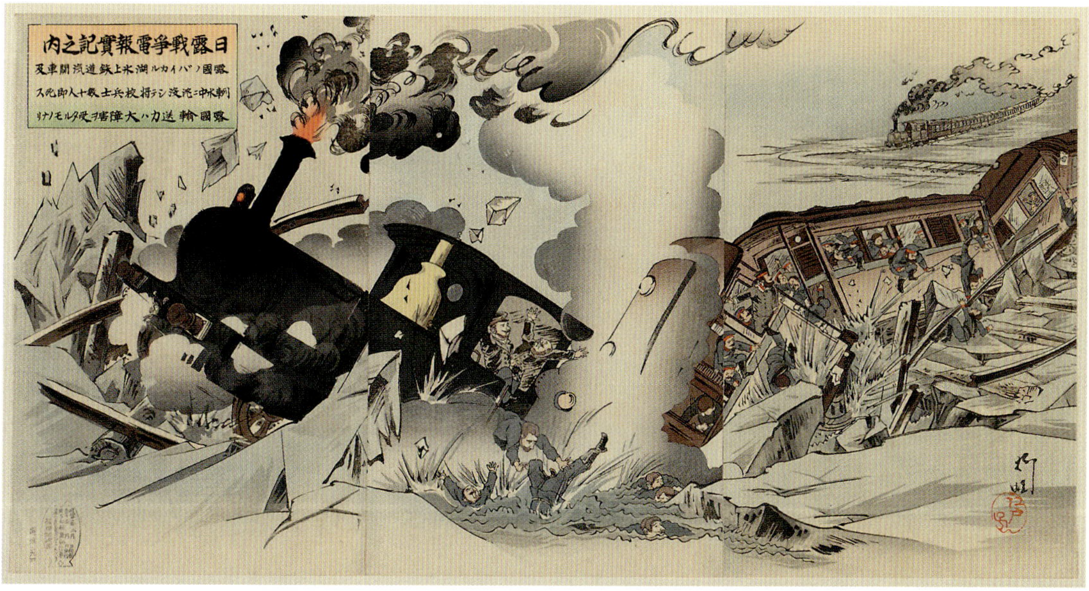

3.7 Utagawa Kokunimasa (Ryūa), 1874–1944. Publisher: Fukuda Hatsujirō. *Telegraphic Record of the Russo-Japanese War: On the Ice of Lake Baikal in Russia, a Steam Locomotive and Its Cars Sank, Killing Tens of Officers and Soldiers. Russia's Transport Capacity Was Greatly Damaged* (*Nichiro sensō denpō jikki no uchi: Rokoku no Baikaru-ko hyōjō tetsudō kikansha oyobi ressha kosui ni chinbotsu shite shōkō heishi sūjūnin sokushi su Rokoku yusōryoku wa dai shōgai o uketaru mono nari*). Japanese, Meiji era, 1904 (Meiji 37), March. Woodblock print (nishiki-e); ink and color on paper. Vertical ōban triptych, 36.4 × 70.3 cm (14 5/16 × 27 11/16 in.). Photograph © 2012 Museum of Fine Arts, Boston, Jean S. and Frederic A. Sharf Collection, 2000.68a-c.

that is, "the wager that the long-term dangers of a technology-intensive society will be avoided by continued progress."[14] Clearly, however, modernity brought even greater risk by introducing ever more spectacular forms of destruction. Viewers were scared, but they could not look away.

The technological tropes of modernity are a clear point of visual contiguity linking war and disaster imagery. For example, the marvel of modern transportation resulted in spectacular train crashes and destruction of railway lines that featured prominently in both. In *Telegraphic Record of the Russo-Japanese War: On the Ice of Lake Baikal in Russia, a Steam Locomotive and Its Cars Sank, Killing Tens of Officers and Soldiers* (1904), Utagawa Kokunimasa presents a horrifying image of a multicar passenger train just as it plunges into a frozen lake in Russia near the battlefront (fig. 3.7). Images of trains violently derailed by earthquakes appeared as early as the Nōbi quake in 1891, although scholars have argued that they must have been fanciful projections since no such inci-

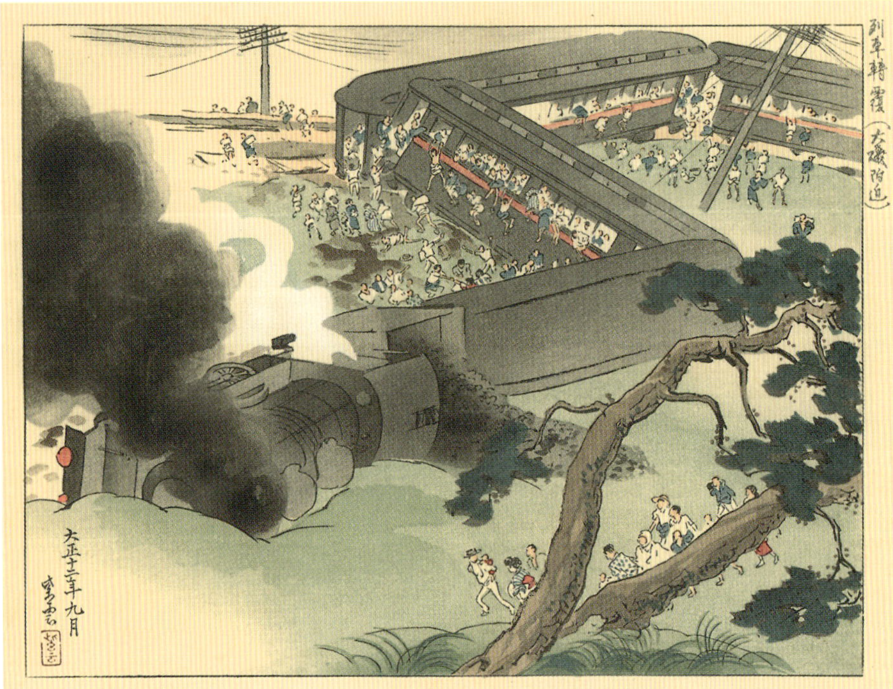

3.8 Kondō Shiun, *Overturned Train (the Road near Ōiso)* (*Ressha tenpuku [Ōiso fukin]*), woodblock print. *Taishō Shinsai gashū* (Collected Prints of the Taishō Earthquake) (Tokyo: Zue Kenkyūkai, 1926). Collection of the author.

dents were ever reported.[15] There were, however, widespread deadly derailings of trains and massive warping of train tracks in the 1923 temblor, which provided a visual feast for artists. Kondō Shiun's (active 1910–30) print (fig. 3.8) in the *Collected Prints of the Taishō Earthquake* (*Taishō Shinsai gashū*), titled *Overturned Train (the Road near Ōiso)* (*Ressha tenpuku [Ōiso fukin]*), depicts a buckled train that has been knocked off the tracks near Ōiso on the coast in Kanagawa prefecture southwest of Tokyo.[16] The front of the locomotive lies on its side spewing dark smoke, a colossal man-made beast felled by the forces of nature. Several other deadly train derailments were widely documented, particularly the dramatic scene at the village of Nebukawa, west of Odawara, where the quake and subsequent tsunami caused the side of the mountain to collapse into the sea, taking with it the village train station as well as a passing commuter train carrying over one hundred passengers.[17] Sharing both the spectacular actuality of landmark historical events and the motifs of death and destruction, images of wars and natural disasters appealed to the same spectators by highlighting the ever-increasing national stakes of modernity.

Disaster Spectators

The pleasure that viewers take in the spectacle of disaster is often predicated on their viewing position—their sense of safety and distance from physical harm. But some took pleasure in spectatorship even while amid disasters, as represented in the strangely bucolic woodblock print by popular *Miyako shinbun* newspaper illustrator Igawa Sengai (1876–1961) *Refuge on the Streetcar Tracks* (*Densha senrojō no hinan*), in which a Japanese family picnics on top of the debilitated electric streetcar tracks as they gaze leisurely on the city in the midst of obliteration (fig. 3.9).[18] The crowns of toppled buildings and smoldering fires are visible in the distance beyond the smoke, as the neatly dressed family of three—with the father casually reclining—prepare tea surrounded by fallen power lines. Inhabitants of San Francisco were known to set up chairs at vantage points in the burning city to watch the encroaching conflagrations, moving with the rhythm of the fires. Similarly, Tokyo residents sat in cautious observation as they waited in the streets with their belongings and tried to assess the movement of the fires (fig. 3.10). Supposedly safe positions, however, could become suddenly imperiled with the changing winds, and participant-spectators could easily become victims. Such was the perilous adventure of disaster spectatorship.

Some images placed remote spectators in the position of participant-spectators through the spectral traces of a figure in the frame that made the viewing action more self-reflexive and called greater attention to the mediation of the image. In one such image, we see the haunting fragment of a figure from behind—the back of a panama hat—which directs our gaze toward a devastated vista and the ruined carcass of the Ryōunkaku (Twelve Stories) (fig. 3.11). Other images highlight the fine line between spectators and witnesses, as the viewer's sight attests to the visible evidence put before him or her, thereby acknowledging the historical tragedy. Cartoonist and *Asahi shinbun* illustrator Okamoto Ippei's simple scene of the destroyed cityscape in *Paintings of the Great Kantō Earthquake* (*Kantō daijishinga*) (fig. 3.12) affirms the signifying power of looking as a mother pushes her young son forward and points sharply toward a smudge of ink. She exclaims in the text below, "This is what's left of your house." "It's a lie! It can't be!" he screams, recoiling in horror from the charred ruins.

The display of charred and bloated dead bodies similarly constituted a form of disaster spectacle that explicitly implicated the spectator by drawing attention to the action of looking. "The Clothing Depot that I saw the day after the disaster was in truth a sea of corpses," wrote celebrated artist and designer Takehisa Yumeji (1884–1934) in his illustrated chronicle that ran in the *Miyako shinbun*. "People who had been killed in war on a battlefield probably would have presented no more tragic a scene. Persons

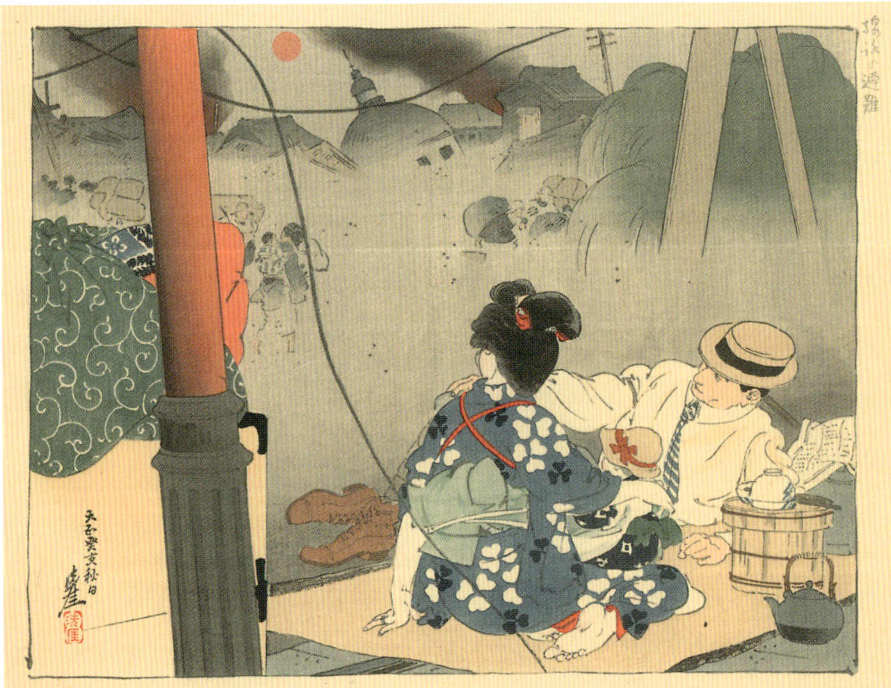

3.9 Igawa Sengai, *Refuge on the Streetcar Tracks* (*Densha senrojō no hinan*), woodblock print. *Taishō Shinsai gashū* (Collected Prints of the Taishō Earthquake) (Tokyo: Zue Kenkyūkai, 1926). Collection of the author.

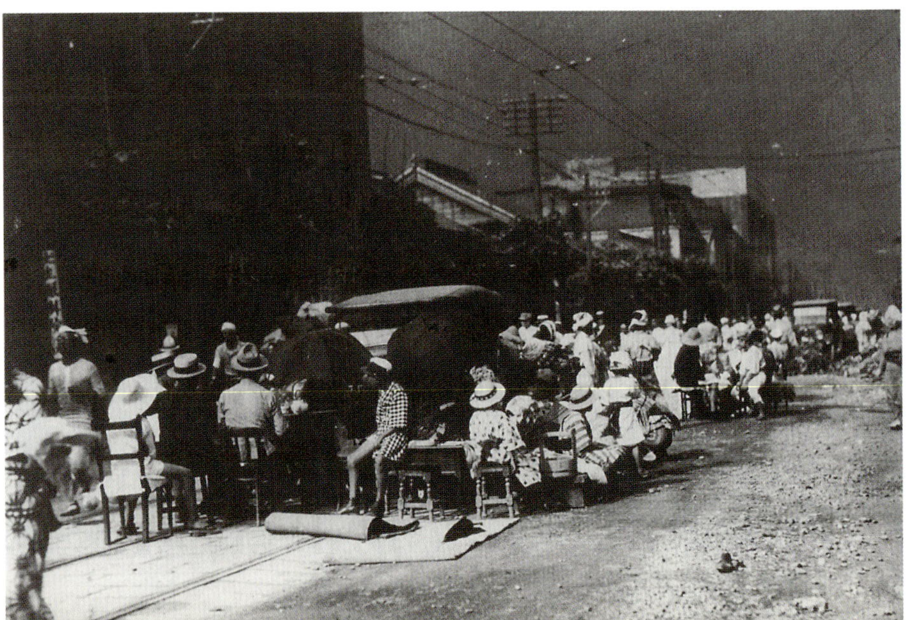

3.10 Artist unknown, *Great Kantō Earthquake: Refugees on Top of the Streetcar Tracks* (*Kantō Daishinsai: Shiden no senrojō ni hinanshita hitotachi*), photograph, 1923. By permission of Mainichi Shinbunsha.

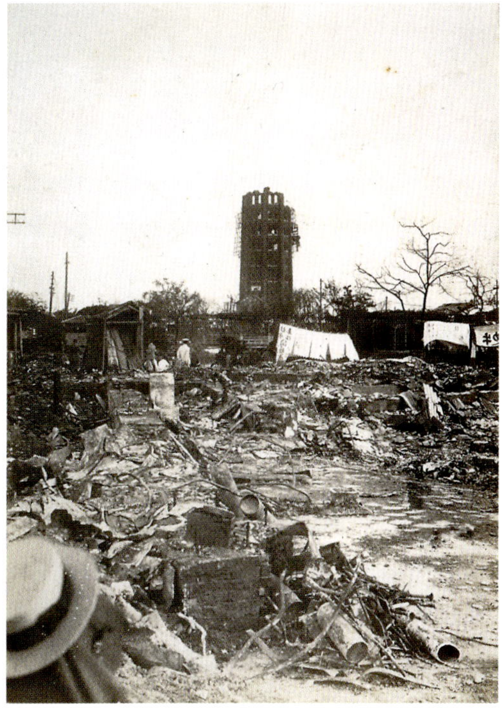

3.11 Naruse Keijirō, Ryōunkaku, photograph, 1923. Collection of Tōkyō-to Edo Tōkyō Hakubutsukan.

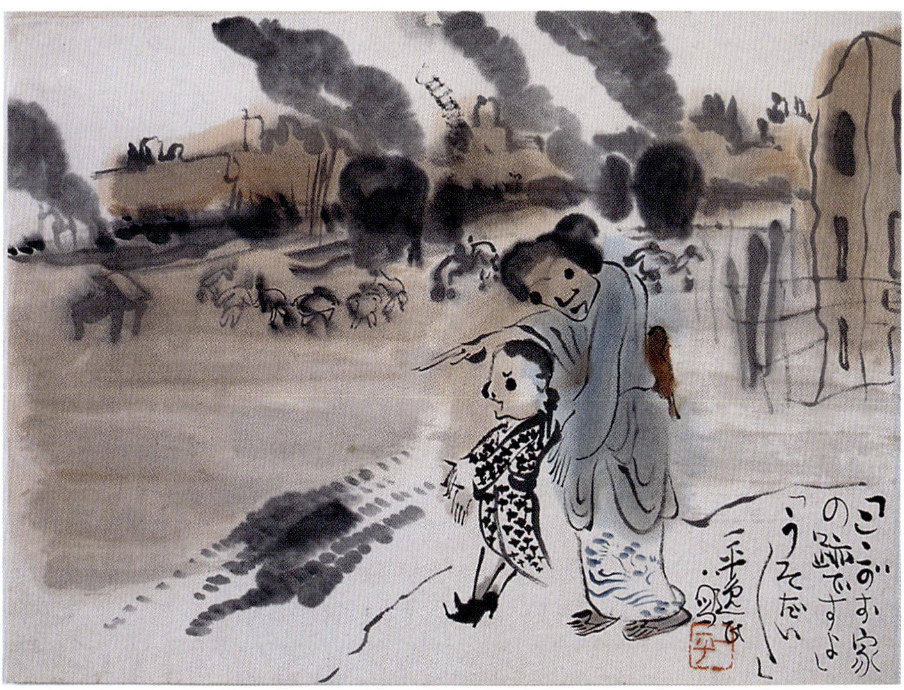

3.12 Okamoto Ippei, "This is what's left of your house." "It's a lie! It can't be!" (*Koko ga oie no ato desu yo. Uso dai! Uso dai!*), ink and color on paper, c. early 1924. *Paintings of the Great Kantō Earthquake* (*Kantō daijishinga*). Collection of Tōkyō-to Edo Tōkyō Hakubutsukan.

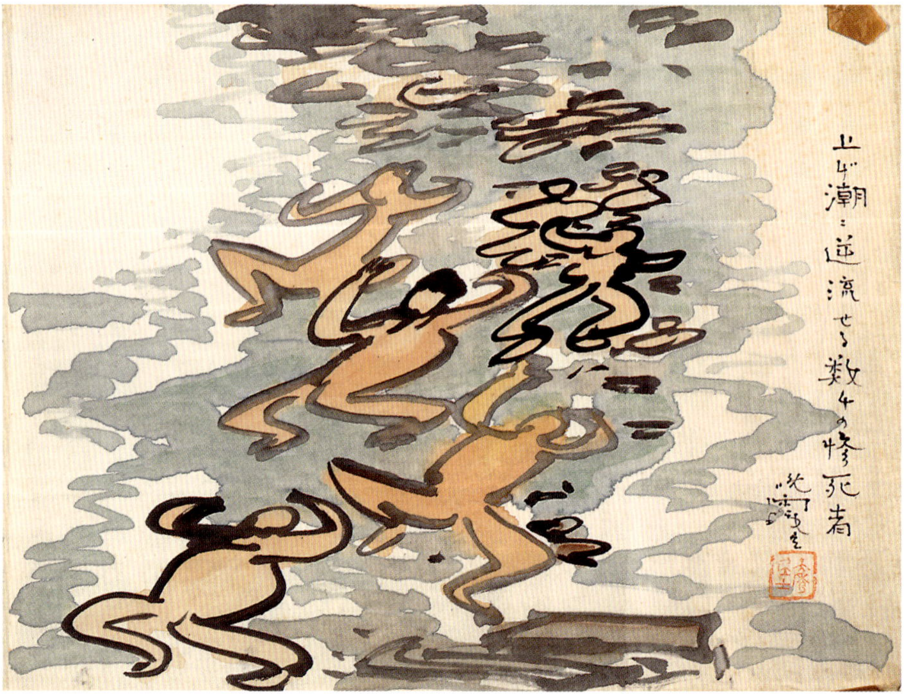

3.13 Maruyama Banka, *Thousands of the Tragic Dead Flowing Upstream with the Rising Tide* (*Ageshio ni gyakuryūseru sūsen no sanshisha*), ink and color on paper, c. 1923. Collection of Tōkyō-to Edo Tōkyō Hakubutsukan.

who had until only a short while earlier been pursuing their lives with no thought of dying were now dead. They didn't want to die, they fervently wanted to live, yet here they were in all their writhing agony; a wave of misery had swept in on top of them."[19] The viewers of such scenes, designed to induce horror, could not remain neutral. As W. E. B. Du Bois said about the grisly activity of "spectacle lynchings" in the United States, the cultural power was "in the looking."[20]

Just as with spectatorship of lynchings, disaster looking was extended publicly in newspapers and privately in personally collected photographs and picture postcards.[21] There was an aggressive quality in looking at and gazing upon the corpses of those killed violently, the marks of their extermination plainly visible. Through the gaze of the spectator, the exposed corpses of the victims, denied their privacy, individuality, or dignity, are converted into objectified, shamed bodies. In view of the Japanese religious and philosophical beliefs associating death and defilement, this act took on even greater significance. Contemplation of the impurity of the nine stages of the decaying corpse (*kusōkan*) was featured in many Buddhist sutras, and this practice was in

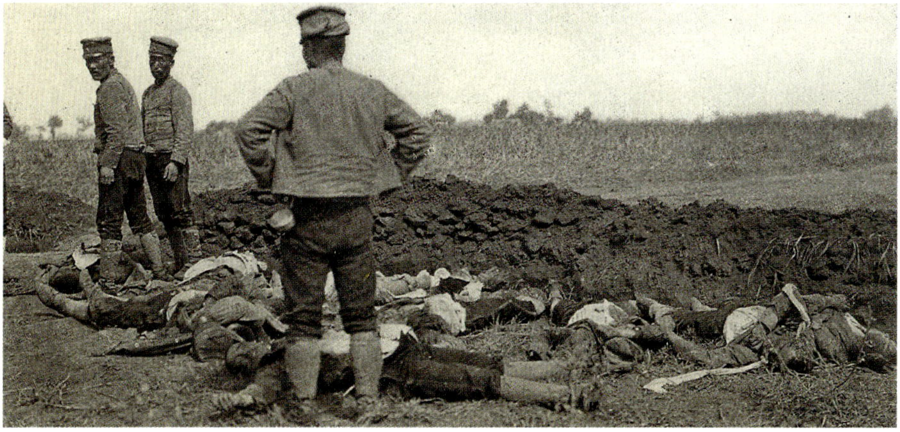

3.14 *Burying Japanese and Russian Dead Together outside Liao-Yang,* photograph. James H. Hare, ed., *A Photographic Record of the Russo-Japanese War* (New York: P.F. Collier, 1905).

widespread practice to help devotees overcome obstacles to enlightenment and to conquer carnal, particularly sexual, desires.[22] Modern photographs and postcards enabled viewers to linger over these degraded bodies with a kind of penetrating stare, unable to look away but lacking moral guidance. Thus, the images capitalized on the voyeuristic and prurient curiosity of the viewer by allowing the eye to caress the bodies. They ultimately evoked an abject visual pleasure in viewing the repulsive yet alluring surface textures of the blistered bodies thrust into view. This same aesthetic sensibility informed artistic renderings of bodies in extremis, with images of floating corpses in the Tokyo waterways permanently etched into people's minds. Pioneering Japanese watercolorist Maruyama Banka's mesmerizing painting *Thousands of the Tragic Dead Flowing Upstream with the Rising Tide* (*Ageshio ni gyakuryūseru sūsen no sanshisha*) (fig. 3.13) captures shimmering corpses floating on the water, dematerializing the bodies into writhing silhouettes that appear trapped under the glassy surface.[23] Their hapless outlines dance wildly with the rhythms of the moving water, caught forever in the animated poses of their agonizing annihilation. These images produce a kind of synesthesia, bringing the visual spectacle of death in touch with the other senses.

Ironically, while images of corpses do appear in Russo-Japanese War photographs and film, strict censorship restrictions did not allow the reveling in "body horror" that was evident in Kantō earthquake representations. Russo-Japanese War volumes, such as *A Photographic Record of the Russo-Japanese War*, include only a few images of soldiers gazing upon the war dead, both friend and foe, in battle trenches and during burial; one example is *Burying Japanese and Russian Dead Together outside Liao-*

Yang (fig. 3.14).[24] This kind of somber documentary photograph still imparts to the viewer what Bill Nichols terms "an indexical whammy" in its display of the spectacle of actuality, thereby creating an effect quite different from that of the staged casualties frequently seen in woodblock prints that aesthetically transform death into a noble enterprise.[25]

Sensation and Subjectivity

From its inception, photography has been considered a spectacular medium because of its ability to reproduce the real.[26] Yet while it is ostensibly the most veristic medium, it is also the most epitaphic. Keenly aware of this fact, publishers selling photographic postcards tried to revivify their images by tinting them to simulate the visual immediacy of the cityscape in crisis. These images, however, hover in the uneasy middle ground between documentary and re-creation, as the visual prostheses lend a surreal, simulated quality to the photographs rather than making them appear more alive.

Eschewing mimesis, other art forms emphasized sensation. A series of fourteen colorful lithographs produced just one month after the quake by Tenshōdō and printed by Shūgadō in Komagata-chō (both in Asakusa) under the title *Pictorial Account of the Great Earthquake in the Imperial Capital* (*Teito Daishinsai gahō*) vividly capture the intensity of the experience.[27] The lithographs share certain characteristics, most notably the motif of people frantically fleeing mammoth firestorms, as seen in the fifth scene in the series picturing the destruction of the Shin Yoshiwara licensed prostitution quarter (fig. 3.15). These images clearly draw on previous conventions of fire and disaster representation seen in the Edo-period works of Ōkyo or the popular lithographs of *Fūzoku gahō.* The 1923 lithograph depicts a cyclone of fire churning up the street, tossing everything in its path into the sky. The flames chase rich and poor alike, as elegantly suited gentlemen in chauffeur-driven cars flee alongside manual laborers and prostitutes. The image elicits emotions of terror mixed with excitement. Unlike the added static flames on tinted postcards, the print artist is able to convey the burning intensity of the blaze that illuminated the night sky, transforming it into a sea of "sheer vermilion," as it was often described. The print gives physicality to the disaster through movement and color.

One can argue that these prints are part of a long-standing East Asian narrative tradition that prefigures cinematic aesthetics—particularly illustrated handscrolls with their continuously unfolding depictions or linked monoscenic "stills"—and thereby express the temporality of narrative cinema.[28] What is particularly striking about this moment in history—the late Taishō era—is that the pictorial narrative, the precine-

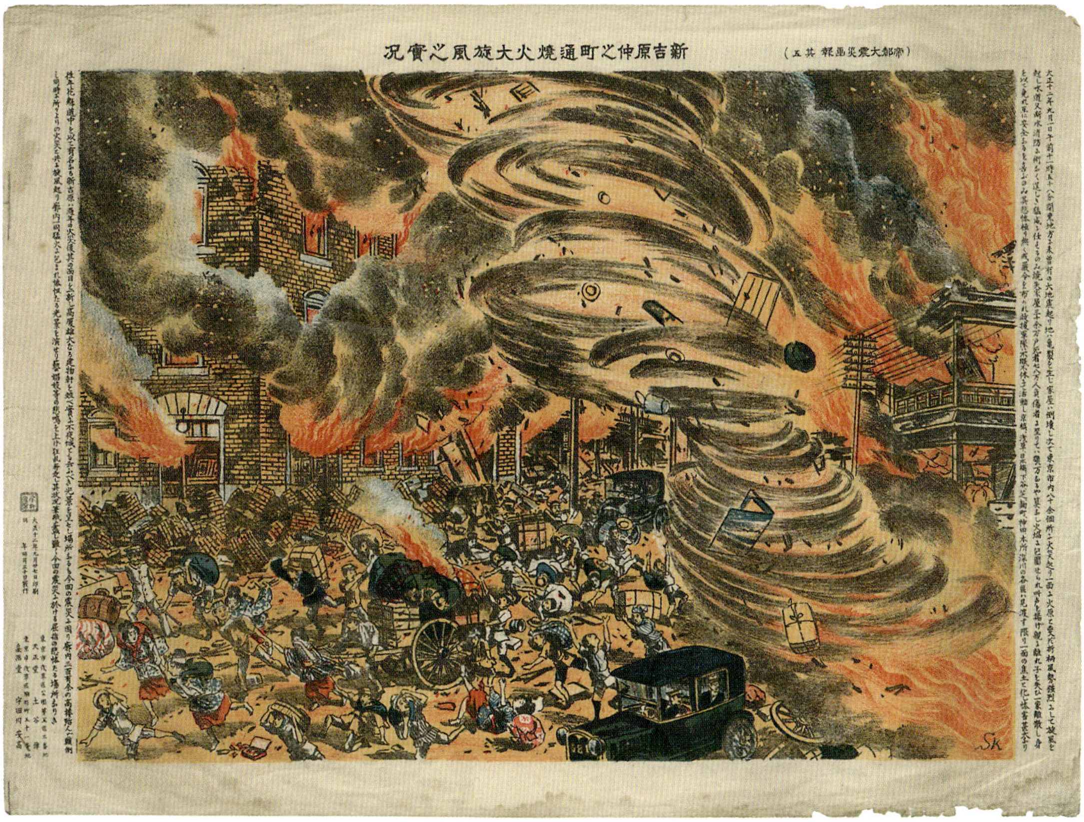

3.15 S.K., *Actual Situation of the Large Whirlwind of Fire on Nakanomachi Street in Shin Yoshiwara* (*Shin Yoshiwara Nakanomachidōri takuhi daisenpū no jikkyō*), lithograph. *Pictorial Account of the Great Earthquake in the Imperial Capital* (*Teito daishinsai gahō*), print no. 5 (Tokyo: Tenshōdō, Shūgadō, 1923). Printed 27 September 1923 and published for sale on 30 September 1923. Collection of the author.

matic, the photographic, and the cinematic all shared the same space and were all valued for expressing different aspects of this momentous event. The term *precinematic* is commonly applied to optical devices such as dioramas, panoramas, stereoscopes, and magic lanterns that simulated visual effects akin to cinema, although they were not necessarily developed in this teleological framework.[29] The year 1923 marked a dividing line in representational history: subsequent global disasters came to rely almost entirely on photographic and cinematic representation (and then later on televisual and digital media as well), precipitating what one might argue is a narrowing—and an impoverishment—of the visual sphere, limiting the ability to express sensation and affect in response to major historical events.

Each lithograph in the *Pictorial Account* series had two captions: the one on the right was the same on all the images, presenting one dramatic omnipresent description of the earthquake, echoing the authoritative, omniscient voice of a documentary narrator:

> On September 1 of Taishō 12 (1923), 11:58 A.M., the Kantō region experienced a massive earthquake of unprecedented size that created deep fissures in the ground and made houses collapse; more than eighty places in Tokyo were transformed into a field of flames, at which point a fierce wind blew up, causing a whirlwind. The water mains ruptured, and there was no way to put out the fires, whose ferocity simply worsened. Over 300,000 houses burned, and the number of dead came to 780,000 victims, but there was no way to calculate the wounded (accurately). The only joy was by families who were dispersed—their children lost and surrounded by flames and screaming as they were separated from their parents—and who barely escaped, only to find each other safe. The misery was so vast that martial law was declared and rescue troops worked without sleep or rest. Across several wards of the city—Kyōbashi, Asakusa, Nihonbashi, Shitaya, Shiba, Kōjimachi, Kanda, Honjo, Fukagawa—as far as the eye could see, everything had been transformed into a vast expanse of scorched earth. The devastation was immense.

Then the left-hand caption offered specifics, as in the following description of Shin Yoshiwara:

> The New Yoshiwara, famous in former years for its colorful processions of courtesans, has changed completely in appearance since the last time there was a fire there. Of the place that once presented a spectacle of tall, grand buildings lined up eave to eave, indeed a Nightless City, nearly all of the over three hundred lofty structures in the quarter collapsed in the quake. At the same time, whirlwinds came up along with fires in various places, and the entire quarter was enveloped in fierce flames, a horrifying spectacle. Screams rose from the geisha and prostitutes, and they dashed about in confusion. The scene was simply beyond description. In this earthquake it was a place of profound misery.[30]

Describing the "actual situation" (*jikkyō*) of the quake (a contemporary term that was commonly used in film along with *jissha* to connote "documentary"), the captions make truth claims while mirroring the image's theatricality; in other words, they tell the viewer that there are ways to represent the truth, and there are ways to represent the truth *more truthfully*.[31] This narrative captioning serves the same function as a documentary film voice-over, in which a "Voice of God commentator," who is heard but not seen, is a surrogate voice for the producer, who in the case of the lithographs also serves as a media proxy for officialdom.[32]

Such spectacular narrations in the guise of objective reportage were complemented by the copious literary descriptions of the earthquake, such as the ostensibly documentary yet highly melodramatic writing of novelist Nagata Mikihiko (1887–1964) in his widely read *Daichi wa furū* (The Earth Shakes), in which he describes what he saw when he visited the earthquake ruins and recounts the horrifying firsthand accounts of survivors.[33] These spectacular expressions also echoed the melodramatic tone of popular ballads sung on the streets of the capital, like the hit "Song of the Great Taishō Earthquake" (*Taishō Daishinsai no uta*) by pioneer *enka* singer-songwriter and political activist Soeda Azenbō (1872–1944):

Ah disaster, disaster
The time: 1923
September 1, noon
Heaven's pillars and Earth's axis pulverized
The ground rumbled, houses shook and quivered
Destruction's mad din horrific
Buildings collapsed, hills crumbled
The land cracked, the sea roared
. . . .
The fire spread and scorched the sky
In a moment a horrific scroll unfurled
Carnage and cries, scenes from a living hell
How pitiful these people
Their greed and fear and misery
. . . .
Those that survive torture by fire and water
Get buried alive when the bridge collapses
Others are swept away when their boat burns around them
The number drowned is beyond count
A wind fans the inferno
Clueless as to escape they sink down into the pond's mud
Trying to keep the fire's flying sparks at bay
Their bodies sink deeper and deeper
But relentless the approach of those tongues of flame
So many are steamed to death
And atop their corpses tread still others
Submerged in smoke and flame
Those who had somehow managed to escape with their lives
Show wounds unbearable to behold
Half alive or more likely half dead

Their wretchedness insect-like

. . . .

We cannot turn our eyes to behold our compatriots
Their corpses in the tens of thousands piled into a mountain
A scene from hell more terrible than hell itself
Older brother, younger brother, parents, children
Men or women
One cannot differentiate faces
Some are festered, others swollen
Heads crushed, stomachs ripped
Bones smashed, guts flowing forth
The sharp stench sticks in the nose
Pile the corpses as they are
Pour on the oil and watch the burning sight
Choke on the smoke from the pyre

. . . .

Ah, horrific calamity
Ah, the horrific calamity[34]

In prewar Japan, songs, particularly *enka* ballads, were intimately associated with the voice of the common folk—the everyman. Singing tales of disaster and loss was a communal mechanism for coping with tragic experiences. It was also a haunting, personal rejoinder to the emotional, gut-wrenching headlines put forth by the mass media. Like the lithographs, the music reenacted the carnage and suffering in spectacular detail, but it also enveloped the bodily senses in horror; it linked sight, smell, and touch in a relentless tale of tragedy.

Disaster is a bricolage of visions and voices in which fact and fiction are entangled. All representations are imbricated with layers of subjectivity, and none can be taken as transparent historical records, even when they seem directly to index actual places and events. Personal narrations of the event, like Azenbō's song and other subjective expressions, formed an important part of the heterogeneous cultural field against which all visual imagery was created and all visible evidence read.

The spectacular theatricality of Kantō earthquake print imagery even bridged the diachronic time of history and the mythic time of legend. Despite the increasing scientific rationalism of the modern era, there was still a place for the folkloric and the miraculous. Modernity did not chase magic or myths from the world. Real-life modern historical events were still saturated with supernatural aspects that could not be depicted in photography but provided a feast for the realm of the spectacular. One lith-

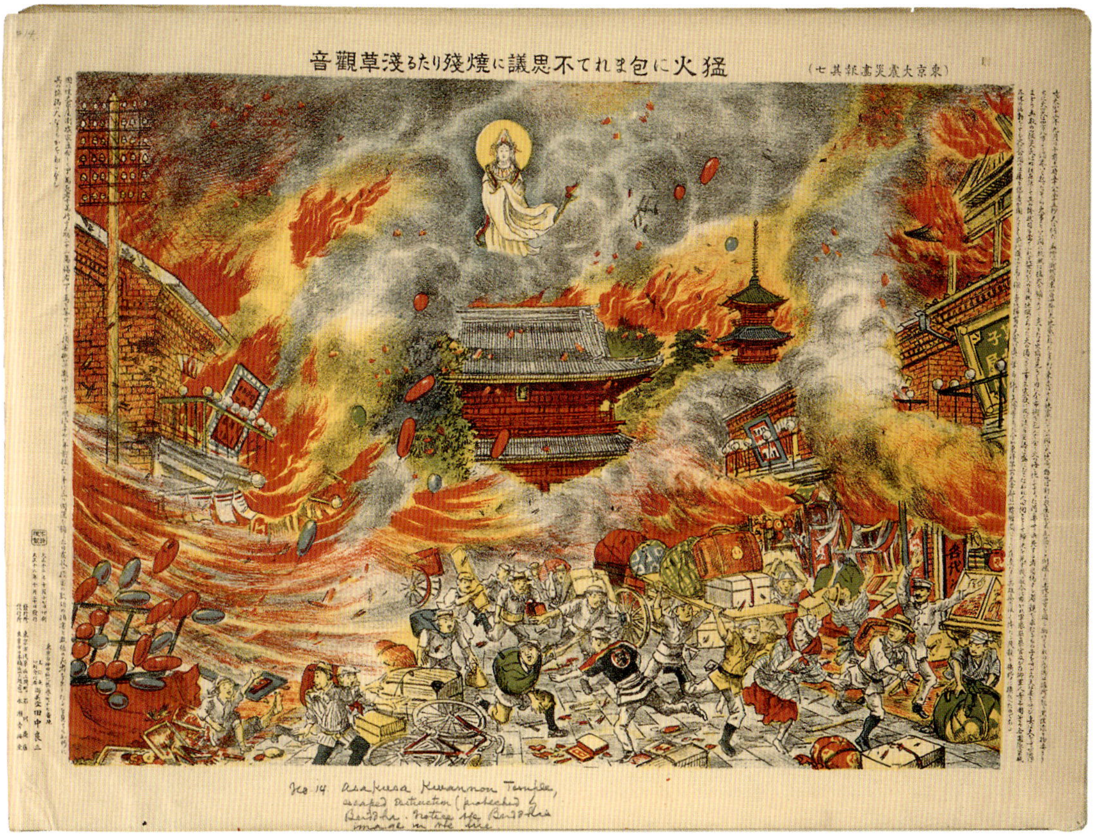

3.16 Artist unknown, *Asakusa Kannon Enveloped by Raging Fires Miraculously Escapes Burning* (*Mōka ni tsutsumarete fushigi ni yakenokoritaru Asakusa Kannon*), lithograph. *Tōkyō daishinsai gahō* (Pictorial Account of the Great Tokyo Earthquake) (Tokyo: Shōbidō, Ishikawa Shōten, 20 October 1923). Prints and Photographs Collection, Library of Congress, Washington, DC.

ograph in another series titled *Pictorial Account of the Great Tokyo Earthquake* (*Tōkyō daishinsai gahō*) (fig. 3.16) pictures the now-legendary tale of the bodhisattva Kannon miraculously appearing over the deity's tutelary temple Sensōji in Asakusa and warding off the firestorm caused by the earthquake, sparing Sensōji's main icon hall (*hondō*), its iconic five-story pagoda, and all the refugees crowded inside.[35] In the print, Kannon, in majestic billowing robes, appears to float over the temple, parting the sea of flames just as Moses parted the Red Sea to enable the exodus of the Jews from Egypt. While the gods may have lost control of the catfish, Kannon still controlled the dispensation of mercy—to those who were worthy.[36] Perhaps not coincidentally, the deity's compassion did not extend to the bustling Nakamise shopping street leading up to the tem-

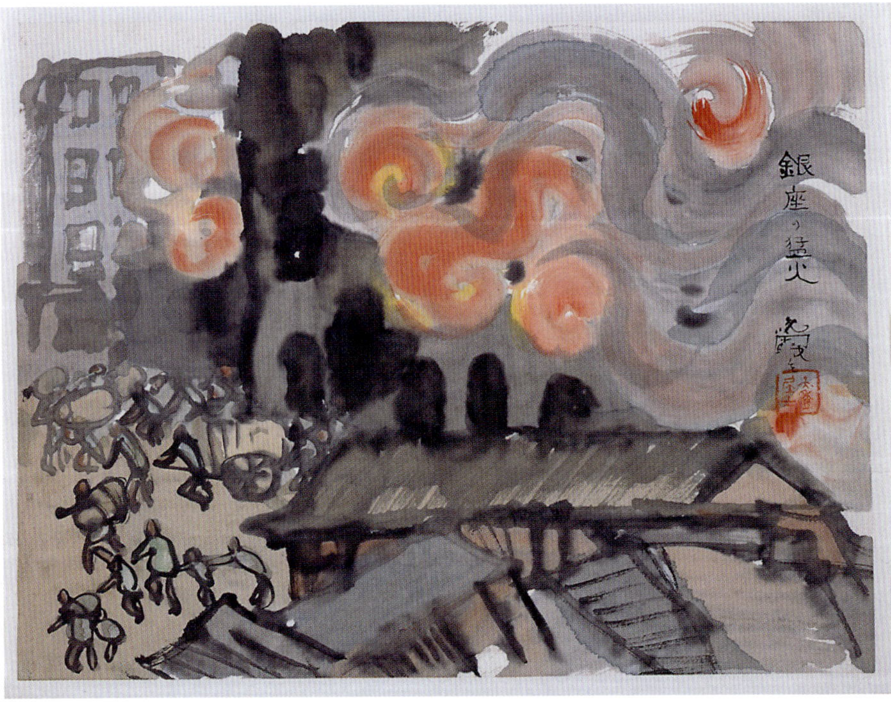

3.17 Maruyama Banka, *Raging Flames of Ginza* (*Ginza no mōka*), ink and color on paper, c. 1923. *Paintings of the Great Kantō Earthquake* (*Kantō daijishinga*). Collection of Tōkyō-to Edo Tōkyō Hakubutsukan.

ple or the lively *shitamachi* entertainment and commercial district (*sakariba*) across Asakusa Park from the temple area, known as Asakusa Rokku (Asakusa Sixth District), which included the landmark Twelve Stories skyscraper, the Hanayashiki (Japan's oldest amusement park, open since 1853, which also had zoo exhibits), cafés, movie theaters, and live entertainment playhouses, all of which were reduced to a burned wasteland. This was a story of divine salvation and punishment.

The material qualities of painting opened up additional spheres of affective expression that captured the spectacle of disaster. Expressive and ethereal brushwork, saturated pigments, and layered inky washes or impasto in oil produced visual effects unlike those of other media. In *Paintings of the Great Kantō Earthquake* (*Kantō daijishinga*), a series of one hundred small-scale paintings by a group of ten well-known artists document the various stages of the calamity and exemplify the expressive possibilities of disaster painting.[37] Selected works in the series were reproduced in black and white and circulated in published albums such as *Pictorial Collection of the Great Earthquake Disaster* (*Daishinsai gashū*).[38]

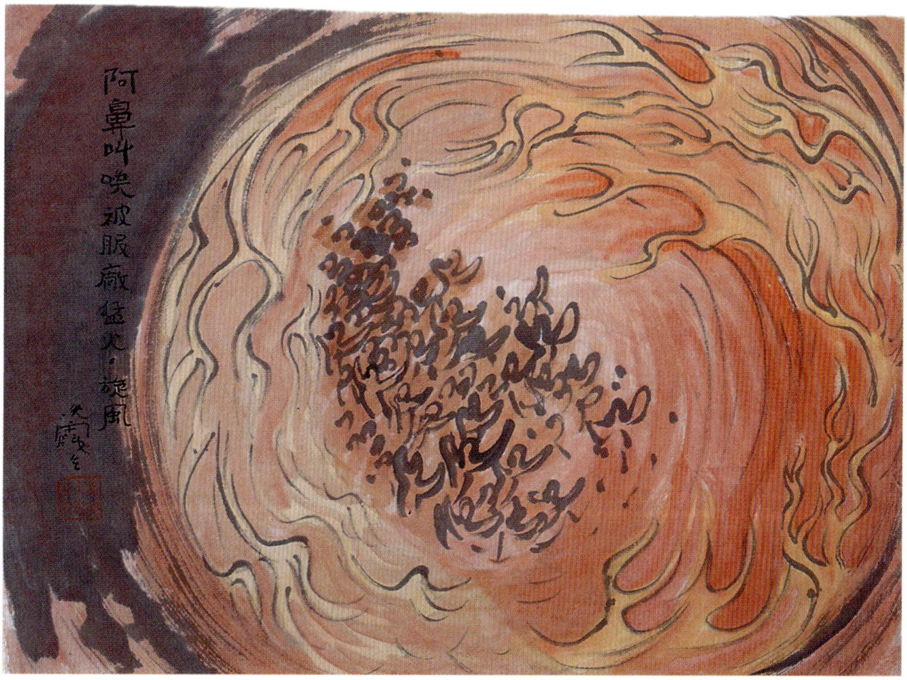

3.18 Maruyama Banka, *Furious Flames of the Army Clothing Depot One of the Eight Hells* (*Abi kyōkan jigoku Hifukushō mōka*), ink and color on paper, c. 1923. *Paintings of the Great Kantō Earthquake* (*Kantō daijishinga*). Collection of Tōkyō-to Edo Tōkyō Hakubutsukan.

One of these paintings, Maruyama Banka's *Raging Flames of Ginza* (*Ginza no mōka*) (fig. 3.17), skillfully evokes the intensity and awesome power of the fire through the abstract use of line and ink wash. The brushwork makes the flames palpable, while the bleeding inky splotches threaten to consume the paper as they spread, much as the conflagration seemed to devour everything around it.

Banka's *Furious Flames of the Army Clothing Depot One of the Eight Hells* (*Abi kyōkan jigoku Hifukushō mōka*) (fig. 3.18) refers to the genre of Buddhist paintings of hell, which were known for their fearsome infernal imagery and their implied morality lessons.[39] The artist outlines a mass of people caught in the middle of an orb of flames—an infernal crystal ball. This dehumanized sphere is all-enveloping, cremating the souls within. The lurking mass of blackness on the edge of the vortex of fire beckons the viewer to the abyss beyond. Such scenes could not be witnessed because participants and bystanders alike would have been incinerated. Banka instead enables viewers to imagine the hellish world of those who perished through emotional identification and abstraction.

Personal lamentations in literature and song were paralleled in expressive renderings of physical and psychological torment by a range of artists experimenting with modernist abstraction. Yamamura Kōka (1885–1942), a 1907 graduate of the Japanese-style painting (*nihonga*) division of the prestigious imperial art academy, the Tokyo School of Fine Arts (Tōkyō Bijutsu Gakkō), and a regular participant in the annual national art salon sponsored by the Ministry of Education, was mainly known for his actor portraits.[40] However, in response to the Kantō quake, he produced a series of thirty dramatic expressive sketches that ran in the evening edition of *Hōchi shinbun* from 30 September until 1 November 1923.[41] The works were selected from sketches the artist had made on the day of the quake, when he walked across the city with his sketchpad from Tokyo to Yokohama and then on to Kamakura and beyond—an arduous pilgrimage driven by fascination and disbelief. In the newspaper series, the images were paired with texts by playwright and theater critic Kawamura Karyō (1884–1954) collectively titled *Daishinsai inshōki: Taishō Musashi abumi* (Record of Impressions of the Great Earthquake Disaster: Taishō Stirrups of Musashi Province).[42] The title refers to the famous two-volume retrospective account of the Great Meireki Fire (Meireki no Taika) that devastated the Edo capital (on the same Kantō plain in Musashi province) in 1657 penned by Asai Ryōi in 1661 under the title *Musashi abumi* (Stirrups of Musashi Province).[43]

The Meireki fire, which claimed close to one hundred thousand lives, had strong geographical links to the 1923 quake, not only because the two disasters devastated the same capital city but also because bodies of victims in the Edo period were transported down the Sumida River to Honjo (an area outside the city that was later incorporated into metropolitan Tokyo), where the dead were buried and commemorated at the temple Ekōin (Hall of Prayer for the Dead) built in 1657. Ekōin was at the heart of the devastation of the Kantō quake and served as a model for commemorating the later disaster. The shogun in the Meireki period, Tokugawa Ietsuna, set aside the land for the temple to bury and commemorate the fire victims, the majority of whom could not be identified and did not have family or relatives, and he erected "The Mound of a Million Souls" (Banninzuka) at the center of the temple. From then on, Ekōin became identified as a place to pray for the victims of disasters (as well as criminals) who were not survived by or claimed by family or relatives. Asai's evocative illustrated account of the fire described a human experience that struck a deep chord with people who had experienced the 1923 quake and imbued Kōka's drawings with historical resonance. Kawamura Karyō's text accompanying Kōka's illustration titled *Site of the Former Honjo Clothing Depot* (*Honjo Hifukushōato*) (fig. 3.19) first quotes Asai, making a visceral connection between the present disaster and the 1657 event:

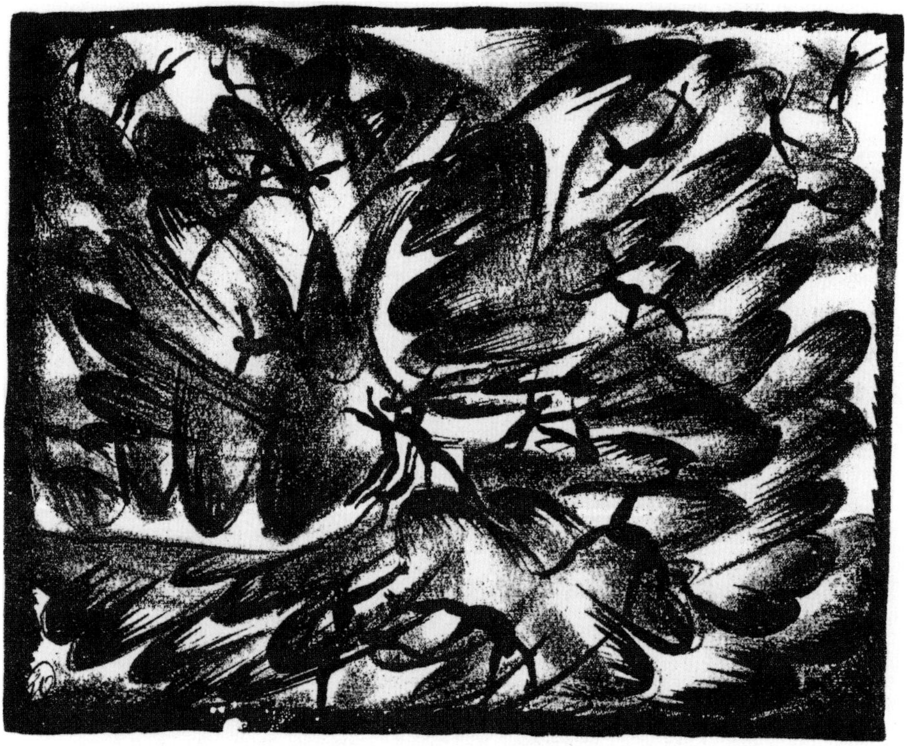

3.19 Yamamura Kōka, *Site of the Former Honjo Clothing Depot* (*Honjo Hifukushōato*), drawing, from the series *Record of Impressions of the Great Earthquake Disaster: Taishō Stirrups of Musashi Province* (*Daishinsai inshōki: Taishō Musashi abumi*), originally in *Hōchi shinbun,* 30 September–1 November 1923, P.M. ed. Reproduced in Takehisa Yumeji, Kawamura Karyō, and Yamamura Kōka, *Yumeji to Karyō, Kōka no Kantō Daishinsai rupo* (Yumeji, Karyō, and Kōka's Reportage of the Great Kantō Earthquake) (Tokyo: Kress Shuppan, 2003), 67. Courtesy of Kress Shuppan.

They vied with one another, carrying with them their miscellaneous household goods, and just as they put them down for a moment to rest at the front gate of the Nishi Honganji Temple a huge whirlwind came up. It started from the main hall of the temple, and at the same time temples in several other places suddenly burst into flames. People's belongings, piled up like a mountain, caught fire, and the people gathered there milled about in confusion, trying to save their lives, leaping into a well or escaping into a ditch. Those on the bottom drowned in the water; those in the middle were pressed down by their companions, while those on top were burned to death. More than 450 of them died here.

So records the work *Musashi abumi* [Stirrups of Musashi Province]. At the place where the Honjo Clothing Depot once stood far more than 30,000 people lost their lives, all due entirely to this whirlwind. Such was the horror of this great fire that they were surrounded by fire from all directions. In the open space where the whirlwind and flying sparks blew

> about—indeed the very scene that night at the Clothing Depot—a great number of people were blown upward and around in the searing wind and whirlpool of flames. Among them one person was dropped from the sky into a rice paddy in Ichikawa some ten miles away. There is a story that this person, now in a bed in the Ichikawa Hospital, can only babble, "The Clothing Depot, the Clothing Depot." A true story about one person whose life was saved from among nine who died that night is even more horrifying: according to it he was borne in flames atop a huge handcart filled to capacity with his goods high above the electric wires. In such a wind as this, great nature, quite beyond our imagination, can dispose of things with huge force, as though such an immense task was nothing at all.
>
> To the extent that human beings can conceive it, just how much were people confused, agitated, what agony did they feel thinking they were safe, when to the place where so many lives had put their trust the unexpected hand of death came to them? All are now burned to cinders. They will be forever silent, but when I think of what was in their hearts at the instant of death I tremble within myself and feel deeply the fear that they experienced.[44]

The sensationalized text, daring the audience to suspend its disbelief, expresses the internalization of the spectator's experience that literally causes his body to shake from within. The experience of the disaster spectacle was suffused with sheer terror.

Illustrator Kōka's highly expressionistic works show clear awareness of the strong currents of artistic modernism in Japan. He was affiliated with the *shin hanga* (new prints) neotraditionalist print movement that was primarily known for revitalizing Edo-period print production practices and themes (pictures of beautiful women [*bijinga*], landscapes, and actor prints) but that also selectively incorporated elements of modernist Western painting, particularly the pictorial effects of light seen in impressionism. From the turn of the century, an increasing number of Japanese artists traveled abroad to study in Europe and America, and they either directly or through reproductions encountered the lively new artistic movements of modernism and the international avant-garde, including the radical cultural developments in revolutionary Russia. Printmaking underwent several renaissances, some spurred on by widespread popular interest in representations of Japan's triumphant wartime victories against China and Russia and others driven by new interest in prints as art, exemplified by the *shin hanga* and *sōsaku hanga* (creative prints) movements, which helped revitalize the content and creative process of printmaking. *Shin hanga* print artists were keenly aware of the avant-garde and the sometimes highly abstract modernist printmaking of their contemporaries in the *sōsaku hanga* movement. While one could argue that representations of disaster in Japan were always stylized, modernist stylizations of disaster took on new meaning after the introduction of photography, when artists self-reflexively imparted subjectivity to events and drew attention to the pictorial style as a means of expressing the aesthetic visuality of the disaster experience.

These modernist images indulged in the construction of vivid form, taking on the challenges of the tangible and intangible elements of disaster. One highly abstracted rendering of urban chaos in the midst of the disaster is Mizushima Niou's (1884–1958) cover for the Japan Cartoonists Association (Nihon Mangakai) publication *Daishinsai gashū* (Pictorial Collection of the Great Earthquake Disaster) (fig. 3.20), which features panicked figures in a turbulent cubistic cityscape menaced by red clawlike arcs of flame and a mammoth, stylized plume of white smoke overhead.[45] In this painting, the modernist techniques of distortion, fragmentation, and radically cropped and angular compositions instantiate the earthquake experience of radical rupture and the realization of civilization's vulnerability. They are also metonyms for the similarly disruptive forces of modernity that fragmented vision and the holism of the object. Disaster images highlighted the extreme, often surreal physical dislocations caused by the force of the quake. Visible throughout the devastated cityscape, the violent rupturing of steel—a material that symbolized modern technological strength and invincibility—epitomized these physical dislocations. The rupture of steel not only simulated the brutal somatic sensation of the shock, but, like modernism, psychologically undermined the premise of a positive, progressive modernity.

Contemporary artists and critics widely noted the parallel disruptive capacities of modernism and the earthquake. Quoting a fellow artist's description of his surreal and spectacular sensory experience, Takehisa Yumeji paired it with his own turbulent expressionist quake landscape (*Expressionist Picture* [*Hyōgenha no e*]) (fig. 3.21):

> The sound—it wasn't just some ordinary bang or slam. It was a more awful sound rather than some great magnificent sound. When it came, the glass in the window seemed to glitter like the pictures that the Triangle group painted using mica. And just as that was going through my mind, the tatami mats rippled and surged toward me, and a hole opened up in the boards in the ceiling above my shins. Everything lost its color and shape, the elements all dissolved, cells split apart—everything was like it must have been at the mythical formation of the islands of Japan back in the chaotic darkness of the age of the gods. Even so, it was strange: people at the time instinctively knew the direction of things, how and where to go to get out of the pitch darkness. They would break a hole through a wall, and when they thrust out just their heads they'd see the sky was blood red, right in broad daylight. This, they thought, was what it was like at the creation of the world. As far as the eye could see, roof tiles undulating like waves. . . . Someone crawling on all fours came over the rooftops that were rolling like waves. The forest at Sannō was swaying back and forth like a thin

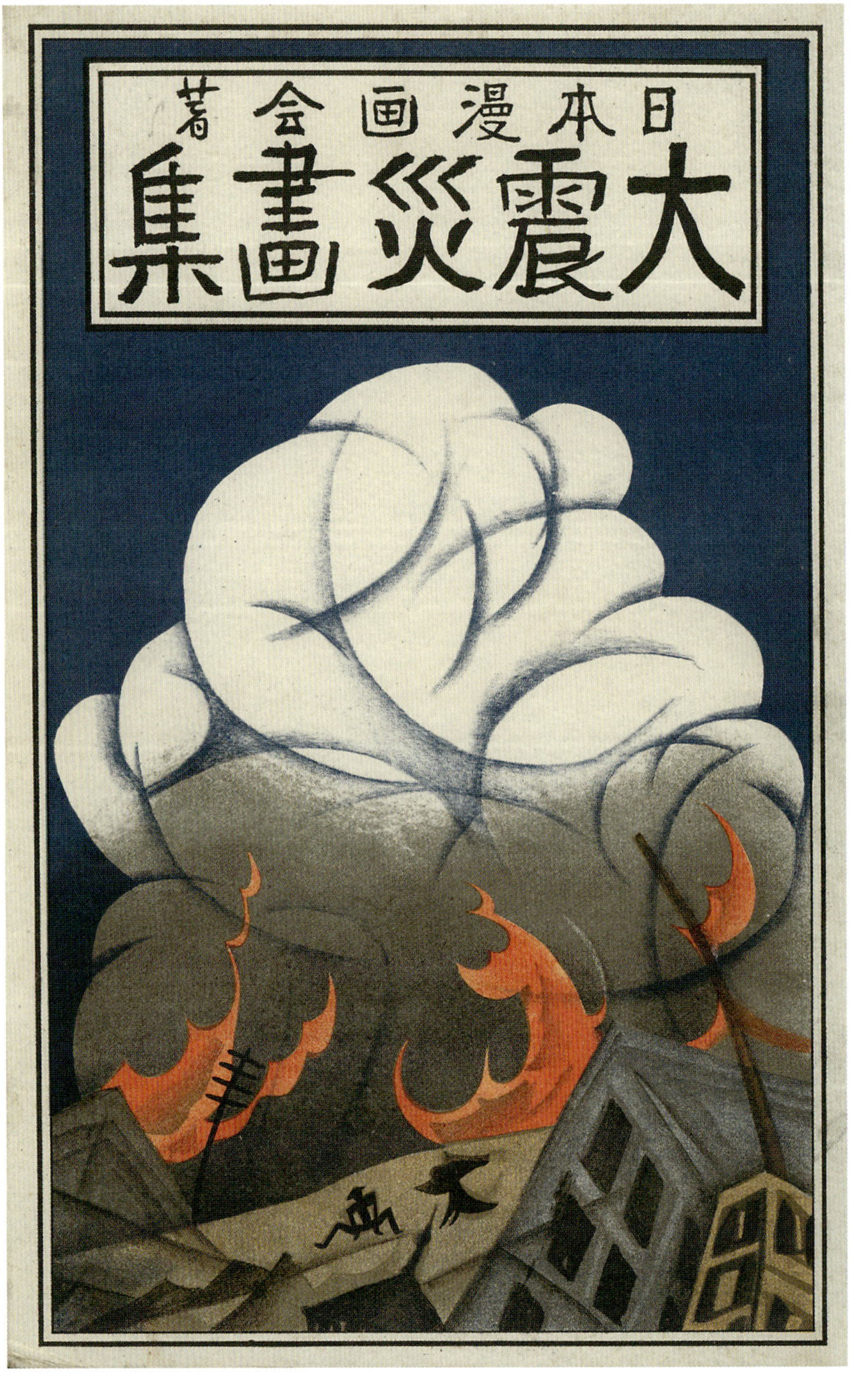

3.20 Mizushima Niou, people fleeing fire in abstract cityscape, cover. Nihon Mangakai, *Daishinsai gashū* (Pictorial Collection of the Great Earthquake Disaster) (Tokyo: Kanao Bunendō, 1923).

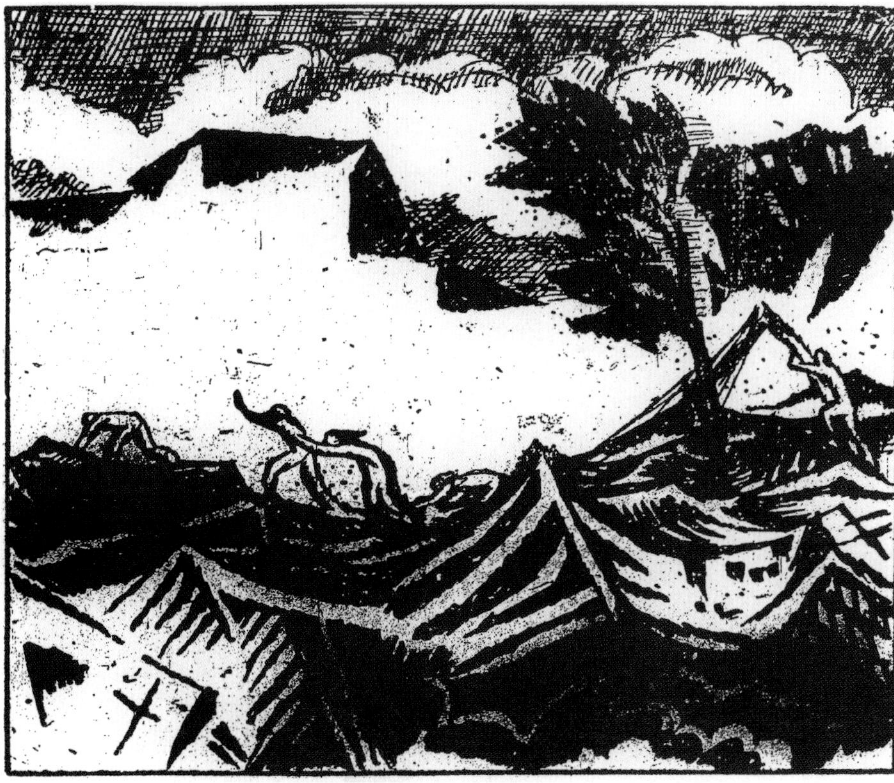

3.21 Takehisa Yumeji, *Expressionist Picture* (*Hyōgenha no e*), drawing, from *True Pictures of the Tokyo Calamity* (*Tōkyō sainan gashin*). *Miyako shinbun,* 14 September–4 October 1923. Reproduced in Takehisa Yumeji, Kawamura Karyō, and Yamamura Kōka, *Yumeji to Karyō, Kōka no Kantō Daishinsai rupo* (Yumeji, Karyō, and Kōka's Reportage of the Great Kantō Earthquake) (Tokyo: Kress Shuppan, 2003), 13. Courtesy of Kress Shuppan.

curtain in one of those second-rate theaters, so it was no wonder that people couldn't walk. I finally understood the pictures of the German Expressionists.[46]

The fragmented cinematic aesthetics of montage used in the film *Teito fukkō* (Reconstruction of the Imperial Capital) further reinforce the analogies between modernism, disaster, and spectacle.[47] The film's title is announced in stylized Japanese cinematic typography (*kinema moji*) followed by a series of atmospheric fade shots, one superimposed on the next: a spinning globe; the waving flag of Japan; an image of the Imperial Palace, serenely set in the distance; the bustling streets of the metropolis. Suddenly the scene goes topsy-turvy, spinning in a vertigo-inducing circle, and a large fire explosion bursts from the center of the vortex and immolates the cityscape

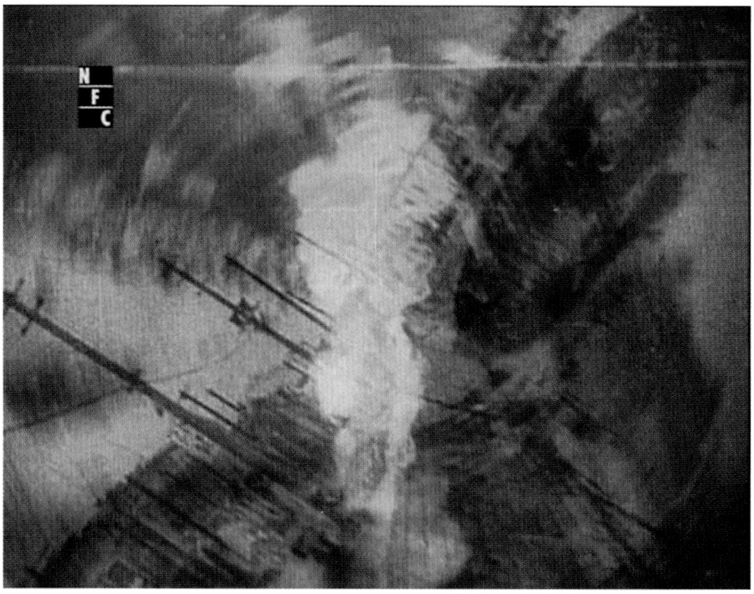

3.22 A world turned topsy-turvy, screen shot. *Reconstruction of the Imperial Capital* (*Teito fukkō*), 1930. Oda Hamatarō, director; Reconstruction Agency (Fukkō Kyoku), producer. By permission of Tōkyō Kokuritsu Kindai Bijutsukan Film Center.

(fig. 3.22). Like the perceptual explosion of modernity and modernism, the earthquake's shocks and blast fractured vision, literally turning the world upside down. In the filmic world, this disruption is temporary, eventually righted through the rational vision of reconstruction that returns the viewer to an even-keeled, holistic state. Modernist critics would argue, however, that the ruptures of modernity endured in the external world.

Spectacular Sites

The spectacular destruction of particular Kantō landmarks dominated the press coverage, imparting special significance to these locales and structures, which were subsequently immortalized in other visual arenas. The dramatically cracked edifice of the Asakusa district skyscraper, the Twelve Stories, captured on the cover of *Hōchi shinbun*'s two-page evening special edition on 5 September (fig. 3.23), was one of many prominent sites pictured, including Sensōji Temple and the Nakamise shopping street in Asakusa, the Nikolai Russian Orthodox Church in Ochanomizu, the Imperial Palace, the Ginza, the Nihonbashi central business district, Mitsukoshi Department

Store in Nihonbashi, Ueno Park and Zoo, the bronze statue of famed military leader Saigō Takamori and the Great Buddha in Ueno Park, the Honjo Army Clothing Depot and the Kokugikan sports stadium in Ryōgoku, the Shin Yoshiwara licensed prostitution district, the Central Meteorological Observatory, and the cityscape of Yokohama, among others.[48]

The Twelve Stories was a popular sight in the urban landscape of modern Tokyo and appeared frequently in commercial images of beautiful women (*bijinga*), touristic promotional images of Asakusa, and even prints of the Meiji emperor and his imperial cortege engaged in sightseeing activities in the capital. Built in 1890, at 225 feet, the octagonal edifice with red brick covering a wooden frame was just slightly shorter than the tallest building in the world at the time, the World Building in New York. The tower housed Japan's first two elevators and had electric lights on every floor, including an additional row of lights above the outdoor roof observation deck.[49] The panoramic view of the entire Kantō plain from the top was a major attraction, and visitors could see Mount Fuji on a clear day. It also offered titillating distant views of the nearby magnificent Shin Yoshiwara pleasure quarters and elegant parading courtesans. The Twelve Stories housed a range of commercial enterprises, including exhibition halls, shops with international goods, restaurants, and bars. A variety of cultural events, such as music concerts, art exhibitions, and beauty contests, also took place in the tower. The roof even served as a launchpad for popular entertainments like the dramatic aerial ballooning of the acrobat Spencer during his celebrated tour of Japan. One of the most famed places in the tower was the store where woodblock prints were made for the popular board game *sugoroku*.[50]

In a district laden with spectacles that included the Panorama Hall (Panoramakan) and its three-dimensional battle scenes as well as a high concentration of movie theaters, the Twelve Stories was an urban landmark that embodied the spectacular. Spectacular decapitation and incineration were therefore the only demise befitting such a symbolic site.

In a large-scale oil painting, renowned Western-style academic painter Okada Saburōsuke (1869–1939) spotlighted the vanishing—yet proudly defiant—body of the Twelve Stories on its funeral pyre as this noble monument to modernity was cremated (*Asakusa Twelve Stories Building on Fire* [*Moetsutsu aru Asakusa Jūnikai*], fig. 3.24).[51] Okada's thick impasto rendition of spectacular incineration focused on the expressive yet intangible visual qualities of fire. His painting was subsequently emblazoned on the cover of the October 1923 issue of the popular women's magazine *Shufu no tomo* (Housewife's Companion) and became known to a broad female readership through its reproduction in the mass media. The building's fiery downfall was

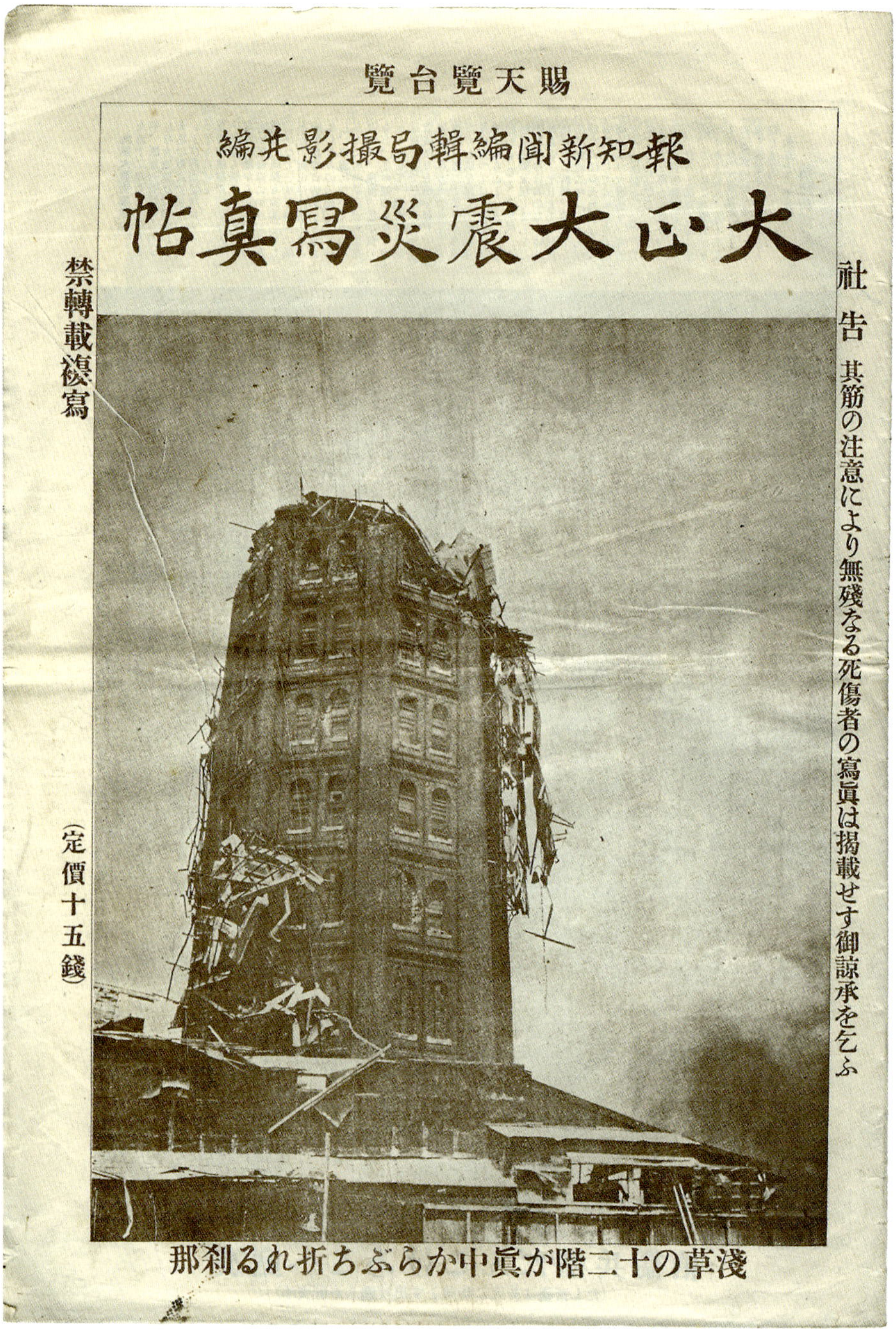

賜天覧台覧

報知新聞編輯局撮影共編

大正大震災寫眞帖

禁轉載複寫

（定價十五錢）

社告 其筋の注意により無殘なる死傷者の寫眞は掲載せす御諒承を乞ふ

淺草の十二階が眞中からぶち折れる刹那

3.23 "The Asakusa Twelve Stories the Moment It Snapped in the Middle" (*Asakusa no jūnikai ga manaka kara buchioreru setsu*), photograph, front page. *Hōchi shinbun*, special edition, 5 September 1923. *Great Taishō Earthquake Photograph Album* (*Taishō Daishinsai shashinchō*). Tano Yasunori Collection. By permission of Yomiuri Shinbun.

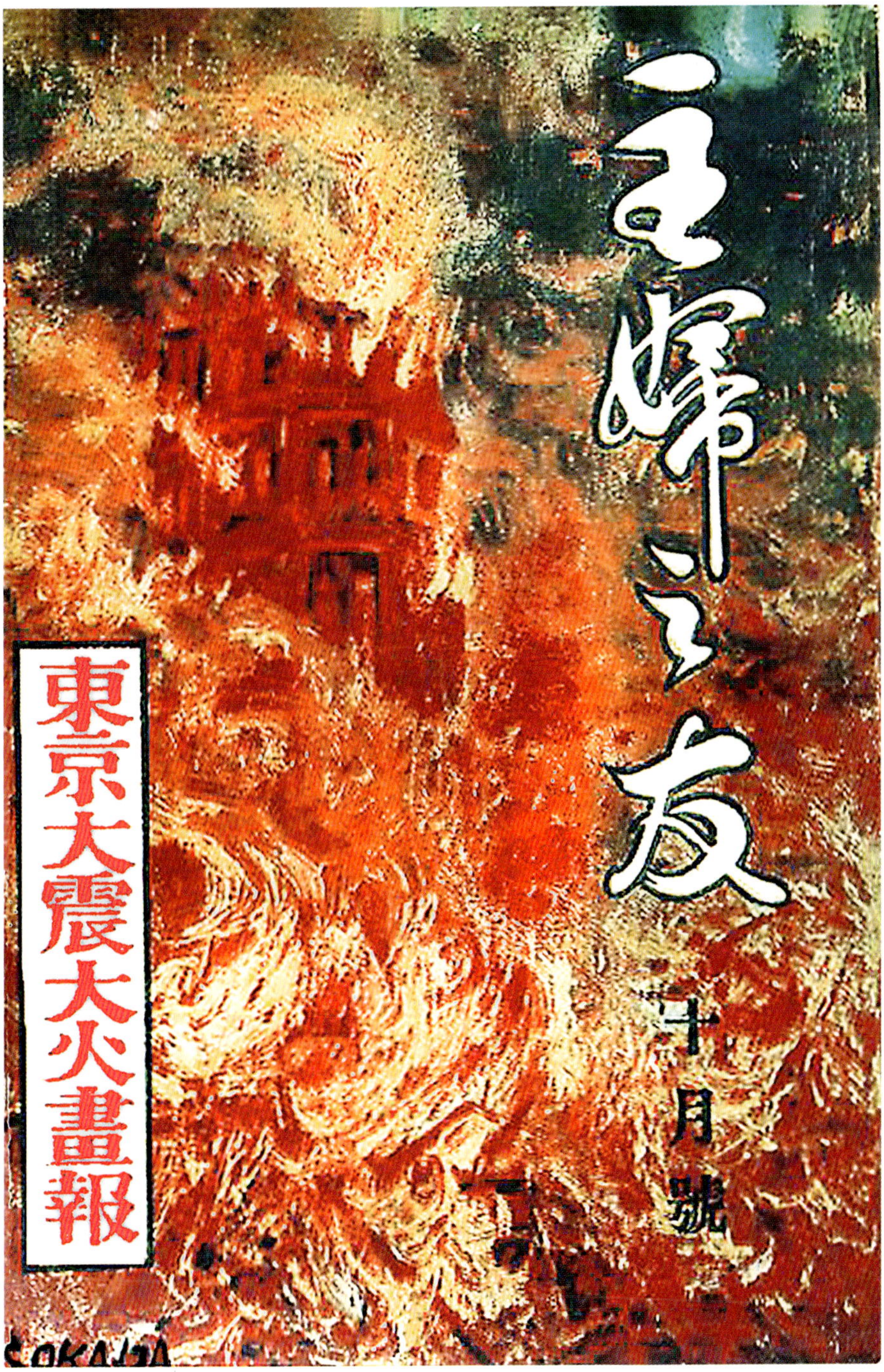

3.24 Okada Saburōsuke, *Asakusa Twelve Stories Building on Fire* (*Moetsutsu aru Asakusa Jūnikai*), print reproduction of original oil painting. Cover of *Shufu no tomo* (Housewife's Companion), October 1923. By permission of Shufu no Tomosha.

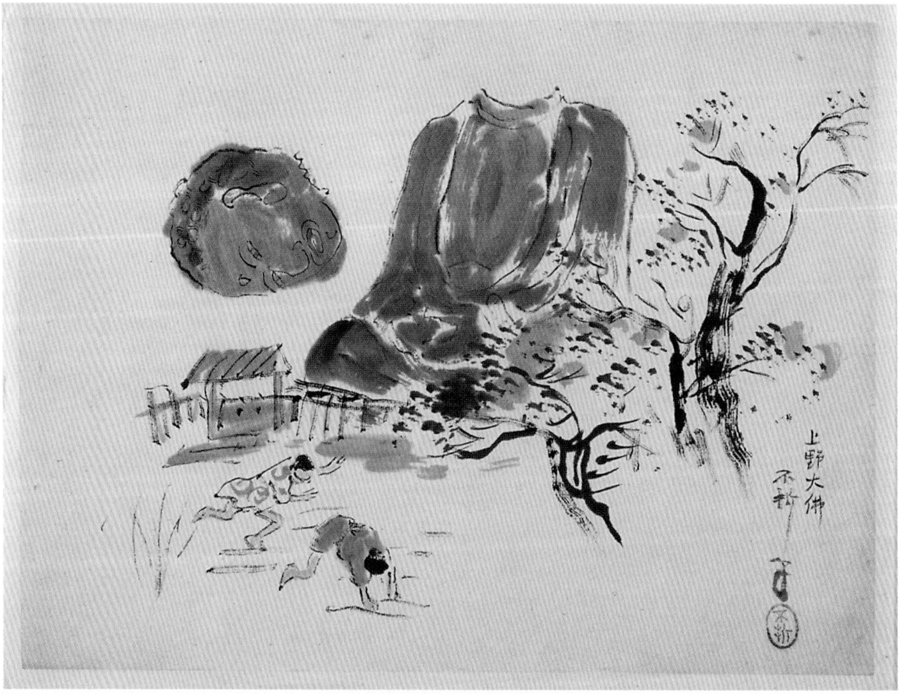

3.25 Nakamura Fusetsu, *Great Buddha at Ueno* (*Ueno Daibutsu*), its head thrown off its body by the force of the quake, ink and color on paper, c. 1923. *Paintings of the Great Kantō Earthquake* (*Kantō daijishinga*). Collection of Tōkyō-to Edo Tōkyō Hakubutsukan.

the subject of numerous works that capitalized on its iconic stature, and the tower loomed in the background of many others, including depictions of the animal stampede out of the Hanayashiki Zoo in Asakusa famously led by a charging elephant.[52] Many artists rendered the structure's agonistic demise as if it were in a superhuman battle for eternity.

Having cracked at midelevation in the quake, the burned, empty shell of the Twelve Stories was one of many literal and symbolic decapitations. Famous buildings throughout the region lost their upper stories, grand cupolas, and majestic domes, leaving the city littered with headless architectural carcasses. Even the gigantic head of the *Daibutsu,* or Great Buddha, statue in Ueno Park was thrown off its body by the earthquake, an image captured by prominent Western-style painter and calligrapher Nakamura Fusetsu (1866–1943) in his ink painting showing people running for cover as the mammoth head topples toward them (fig. 3.25).[53] Cast in 1660, the bronze Buddha stood in the precinct of Kaneiji, the personal Tokugawa family temple where several successors of the founding Tokugawa shogun, Ieyasu, are entombed. Its beheading

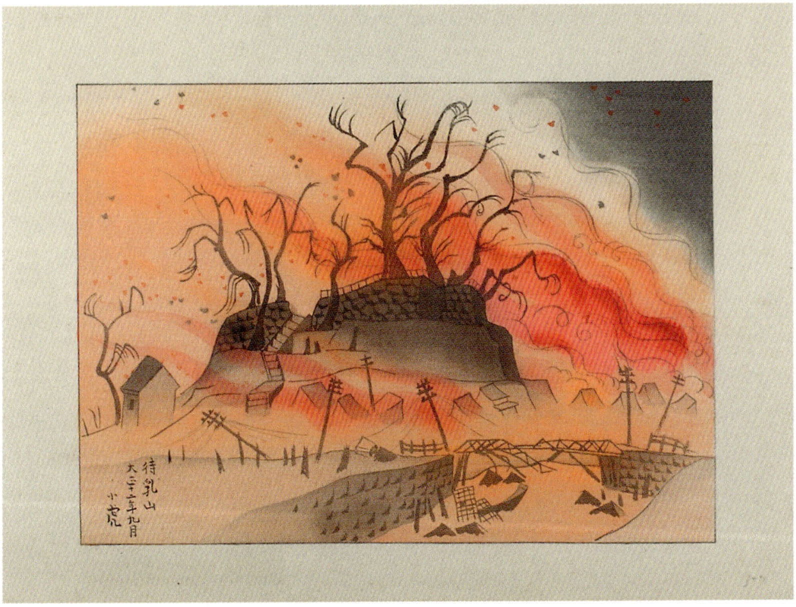

3.26 Kawasaki Shōko, *Matsuchiyama*, a small hillock near Ueno, in the *meisho-e* tradition, woodblock print. Nishizawa Tekiho et al., *Taishō shinkasai mokuhangashū zen sanjū rokkei* (Collection of Woodblock Prints of the Taishō Earthquake—Thirty-Six Scenes) (Tokyo: Gahōsha, 1924). Collection of Machida Shiritsu Kokusai Hanga Bijutsukan.

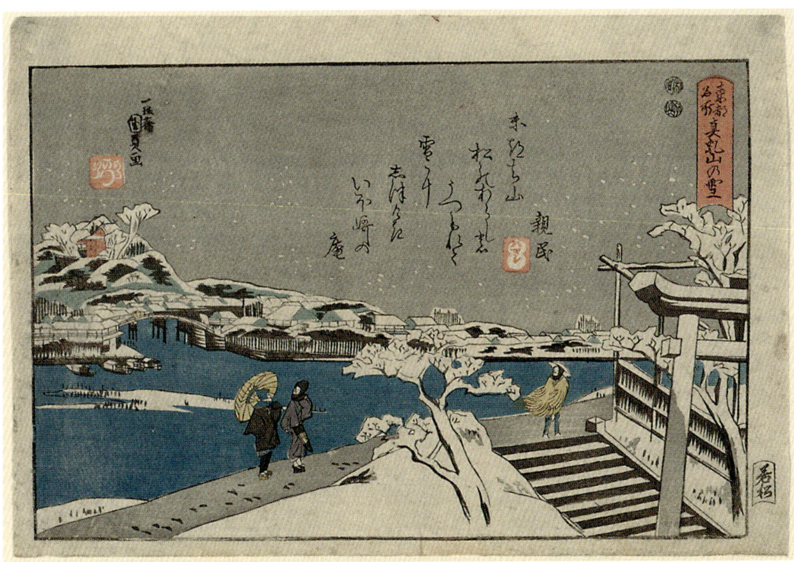

3.27 Isshusai Kunikazu, active about 1848–68. Publisher: Wakamatsuya Gensuke. *Snow at Matsuchiyama* (*Matsuchiyama no yuki*), from the series *Famous Places in the Eastern Capital* (*Tōto meisho*). Japanese, Edo period, 1847–52 (Kōka 4–Kaei 5). Woodblock print (nishiki-e); ink and color on paper. Horizontal ōban. Photograph © 2012 Museum of Fine Arts, Boston, gift of Louis W. Black, 51.26.

could be seen as one more indicator of the end of the Tokugawa era, a theme put forward by many cultural critics and commentators as they lamented the erasure of Edo-period structures in the city. In a sense, the disaster accelerated the process of destruction already begun by the forces of modernity.

As with the Twelve Stories, buildings and sites across the city were imbued with a memory and a life force in many images, even taking on subjective qualities of personhood in the spectacular representations of their destruction. In *nihonga* painter Kawasaki Shōko's (1886–1977) depiction of Matsuchiyama during the Taishō earthquake (fig. 3.26), a small hillock near Ueno along the Sumida River, the artist drew on the genre of *meisho-e* (pictures of famous sites), an older Edo-period pictorial tradition of representing popular attractions in the capital and throughout the country.[54] The *shin hanga* movement had begun revitalizing the *meisho-e* genre as *shin-meisho-e* (pictures of new famous sites) around 1910 to highlight the new urban locales associated with Japan's modernity. Some sites crossed over from Edo to Tokyo to represent the reinvention of the city; others were entirely new locations. Matsuchiyama was well known from the Edo period as a place where one could get a good view of Mount Tsukuba. It was also a popular landmark along the river journey to the Shin Yoshiwara prostitution district as one went through the Sanya Canal. Matsuchiyama had a small subtemple of Asakusa Sensōji called Honryūin (also known as Matsuchiyama Shōden), which was a popular religious site for those seeking marital harmony and business prosperity, and it attracted many believers from the entertainment businesses in Asakusa, including numerous geisha and performers in the *karyūkai* ("the flower and willow world").

A comparison of Kawasaki Shōko's rendering of Matsuchiyama with an earlier Edo rendition of the same location, a print in *Tōto meisho* (Famous Places in the Eastern Capital) by printmaker Isshusai Kunikazu (Utagawa Kunikazu, active 1849–67) from around 1850 (fig. 3.27), reveals how much Shōko dramatized his scene. Isshusai's conventional genre scene of figures in a snowy landscape with the Matsuchiyama hill, Imadobashi Bridge, and temple in the background is peaceful and picturesque. Shōko's print focuses on the fires that engulfed Matsuchiyama from the directions of the Shin Yoshiwara and Asakusa Senzokuchō. It features in the foreground the Imadobashi Bridge that crossed the Sanya Canal, which fell and tragically killed many people during the quake. His burning trees on the prominent hill in the center have an anthropomorphic, agonistic quality, as they seemingly writhe in pain amidst the conflagration in empathy with the quake's human victims. The famous sites of the capital were now infamous for their spectacular destruction and state of ruin.

Cinematic Spectacle

Motion was critical to the drama of visual spectacle, and there is no medium more closely associated with motion than cinema, or "motion pictures" (*katsudō shashin*). Cinema shared photography's claims to the spectacle of actuality. And photography gestured toward the temporal flows of cinema. In the 1880s, the early chronophotographers Muybridge and Marey, in their famous time-release experiments, took up the challenge of depicting movement. By 1923, some photographs of the earthquake were able to capture the sense of instantaneous arrest of a live-action film still. One stunning news image of the Marunouchi district from the *Ōsaka mainichi shinbun* comes very close to approximating the cinematic (fig. 3.28). It snaps the blurry figure of a fleeing man in a white jacket as he turns his head to gaze back at the encroaching flames. The indistinctness of the figure, his action stance midstride, and the sense that he is about to run off the edge of the picture all mimic the aesthetics of the freeze frame. This impression of an action shot is reinforced by the two men in the middle ground who are snapped leaping midair across twisted hoses that spout water into the smoky blaze in the far distance. The running figure highlights the spectator's position in the path of escape and suggests a temporal and spatial continuation of the scene beyond the bounds of the image. The snapshot appears to be candid. It is a fragment of an ongoing narrative that one imagines continuing to unfold. According to David Campany, photography's focus on the "decisive moment" emerged in the shadow of cinema, which, "in colonizing the popular understanding of time, implied that life itself was made up of distinct slices and that still photography had the potential to seize and extract them." Hence, "the frame cuts into space and the shutter cuts into time, turning the photographic act into an event in itself." By fixing this random yet decisive moment, the photographer crystallizes the emotional impact of the spectacle of photography, producing an image that is fragmentary yet summative.[55]

Taken together, static snapshots begin to form a montage of spectacular actuality that parallels the numerous contemporary cinematic representations of the earthquake and produce similar emotive vibrations.[56] According to Elizabeth Cowie, "Documentary film is associated with the serious. . . . Yet, for all its seriousness, [it] nevertheless also involves more disreputable features of cinema usually associated with entertainment film, namely, the pleasures and fascination of film as spectacle. In documentary film these pleasures arise not through make-believe or fictional enactment but by the re-presentation of actuality."[57] The representation of disaster was both inherently sensational and perversely pleasurable. Filmmakers consciously enhanced these

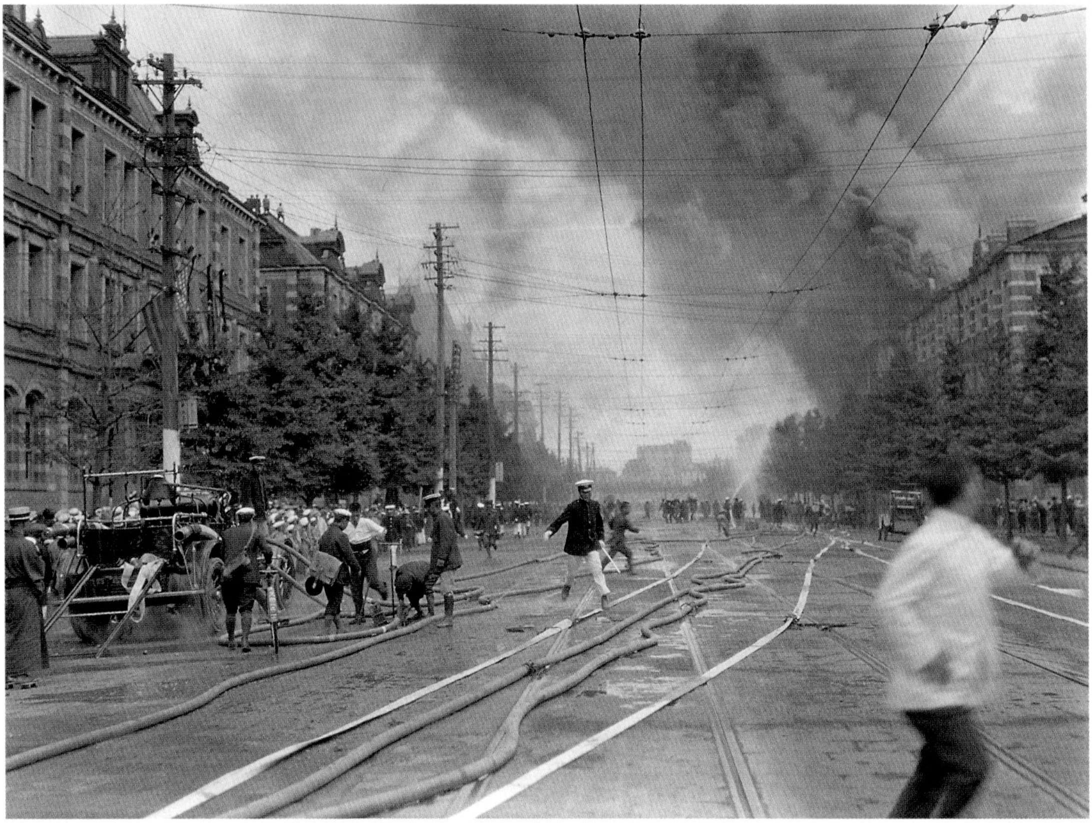

3.28 Artist unknown, *Marunouchi,* a sense of captured motion, photograph. *Ōsaka mainichi shinbun,* 1923. By permission of Mainichi Shinbunsha.

features through the manipulation of time, space, and affect in filming and editing to heighten the spectacular qualities of the viewing experience while still providing authentic reportage of the event.

One landmark documentary work, *Kantō taishin taika jikkyō* (Actual Conditions of the Great Kantō Earthquake and Fire), is a sixty-four-minute silent film produced by the studio Tōkyō Shinema Shōkai for the Ministry of Education's social education division (Monbushō Shakai Kyōikuka), using footage shot by cameraman Shirai Shigeru. Shirai had moved to the production studio in 1921 after working in commercial filmmaking for Shōchiku Kinema Kenkyūjo and at the Shōchiku Kamata film studio.[58] Tōkyō Shinema Shōkai was an independent studio that handled many educational films (*kyōiku eiga*) and documentary projects (*jissha eiga, kiroku eiga,* or *bunka eiga*) as well as American film magnate David O. Selznick's imported newsreels.

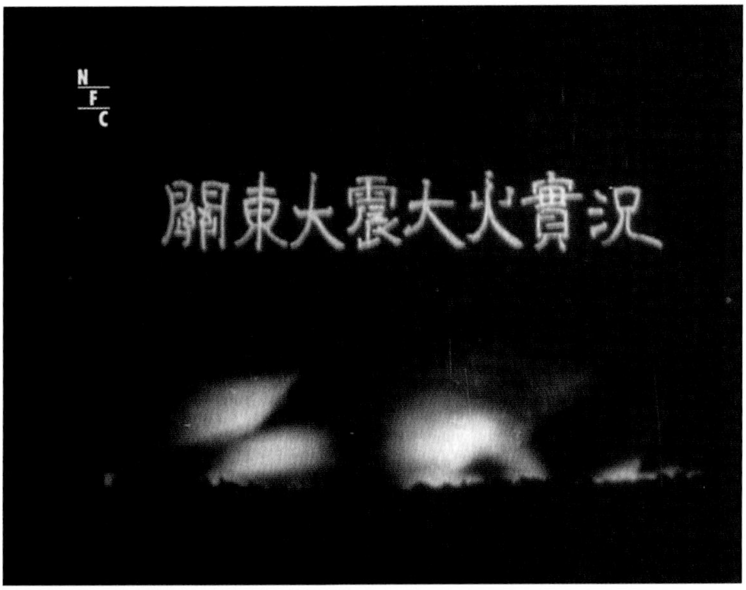

3.29 Title screen, *Actual Conditions of the Great Kantō Earthquake and Fire* (*Kantō taishin taika jikkyō*), 1923. Produced by Tōkyō Shinema Shōkai and the Ministry of Education. By permission of Tōkyō Kokuritsu Kindai Bijutsukan Film Center.

This film opens with a title screen that could easily be mistaken for that of a fictional disaster movie, showing the title characters illuminated from behind by a theatrical burning cityscape on the horizon (fig. 3.29). The title quickly fades to reveal the front pages of newspapers with dramatic headlines bewailing the massive devastation of the Kantō region, a technique for underscoring the historic and newsworthy nature of the event. Echoing other pictorial representations, the film then switches to a map of the city with large portions blotted out. The following intertitle reads, "Taishō 12, 1 September, 11:58 A.M." "A violent quake strikes the entire imperial capital. The earth splits apart; houses collapse—a view of the Marunouchi area." Here the film shifts to the first images of people, who are shown gingerly stepping over enormous fissures in the street in the Marunouchi financial district. The close-up low angle of the camera magnifies the ground. Yet contrary to the dramatic buildup of the opening sequences and the visible evidence of the earth's massive dislocation, the film's calm tone is shocking for its lack of alarm. It might just be another day in the bustling capital—if the ground were not ripped apart. This uncanny combination of calm and turmoil pervades the film.

Despite its editing, the film lacks a coherent style; it is a collage of fragments shot at a variety of locations and from greatly varying viewpoints. This seeming randomness does convey the disturbing temporal dimensions of disaster and its ultimate unknowability at the time. No sooner does the major seismic event appear to be over than it is quickly revealed to have become an enormous, spreading conflagration, producing a sense of impending doom.

Like an instrument of scientific reconnaissance, the camera shows Ochanomizu's steel train tracks torn in half, dangling limply from a hillside, and then cuts back and forth to distance and close-up shots that probe the terrestrial violence. The film then zeroes in on collapsed storefronts in local neighborhoods and follows a seemingly random assortment of people as they matter-of-factly assess damaged buildings. It quickly shifts locales again to record the crumbling exterior of the famed Marunouchi office building and the nearly desolate streetscape of the financial district. The resulting disorientation mirrors what one imagines was the sensation on the ground.

An intertitle then abruptly announces, "As one is watching the fierce fires rise from all directions, the city becomes an ocean of fire," at which point it presents the first image of the inferno: an enormous plume of smoke in the air that indicates the disaster is still unfolding. Fire brigades are seen heroically administering the little available water to burning buildings. They also seem oddly calm in contrast to the frantic tenor of the photograph of Marunouchi described earlier; no one is running. The halting movement of the camera throughout the film adds to the authentic feeling of newsreel footage. Slowly panning across smoldering wreckage or displaying a panoramic view of encroaching fires and masses of refugees fleeing along the city's major thoroughfares, the camera remains deliberate and dispassionate.

There are notable intrusions into the detached gaze of the camera eye. A man wearing a loosely tied *yukata* robe casually saunters across the frame in Kanda, staring directly at the camera as he passes by (fig. 3.30). In his nonchalance and lack of worldly possessions, he stands in marked contrast to the other figures around him, particularly the man immediately behind him who is struggling with a massive handcart full of household goods. His extended gaze temporarily ruptures the fourth wall, drawing attention to the camera eye as spectator or interloper. This effect is repeated at several junctures in the film. In his memoirs, cameraman Shirai recounts his sense of being an interloper and describes the hostility he encountered as he tried to record the event, several times fearing for his life. At one point, surrounded by an angry mob, he and his assistant had to climb on top of a car and plead for their lives, explaining that their purpose in filming was to disseminate information about the disaster to the rest of the country to expedite relief, not to capitalize on people's misery. Later, claiming concern for maintaining public peace, the authorities took him

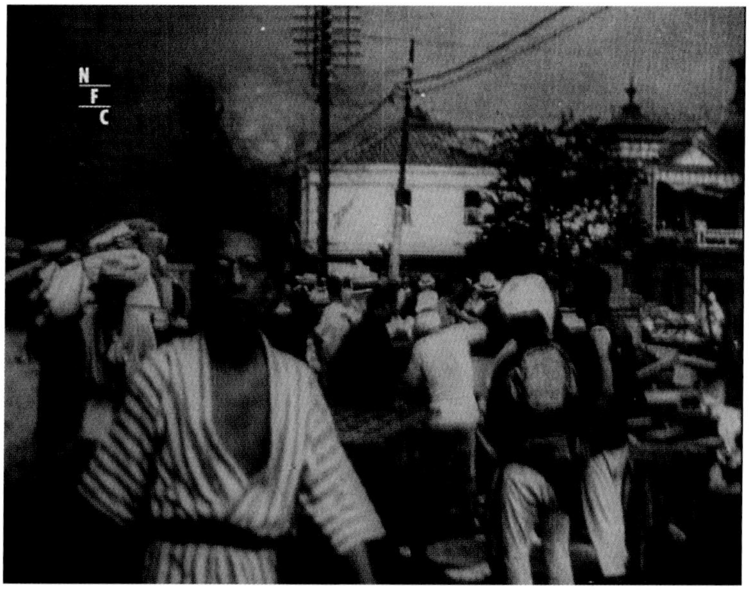

3.30 Rupturing the fourth wall, staring man's challenge to the camera, *Actual Conditions of the Great Kantō Earthquake and Fire* (*Kantō taishin taika jikkyō*), screen shot, 1923. Produced by Tōkyō Shinema Shōkai and the Ministry of Education. By permission of Tōkyō Kokuritsu Kindai Bijutsukan Film Center.

into custody and eventually confiscated portions of his film, a harrowing experience others did not survive.[59]

Throughout the film, numerous people turn their heads to see what the camera eye is seeing, wondering what is so important that it must be immortalized on film. They are either blithely unaware that they, too, are the object of the gaze or deeply suspicious of the camera's presence. The film is consistently self-reflexive in rendering the spectator's position, positioning us behind crowds watching billowing smoke and encroaching fires; in one scene, a mother facing away from the camera in the foreground rocks from side to side to calm the infant on her back, the agitation of her nervous rhythm conveying the collective anxiety and vulnerability. This imagery produces a sense of critical distance and implies that the cameraman/producer is not a participant.

Unlike the frenzied drama of other spectacular representations of disaster, the film's scenes of incineration often feel as if they are in slow motion. The camera rests on the structure of a burning storehouse and lingers as all four walls burn and fall. Yet intensity is subtly communicated in other ways through camera work, particularly in the claustrophobic impingement on the viewer's space as people crowd closer to the lens.

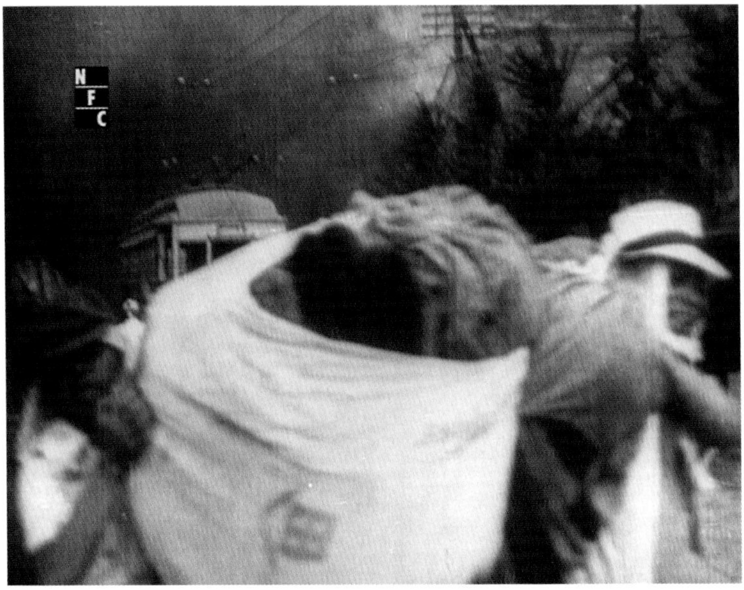

3.31 The limits of the picture frame, *Actual Conditions of the Great Kantō Earthquake and Fire* (*Kantō taishin taika jikkyō*), screen shot, 1923. Produced by Tōkyō Shinema Shōkai and the Ministry of Education. By permission of Tōkyō Kokuritsu Kindai Bijutsukan Film Center.

At one point, two men carrying a heavy load of household possessions heave into the frame, and as they move across the field of view, their cargo covers three-quarters of the screen, nearly obscuring the incinerated streetcar in the background (fig. 3.31). With their belongings pushed beyond the boundaries of our vision, they also make us keenly aware of the limits of the film frame. Although the camera could follow them, its gaze remains steady, leaving the area beyond the margins out of sight.

The film captures famous sites and panoramic vistas of destruction around the city, such as the remains of the broken Twelve Stories, which still smolders in the distance across Asakusa Park, and a devastated wasteland as far as the eye can see in Honjo. While the early films of the quake presented authentic footage in the "documentary mode," laying claim to a kind of truth-telling facticity, they were still retrospective in viewpoint and presented motivated visual narratives of the disaster. The "reconstruction" (*fukkō*) explicit in later films was always implicit in the earlier works: What did this event say about the Japanese people and the future of the nation? How would they recover and rebuild? Produced under official supervision, the films were part of a multiyear fund-raising campaign and were designed to educate and mobilize the populace, although they could not help but be entertaining as well, a fact that social reformers feared and debated.

In fact, certain filmmakers, like their artist colleagues, overtly catered to the audience's desire for sensational aesthetics, injecting theatrical imagery and textual interventions that made their works highly entertaining. The seven-minute silent short *Fukkō no Tōkyō* (Reconstruction of Tokyo) dramatized its intertitles with wildly colored backgrounds and lettering and drew attention to iconic motifs in the landscape through portentous tinted images shot at exaggerated angles: a street clock looming overhead, stopped at exactly 11:58 A.M.; or an enormous plume of smoke in the sky sublimely illuminated from behind. The thirty-one-minute silent film *Teito Daishinsai Taishō 12-nen 9-gatsu tsuitachi* (Great Earthquake of the Imperial Capital, 1 September 1923), produced on 26 September, little more than three weeks after the quake, and possibly reedited by the Ministry of Education and Nikkatsu, focused on the intensity of crowds (fig. 3.32). In contrast to the images of a vast empty cityscape in ruins, the scenes in this film, teeming with humanity and tinted an alarming red, express the magnitude of disaster in a social register. Through cunning artifice, the visual claustrophobia of the packed frame makes the viewer physically uncomfortable. The popular singer-songwriter Soeda Azenbō described in his memoir *A Burnt-Out Diary* the feeling of masses of terrified people impressing themselves upon his body as he narrowly escaped the quake and conflagrations:

> Satsuki Shō, Misao, and I joined the stream of fleeing refugees. We advanced toward Ueno. Well, not really advanced, we were pushed there. After somehow managing to get to the Shimokurumazaka bus stop, the crowd came to a standstill in a human traffic jam. Worse yet, the people in back were being pushed by the people to the rear of them. The crush was unbearable. There was no way to go forward or move back. I could see Misao's face in the crowd some ten feet behind me, but had lost sight of Satsuki. The fire gradually came nearer and nearer. I felt terrified. A space near the train tracks; there was a little opening there. Maybe that would provide some safety. Each and every face was completely clouded by fear. At last, Satsuki and Misao came along. Together we followed the railway tracks until we came out at a place near Uguisudani. We saw the weird sight of the flames rising from the huge tiles of the Honganji Temple. Soon the sky began to lighten as night gave way to dawn and the sun appeared above the flames. The second of September. The fire grew ever fiercer. I stood on the tramline tracks and stared unmoving at the deep red of the rising sun. The culture of half a century had been so easily destroyed. How flimsy science. How weak humanity. How small I am. These are the things I thought. I turned to look toward the hills of Ueno and saw a mountain of people there. Some appeared at the edges of precipices

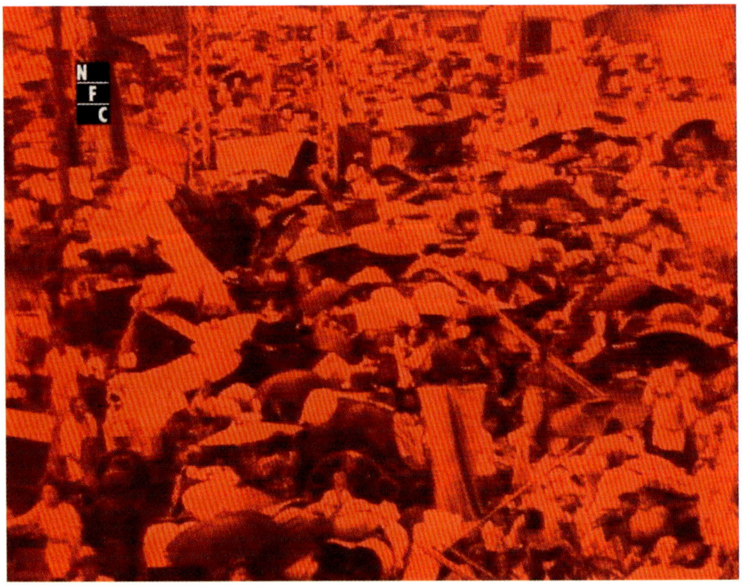

3.32 The social aspects of disaster, oceanic crowds, *Great Earthquake of the Imperial Capital, 1 September 1923* (*Teito Daishinsai Taishō 12-nen 9-gatsu tsuitachi*), screen shot, 1923. Reedited by the Ministry of Education and Nikkatsu. By permission of Tōkyō Kokuritsu Kindai Bijutsukan Film Center.

on the verge of falling, others clung to the edge. People clustered below the ridge near the railroad tracks. Some sat on bundles of belongings, others leaned against their bags. Some just looked dazed and vacant, while others wept, maintained a stony silence, or kept repeating to themselves, "What do we do now?" "What will happen?" Such people filled the area to overflowing.[60]

As Jeffrey Schnapp and Matthew Tiews describe, crowds became the new protagonists of modern history: "Heterogeneous and unstable, they arise as the result of the promiscuous intermingling and physical massing of social classes, age groups, races, nationalities, and genders along the boulevards of the industrial metropolis. They can no longer be conceived of as the passive subjects of history."[61] Their historical political actions have demonstrated around the world that crowds can make and unmake government through their mass assemblies, a lesson particularly vivid in Japan, where the violent nationwide Rice Riots of 1918 (Kome Sōdō), involving over two million protesters, shocked the government, required over one hundred thousand troops to subdue, and eventually caused the resignation of Prime Minister Terauchi Masatake. And as images showed after the Kantō quake, crowds embodied the collective terror and victimization of disaster as well as the mob savagery it unleashed. Reading Azenbō's account against the many printed and cinematic images of the heaving, oceanic masses, the viewer is able to reexperience the bodily sensations of the quake and its fearful aftermath. In such images, the thrill of the incipient action of the crowd as spectacle draws one into the aesthetics of death and destruction, hyperstimulating the senses to allow the viewer to indulge in the visual pleasure of terror.

Crowds, weapons, trains, skyscrapers, photography, cinema—these modern visual spectacles were a constant reminder of modernity's disruptive capacity in the name of progress. The physical ruptures of the Great Kantō Earthquake reiterated and magnified this disruption. And as much as the modern mass media might try to control the spectacle of actuality, they could not, for preexisting forms of artistic expression contributed an important sensory and affective register to the visual experience of disaster that inextricably linked the documentary and the subjective imagination. This aestheticized visual spectacle was pleasurable because it evoked the sensorium. At the same time, it morally implicated the spectator as witness and voyeur by unequivocally asserting the relationship between seeing and power. Modernists felt the congruity between modernity and disaster acutely, visually articulating the parallel disruptive aspects through subjective expression and abstraction. Their new visual language found a perfect analogy in the fracturing effects of the earthquake. Yet this association between modernity's profound dislocations and catastrophe provoked deep anxiety about the future of the nation.

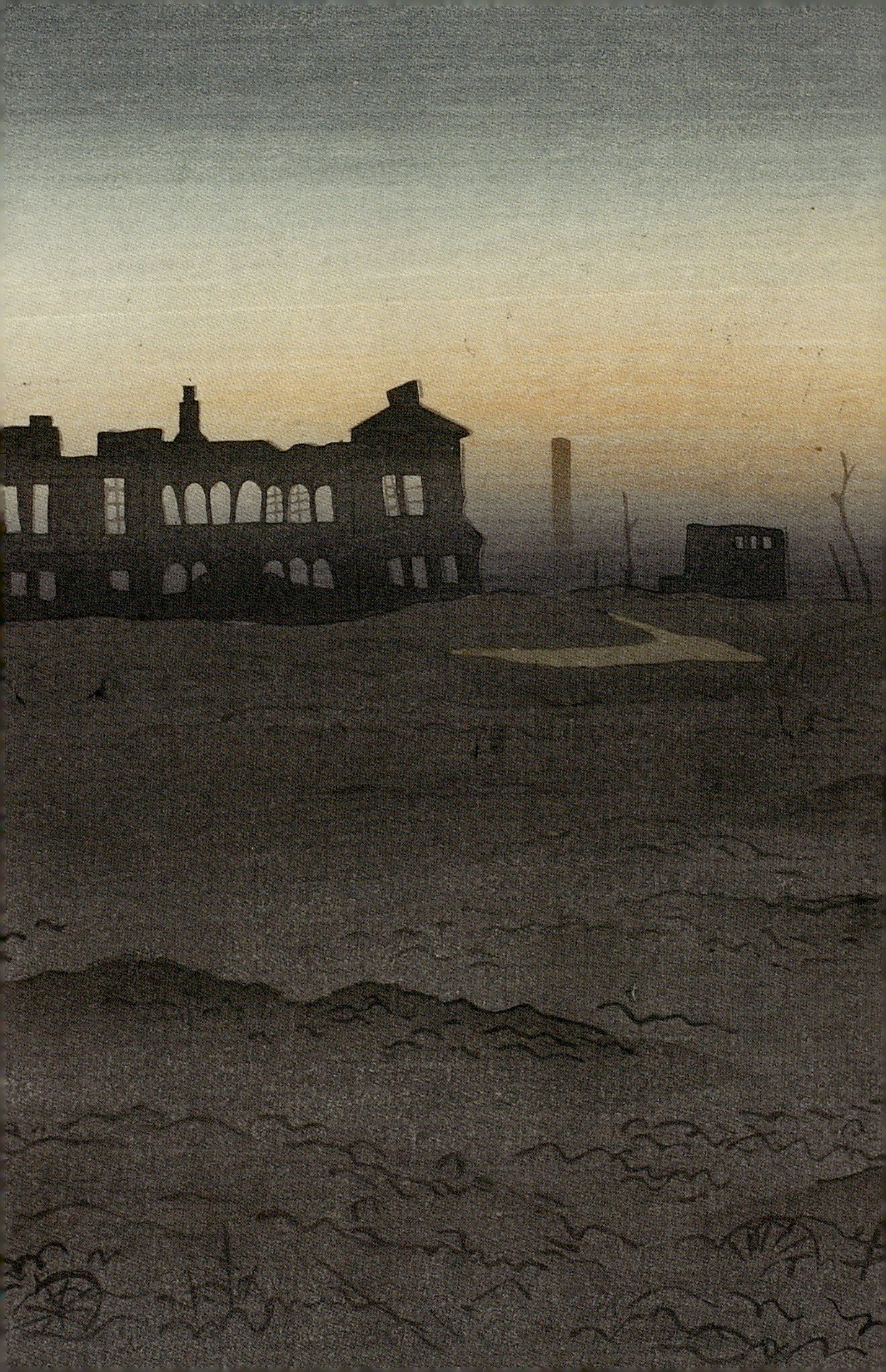

4

THE SUBLIME NATURE OF RUINS

Around ten o'clock
On a lonely night
The moon in the sky
Is all alone.
Its yellow face
Looks down upon
Tokyo
Which was completely burned.
All alone,
It stares.

"YAKETA TŌKYŌ" (BURNED-OUT TOKYO),
BY THIRD GRADER KAWAKAMI YOSHINO, C. 1923

Sketching the Sublime

The burned expanse, or *yakenohara,* that had been Tokyo became an enduring image of the earthquake experience that was seared into the minds of all who had survived. Like the solitary moon staring dejectedly down at the ruins (fig. 4.1), Japanese people felt compelled to gaze upon their surroundings to record the image for posterity. While this activity was undoubtedly a form of disaster tourism, it also served as a kind of psychological catharsis and nascent memory work. This function is particularly evident in the copious drawings people made of the charred remnants and rubble piles left in the quake's path. Prominent artists from every stylistic school took to the streets in the immediate aftermath of the quake to create an enduring testament to this unfathomable experience.

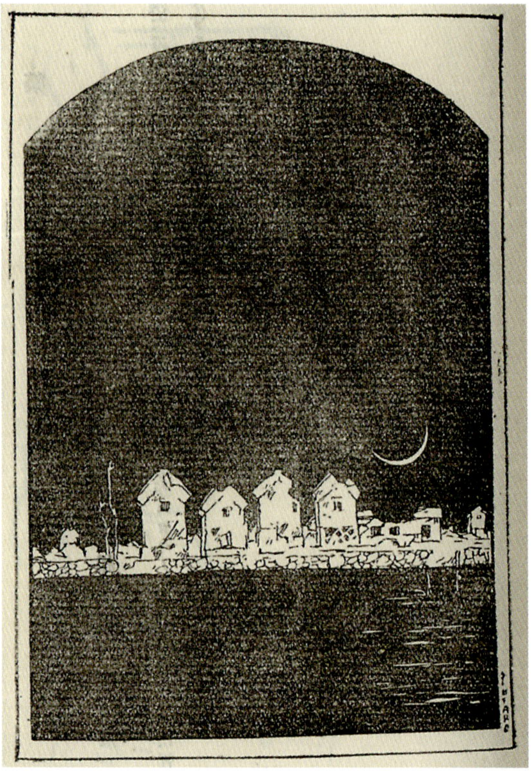

4.1 Iwata Sentarō, *In the Burned Ruins* (*Yakeato nite*), drawing. Illustration for Iwata's short essay "Yakeato nite" (In the Burned Ruins), in Reimeisha Henshūbu, *Shinsai gafu: Gaka no me* (Earthquake Sketchbook: The Eye of the Artist) (Tokyo: Reimeisha, December 1923), 49.

Sketches are generally private modes of expression and have an informality born of their more personal and exploratory mode. But sketches of the earthquake had a public dimension as well. Western-style academic painter Kanokogi Takeshiro (1874–1941), for instance, memorialized the silent, gnarled topography of the city in the hundreds of sketches he produced in the month after the quake, which he exhibited and later used as preparatory drawings for more monumental public tableaux (fig. 4.2).[1] In addition to being publicly exhibited, sketches were widely reproduced in newspapers, magazines, and books.[2] Perhaps one of the most famous examples of published sketches was by the enormously popular artist-illustrator Takehisa Yumeji, which ran with his personal commentary for twenty-one issues of the *Miyako shinbun* (14 September to 4 October) under the series title *True Pictures of the Tokyo Calamity* (*Tōkyō sainan gashin*).[3]

Yumeji's final installment in the series, *Ancient Babylon* (*Babiron no mukashi*) (fig. 4.3), is particularly portentous:

> Even now the conductor on a city bus will say, "Anyone for the Shirokiya Department Store stop?" There is nothing left of the Shirokiya. From atop a horse-drawn carriage a gentleman tells a youth from the country, "This used to be called the Ginza. It was once a really busy

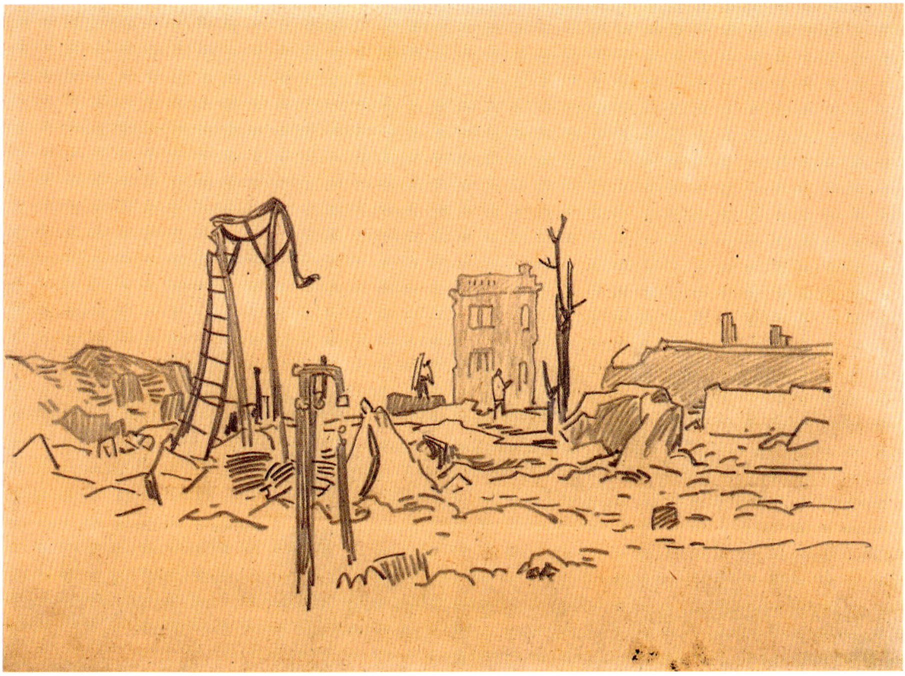

4.2 Kanokogi Takeshiro, *Sketch—Burned-Out Area* (*Suketchi [yakeno]*), charcoal on paper, 1923. Collection of Tōkyō-to Gendai Bijutsukan.

street." Another voice says, "This is where the famous kabuki actor Morita Kanya used to pack 'em into the theater (he played in 'Okuni and Gohei')."

In front of the Imperial Theater (Teigeki) they used to sell hand towels (*tenugui*) with the design favored by the actor Kikugorō for five *sen* apiece. Who would have thought that they would now be secretly selling those awful picture postcards? There are salesmen who take double advantage of people who have a vain streak by palming off on them poor-quality rings that may have been taken off the fingers of corpses or that have been treated with tobacco smoke to make it appear that they were dug out of the ashes.

People who are stark naked and have barely escaped with their lives two or three times don't seem to feel that life is dear to them any longer. I hear that in Asakusa Park not a night goes by when there aren't four or five violent fights. Whether it's because of wine, or women, or the question when does one die, what do we know about people? Only the deity Kannon knows. And so, there are fights.

Tokyo has gone back to Ancient Babylon.[4]

Yumeji's vivid textual and visual description of Tokyo as "Ancient Babylon" in a physical and moral state of ruin uses the ancient capital of Mesopotamia, one of the

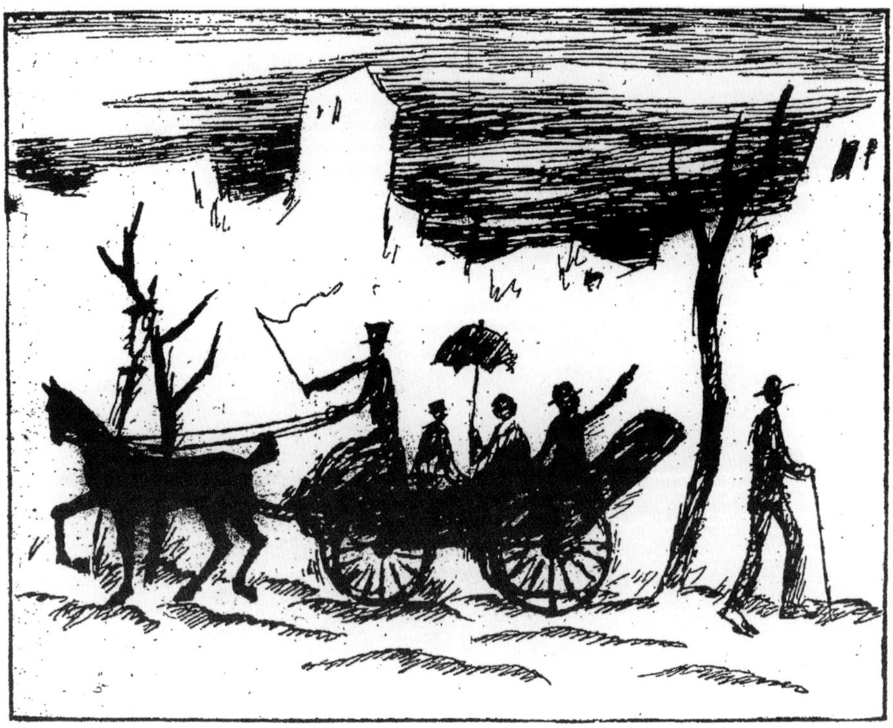

4.3 Takehisa Yumeji, *Ancient Babylon* (*Babiron no mukashi*), from *True Pictures of the Tokyo Calamity* (*Tōkyō sainan gashin*), drawing, originally in *Miyako shinbun,* 14 September–4 October 1923. Reproduced in Takehisa Yumeji, Kawamura Karyō, and Yamamura Kōka, *Yumeji to Karyō, Kōka no Kantō Daishinsai rupo* (Yumeji, Karyō, and Kōka's Reportage of the Great Kantō Earthquake) (Tokyo: Kress Shuppan, 2003), 43. Courtesy of Kress Shuppan.

most famous cities of antiquity, as an emblem of a civilization in ruins. In biblical literature, Babylon represents the misguided belief in false idols and sinful indulgence in luxury that caused the city's eventual downfall. As a monument built to the glory of man rather than the glory of God, the city's Tower of Babel was a symbol of humanity's hubris and its eventual punishment by divine power.

The two great cities of the Kantō region, Tokyo and Yokohama, were filled with places of "magnificent destruction," scenes of ruin that produced both horror and fascination, providing abundant material for moralistic philosophizing about the nation and the individual and stimulating a range of emotions.[5] Japan's imaging of the ruins of its modernity revealed a profound ambivalence toward the national ideology of progress and modernization. Having watched modernity relentlessly destroy the past in just a few decades, the Japanese indulged in this evanescent cult of ruin as a means of probing the pervasive metaphysical anxieties of the modern.

This symbolic evocation of ruin was common to disaster experiences around the world. For example, one account of the San Francisco earthquake noted, "It was with grief and horror that the mass of the people gazed on this steady march of the army of ruin. They were seemingly half dazed by the magnitude of the disaster, strangely passive in the face of the ruin that surrounded them, as if stunned by despair and not yet awakened to a realization of the horrors of the situation."[6] Yet ruins connote more than devastation and loss; they also bear witness to the relationship between life and death—the dialectic of destruction and construction—and thereby prompt meditations on this tentative balance. Ruins represent the sublime power of nature in all its terrifying magnificence; they also embody the fragility of modernity's gamble.

Visions of ruin, such as Kiriya Senrin's *Burned Ruins of Ginza Street* (*Ginza dōri no yakeato*) (fig. 4.4), draw in the viewer with their sensual stillness and evoke the sublime in their silent aesthetics of horror.[7] Historically, horror has been a fundamental aspect of the sublime. In the writings of the eighteenth-century romantics, the sublime is produced by a combination of aesthetic joy and emotional horror and despair. In recognizing the cruel extremes of nature, the sublime merged nature's majesty with its fearfulness. In contrast to beauty, sublimity could produce pleasure from darkness, uncertainty, and confusion. In Edmund Burke's 1757 tract *A Philosophical Enquiry into the Origin of Our Ideas of the Sublime and Beautiful,* he says, "Whatever is fitted in any sort to excite the ideas of pain and danger, that is to say, whatever is in any sort terrible, or is conversant about terrible objects, or operates in a manner analogous to terror, is a source of the *sublime;* that is, it is productive of the strongest emotion which the mind is capable of feeling."[8] By distinguishing beauty from the sublime, Burke called attention to the physiological pleasure produced by the aesthetics (or anti-aesthetics) of pain, death, destruction, ugliness, and the darker aspects of human existence. He emphasized the aesthetic duality of the sublime, which simultaneously produced the sensations of fear and attraction. Building on Burke's ruminations, Immanuel Kant, a philosopher who was highly influential in Taishō Japan among prominent neo-Kantians such as Nishida Kitarō (1870–1945), identified the sublime's ability to produce "a pleasure in the way that nature's capacity to overwhelm our powers of perception and imagination is contained by and serves to vivify our powers of rational comprehension."[9] Thus, the incomprehensibility of overwhelming conditions—such as those after a major natural disaster—could produce a psychological transcendence that crystallized one's ability to think rationally.

The visual and symbolic allure of images of ruin is borne out by the many such images in the numerous collections of artists' sketches published in the months following the quake.[10] The popular paperback *Shinsai gafu: Gaka no me* (Earthquake

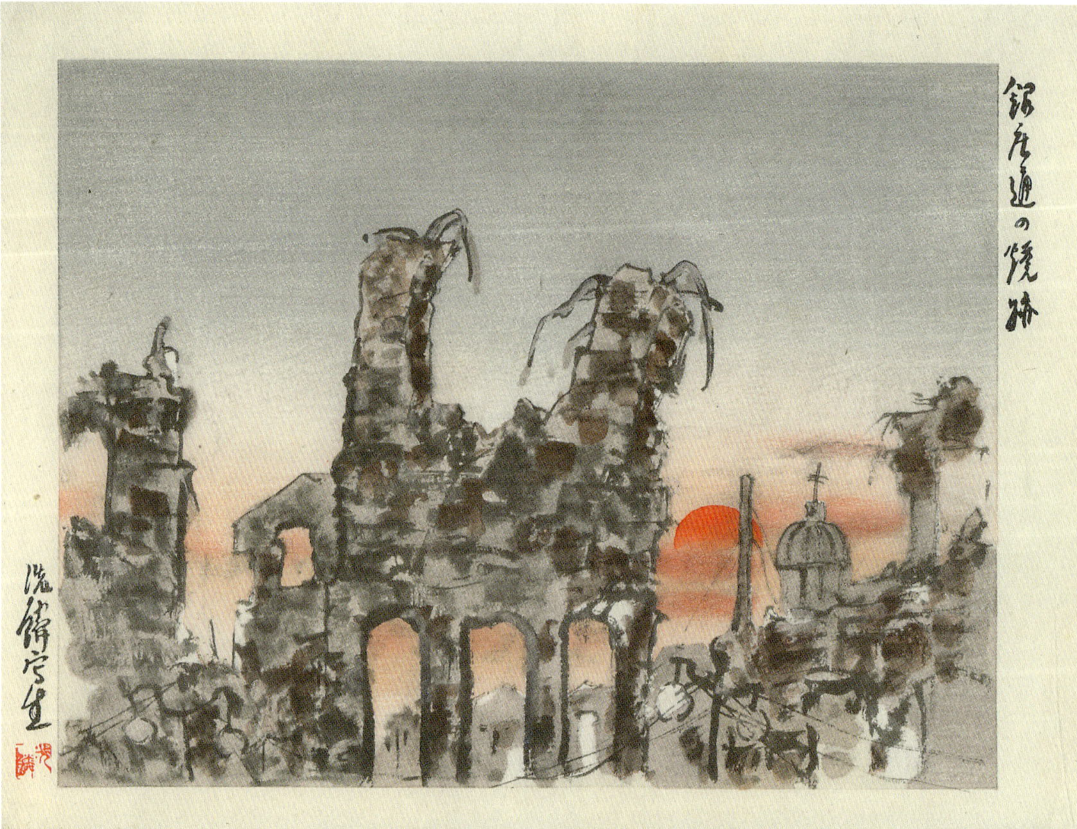

4.4 Kiriya Senrin, *Burned Ruins of Ginza Street* (*Ginza dōri no yakeato*), woodblock print. *Taishō shinsai gashū* (Collected Prints of the Taishō Earthquake) (Tokyo: Zue Kenkyūkai, 1926). Collection of Dr. Gisela Rabitz and Dr. Albrecht Rabitz.

Sketchbook: The Eye of the Artist) included images and text by forty-four well-known artists and cartoonists, who felt compelled to publish their responses to the quake in visual form to preserve for posterity their "feelings of horror" (*obiyakasareta kokoromochi*). Quickly selling out the first run, the book went into a second printing within just one month. Among its many images of ruin, the book features Shimizu Miezō's (1893–1962) sketch of the ruins of the Ichimuraza kabuki theater in Asakusa (fig. 4.5) that had been destroyed by fire on numerous occasions, reaching all the way back to the Great Meireki Fire of the seventeenth century.[11] Keenly aware of the picturesque qualities of the ruins around him and their evocative visual aesthetics, Shimizu explained his choice of subject in the simplest terms: "Of all the theaters that I saw, I thought this one would make the best picture."[12] Such "mournful skel-

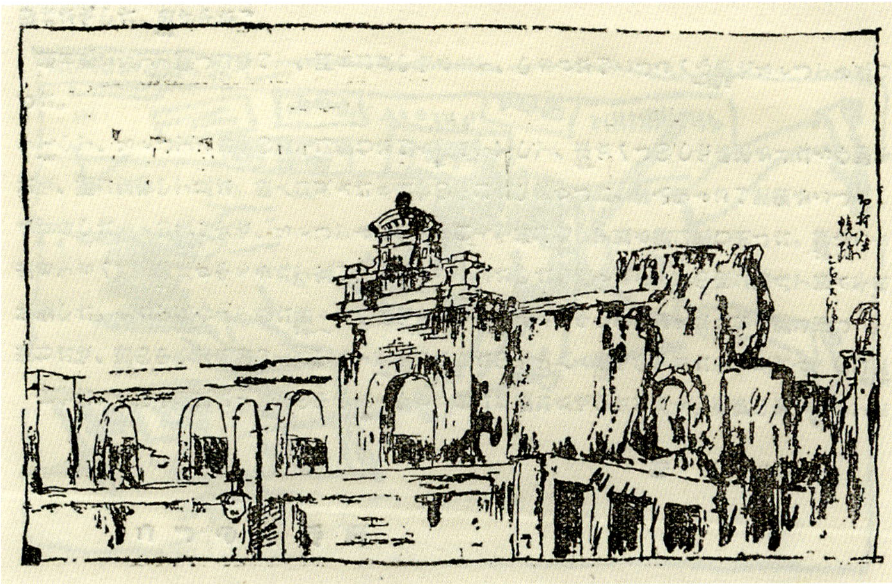

4.5 Shimizu Miezō, *Burned Ruins of the Ichimuraza Theater* (*Ichimuraza no yakeato*), drawing. Reimeisha Henshūbu, *Shinsai gafu: Gaka no me* (Earthquake Sketchbook: The Eye of the Artist) (Tokyo: Reimeisha, 1923), 89. Private collection. Courtesy of Shimizu Toshimoto.

etons," to borrow a term from the French writer Comte de Volney, evoked the nostalgia of time and place, lamenting the passing of all that man had built and the eclipse of humanity's glory, leaving only the perverse pleasure of a melancholic memento mori.[13] But in modernity's relentless cycle of creative destruction, only hastened by disaster, these sublime monuments were not permitted to linger long for public contemplation. With most of them slated for imminent demolition or facing an unknown fate, these hobbled structures were double symbols of ephemerality, and the images of their ruin were eventually all that remained to testify to this fleeting moment of horror and fascination.

In their sublimity, these images conjure up a latter-day vision of William Wordsworth's "Tintern Abbey" (1798), his ode to a de rigueur tourist site in the Welsh countryside along the Wye River, its ruin calling him to memory by inspiring a spiritual communion with nature. Ruins were a desirable destination. Artists from the Renaissance on, notably Nicolas Poussin and Giovanni Battista Piranesi, immortalized the ruins of Rome and classical antiquity, which became popular landmarks on the Grand Tour of Europe. In fact, people have been so drawn to ruins over the centuries that they have often created them artificially if none occurred naturally.

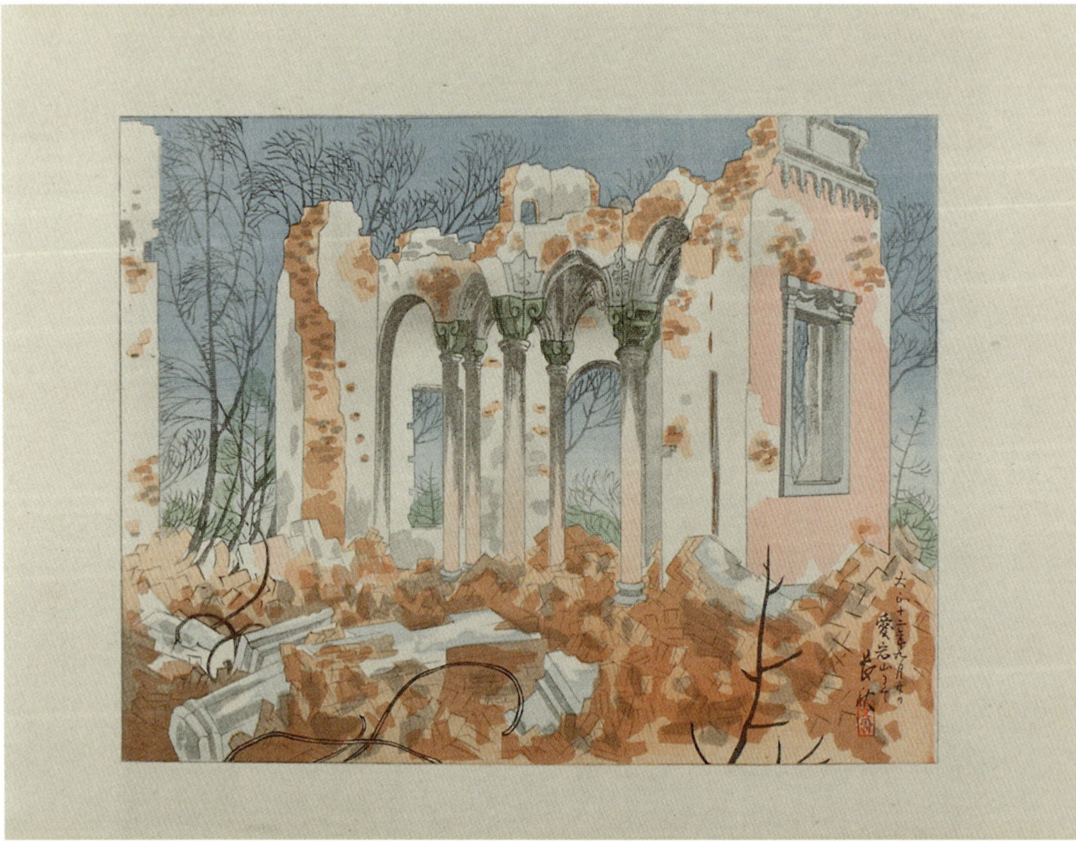

4.6 Isoda Chōshū, *Atagoyama*, ruins of a Western-style building near Shinbashi, woodblock print. Nishizawa Tekiho et al., *Taishō shinkasai mokuhangashū zen sanjū rokkei* (Collection of Woodblock Prints of the Taishō Earthquake—Thirty-Six Scenes) (Tokyo: Gahōsha, 1924). Collection of Machida Shiritsu Kokusai Hanga Bijutsukan.

In the West, from the eighteenth century on, it was common to design ruins as pastoral garden follies that would contribute to picturesque vistas in the landscape.[14] Ruins provide an enticing scopic sensuality in their decay and disorder. In a woodblock print of modern ruins at Atagoyama near Shinbashi (fig. 4.6), Isoda Chōshū (1880–1947), acclaimed historical figure painter and founding member of the dynamic young *nihonga* artists group, the Kōjikai, displays the fragments of a stalwart Western-style building, revealing its interior arches with decorated capitals and showing it surrounded by scattered debris as if nature were reclaiming the structure, mirroring the climbing ivy that enveloped Wordsworth's revered abbey. Here, he could show collapsed columns, symbolic of the passing of time, that had been created in an instant

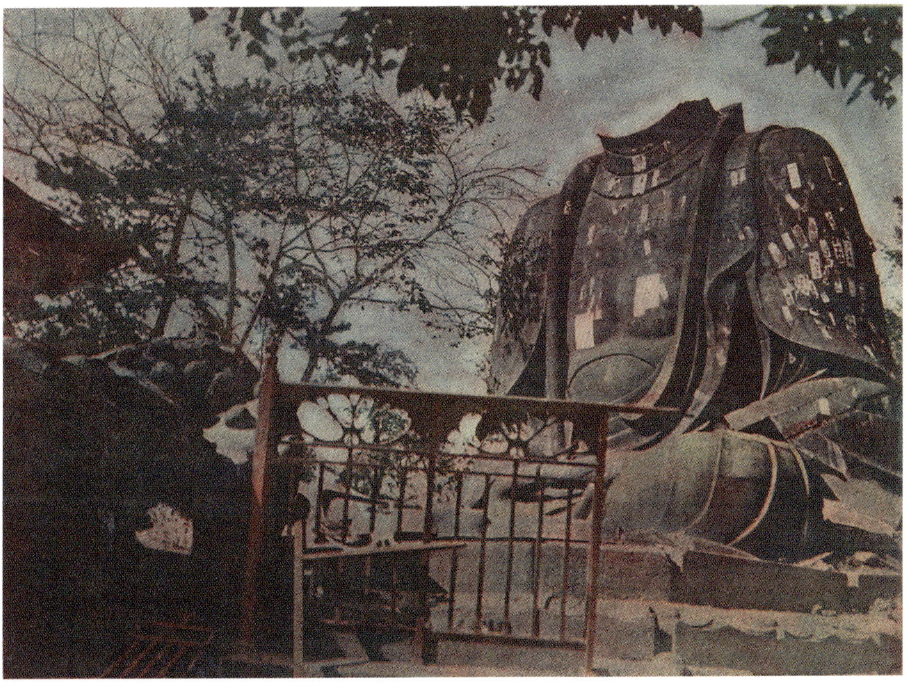

4.7 *The Ueno Daibutsu That Has Dropped Its Head (Kubi o otosaretaru Ueno no daibutsu)*, photograph, 1923. *Rekishi shashin* 1, no. 213 (October 1923).

by the earthquake, prompting a sudden awareness not only of the fearful power of nature but also of the horrific possibility that modernity itself could suddenly degenerate into ruins.

Famous Sites of Ruin

The postquake "pictures of new famous sites" (*shin-meisho-e*) immortalized locales for both their spectacular destruction and sublime state of ruin, putting a dark twist on this touristic genre. From the literally headless Great Buddha at Ueno (fig. 4.7) to the symbolically headless Nikolai Russian Orthodox Church with its collapsed dome, decapitated ruins were pictured throughout the devastated Tokyo cityscape. The Nikoraidō (as the Nikolai Russian Orthodox church in Ochanomizu was commonly called in Japanese) was one of the most iconic modern sites of ruin in the capital. Designed for the congregation of Ivan Dimitrovich Kasatkin (1836–1912) by Russian architect Mikhail A. Shchurupov and built in 1891 by British architect Josiah Conder using local craftsmen, the Nikoraidō was one of the few Byzantine-style churches in

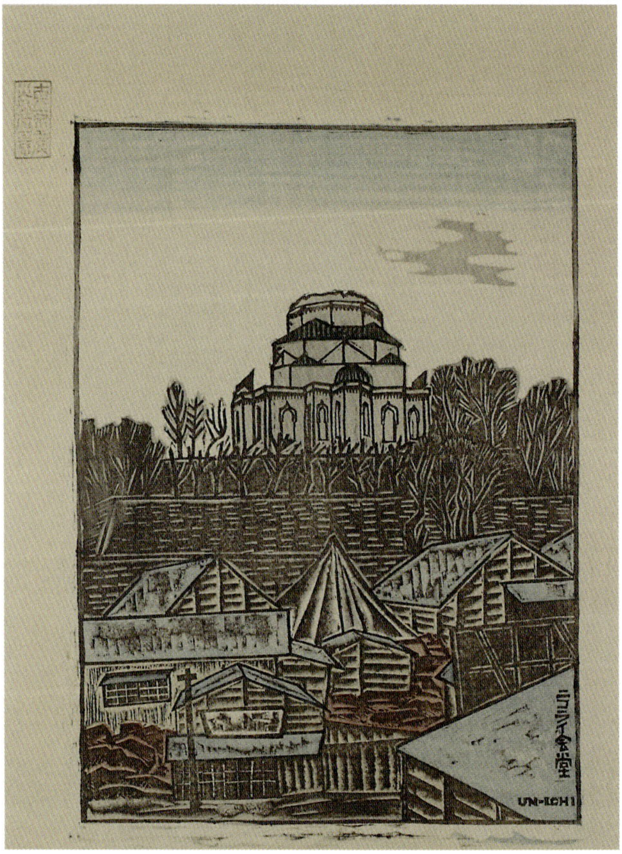

4.8 Hiratsuka Unichi, *Nikolai Church* (*Nikorai Kaidō*), in the series *Scenes after the Tokyo Earthquake* (*Tōkyō Shinsai ato fūkei*), woodblock print, 1925. Collection of Machida Shiritsu Kokusai Hanga Bijutsukan.

Japan. Standing at nearly 125 feet high at the dome, it was (and still is) a distinctive architectural symbol in Tokyo.[15] The church stood high on a promontory, which contributed to its regal impression. It presided over the area, often photographed from below to emphasize its majesty. When the dome fell and the interior was burned during the quake, a striking ruin remained, making the structure as symbolic in destruction as it had been in its grand position above the city. Visualizations of the destroyed church, stripped of its elegant dome and steeple, serve as visual lamentations. A series of twelve woodblock prints by skilled woodcarver and leading figure in the creative prints movement Hiratsuka Unichi (1895–1997), *Scenes after the Tokyo Earthquake* (*Tōkyō Shinsai ato fūkei,* 1925), present a forlorn and eerily silent city, devoid of human beings, stony reminders of their absence (fig. 4.8).[16] The famous church silhouette stands amid burnt trees as it towers over a cluster of makeshift tents and barracks. The roughness of Unichi's woodblock carving style replicates the effect of chiseled stone, contributing to the image's epitaphic solemnity. Two different printings of Unichi's work show his

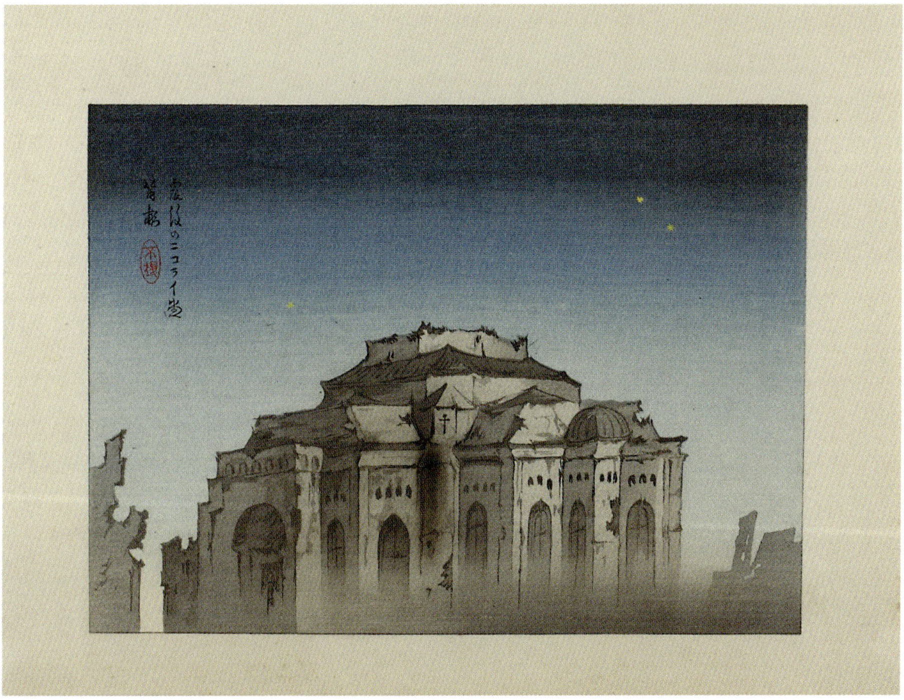

4.9 Nishizawa Tekiho, *Nikolai Church after the Earthquake* (*Shingo no Nikoraidō*), woodblock print. Nishizawa Tekiho et al., *Taishō shinkasai mokuhangashū zen sanjū rokkei* (Collection of Woodblock Prints of the Taishō Earthquake—Thirty-Six Scenes) (Tokyo: Gahōsha, 1924). Collection of Machida Shiritsu Kokusai Hanga Bijutsukan.

experimentation with inking, with images becoming even more somber and lifeless as they are almost entirely denuded of color.

Unichi's print contrasts with the image of the Nikoraidō in *nihonga* artist Nishizawa Tekiho's (1889–1965) print (fig. 4.9), which is more haunting and ethereal, bathing the church in the ghostly glow of a vivid blue twilight sky with twinkling stars.[17] These images, and the ones to follow, bring up a consistently problematic issue, posed by Mark Dery in the subtitle to his article "Terrible Beauty" in which he considers the status of visual culture in the post-traumatic landscape after 9/11: "Does our humanity falter if we acknowledge an esthetic sublime in the visual façade of tragedy?" In the article, he notes "the moral vertigo we feel when we gaze, rapt, at images of spectacular tragedies and simulated horrors, viewing the real and the recreational alike through esthetic eyes."[18] Yet, as a fundamentally visual experience, can disaster and its aftermath ever be divorced from the aesthetic, as uncomfortable as that fact may be? The aesthetics of catastrophe inevitably stimulate our senses while evoking our emotions and empathy. The imaging of disaster does not allow the viewer to remain dispassionate about the tragedy of an earthquake or ignore its ocular dimensions. The aestheticization process is not only in the eye of the viewer, but in the perspective of the producers as well, even in the midst of the disaster. As Maruyama Banka noted in *Wondrous Cloud* (*Kaiun*), his vivid description of the massive and terrifying plume of smoke that rose above Tokyo after the earthquake, "This time the cloud of smoke that appeared was strikingly large; it was said to be over eight miles high. The spectacle of the setting sun reflecting on it produced a kind of beauty (*bikan*) within the gruesomeness of it all."[19] Horror and the sublime, shock and awe, went hand in hand.

Kawasaki Shōko's woodblock print of the ruins of the Honjo Army Clothing Depot and the Yasuda Zenjirō Mansion (fig. 4.10), set against the backdrop of a radiant sunset, suggests the artist's awareness of the long Western tradition of romantic ruins, such as those associated with Germanic painters Caspar David Friedrich (1774–1840) and Arnold Böcklin (1827–1901).[20] The image is a peaceful, sublime representation of the site of the most horrific loss of human life of the earthquake. In the work of the romantics, the site of ruin was a space for contemplating man's mortality and his purpose in life. Dramatizing the barren scene through evocative illumination, Shōko transforms the depot into an affecting landscape of existential reverie.

Such ruins were profoundly emotional and spiritual, a point underscored in the highlighting of religious motifs. Yumeji's simple line drawing of the empty shell and ruined tower of a church graced the cover of the popular poetry collection *Aa Tōkyō* (Ah Tokyo!) (fig. 4.11), a melancholic work printed soon after the earthquake and so resonant with the mood of Japanese readers that it was reprinted five times.[21] Emblematic of the romantic landscape, the ruined house of worship beckons to a higher power, infusing spirituality into the devastated scene. Disaster tests people's faith, but it can also affirm their spiritual strength and sense of community. In the burned ruins of Japan's cities, telephone and electricity poles, knocked askew by the force of the temblor, remain defiantly standing, visually transformed into symbolic crosses that spotlight the existence of the divine, whether it be the God of Christianity or the Buddhist deity of compassion, Kannon. Such was the ecumenical religiosity of Japan that the Christian concept of *sic transit gloria mundi* (thus passes the glory of the world) easily united with the Buddhist belief in *mujō* (the transience of all worldly things). In Kawabata Masamitsu's (active c. 1910–40) sketch *Evening in Yokohama* (*Yokohama no yoru*) (fig. 4.12), the tilted cross of a telephone pole doubles as an altar and a tombstone, unambiguously tapping into these spiritual symbols of faith and mortality amid the beacons of light piercing the night sky.[22] As a member of the prominent artistic and literary association the White Birch Group (Shirakabaha), Kawabata and many

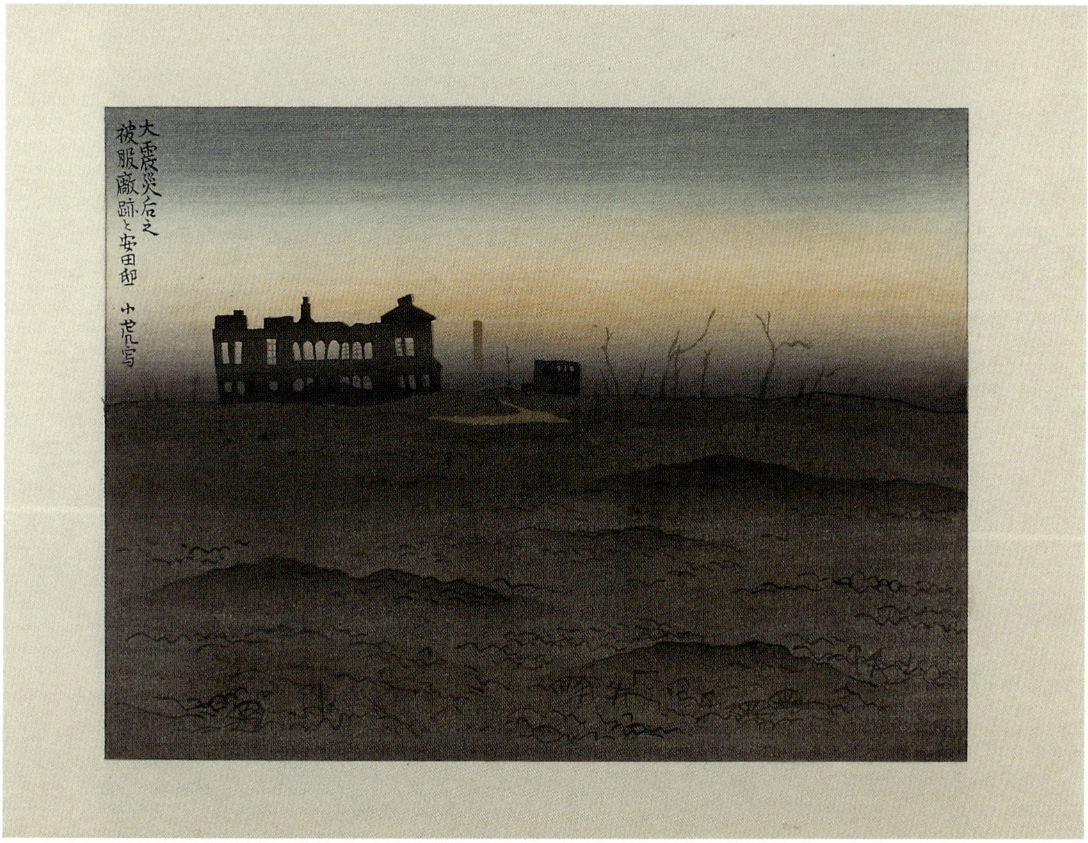

4.10 Kawasaki Shōko, *Former Clothing Depot and the Yasuda Mansion after the Great Earthquake* (*Daishinsainochi no Hifukushōato to Yasudatei*), woodblock print. Nishizawa Tekiho et al., *Taishō shinkasai mokuhangashū zen sanjū rokkei* (Collection of Woodblock Prints of the Taishō Earthquake—Thirty-Six Scenes) (Tokyo: Gahōsha, 1924). Collection of Machida Shiritsu Kokusai Hanga Bijutsukan.

of his elite cohort educated at the Peers' School (Gakushūin) were drawn to the utopian Christian humanism espoused by the maverick Japanese theologian Uchimura Kanzō (1861–1930).

Kawabata's imagery directly recalls the sublime landscapes of Friedrich, whose Gothic ruins, gnarled trees, and foggy vistas mixed spiritual mysticism and German nationalism. In his famous painting *The Abbey in the Oak Wood* (*Abtei im Eichenwald,* c. 1809–10), Friedrich explored the powerful contending forces of life and death—man and nature—emphasizing the regenerative power of nature in contrast to mankind's mortality and fear of death.[23] Many other works by the artist express a sense of impending doom and romantic martyrdom that parallel the pervasive theme of patri-

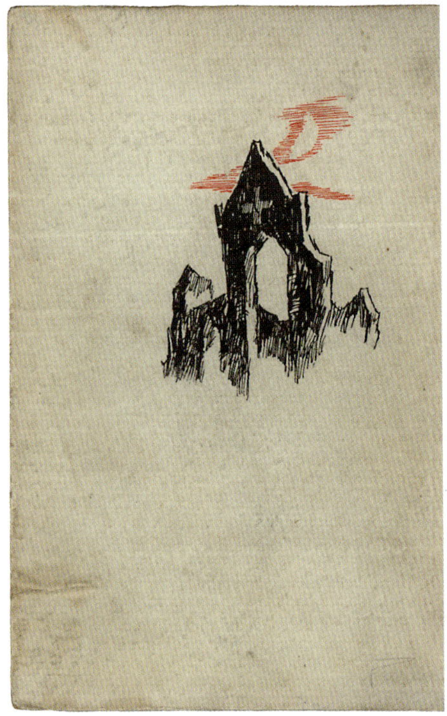

4.11 Takehisa Yumeji, house of worship in ruins, cover. Saijō Yaso, *Ah Tokyo! Tales of Tragedy Poetry Collection* (*Aa Tōkyō: Sanbun, shishū*) (Tokyo: Kōransha, 1923).

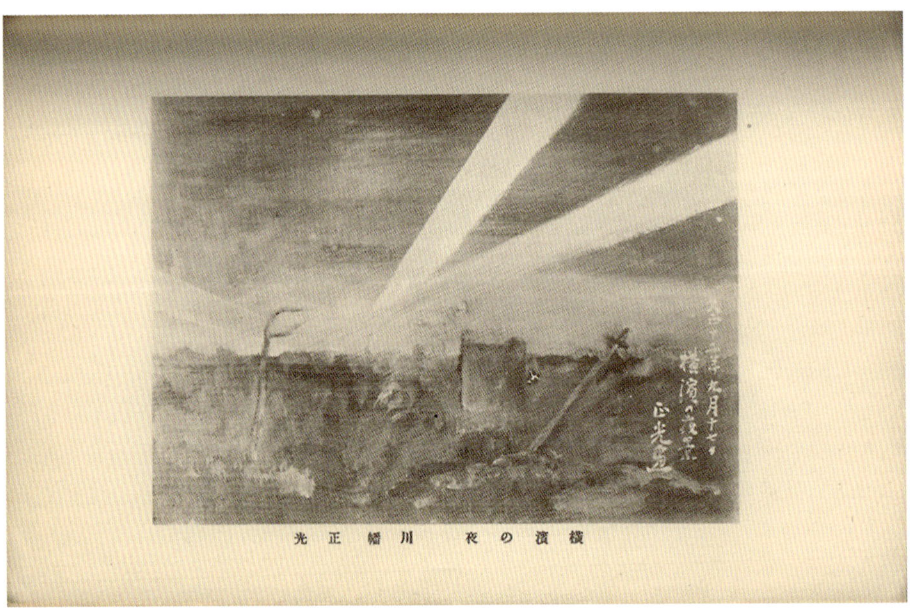

4.12 Kawabata Masamitsu, *Evening in Yokohama* (*Yokohama no yoru*), drawing. Risōsha, ed., *Shinsai suketsuchi gashū* (Collection of Sketches of the Earthquake) (Tokyo: Gahōsha, December 1923).

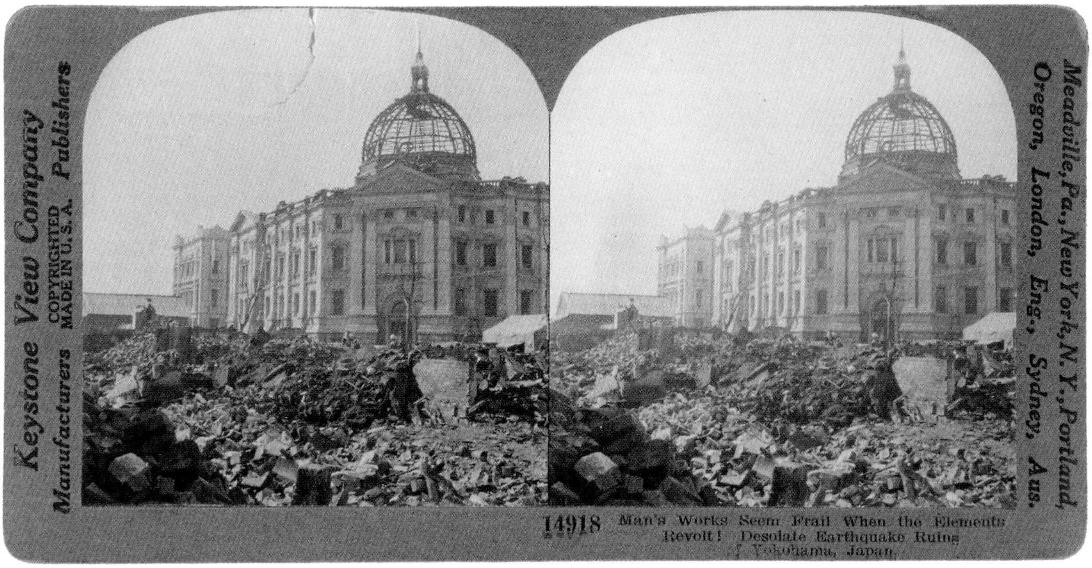

4.13 Yokohama Specie Bank, "Man's Works Seem Frail When the Elements Revolt! Desolate Earthquake Ruins, Yokohama, Japan," stereoscopic photograph, 1923. Produced and sold by the Keystone View Company. Prints and Photographs Collection, Library of Congress, Washington, DC.

otic sacrifice associated with the landscape of ruin in the Japanese imperial capital after the Kantō quake. Those who died for the Japanese nation in the disaster, even if they were merely "accidental" victims of a "random" natural event, became martyrs to the shared tragedy and grief that would make the survivors—and the nation—stronger.

The Ruins of Modernity

While sketches and prints dramatized the emotional symbolism of ruins, mass-market photographs explicitly called into question the modern Japanese national ideology of progress epitomized by the nation-building slogan "civilization and enlightenment" (*bunmei kaika*). More often than not, the photographs of ruins were of "famous sites" with connections to the establishment of the modern Japanese nation-state in the Meiji period. With all the technological advances of the modern era, they seemed to say, humanity still could not match the ultimate force of nature.

Stereoscopic photographs of the earthquake produced and sold by the Keystone View Company in the United States featured the ruins of many stalwart Japanese Western-style edifices, such as the Yokohama Specie Bank (fig. 4.13), their elegant domes that had heralded the nation's technological achievements now reduced to spindly steel

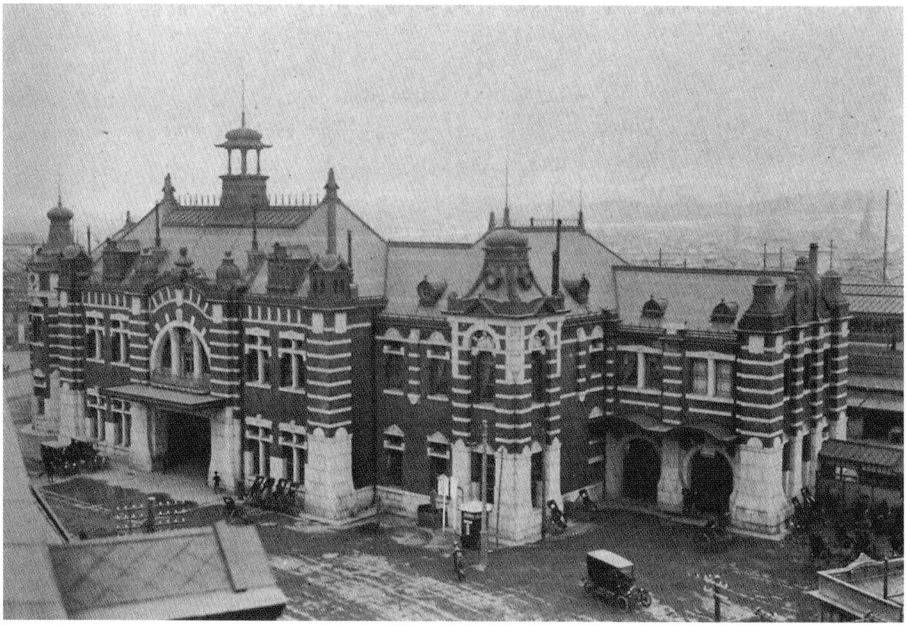

4.14 *Shinbashi Station in Perfect Face before the Earthquake*, photograph. Tokyo Municipal Office, *The Reconstruction of Tokyo* (Tokyo: Tokyo Municipal Office, 1933), 21.

skeletons. These popular images of the ruins of modernity implicitly questioned the permanence of Japan's advances, bearing melodramatic captions such as "Man's works seem frail when the elements revolt! Desolate earthquake ruins Yokohama, Japan" or "Heart of ruined Tokyo—Man's works toppled and crushed by earth's giant tremors."[24] People wondered, was modernity only skin deep? Such thoughts echoed the widespread assessment of the San Francisco quake twenty years earlier, when commentators wrote, "We are fond of boasting in this latter day of man's marvelous success in subduing the forces of Nature; and, while we are in the midst of our exultation over our victories, Nature tumbles the rocks about somewhere within the bowels of the earth, and we have to learn the old lesson that our triumphs have not penetrated farther than to the very outermost rim of the realms of Nature."[25] With the memories of San Francisco still fresh for people around the world, images of Japan's modern cities in ruin were a horrific reminder of this possibility for all nations and had ominous implications for the future of major metropolises across the globe.

These images not only implied a humbling of man's achievements and showed their fragility in the face of the indomitable forces of nature but also implicitly questioned national leadership that was predicated on the ethos of continuous progress.

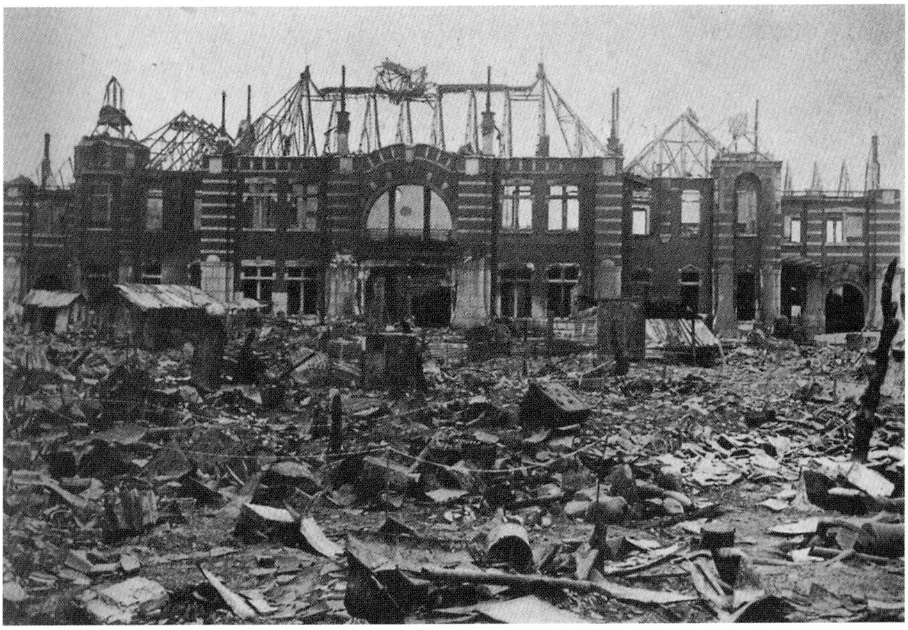

4.15 *Changed Scene of the Shinbashi Station after the Disaster*, photograph. Tokyo Municipal Office, *The Reconstruction of Tokyo* (Tokyo: Tokyo Municipal Office, 1933), 21.

This questioning was blatant in the relentless visual repetition of famous modern sites ruined in the quake, paired in suggestive "before and after" comparisons. Seeing this parade of images placing structures shoulder to shoulder with their eviscerated twins, viewers began to question the permanence of the nation's progress that had been visually signified by and embodied in the transformed cityscape and its monumental new Western-style edifices. The spectacular destruction of gleaming locomotives and the sublime ruins of the palaces of modern transportation, like the stately brick and stone structure of Shinbashi Station (figs. 4.14 and 4.15), was a horrific reminder of this impermanence.

Just as the Asakusa Twelve Stories loomed large in the public imagination as the ultimate disaster spectacle, the structure was also the ultimate modern ruin. Seeing such signature buildings in ruins suddenly made the unstoppable forward vector of the modern seem reversible. The skyscraper featured prominently in the panoply of dramatic before-and-after images that circulated in the mass media and was a particular favorite for commemorative postcards, where its unmolested prequake visage was often shown dramatically eclipsed by its spectacularly destroyed double (fig. 4.16). Photodocumentary images of the tower after the quake even imbue this infamous site

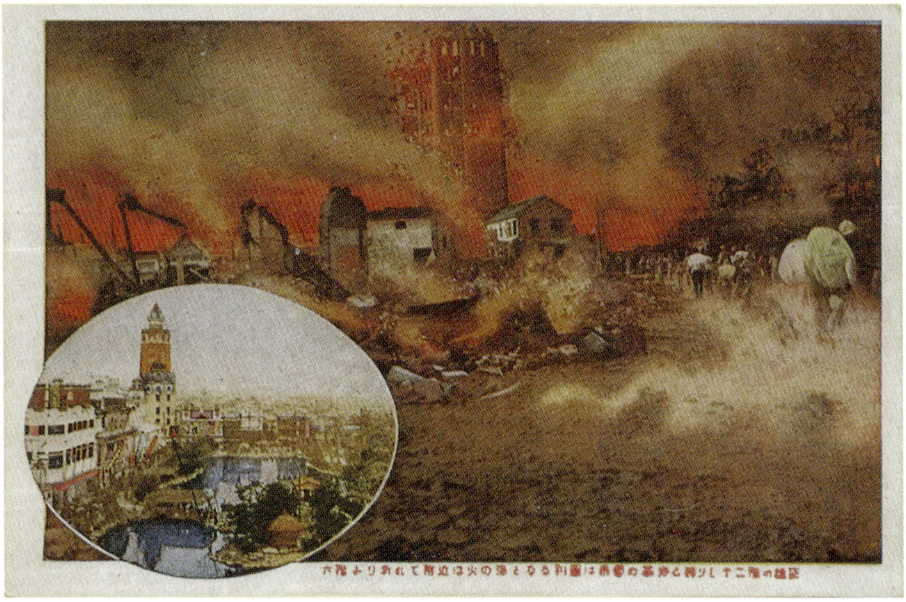

4.16 *Broken at the Sixth Floor, the Neighborhood Is a Sea of Fire. The Separate Illustration Is the Magnificent Figure of the Twelve Stories, the Imperial Capital's High Tower and Pride* (*Rokkai yori orete fukin wa hi no umi to naru. Betsuzu wa teito no kōtō to hokorishi jūnikai no yūshi*), postcard, c. 1923. Ishii Toshio Collection. By permission of Japan Publishing Copyright Association.

of ruin with a human affect, as the building appears to shrug helplessly as steel rebar droops pitifully from the sides of its forlorn, decapitated body. The anthropomorphization of the Twelve Stories recalls descriptions of the Call building, the famous San Francisco skyscraper destroyed in the 1906 quake. One observer noted, "The Call building stood proudly erect, lifting its whited head above the ruin like some leprous thing and with all its windows, dead, staring eyes that looked upon nothing but a wilderness."[26] The Asakusa tower was similarly depicted with its lofty yet broken edifice staring vacantly over a vast wasteland of destruction; whether it stood imposingly close or off silently in the distance, the monument was a haunting specter of the fleeting grandeur of Japanese modernity.

The ruined tower already felt like a relic, and it was soon to become just a memory, demolished with dynamite and reduced to a nondescript pile of rubble. The final destruction of the remains of the iconic tower was captured on film and in photographs. Like the shooting of a wounded animal, the demolition was presented as a kind of mercy killing, putting the crippled building out of its misery and consequently closing the door on a nostalgic stage of Japan's modern past.[27] Building carcasses were

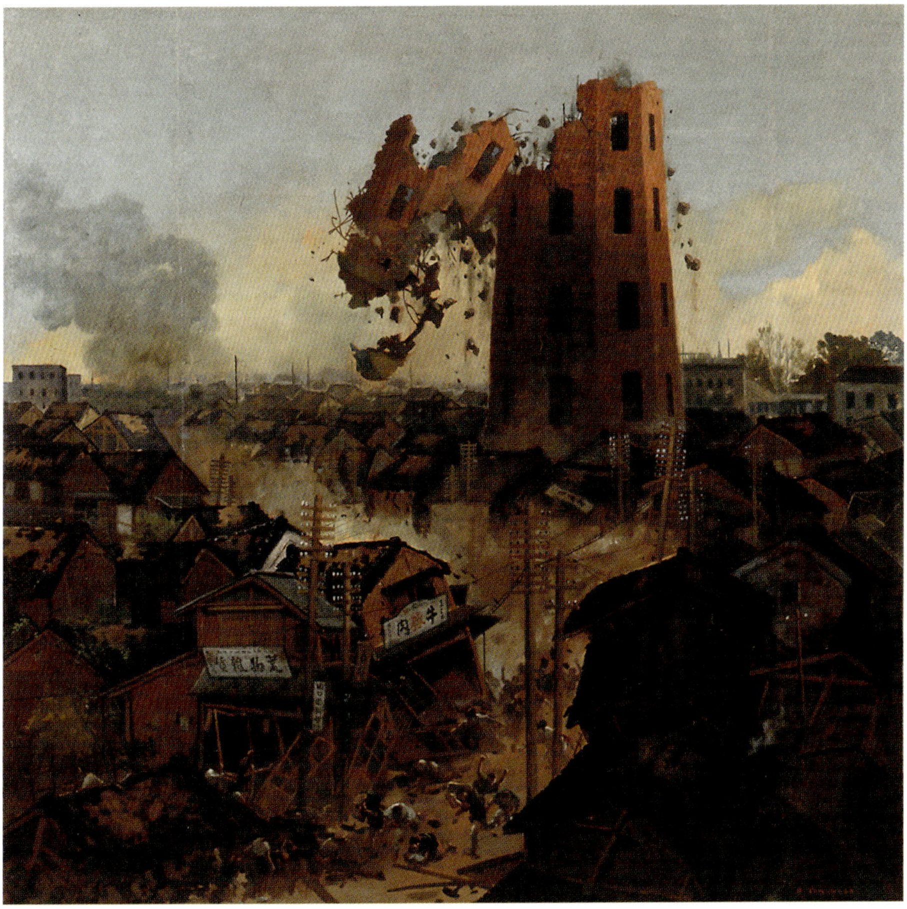

4.17 Tokunaga Ryūshū, *Collapse of the Twelve Stories* (*Jūnikai no hōkai*), oil on canvas, c. 1923. Earthquake Memorial Hall. Tōkyō-to Irei Kyōkai. Photograph courtesy of Art Restoration Studio 21.

akin to those of the animals that lay dead in the streets, particularly the often-pictured horses, which paralleled images of cattle that were killed in San Francisco, many shot by the police to prevent dangerous stampeding. Such imagery raised the persistently ambiguous question of what constituted death by "natural" causes.

Perhaps the best way for contemporary viewers to understand the horror and shattered confidence in seeing the Twelve Stories, a marvel of modern technology, crack and fall to the ground with people still inside (fig. 4.17) is to compare it to the more recent destruction of the World Trade Center on 9/11. While the contexts for the destruction of the Twelve Stories and the World Trade Center differed in significant ways, both were monuments to man's achievements whose loss shattered national confidence and security, and both acquired great status as symbols of trauma. Moreover, the obvious evocation of the Gothic cathedral ruin at Ground Zero of the World Trade Center, making it a symbol of the sublime with direct ties to the spirituality of the romantics, shows how pervasive this symbolism has been through the ages and around the world. Once again, religiosity and spirituality emanated from the devastated landscape of modernity.

Images of the imperial capital in ruins presented serious problems for Japan's ongoing project of modernity. Identification as an "earthquake nation" cast doubt on the country's ability to match the level of civilization and enlightenment of its world peers, which threatened to jeopardize the nation's efforts to jump-start reconstruction and ensure its future. Even Frank Lloyd Wright, a strong Japan supporter who had worked in the country for extended periods, read deep meaning into this event and registered concern for Japan's future. On 26 September 1923, in an act of self-congratulation as much as social concern, he wrote to his teacher Louis Sullivan about the miraculous survival of his major Tokyo project, the Imperial Hotel:

> I have received another cable from Tokyo reading—"Imperial stands square and straight," signed Hayashi, who was the manager that came to America to get me and signed Endo, who was my Japanese right bower in my office in Tokyo.
>
> Corroboration comes now from every side.
>
> The Yankee Skyscrapers are some of them standing badly wracked and some with the top floors shaken down—all visibly seriously damaged, probably murdering thousands trapped in them—unable to get out.
>
> Several are completely gutted by fire also.
>
> Of course it is impossible to tell how the strain has really affected the connections still hidden by whatever of "architecture" is left clinging to their bones.
>
> People die of panic and fright and elevators don't run in an earthquake. You can imagine the piles of dead on the stair landings—stairs four feet wide—10 stories high—half flights.
>
> And this congestion on the basis of eight or ten stories, where it was already too great for safety or comfort on the basis of three stories, was a crime, but a tribute to Yankee salesmanship.
>
> I am opposed to the tall building from now on, in the Pacific Basin, even if the human scale of things and safety and convenience are to be sacrificed to the ubiquitous American Landlord. I have written something outlining these views which will appear somewhere perhaps, and I am sending it to Tokyo to try and head off the propaganda which will try to rebuild Tokyo as a modern American city.[28]

If architects like Frank Lloyd Wright had had their way, Tokyo would not have been

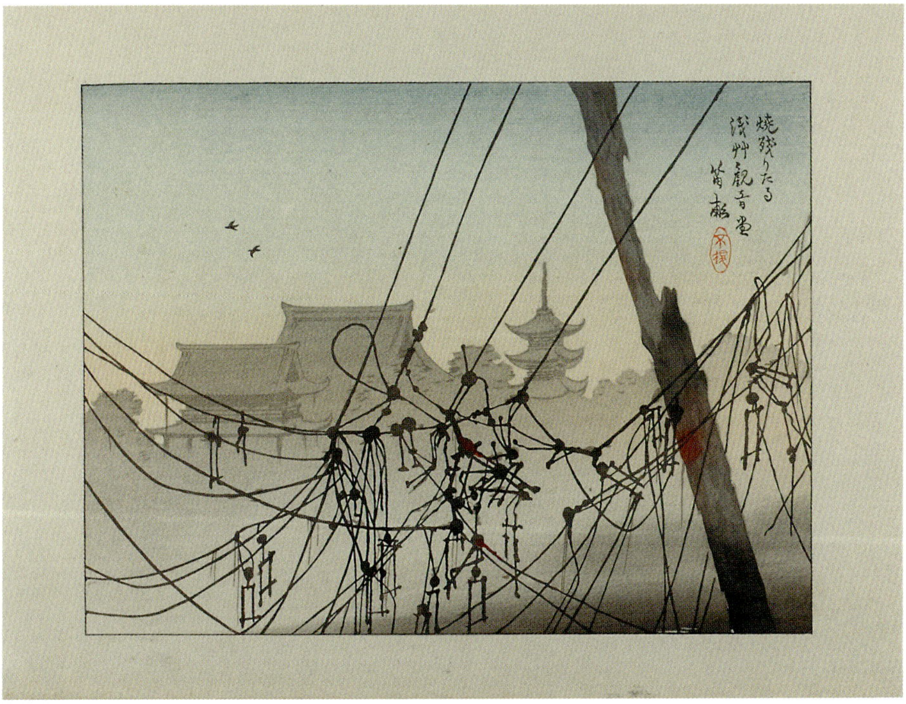

4.18 Nishizawa Tekiho, *Asakusa Kannon Hall That Escaped the Fire* (*Yakenokoritaru Asakusa Kannondō*), woodblock print. Nishizawa Tekiho et al., *Taishō shinkasai mokuhangashū zen sanjū rokkei* (Collection of Woodblock Prints of the Taishō Earthquake—Thirty-Six Scenes) (Tokyo: Gahōsha, 1924). Collection of Machida Shiritsu Kokusai Hanga Bijutsukan.

rebuilt as the megalopolis that it currently is. Nature would have been obeyed. In fact, the survival of the prominent wood Buddhist temple Sensōji in Asakusa (also known as Asakusa-dera) when so-called modern buildings of brick tumbled to the ground seriously undermined the fundamental assumption of Japanese modernity that the West knew best. The significance and long-term implications of this revelation were not lost on the artistic community.

In his contribution to the *Collection of Woodblock Prints of the Taishō Earthquake—Thirty-Six Scenes,* Nishizawa Tekiho pictures the Kannon hall at Sensōji (fig. 4.18) amid a tangle of downed power lines in a way that romantically allegorizes the relationship between past and present, using the abstract, tangled mass in the foreground as a shattered lattice through which the viewer gazes at the majestic, resilient temple structure in the distance, awesome in its hazy luminosity.[29] Tekiho here presents an image of serenity and contemplation as a morality tale that celebrates the resilience of traditional Japanese wooden architecture and cautions against the progressive ambitions of the nation. This kind of imagery resonated with a raging contemporary discourse on earthquake-resistant building practices in Japan. On one side were those who advocated carpenter-built, traditional joinery Japanese-style buildings; on the other were those who championed masonry and steel-reinforced concrete construction of Western-style structures.

In this context, ruins took on an added level of resonance in Japan, paradoxically becoming associated with modernity rather than antiquity. Before the introduction of stone and brick, Japanese disasters generally did not produce ruins because wood structures usually burned to the ground, leaving only debris. In *Ansei-Era Observations,* only one image from the third volume shows stone ruins of the destroyed base and fortifications of Edo castle. The rest of the images from the 1855 quake merely show debris from toppled or burned structures.

By the time of the large-scale Nōbi earthquake near Nagoya in October 1891, however, masonry architecture was already being introduced into Japan, and several publications used images of collapsed brick walls on their covers that highlighted the appearance of ruins while implicitly indicting Western-style structures for their inability to withstand earthquake conditions. This launched an ongoing debate on the benefits of wood versus brick and stone in earthquakes, which Gregory Clancey has discussed in detail: "Stone and brick were also actors in a story about change, which was also the story of 'civilization' (*bunmei* in Japanese, which in the Meiji period was commonly used synonymously with Western civilization). Masonry was perceived to have replaced wood (to have driven carpentry up into the roof or down into the floor) wherever civilization occurred. Indeed, the presence or absence of civilization in European eyes was marked not only by the presence or absence of agriculture, but by masonry, and more especially by ruins."[30] Ironically, ruins were a sign of civilization and enlightenment right at the moment when their appearance undermined the possibility of sustaining these very ideals.

For European observers in the Meiji period, "Japan's lack of masonry ruins was, among other factors, equated with an absence of memory, a contempt for the serious, a disregard of the solid."[31] Through their connection with history, ruins were thought to have an inherent educative value, not to mention a moral dimension. They figured in scenes that either waxed nostalgic about the past or constituted moralistic commentary on the decay of the present. Moreover, as Clancey also notes, while Western thinkers separated wood and stone into binary categories of "organic" and "inorganic" or "ligneous" and "lithic," respectively, which designated wood as living and subject to change while considering stone enduring and permanent, there was no such division in the Shinto-inflected Japanese thought that both wood and stone housed animistic spirits.

In the culture of Meiji building practices, however, some of these Western philosophical beliefs began to engender a preference for brick and stone.[32]

As oxymoronic as the "modern ruin" might seem, running counter to the notion of the gradual "natural" decay to ruins, the term pointed to a terrifying, accelerated genesis of ruin caused by the sudden and violent disruptions of modernity. The modern landscapes featured in the famous sites of ruin underscored modernity's relentless path of destruction even before the quake, as many of the sites no longer featured Edo-period attractions but rather highlighted brick and masonry structures that were landmarks of a thriving modern Japan. Iwata Sentarō's (1901–74) short illustrated text "Yakeato nite" (In the Burned Ruins) recounts a simple conversation between two acquaintances about this tension in the modern urban landscape: "Everything has been burned to the ground," one man says. His companion replies, "Ah, it feels good. If anything, the complete burning feels good." The second man then expresses his nostalgia for the low city of bygone days, the vestiges of old storehouses that lined the riverbanks, the old wooden bridges over small rivers, the things that gave the low city its character, which had already all but disappeared. He concludes, "That thing called the low city is probably already gone." The first man rejoins, "But . . . now a new beautiful Tokyo will be born, isn't that great?" Incredulously, the second retorts, "You really think so? It doesn't look to me like it's going to be much of a change from what we've had up till now. Just that old things crumble, things start afresh, it's all boring." The first man cuts him off: "That's your crude way of thinking. Listen, old things have just simply become victims."[33]

The creative destruction of modernity was the background to Tekiho's picture of Nihonbashi Bridge in the print series (fig. 4.19). Although the Nihonbashi district had been a flourishing center of Edo commercialism in Tokugawa times, the place Tekiho depicted instead spotlighted the landmark Renaissance-style stone bridge that was built there in 1911 and that still stands today. The lionlike sculptures adorning the ends of the bridge are *kirin* (Chinese-inspired chimerical animals) that were thought to be auspicious. Here, somewhat ironically, surrounded by the debris of the quake, they celebrate Tokyo's prosperity and symbolize defense of the city. The sculpted figures themselves were symbols of Japan's modernity, as they were fashioned by artists affiliated with the new Western-style art academy, the Tokyo School of Fine Arts (Tōkyō Bijutsu Gakkō).

Tekiho similarly immortalized the destroyed façade of Mitsukoshi Department Store in the Nihonbashi area (fig. 4.20). An industry innovator, Mitsukoshi was founded in the seventeenth century as Mitsui Echigoya dry goods. Its storefront was a famous landmark of the Edo period and was prominently featured in contemporary

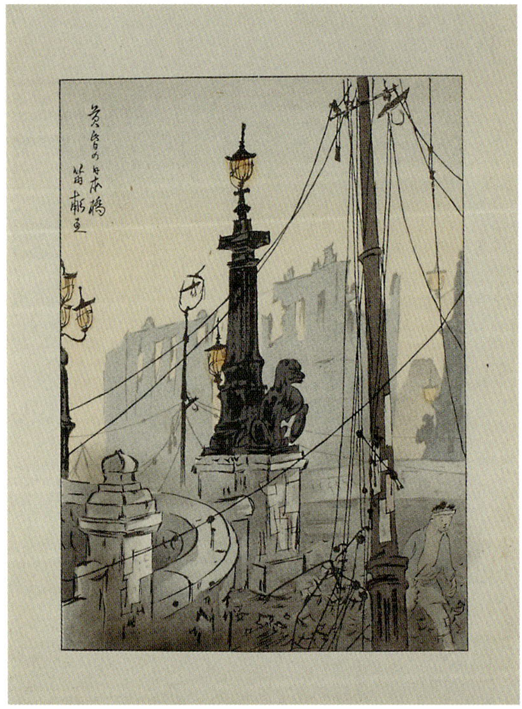

4.19 Nishizawa Tekiho, *Nihonbashi Bridge at Dusk* (*Tasogare no Nihonbashi*), woodblock print. Nishizawa Tekiho et al., *Taishō shinkasai mokuhangashū zen sanjū rokkei* (Collection of Woodblock Prints of the Taishō Earthquake—Thirty-Six Scenes) (Tokyo: Gahōsha, 1924). Collection of Machida Shiritsu Kokusai Hanga Bijutsukan.

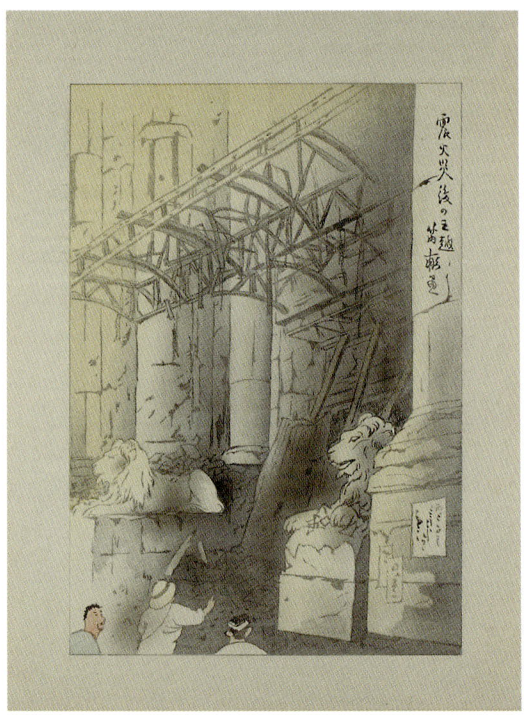

4.20 Nishizawa Tekiho, *Mitsukoshi Department Store after the Earthquake* (*Shinkasaigo no Mitsukoshi*), woodblock print. Nishizawa Tekiho et al., *Taishō shinkasai mokuhangashū zen sanjū rokkei* (Collection of Woodblock Prints of the Taishō Earthquake—Thirty-Six Scenes) (Tokyo: Gahōsha, 1924). Collection of Machida Shiritsu Kokusai Hanga Bijutsukan.

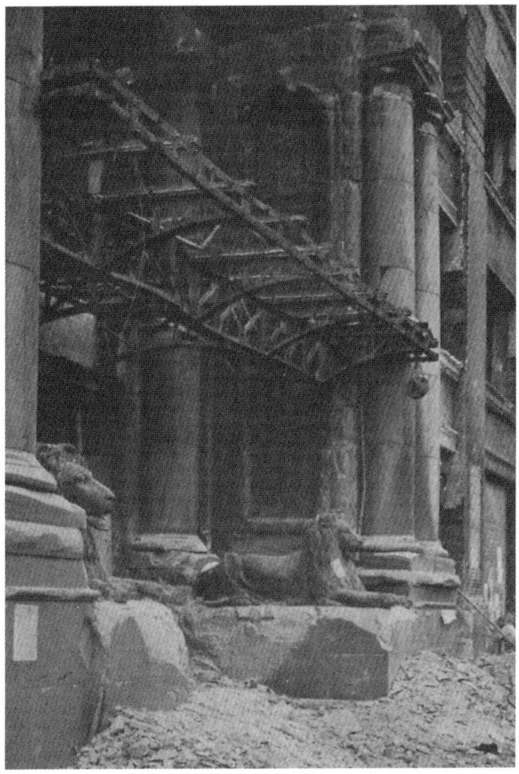

4.21 Mitsukoshi Department Store, photograph, 1923. Tokyo Municipal Office, *The Reconstruction of Tokyo* (Tokyo: Tokyo Municipal Office, 1933), 50.

woodblock prints. Not surprisingly, the postquake image featured the ruins of the store's stylish and popular Western-style edifice, which had been built in 1914 and expanded in 1921.

This scene, among many others, reveals the close visual connection between photographic (fig. 4.21) and print representations, as well as their radically different affect. Both images show Mitsukoshi's signature lions flanking the entrance. The lions, modeled after the ones at the base of the statue of Admiral Nelson in Trafalgar Square in London, symbolized elegance, bravery, and magnanimity, and they stood to protect the entrance. Despite the lions' best apotropaic efforts, however, 93.3 percent of buildings in Nihonbashi were destroyed in the quake, including Mitsukoshi and its annex in Marunouchi.[34] Both emphasize chipped and eroded columns and highlight the building's evacuated sculptural niches, evoking ancient classical ruins, and they invite the viewer to complete their form. But Tekiho has made a few suggestive changes. His foregrounding of the figures and addition of light color in the woodblock subtly transform the more somber and elegiac photographic image into an active scene that curiously resembles images of archaeologists discovering ancient ruins. This conveys a sense of

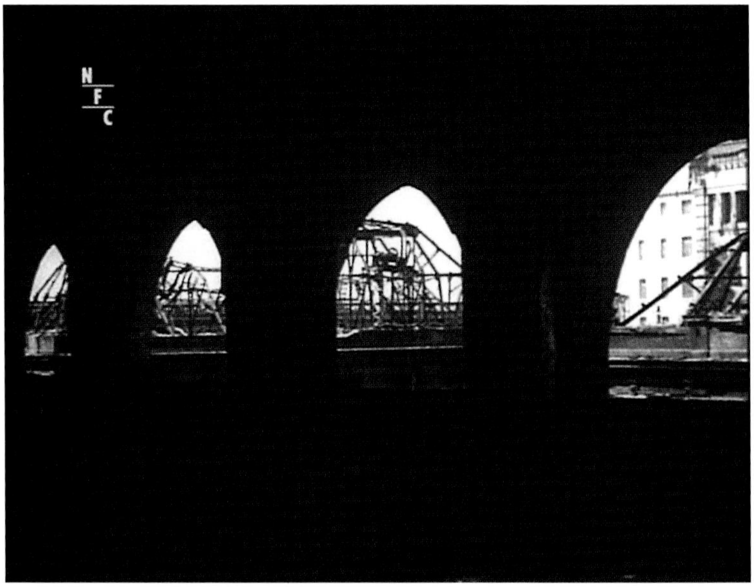

4.22 Ruins of the Mitsui Building, *Great Earthquake of the Imperial Capital, 1 September 1923* (*Teito Daishinsai Taishō 12-nen 9-gatsu tsuitachi*), screen shot, 1923. By permission of Tōkyō Kokuritsu Kindai Bijutsukan Film Center.

adventure in the ruins and allows the viewer to experience vicariously the pleasures of disaster sightseeing.

Cinema, through its modern vision of ruins, brought the sensibilities of classical antiquity and romanticism into the technological age. The film *Great Earthquake of the Imperial Capital, 1 September 1923,* accentuates the viewing experience by showing the skeletal steel frame of the Mitsui Building's roof, naked and spindly without its dignified sheathing, through the Gothic arches of the Mitsukoshi Department Store (fig. 4.22). This framing produces a curious double layer of modernity in ruins, allowing the viewer to glimpse the burnt shell of finance through the broken screen of retail capital.

In the film *Reconstruction of the Imperial Capital,* the viewer sees the vast expanse of destruction from the air and is then taken on a driving tour of the ruins, accompanying the camera as it speeds through the landscape, accelerating the shocking effect of the desolation. From ground level, ruins are visible through other ruins, layer upon layer of empty shells. In some places, the ground still smolders. Like the adventurers in Tekiho's print, we too come upon the vestiges of Mitsukoshi, and then are deposited in the mangled debris in front of the Twelve Stories. Yet unlike the single images in prints and photographs that allow the viewer to ruminate on a scene, the fast-moving cinematic images thrust us forward through time, allowing little occasion for extended contemplation. We are left with ruins that call out for reconstruction, the subtext of the cinematic narration.

Redeploying Ruins

Looking into the sea of ruins after the Kantō quake, the Japanese nation was forced to contemplate its fate. The stoic statue of the martyred Joan of Arc, still standing erect in battle armor in the devastated landscape of Kanda, called the nation to attention and united it in its determination to endure the disaster. The statue communicated a message of unity and rebirth that echoed the efforts of state officials and public intellectuals to mobilize the populace to provide relief and aid for the reconstruction of the imperial capital. Capitalization was critical to reconstruction, and investors needed to have confidence in Japan's future. As reminders of lost loved ones, income, and valuable property, ruins were unequivocally emblems of loss. And in the shadow of the sublime lurked the fear of future devastation. One group that capitalized on this horrifying specter was the insurance industry.

In a full-page advertisement for the Yachiyō Life Insurance Company (Yachiyō Seimei Hoken Kabushiki Kaisha) (fig. 4.23) on the back of the *Asahi shinbun*'s special earthquake pictorial issue (25 September 1923), a young mother with an infant strapped to her chest and all of her worldly belongings slung over her shoulder shelters a small child at her feet and stands staring at a vast expanse of ruin and a menacing plume of smoke in the distance. She embodied the pervasive anxiety people felt after the quake, unsure how they could rebuild and worried that major aftershocks might continue for years. Ruins were a symbol of transience; insurance offered the semblance of stability in an unstable world. Yachiyō's advertising copy promoted the commodity of "peace of mind" that life insurance offered, while stylistically mimicking the emotive sensibilities expressed in poetic venues. "Bitterness flows, on the ground reduced to ashes, your house is burned your husband dead, as you stand still again the tears come. Forlorn and desolate, in the abyss of despair, night falls, there are no stars, no moon. But if you look up, the blessings of your dead husband, [like] the smile of Daikoku, the God of Wealth, flood over your young children." To reinforce the connection, a small roundel above the company's name, perhaps part of the corporate logo, contains Daikoku's grinning face beaming out from the page. While the romantics wrestled with the existential angst of human mortality and stood in awe of the sublime powers of nature, twentieth-century corporations sought to mitigate

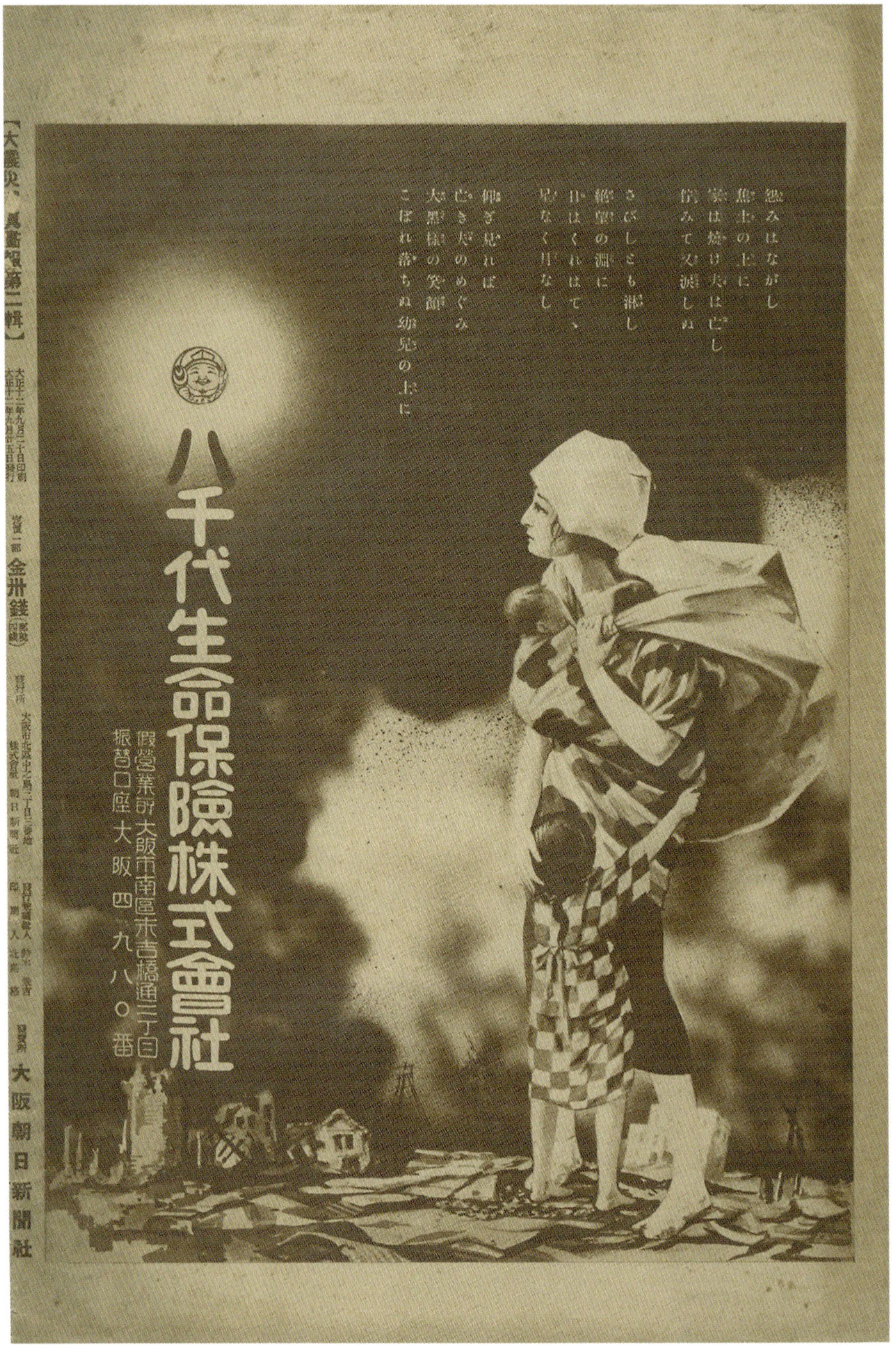

4.23 Yachiyō Life Insurance Company (Yachiyō Seimei Hoken Kabushiki Kaisha), advertisement, back cover. Ōsaka Asahi Shinbunsha. *Daishinsai shashin gahō* (Photographic Pictorial of the Great Earthquake), vol. 2 (Osaka: Ōsaka Asahi Shinbunsha, 25 September 1923).

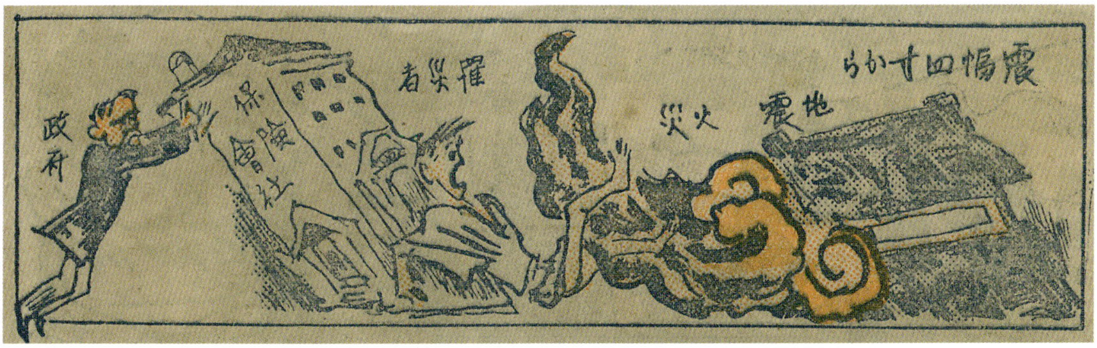

4.24 Ogawa Jihei, "From an [earthquake] amplitude of 4 *sun*" (*Shinpuku yonsun kara*), cartoon. *Jiji manga* 132, 7 October 1923, 2. Collection of the author. By permission of Sangyō Keizai Shinbunsha.

these emotions through compensatory measures and commodified safety nets, aiming to reduce the impact of devastating experiences and their philosophical implications to an equation of material loss.

Initial estimates placed the value of property insured by quake victims at 2.2 billion yen, which, according to economic historians, far exceeded available insurance assets and would have instantly bankrupted the entire industry. Insurers were quick to point out, however, that the exemption clause in section 17 of the standard insurance policy used by all fire and marine companies specified that no payments would be made for fire damages due to earthquakes and volcanic eruptions. Nevertheless, pressure from policyholders and the government made this position socially and politically untenable, and insurers were eventually obliged to pay 10 percent of the policy amounts as "gift money" to express their sympathy to the victims and to facilitate reconstruction. In return, they received government loans to help cover these payments.[35]

The insurance companies determined this arrangement to be the only way to maintain confidence among their clientele. Satirist Ogawa Jihei (1887–1925) parodied this precarious support structure in *Jiji manga* (fig. 4.24) under the caption "From an [earthquake] amplitude of 4 *sun*" (*shinpuku yonsun kara*) showing (from right to left) the earthquake causing the fire and knocking over the victims, who topple the insurance companies, which are propped up by the government, symbolized by Prime Minister Yamamoto.[36] Some contemporary newspapers criticized the payment scheme for insured losses as an aid for the rich rather than the poor, something that insurance executives also recognized but confided only to their private diaries. Ultimately, fire insurers paid the most to the most powerful pressure groups, business owners, and the poor benefited little.

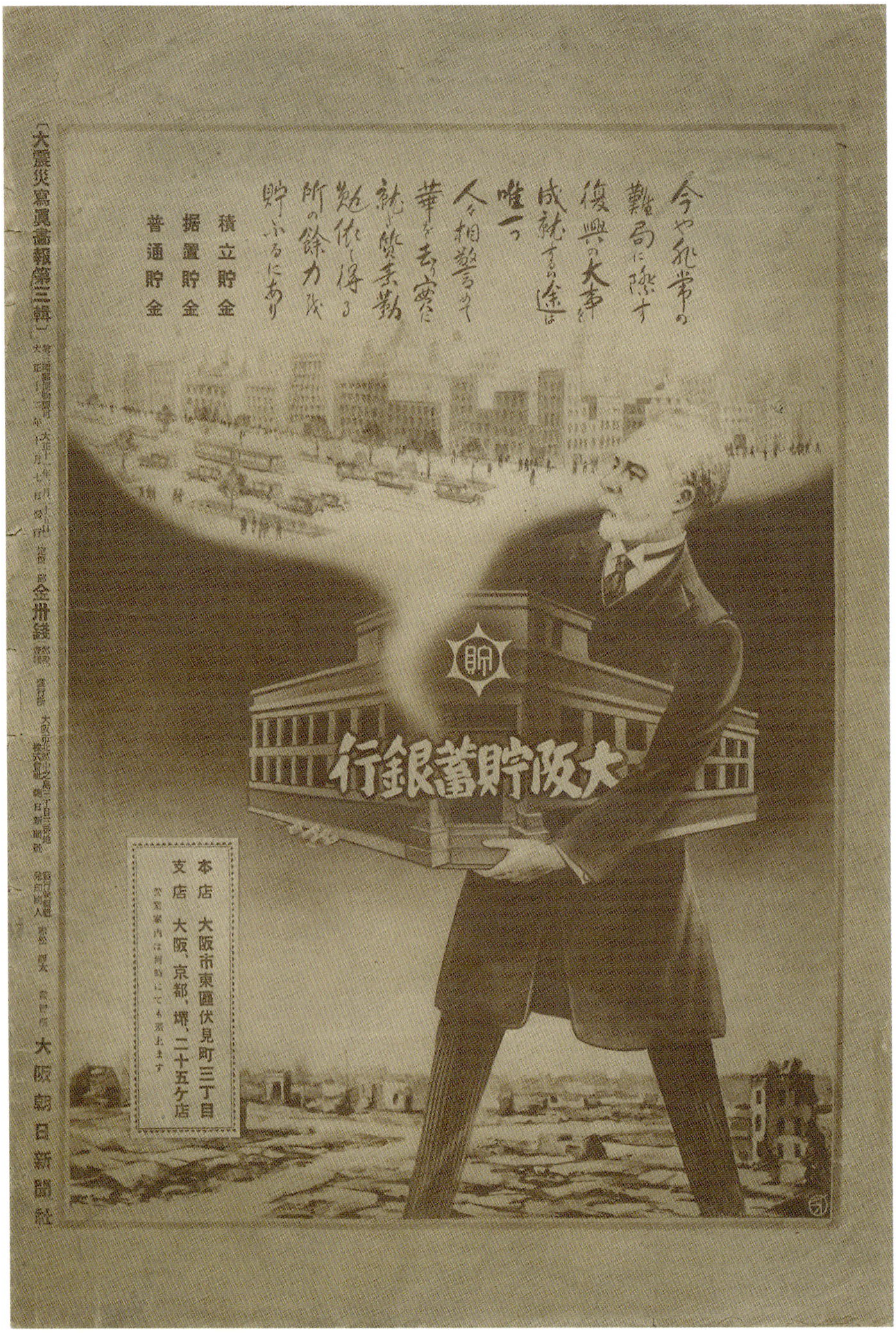

4.25 Osaka Savings Bank (Ōsaka Chōkin Ginkō), advertisement, back cover. Ōsaka Asahi Shinbunsha, *Daishinsai shashin gahō* (Photographic Pictorial of the Great Earthquake), vol. 3 (Osaka: Ōsaka Asahi Shinbunsha, 7 October 1923).

Appeals to permanence were also seen in other commercial sectors, where steel-reinforced concrete (which was determined to be one of the most earthquake- and fire-resistant forms of construction) was heavily promoted.[37] Similarly, the AllSteel Company ran advertisements for its metal furniture pointing out that it would not burn. And the Takeuchi Safe Company's advertisement promoted its fire-rated safes' ability to secure valuables.[38] At the same time, banks deployed images of ruins—now the proverbial "before" scene—to instill a passion for savings to capitalize the reconstruction. In one advertisement (fig. 4.25), a distinguished banker in suit and tie hoists the impressive, sturdy-looking Western-style edifice of the Osaka Savings Bank (Ōsaka Chōkin Ginkō) as a metaphoric weapon in combating the effects of the disaster. The figure stands towering over the desolate ruined landscape of Tokyo, and a cloud emerges from the bank's entrance to create a bubble containing the image of a gleaming new modern metropolis with tree-lined streets, buses, and automobiles. The text exhorts viewers to be frugal and save their money ("strength") so that these economic resources can be available for investment in the reconstruction of the capital: "In this time of serious crisis, the only road for us to take is to complete the great task of rebuilding. People must caution one another: the flower has been left behind, but the fruit will (yet) come if we can rely on frugality and diligence, and this lies in conserving our remaining strength."[39]

Ruins were anathema to the reconstruction project because they needed to be obliterated before recovery could begin. Many were cleared within five months of the quake. If successful, reconstruction would wipe away the conflicted memories embodied in ruins and replace them with a coherent commemorative narrative of the tragedy. But reconstruction was not an instantaneous process, and the city would not return to its former stature for nearly a decade after the quake. In this interim "reconstruction period," new tropes would emerge to replace the rapidly disappearing ruins, and the aesthetics of the sublime would give way to new symbols of social solidarity, conflict, and transformation.

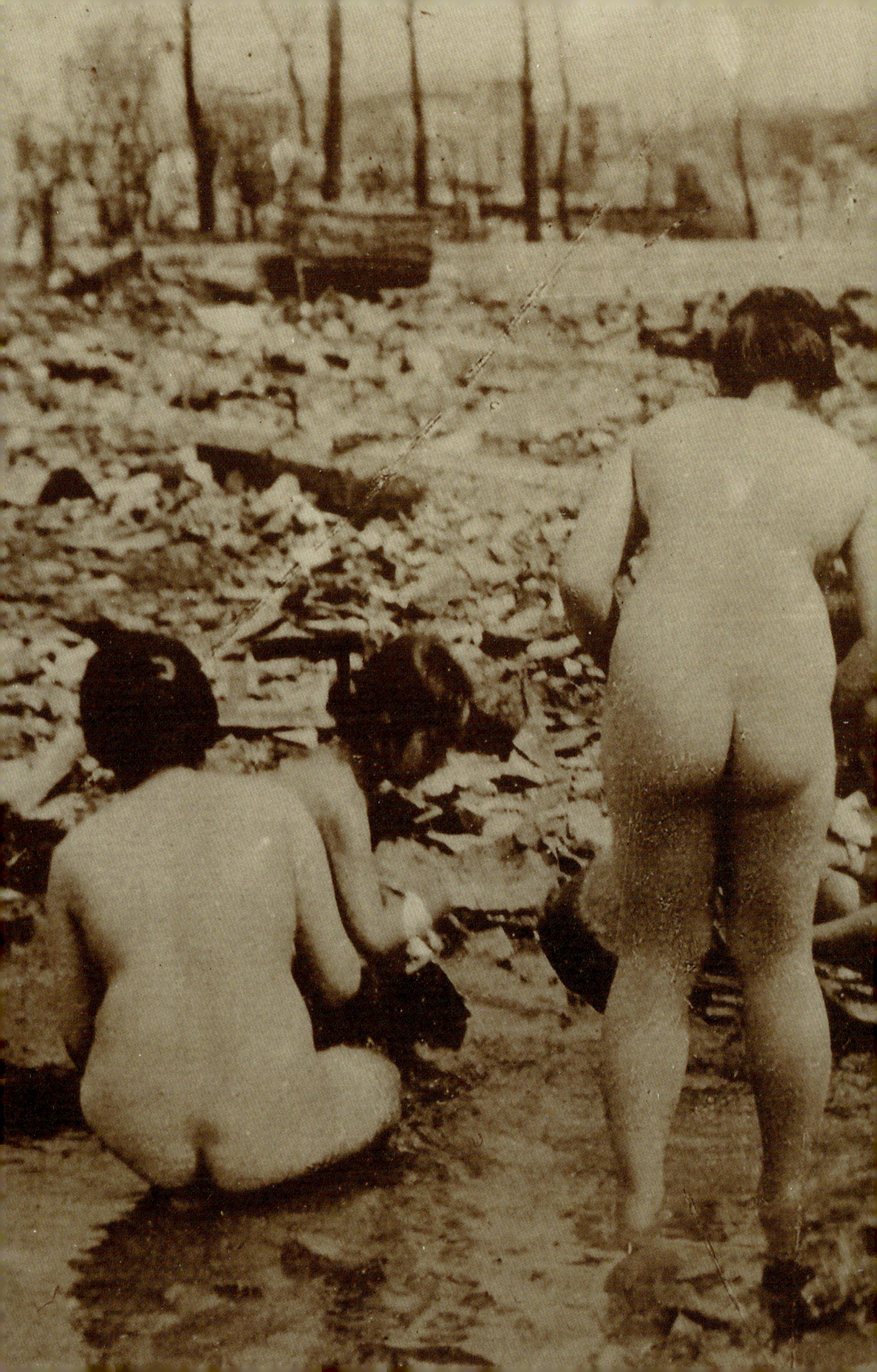

5

RECLAIMING DISASTER

Altruism and Corrosion

The dominant public vision of the Great Kantō Earthquake was an affirmative image of collective suffering and sacrifice that downplayed divisiveness. Unity was the order of the day, and tenacity in the face of adversity would prove Japan's moral backbone. The Kōdansha editors, in the preface to *Taishō Daishinsai daikasai* (The Great Taishō Earthquake and Conflagration), voiced the common message emphasizing the silver lining in the cloud of tragedy: "Painful suffering makes people great. Such an enormous disaster gives us profound and innumerable lessons. If we can properly report this experience to future generations all over the world, and if we can make them reflect upon our experience, the disaster will turn into wealth and happiness. We believe that this subtle providence is also with us."[1]

There were considerable stakes in claiming the disaster for social and political purposes. Hence, heroic images that were part of this narrative of resilience, and its aim of galvanizing the nation for the task of reconstruction, did not go unchallenged. They were countered by a range of social critiques, both subtle and explicit, some in the form of general moral pronouncements about the earthquake as divine punishment and others leveling more pointed criticisms of public and private conduct, economic inequities, or the official mishandling of the reconstruction process. Perceived as a potentially transformative moment in Japan's modern history, the period immediately after the quake echoed with the popular phrase "under these circumstances!" (*kono sai [da kara]!*), which served as the preface to a wide array of reform

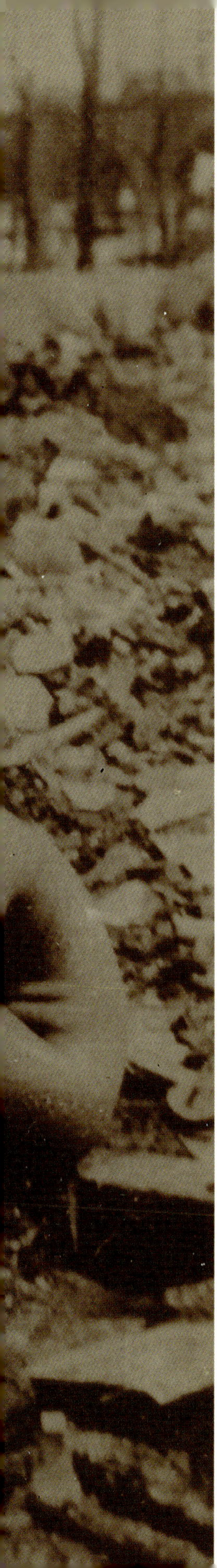

proposals.[2] This period even saw a new generation of artists inspired by global movements in modernism take to the streets of the ruined cityscape to facilitate social change. As visible representatives of modernity's radical cultural upheaval—seen by many as commensurate with the ruptures caused by the earthquake—these artists were alternately commended and vilified for their new social visions.

When people confront a traumatic past, "the desire to bring forth 'positive' forms of identification (heroic memory) can occur at the expense of working through a painful and difficult past," writes Timothy Brown.[3] Three interrelated emblems of the Kantō earthquake are poignant lenses for viewing the tension between the heroic and painful past of the disaster: the refugee, the catfish, and the barracks. They reveal attempts to portray postquake Japan as what the sociologist Allen H. Barton describes as a kind of "altruistic community," in which communal suffering generates "democracies of distress" or "cities of comrades," prompting people to perform extraordinary acts of mutual kindness. At the same time, these images divulge a sinister underside of the experience, in which fault lines in the community are forced open and preexisting social tensions are exacerbated, forming the kind of "corrosive community" that sociologist Kai Erikson has documented.[4] Disaster produces a corrosive community by precipitating intracommunity conflict that reveals or deepens social divisions. It also often blurs the lines between victims and perpetrators.

Refugees: Heroes and Vigilantes

ALTRUISTIC SOLIDARITY

Kanokogi Takeshiro's monumental oil painting *September 1, 1923* (*Taishō 12-nen 9-gatsu tsuitachi,* 1923) (fig. 5.1) highlights the galvanizing dual tropes of catastrophic loss and heroic perseverance through the figure of the refugee. Although depicting a contemporary scene of the quake, Kanokogi clearly drew on his academic training at the Académie Julian in Paris and his familiarity with the important Western Salon genre of history painting to allegorize and heroicize his representation of fleeing refugees, underscoring the moral and didactic virtues of perseverance in the face of extreme hardship. And like many history paintings in the Western canon, the image conveyed the universality of human suffering as the artist sought to elicit public sympathy.

On a more personal level, *nihonga* master Ikeda Yōson (1895–1988) focused on the individual experiences of loss and desperation in *After the Disaster* (*Saika no ato*) (fig. 5.2), his 1924 painting in ink and mineral pigments on a paper folding screen.[5] Yōson's affecting image of survival exhibits the artist's diverse influences, which ranged from Edvard Munch and Francisco Goya to *ukiyo-e* printmakers and *yamato-e*–style classical Japanese

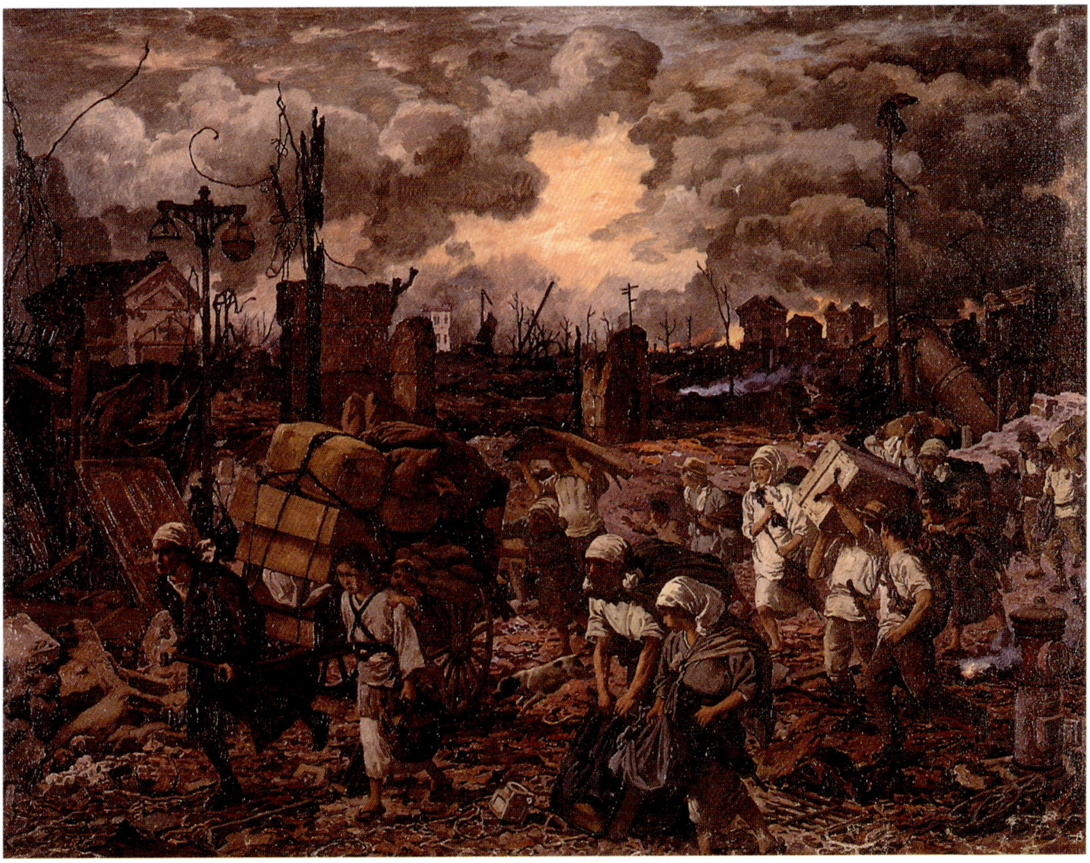

5.1 Kanokogi Takeshiro, *September 1, 1923* (*Taishō 12-nen 9-gatsu tsuitachi*), oil on canvas, c. 1923. Collection of Tōkyō-to Gendai Bijutsukan.

painters. Unlike Kanokogi's activated figures that purposefully make their way through the landscape of ruin, Yōson's are very still, their shoulders slumped dejectedly and their direct outward gaze seeming to beseech the viewer. The static composition of Yōson's scene, its peaceful stillness, conveys a kind of melancholy that seems neither heroic nor triumphant. It also focuses on the basic family unit—a mother, a father, and two children. Kanokogi's refugee hordes represent the masses of Japanese society and humanity in general. Yōson provides no road to recovery; the figures are not on a path. They simply stand in silent disbelief under the dim light of the crescent moon.

Despite the shattering social effects of disaster, earthquake imagery averred that traumatic collective experiences also provided chances to reaffirm community and leadership, particularly across class lines. In a print depicting the bronze statue in Ueno Park of renowned samurai military leader and patriot Saigō Takamori (1828–77) walk-

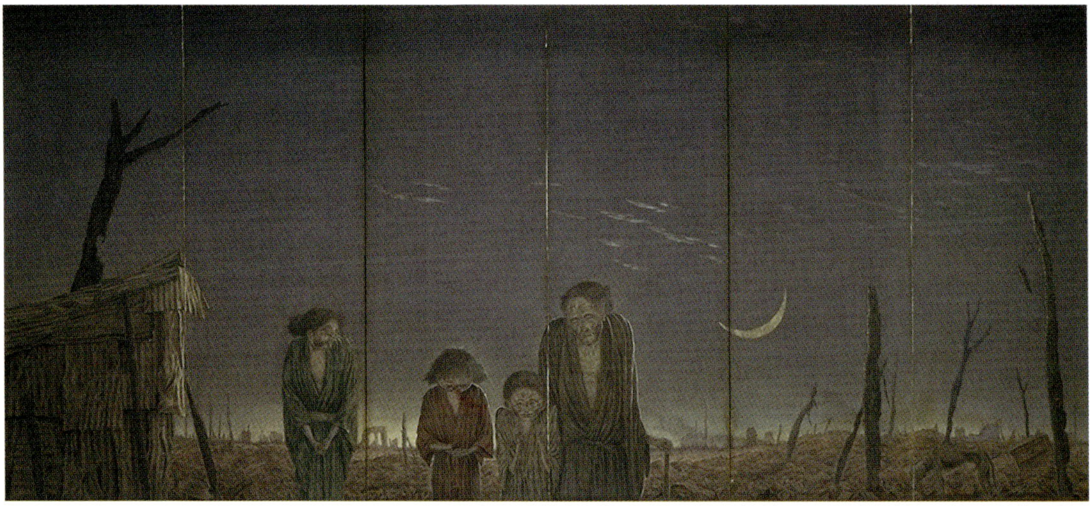

5.2 Ikeda Yōson, *After the Disaster* (*Saika no ato*), ink and mineral pigments on silk, 1924. Collection of Kurashiki Shiritsu Bijutsukan.

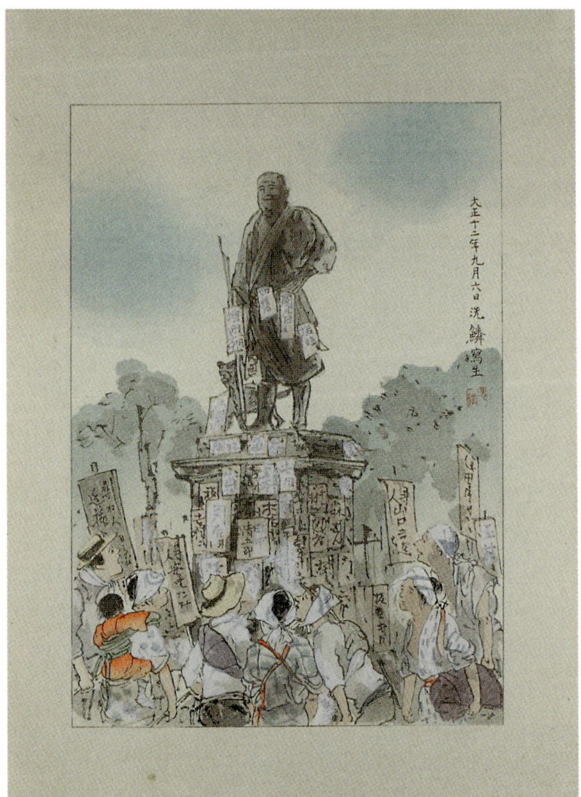

5.3 Kiriya Senrin, *Bronze Statue of Saigō* (*Saigō no dōzō*), woodblock print. Nishizawa Tekiho et al., *Taishō shinkasai mokuhangashū zen sanjū rokkei* (Collection of Woodblock Prints of the Taishō Earthquake—Thirty-Six Scenes) (Tokyo: Gahōsha, 1924). Collection of Machida Shiritsu Kokusai Hanga Bijutsukan.

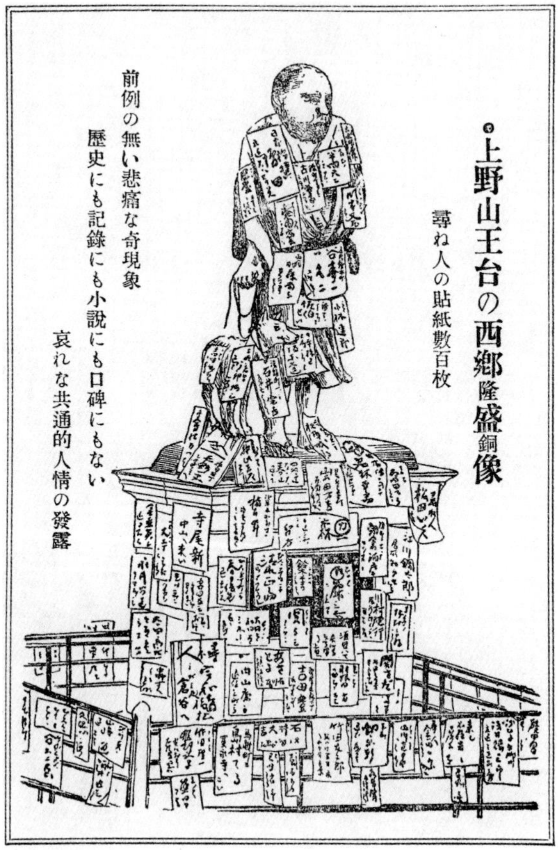

5.4 Miyatake Gaikotsu, *Saigō Takamori Bronze Sculpture at Ueno Sannōdai [Square].* (*Ueno Sannōdai no Saigō Takamori dozō*). Miyatake Gaikotsu, *Shinsai gahō* (Earthquake Disaster Pictorial) (Tokyo: Hankyōdō, 25 September 1923), 1:5. Reproduced in *Miyatake Gaikotsu chosakushū* (Collected Works of Miyatake Gaikotsu), ed. Tanizawa Eiichi and Yoshino Takao (Tokyo: Kawade Shobō Shinsha, 1986), 3:469. Courtesy of Kawade Shobō Shinsha.

ing his dog (fig. 5.3), Kiriya Senrin narrates a valiant saga of makeshift communication networks constructed in the wake of the disaster when most power and telephone lines were damaged.[6] People improvised a message center on the statue, posting the names of their missing loved ones (*tazunebito no harigami*) across Saigō's sturdy body and the plinth below as they tried desperately to reconstitute family and community. The fluttering sheets of paper represent the horrifying unknown, and they communicate unutterable loss. This image was immortalized across the media spectrum.[7] A beloved symbol of national patriotism and traditional military strength from the Meiji restoration period, Saigō was instrumental in Japan's return to imperial power. What better body upon which to begin reconstituting the Japanese citizenry than that of a patriotic martyr to the nation? However, variations on this theme express subtly different messages. One illustration in the series *Shinsai gahō* (Earthquake Disaster Pictorial) by journalist and social critic Miyatake Gaikotsu (1867–1955) shows Saigō's figure covered up to his neck with notices as if he is wearing a cloak of penance (fig. 5.4). His hunched, forlorn body appears about to be swallowed up by the collective grief. The tenor of melancholy and solidarity is echoed in the accompanying text: "Saigō Takamori Bronze Sculpture at Ueno Sannōdai [Square]: Hundreds of pasted missing persons notices; an unusual phenomenon of unprecedented sorrow. Not in history, in chronicles, in novels, or in legend has there been such an expression of compassionate common humanity."[8] Not one to celebrate national heroes, Miyatake did not express his empathy for the nation-state but for the suffering people.

The *New York Times* headline on 4 September, "Princes and Statesmen Lie Dead with the Poor," crystallized a common view at the time: nature pays no heed to class. The quake's purported eradication (at least temporarily) of class divisions was heralded in a caption on a stereoscopic photograph that read, "Calamity obliterates class—The Empress of Japan among her earthquake stricken people." The widely circulated image (fig. 5.5) showed the empress visiting injured and orphaned children in the capital.[9] This scene symbolically suggested that the state was serving as surrogate parents for the orphans and that the nation was one large compassionate family. Images of the empress visiting her injured subjects in makeshift shelters harkened back to previous imperial acts of consolation that garnered the spotlight during the Nōbi earthquake in 1891 and continued through both the Sino- and Russo-Japanese wars, engendering widespread appreciation for benevolent imperial rule.[10] Together with the innumerable images of the prince regent surveying the damaged cityscape and consulting with government officials such as Prime Minister Yamamoto and Home Minister Gotō, the empress's presence among refugees reinforced the image of state compassion.

The first Imperial Edict on Reconstruction (Teito fukkō ni kansuru shōsho), issued on 13 September 1923, less than two weeks after the quake, read:

> It is Our great fortune that, through the divine help of Our forefathers and the loyal cooperation of Our people, we have been able to maintain Our national repose and security despite the occurrence of the world war of unexampled dimensions. . . . We confidently trust that Our loyal people, who always courageously proffer their services and make sacrifices for the public welfare, ardently desire, with Us, to enjoy the best facilities and security obtainable. . . . At this moment of catastrophe unparalleled in our history, Our stricken heart goes out in abundant compassion to the people.[11]

The didactic message was clear. The nation would unite in tragedy and its people would together face the task of reconstruction. The population would act selflessly and support the divinely ordained, benevolent leadership of the imperial household. A marked surge in samaritanism reported after the quake visibly manifested the post-traumatic

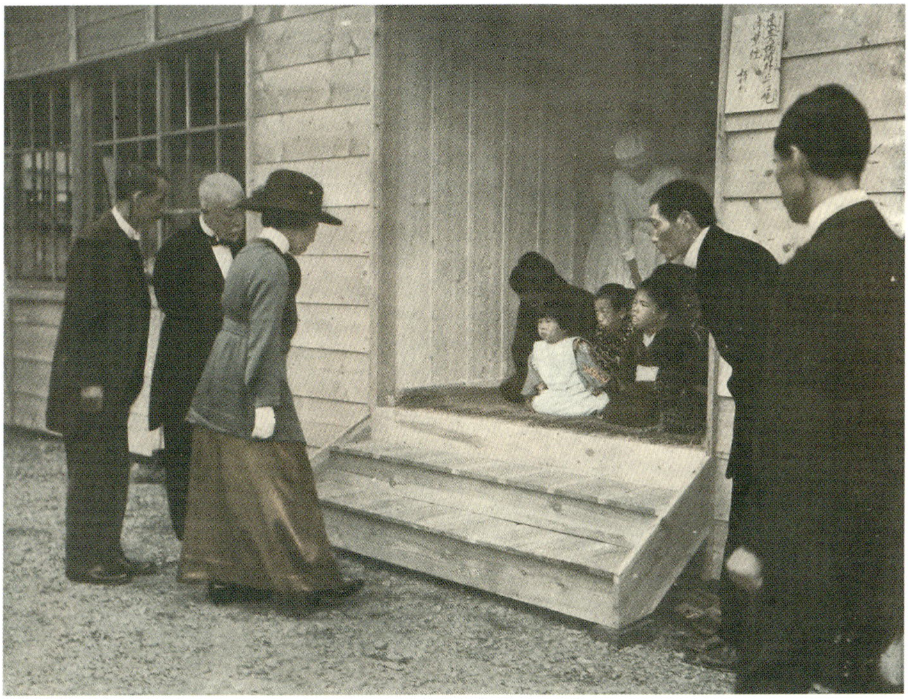

5.5 *Her Majesty the Empress Granting a Comfort Visit to Sympathetic Orphans* (*Awarenaru koji o mimai hasetamau Kōgō Heika*), photograph, 1923. *Rekishi shashin* 2, no. 204 (November 1923).

altruistic community, in which people feel compelled to offer help to their compatriots who are suffering.[12] As the Tokyo municipal government narrative later recounted, "Indeed, one touch of Nature made the whole nation kin; for sympathy and brotherly love were everywhere to the fore, whether in the country or in town, among private individuals or in public bodies, who were, therefore, all absorbed in the common and universal aim of relieving the sufferers."[13] This altruism, often highly dramatized and even sometimes fictional, was promoted in the many literary compilations of "heroic stories and excellent tales" (*bidan kawa*) and "expressions of human warmth" (*ninjōbi no hatsurō*) that recounted individual acts of bravery and kindness.[14]

VIGILANTE VIOLENCE

Government and quasi-official publications featuring altruistic stories of Japanese heroism, humanitarianism, and solidarity tried to drown out narratives or visualizations that emphasized the chaos and savagery that took place in the wake of the devastation.

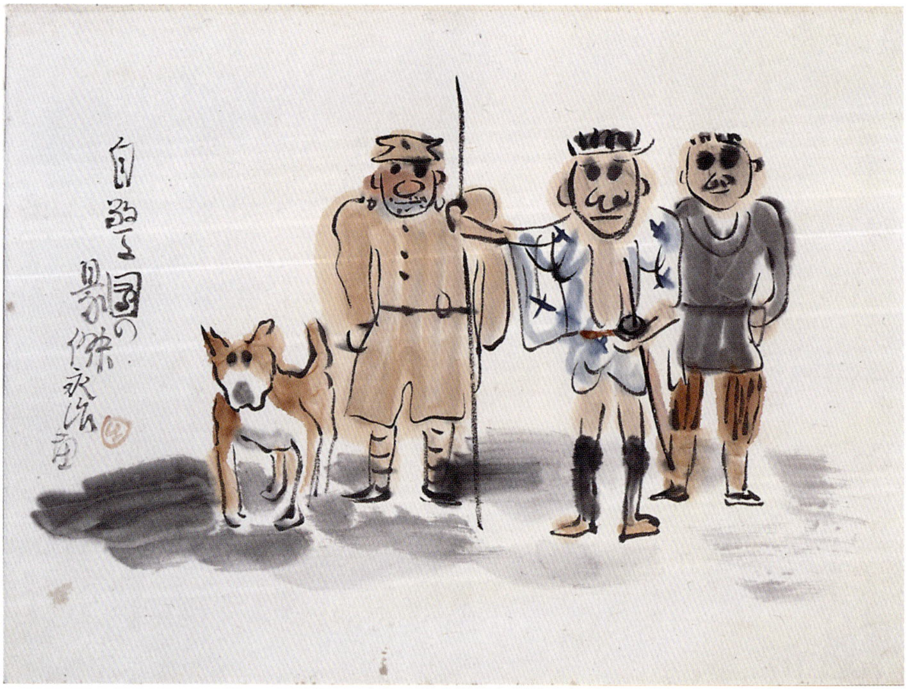

5.6 Ikeda Eiji, *Vigilante Heroes* (*Jikeidan no gōketsu*). *Paintings of the Great Kantō Earthquake* (*Kantō daijishinga*), ink and color on paper, c. 1923. Collection of Tōkyō-to Edo Tōkyō Hakubutsukan.

Not only was the news of deep-seated antagonism toward Japan's colonial subjects and the widespread vigilante rampages against foreign nationals residing in Japan initially suppressed, but food riots, looting, and class conflict were also not widely represented: effective reconstruction required that the Japanese masses be seen as united and good.[15] As we have seen, atrocities were in a sense hidden in plain sight, with the bodies of non-Japanese victims of vigilante violence being mingled with other casualties. Furthermore, the anonymous imagery and the labeling of mass death by neighborhood served to equalize the dead, even the murdered, yet it also symbolically spotlighted the areas that suffered the greatest devastation, subtly pointing to the preexisting social inequities that contributed to their decimation. Still, official pronouncements downplayed these inherent biases, instead emphasizing the arbitrary and class-blind impact of the earthquake and shifting blame for the disproportionate suffering due to human discrimination onto the uncontrollable whim of nature.

Sinister images of local vigilante groups, self-appointed security posses that dominated the anarchic days after the quake, exposed another, more corrosive side of the

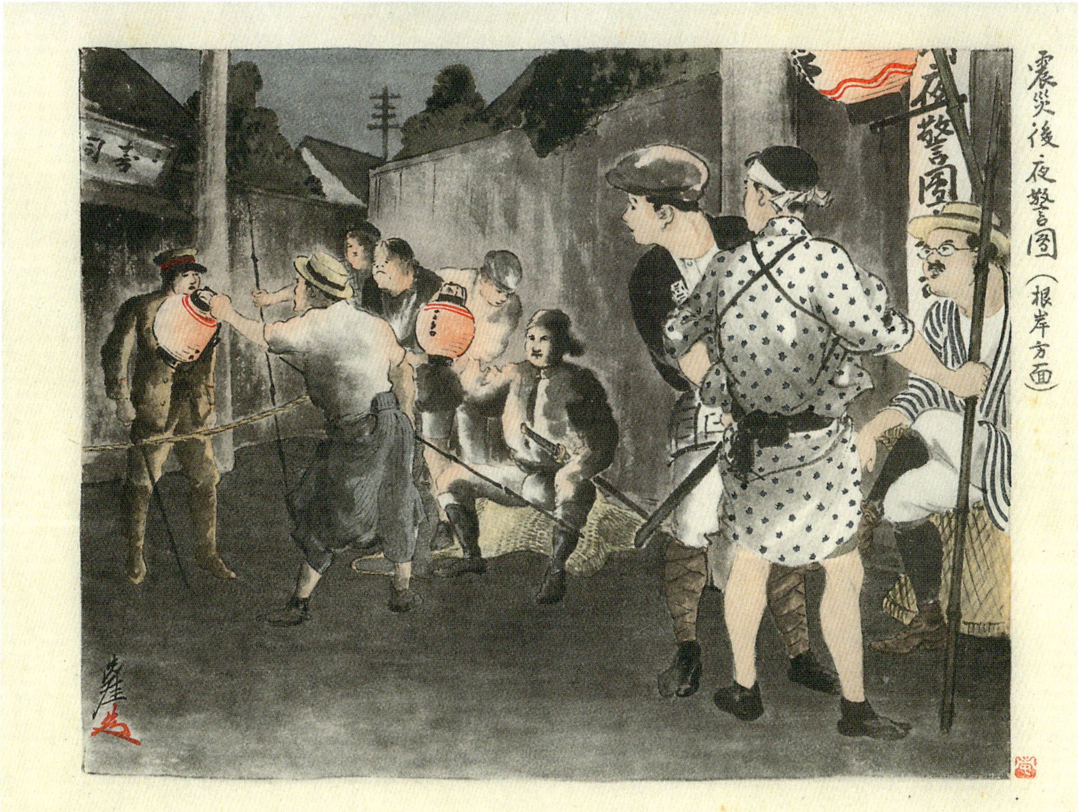

5.7 Igawa Sengai, *Nightwatch Vigilante Groups after the Earthquake (near Negishi)* (*Shinsaigo jikeidan [Negishi hōmen]*), woodblock print. *Taishō shinsai gashū* (Collected Prints of the Taishō Earthquake) (Tokyo: Zue Kenkyūkai, 1926). Collection of Dr. Gisela Rabitz and Dr. Albrecht Rabitz.

refugee community: savagery and lawlessness. *Tokyo Puck* cartoonist Ikeda Eiji (1889–1950) painted figures armed with spears and standing at attention, challenging the viewer as they might have challenged people walking along the roads (fig. 5.6).[16] His fellow series contributor Nakazawa Hiromitsu (1874–1964) created four paintings of security squads harassing Koreans and other fleeing refugees under the title *Rumors and False Reports* (*Ryūgen higo*), using the contemporary term for the malevolent gossip that spurred the vigilante violence and other heinous acts.[17] Many commentators noted the vigilantes' menacing demeanor, aware that the gangs often killed first and asked questions later.

Igawa Sengai's *Nightwatch Vigilante Groups after the Earthquake (near Negishi)* (*Shinsaigo jikeidan [Negishi hōmen]*) (fig. 5.7) depicts an atmospheric night scene with

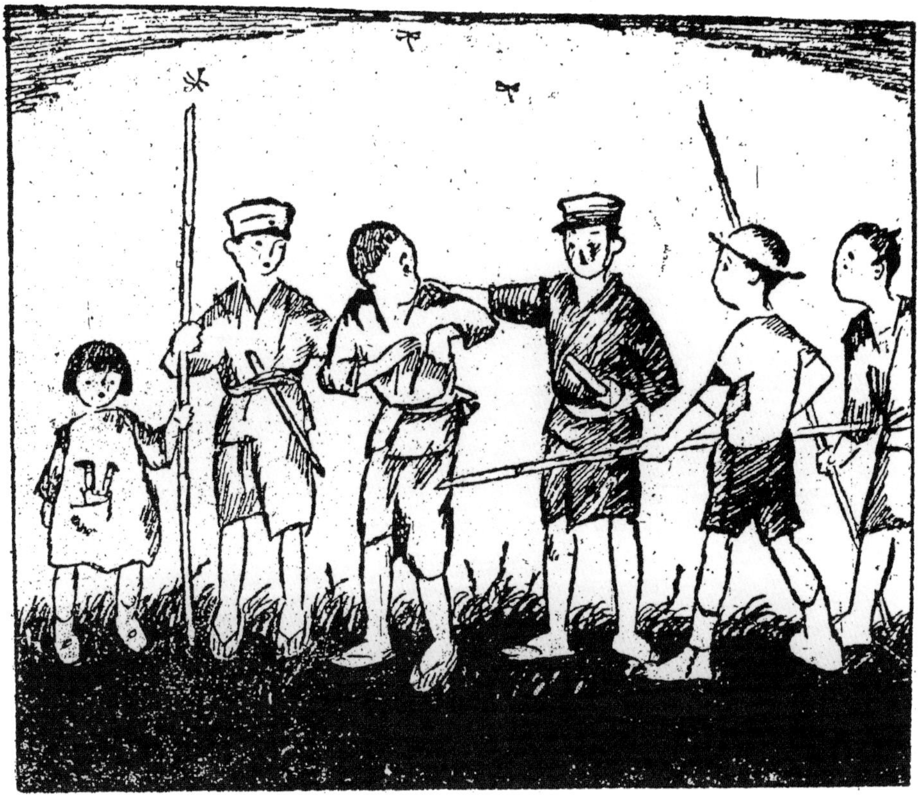

5.8 Takehisa Yumeji, *Vigilante Game* (*Jikeidan asobi*), drawing, from *True Pictures of the Tokyo Calamity* (*Tokyo sainan gashin*), originally in *Miyako shinbun,* 19 September 1923. Reproduced in Takehisa Yumeji, Kawamura Karyō, and Yamamura Kōka, *Yumeji to Karyō, Kōka no Kantō Daishinsai rupo* (Yumeji, Karyō, and Kōka's Reportage of the Great Kantō Earthquake) (Tokyo: Kress Shuppan, 2003), 15. Courtesy of Kress Shuppan.

neighborhood men gathered at a checkpoint. The men all turn suspiciously as their lanterns reveal someone approaching. The tension is palpable, the mood sinister.[18] Later literary and news accounts of the quake expressed shock over these brutal actions and indignation about the subsequent cover-up, but censorship and fear of reprisals precluded widespread critical discussion of the events or direct criticism of the government's lack of response. Even after the government began arresting and publicly prosecuting selected vigilantes for their crimes a month or two after the quake, many rallied to the men's defense, seeing them as patriots who acted bravely to protect the public. Extensive testimony demonstrated that many vigilantes had acted on rumors from government officials, who secretly circulated warnings about Korean sedition while publicly advising restraint. Ultimately, vigilante acts and state actions for "public security" could not be so easily disentangled, and the massacre was excused in light of the extreme circumstances. As Jin-hee Lee has shown, the postquake trials and discourse about social participation in violence vividly revealed the capacity of both the state and its people to inflict violence on colonial subjects within the empire with impunity, since these colonial subjects were not seen as part of the "Japanese public" and stood outside of the ethnically bounded notion of the public good.[19]

Despite the initially suppressed and ultimately conflicted public opinion about vigilante actions, early visual reminders provided a window on the Janus face of the disaster community. Takehisa Yumeji's seemingly innocent sketch of children playing the Vigilante Game (*Jikeidan asobi*) (fig. 5.8) that ran in the *Miyako shinbun* on 19 September shows six adolescents, five boys and a younger girl, engaged in a game that simulated the brutal policing responsible for the massacres. All the children have weapons; even the small girl to the far left grips a bamboo stick along with her older male companion. One boy in the center, highlighted through his darkened clothing, reaches out and grabs the central figure, the son of a local tofu-store proprietor who is cowering in fear, and screams, "Man-chan, somehow your face doesn't look Japanese!" The others stand with weapons poised. "Children in the suburbs have begun to play the vigilante game," comments Yumeji.

> "Let's make Man-chan the enemy!" "No, I don't want to!" cries back Man-chan, flinching, "'cause that means you're gonna stab me with a bamboo spear, right?!" "Nothing like that's gonna happen. We're just pretending!" But even though he said this, Man-chan would not agree, so the neighborhood bully came out. "Man-chan! If you don't play the enemy, I'm going to beat you up!" he said threateningly, and Man-chan was obliged to be the enemy; he was chased around and beat on until he cried. Children like war, but nowadays you see adults get full of themselves and go around imitating policemen and soldiers. They brandish sticks and bother passersby like Man-chan. Excuse me, let's try using a hackneyed advertising slogan here like, "Hey kids! Let's stop carrying around sticks and playing at being a vigilante."[20]

Published volumes of children's drawings commemorating the quake contain vivid representations of vigilante groups and uniformed police subduing "offenders" with swords and sticks. Like Yumeji's drawing, one child's work renders one figure in darker hues, but in this case, the presumed offender is spotlighted rather than the vigilante (fig. 5.9). Unlike poor Man-chan, his face is distorted in a fearsome scowl and he appears to be holding a black object, perhaps one of the bombs that Koreans were rumored to be detonating throughout the city. The level of apprehension that children had absorbed from the rampant fear in the general populace radiates from the image,

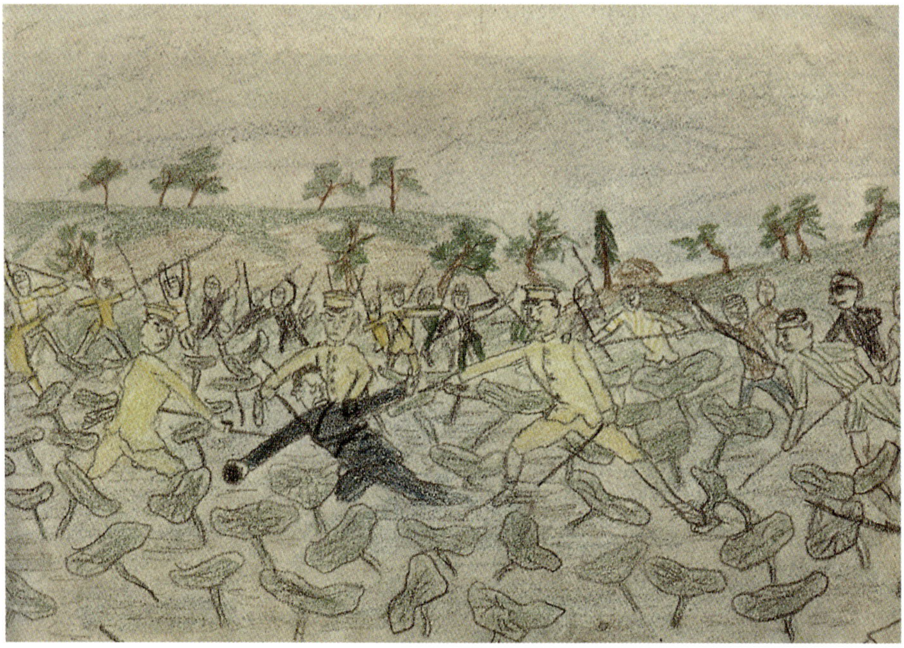

5.9 Yamazaki Iwao (primary school student), a child's response to rumors about Koreans after the earthquake, pencil and crayon on paper, c. 1923. Reconstruction Memorial Museum. Collection of Tōkyō-to Irei Kyōkai. Photograph courtesy of Kanagawa Daigaku Nihon Jōmin Bunka Kenkyūjo Himoji Shiryō Kenkyū Sentā.

and the tenor of the drawing shows that children were not disabused of the misinformation about Korean culpability in the quake. While sometimes feeling sympathy toward the Korean victims after witnessing the horrific vigilante violence against them, children still harbored intense fear about Koreans generated by the rumors.[21] This ambivalence was publicly expressed in the exhibition of these drawings in February 1924 and memorialized in the permanent exhibit of children's drawings on a wall in the earthquake memorial museum.[22]

Private sketches have revealed deep concerns among the artistic community about the savagery the earthquake unleashed in the Japanese people. Yanase Masamu (1900–45), a young painter and graphic artist who collaborated with several avant-garde art groups and had just joined the leftist political movement at the time of the disaster, was himself detained after the quake along with many leftist intellectuals labeled as potential subversives. He was questioned and tortured but eventually released, narrowly evading the terrible fate of others in the movement. He immediately left the city and returned to his hometown in Kyūshū.

Among Yanase's many sketches of regenerating refugee life in the aftermath of the earthquake are several disturbing images of police officers, one viciously bayoneting a dog. While martial law was reassuring to many, it was menacing to others, and sanctioned officials often behaved no better than the vigilantes, as the Ōsugi Sakae and Kameido assassinations and Korean massacre demonstrated. Yanase later wrote in his autobiography that his experiences during the earthquake transformed his artistic mission to working for the organized proletarian liberation movement.[23] Yanase's colleague and fellow founder of the artists' group Mavo, Murayama Tomoyoshi (1901–77), also an ardent leftist, was warned by friends to take off his high-collared Russian-style *rubashka* shirt, considered fashionable in the Taishō period among left-leaning intellectuals, for fear that he might be murdered on the spot as a foreigner. Just as Koreans and Chinese were summarily executed if they failed simple pronunciation tests, others found that their appearance, clothing, or personal adornment could signal insider or outsider status and could make the difference between life and death.

Texts and images underscore the ambivalent identity of quake refugees as victims and victimizers. *Miyako shinbun* political cartoonist and illustrator Shirota Shūichi (1880–1958) drolly captured the cocky yet ragtag manner of the vigilantes in his band of misfits brandishing oversized weapons and ill-fitting galoshes as they strike a defensive pose on the streets of Tokyo in *The Result of the Great Earthquake Disaster* (*Daishinsai no sanbutsu*) (fig. 5.10). In the accompanying text, Shirota described the anxieties of families praying as they waited in the dark, fearing for their safety amid the circulation of frightening rumors. In his narrative, a megaphone calls to the adult men from the street to gather and guard the neighborhoods. The husband goes despite his wife's fearful protests. He is one of the self-appointed national strongmen. The men keep coming like the spectacle of the Night Parade of a Hundred Demons (Hyakki Yakō), when supernatural monsters and spirits (*yōkai*) are believed to take to the streets during summer evenings. And as the legend goes, anyone who comes across the demon procession will soon breathe his last breath.[24]

More explicit in impugning the moral character of the vigilantes, Miyatake Gaikotsu dedicated himself to debunking the "rumors and false reports" (*ryūgen higo*) that circulated after the quake and fueled vigilantism, including far-fetched tales suggesting that pregnant Korean women carried bombs in their bellies. In *The Dissoluteness and Violence of the Vigilantes* (*Jikeidanin no hōjū to bōkō*) (fig. 5.11), Miyatake divides the rectangular frame of the image on the diagonal, using the upper portion to show a group of men passed out among myriad empty bottles of alcohol, and the lower to show these thugs beating a woman with a stick and grabbing a pouch of salvaged valuables off the neck of another refugee.[25]

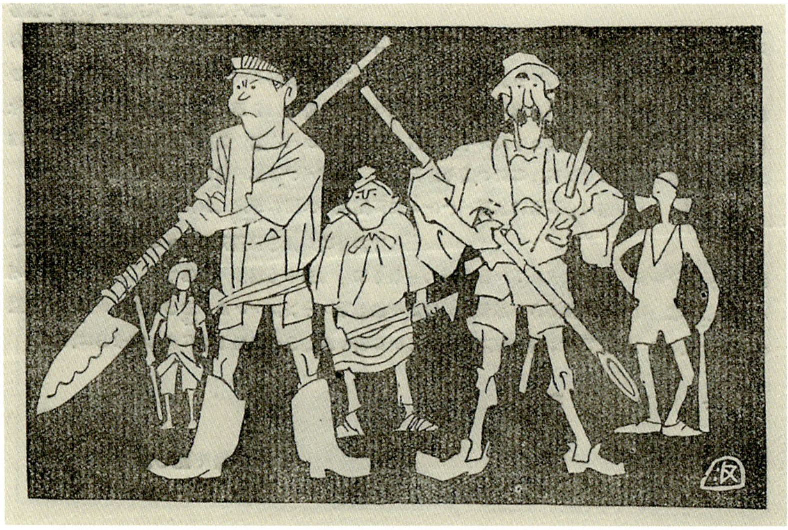

5.10 Shirota Shūichi, *The Result of the Great Earthquake Disaster* (*Daishinsai no sanbutsu*), refugees as victims and victimizers, drawing. Reimeisha Henshūbu, *Shinsai gafu: Gaka no me* (Earthquake Sketchbook: Eye of the Artist) (Tokyo, Reimeisha, 1923), 77. Private collection.

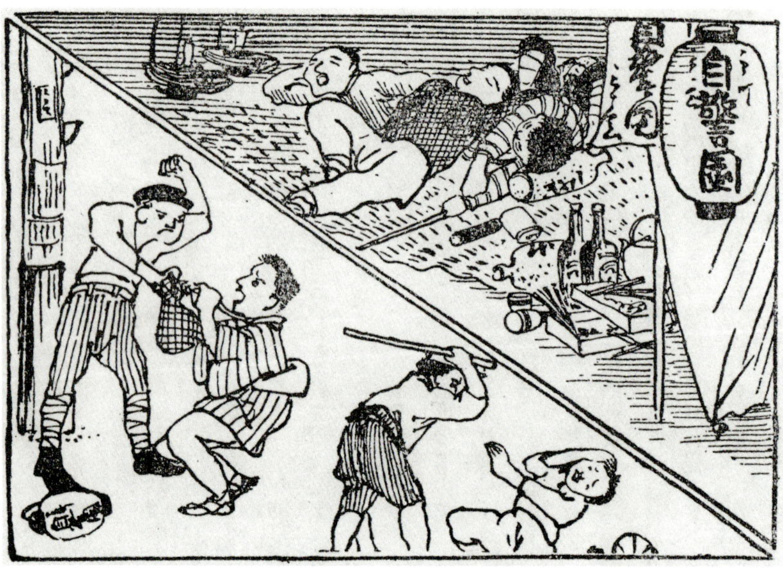

5.11 Miyatake Gaikotsu, *The Dissoluteness and Violence of the Vigilantes* (*Jikeidanin no hōjū to bōkō*). Miyatake Gaikotsu, *Shinsai gahō* (Earthquake Disaster Pictorial) (Tokyo: Hankyōdō, 10 October 1923), 2:37. Reproduced in Miyatake Gaikotsu, *Miyatake Gaikotsu chosakushū* (Collected Works of Miyatake Gaikotsu), ed. Tanizawa Eiichi and Yoshino Takao (Tokyo: Kawade Shobō Shinsha, 1986), 3:499. Courtesy of Kawade Shobō Shinsha.

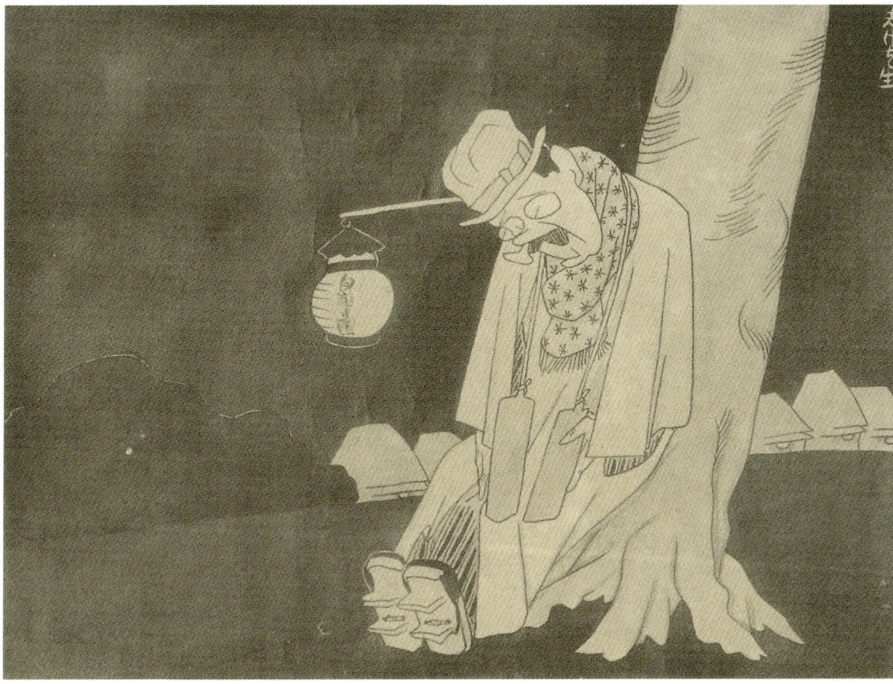

5.12 Miyao Shigeo, *Vigilante* (*Jikeidan*), drawing. Nihon Mangakai, *Daishinsai gashū* (Pictorial Collection of the Great Earthquake Disaster) (Tokyo: Kanao Bunendō, 1923).

Some artists sought to add levity to the overwhelming situation, lampooning the sinister vigilantes while questioning the supposed moral fiber and vigilance of these defenders of justice. One whimsical drawing by Miyao Shigeo (1902–82), a student of premier Taishō-era satirist Okamoto Ippei (1886–1948) and a cartoonist for *Tōkyō maiyū shinbun*, shows a hapless man leaning against a tree taking a nap, his various implements of surveillance dangling limply from his body (fig. 5.12). He is clearly snoozing on the job after a few too many nips at the bottle.[26]

For his part, pioneering cartoonist and political critic Kitazawa Rakuten (1876–1955), considered the father of modern *manga*, satirized the new, purportedly more conscientious police who were now doing things by the book rather than following the orders of their senior officers, which had led to the unimpeded murder of Ōsugi Sakae and his family by Lieutenant Amakasu Masahiko and junior officers (fig. 5.13).[27] The mocking text reads, "Since the Amakasu Incident: 'Hurry up. It's the order of a senior officer' [says the portly senior police officer on the right to his subordinate on the left]. 'Yes, please wait a moment.' Then secretly [he] looks the order up in the *Compendium of Laws* (*Roppō zenshō*)."[28]

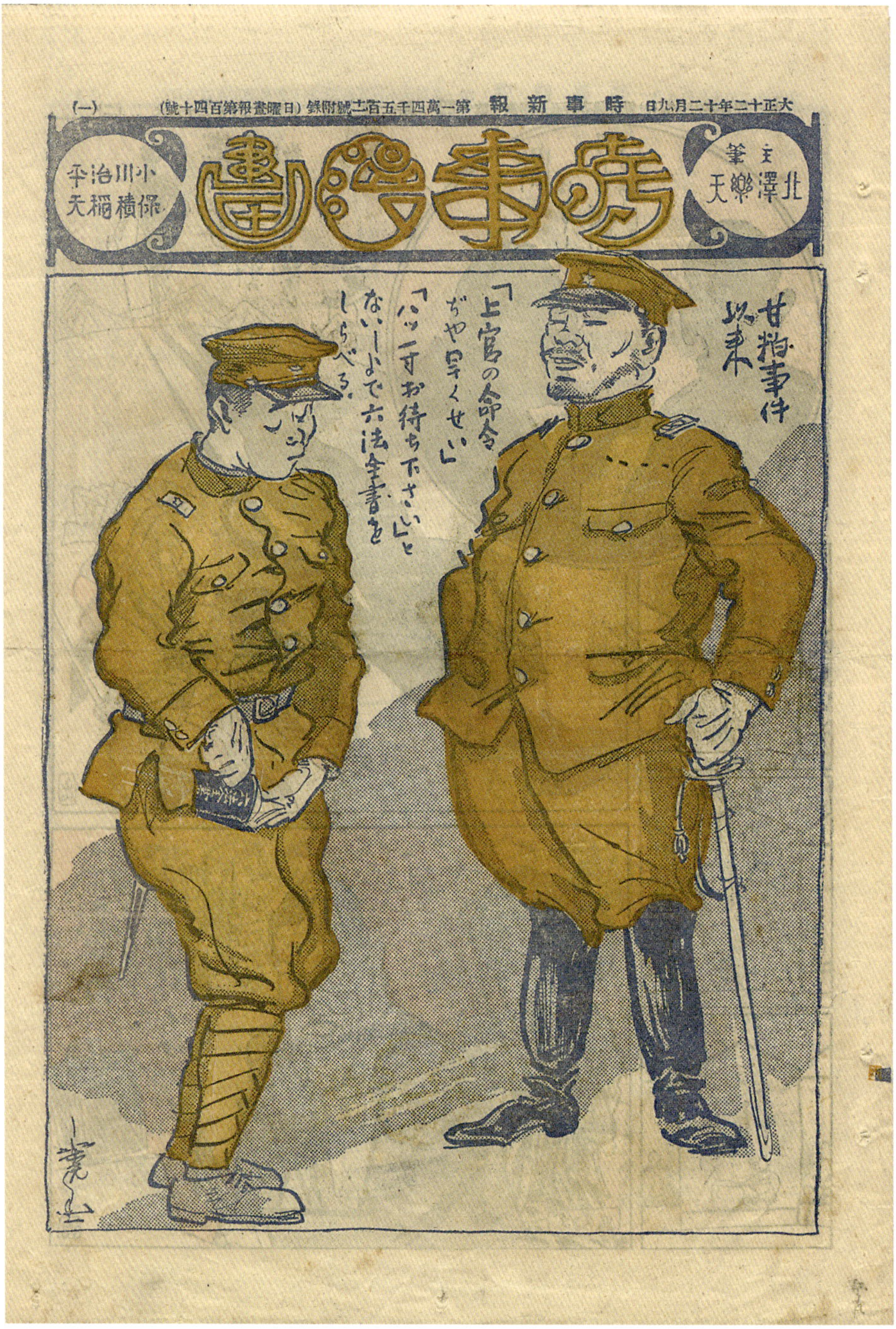

5.13 Kitazawa Rakuten, "Since the Amakasu Incident" (*Amakasu jiken irai*), cartoon, cover. *Jiji manga* 140, 9 December 1923. Collection of Ohio State University Billy Ireland Cartoon Library & Museum. By permission of Sangyō Keizai Shinbunsha.

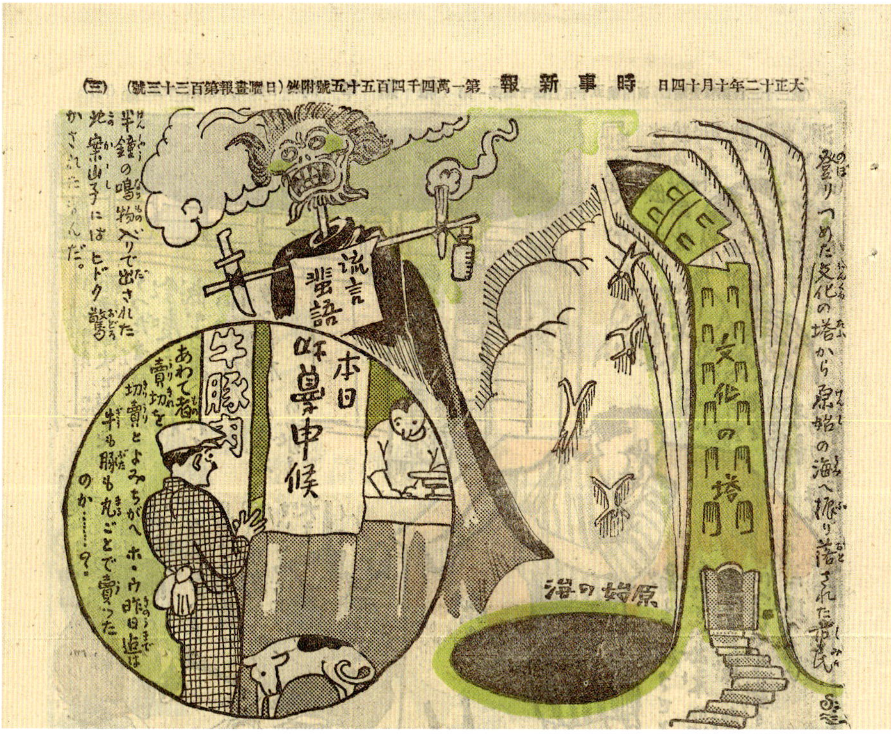

5.14 Arita Shigeru, "The Citizens Who Have Been Shaken Down from the High Cultured Tower into the Primitive Ocean" (*Noboritsumeta bunka no tō kara genshi no umi e furiotosareta shimin*), cartoon. *Jiji manga* 133, 14 October 1923, 3. Collection of Ohio State University Billy Ireland Cartoon Library & Museum. By permission of Sangyō Keizai Shinbunsha.

Nevertheless, all critical responses took a back seat in the popular press to the larger story of heroism and communal solidarity, which was bolstered by the implicit understanding that "those" people were in the end outsiders and subversives. The official earthquake report published by the Home Ministry in 1926 recounts:

> The nerves of the terror-stricken citizens, already frayed by the painful ordeal they had undergone, was further taxed by various wild rumours of a threatening character, heralding the possible visitation of another violent earthquake, or reporting the uprising of recalcitrant elements, the prevalence of arson, jail breaking, robbery, theft and so on. These wild rumours spread all over the city and throughout the suburbs, adding fuel to the general uneasiness, and a veritable reign of terror ensued. The police force, completely exhausted with their desperate efforts for the protection and relief of refugees lasting for days and nights, were not equal to the task of controlling the spread of such injurious groundless rumours and pacifying the general excitement of the multitudes.[29]

The text excuses the vigilante murder of thousands of innocent Koreans and Chinese because of the rumors of recalcitrant and criminal elements, absolves the terror-stricken citizens of their murderous racist rampage because of their frayed nerves, and not only forgives the police for their inability or unwillingness to protect the murder victims but also fails to indict them for their widely known role in instigating the violence and their participation in the murder of prominent leftist political figures.

With the destruction of the metropolis, residents of the capital had not only been reduced to a primitive state, but their "civilized" impulses had been stripped away, bringing out their savage tendencies.[30] *Jiji manga*'s comic montage in its 14 October issue (fig. 5.14) shows an image of the Twelve Stories, labeled "tower of culture," cracking in two. The caption reads: "The Citizens Who Have Been Shaken Down from the High Cultured Tower into the Primitive Ocean." Out of the ocean seems to float a terrifying skeleton-faced scarecrow with a large hunting knife and a torch on either side of its hideous scowling visage. The scarecrow bears the sign "rumors and false reports" (*ryūgen higo*). The caption to the far left exclaims, "The scarecrow [rumors] being issued by the sirens dreadfully surprised us," poignantly underscoring the shock of seeing officials and neighbors join vigilante death squads.

Corrosive Catfish and Divine Punishment

THE ART OF SOCIAL CRITIQUE

The familiar allegorical symbol of the catfish (*namazu*) was revived after the Kantō quake to personify natural or divine discontent and was frequently invoked to express the common perception of the earthquake as "divine punishment" (*tenken* or *tenbatsu*) for the extravagances and materialism of modern life. Moralizing criticism of modern Japanese society and the government abounded. The earthquake was considered an expression of the "anger of the earth," a way of humbling human beings, and a charge to return to basic values. According to Janet Borland, a survey of popular explanations of the earthquake conducted in 1924 acknowledged that while there were various opinions, most people viewed the earthquake as either an act of "divine punishment," penalizing them for their sins; a "divine warning"; or a "divine favor," presenting a "golden opportunity" for national reform and renovation.[31] These readings of the event picked up on themes seen in earlier Ansei-period catfish prints. The labeling of natural disasters as divine punishment was actually common throughout history and across the world, and they were also often linked to politics. After the great earthquake that destroyed Portugal's capital city of Lisbon in 1755, Jesuit religious leaders decried the event as divine punishment for the sins and worldly iniquities of the city. This spurred

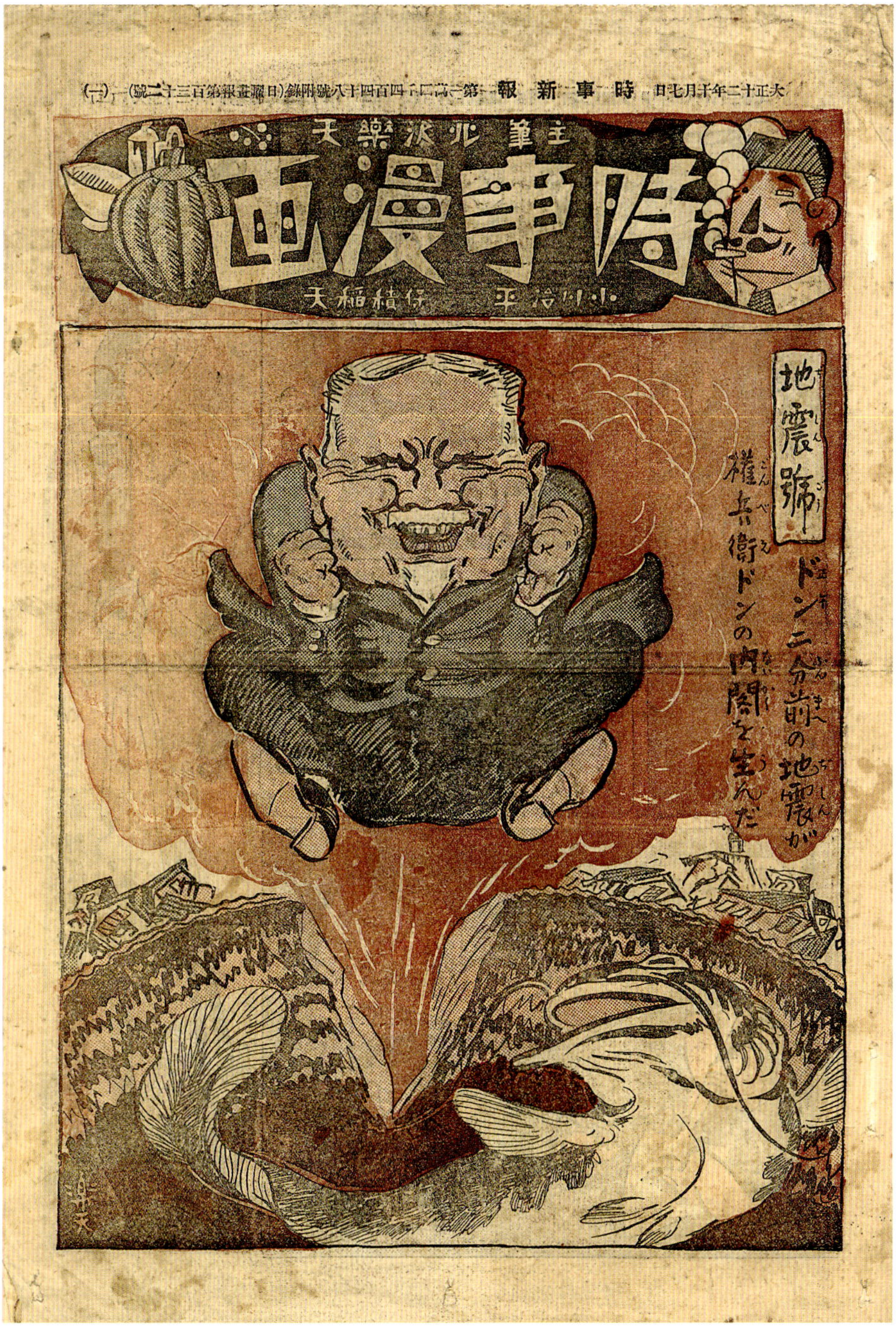

5.15 Kitazawa Rakuten, "Earthquake Edition" (*Jishingō*), cartoon, cover. *Jiji manga* 132, 7 October 1923. Collection of Ohio State University Billy Ireland Cartoon Library & Museum. By permission of Sangyō Keizai Shinbunsha.

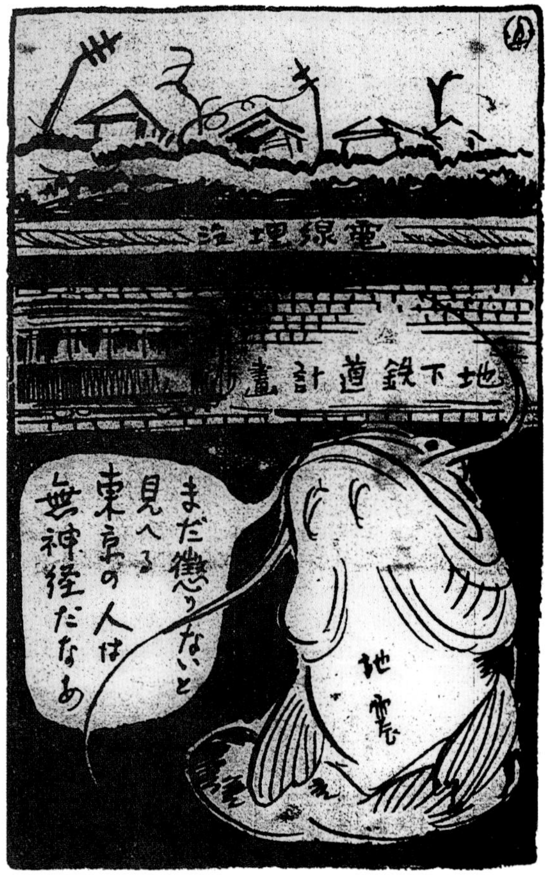

5.16 Kitazawa Rakuten, "They just don't seem to learn from experience. Tokyoites are really callous" (*Mada korinai to mieru. Tōkyō no hito wa mushinkei da naa*), cartoon. *Jiji shinpō*, 17 October 1923, P.M. ed., 3. By permission of Sangyō Keizai Shinbunsha.

an intense power struggle between secular political leaders and religious sects for control of the city.[32]

Humor was a powerful yet entertaining medium for leveling some of this extremely harsh social criticism. Kitazawa Rakuten used the catfish image repeatedly in the popular satirical cartoon supplement he edited for the Sunday edition of the newspaper *Jiji shinpō* titled *Jiji manga* (Cartoons of Current Affairs). Rakuten's cover on 7 October (fig. 5.15) showed the subterranean catfish causing a massive fissure in the ground that "gave birth" to the short-term emergency cabinet of Prime Minister Yamamoto. Yamamoto, in a quasi-fetal pose like that of a newborn baby, springs forth from the seismic fissure in a cloud of smoke. "Earthquake Special Issue: Boom! (*don*) The earthquake at two minutes before [noon] gave birth to the Cabinet of Boss (*don*) [Yamamoto] Gonbei." The text and image play on the homophone *don*—which both represents the booming crash of the earthquake and is slang for a powerful political boss. Yamamoto had already served as prime minister from 1913 to 1914, and he was asked to return to office after the sudden death of Katō Tomosaburō days before the quake, but he had not even begun to assemble his cabinet when the earthquake struck. He served until 7 January 1924, and his political fortunes became tied to the temblor. According to Rakuten's catfish, disaster not only caused massive disruption but also produced a new locus of political power within the conflict-ridden landscape of Japanese party politics.

Catfish imagery also appeared in the regular *Jiji shinpō* paper. One cartoon (fig. 5.16), probably also by Rakuten, shows an animated catfish labeled "earthquake" rearing its head from the subterranean level beneath the buried electrical power lines and the planned subway construction to view the scorched earth above. It exclaims, "They just don't seem to learn from experience; Tokyoites are really callous . . ."[33] Here the catfish presumably seeks to punish the city dwellers again for disturbing his lair with their intrusive underground urban development.

DIVINE PUNISHMENT'S SOCIAL DISCONTENTS

Many public figures lobbed social indictments under the label of divine punishment. Prominent businessman and founder of Japan's first bank Shibusawa Eiichi (1840–1931) was among the most visible and vocal believers in this view. Although his stance was seemingly counter to his professional standing as a dynamic entrepreneur and pillar of the "civilization and enlightenment" movement, he believed that the divine punishment was directed specifically toward the nation's capital as the center of Japan's westernization, urbanization, capitalism, and industrialization. He wrote in the newspaper *Yorozu chōhō* on 13 September, "This earthquake was not only a natural disaster unparalleled throughout history, but also Heaven sent. Since the Meiji Restoration, Tokyo had become the center of politics, economics and other elements and played a leading role in shaping the modern nation. The modern political world, however, became a place of fighting. The economic world became greedy, and morality and ethics reached the point that Arishima's committing a double suicide with his lover was praised. In this context, this earthquake was no coincidence."[34]

A few weeks later, in his own calligraphy, Shibusawa penned a terse elegy for the introduction to Kōdansha's landmark volume that directly blamed modern immorality for the decimation of the capital: "With faithfulness and truth, you gain moral principles; with conceit and lavishness, you lose them" (*chū shin o motte kore o e, kyō tai o motte kore o ushinau*).[35]

Novelist Nagai Kafū (1879–1959), a long-standing sentimentalist who yearned for the Tokugawa days of Edo, also suggested that Tokyo's evil ways had unleashed the earth-

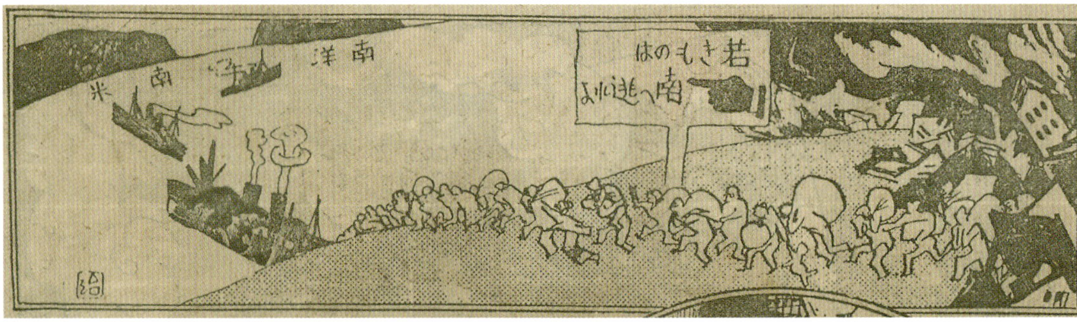

5.17 Ogawa Jihei, "Young People, Escape to the South!" (*Wakakimono wa minami e nigereyo*), cartoon. *Jiji manga* 132, 7 October 1923, 4. Collection of the author. By permission of Sangyō Keizai Shinbunsha.

quake as a form of punishment. Writing in his diary after the disaster, he described the city in the Taishō period before the quake as a "sham hallway, a grand façade with nothing behind it, a device for deceiving the foolish."[36] Vociferous social critic and Tenrikyō priest Okutani Fumitomo weighed in as well. Convinced that divine anger had fallen upon the hedonistic, consumption-driven capital, but clearly indifferent to the class implications of the sites hit hardest by the disaster, he wrote,

> If you peruse a map of the devastated areas of Tokyo, those places completely burnt to the ground include Nihonbashi, Kyōbashi, Kanda, Fukagawa, Honjo and Asakusa. In other words, the amusement districts. Generally speaking, these areas were extremely prosperous in business and industry, and, on 1, 2, and 3 September, they were burnt to the ground. Even the main road of Ginza which prided itself on being the jewel of the urban atmosphere has been reduced to a dreamscape. Mitsukoshi, the number one department store in Japan where crowds flocked to satisfy their vanity, has even become a miserable wreck. On the other hand, the residential areas of Koishikawa, Ushigome, Kōjimachi, Akasaka and Azabu mostly escaped the earthquake and fires. . . . That the entertainment districts were completely destroyed must be an expression of divine will.[37]

Implicit or explicit accusatory classist vitriol was not uncommon in other disasters. In San Francisco, for example, upstanding citizens such as author Charles Caldwell Dobie showed little sympathy for the city's disenfranchised victims, writing in what now seems like shockingly pejorative language, "The square was swarming with refugees from 'South of Market' districts. This was the cheap, poor quarter of the town and many of the wretches who had fled from the flames looked as if they had not faced the morning sunlight for years. They were like rats startled out of their holes, this beer-sodden, frowsy crew of dreadful men and still more dreadful women. A breed that has passed out of American life completely—red faced, bloated, blowsy."[38] Class bias, economic inequities, racism, and political factionalism continue to haunt disasters and their aftermaths, as clearly seen in three recent catastrophic mega-events of the new millennium: the 2004 Indian Ocean tsunami, Hurricane Katrina in 2005, and the 2010 earthquake that devastated Haiti.

Not only did the working class in the low city bear the brunt of the earthquake, but poor displaced quake refugees, a good portion of whom were rural migrants to the city, were encouraged to continue their migration and head abroad, as Ogawa Jihei noted in his pointed cartoon captioned "Young people, escape to the south!" and showing a large hand pointing toward ships in the ocean headed for the South Seas and South America (fig. 5.17). Throughout the 1920s and 1930s, the Japanese government encouraged people to emigrate abroad to relieve domestic overpopulation, particularly pushing them to homestead in the expanding colonial empire. As the capital sought to recover from the quake, authorities thought they could relieve some of the city's housing and social welfare problems by offloading its population to the colonies.

Seeking to elide class issues, home minister and leader of the reconstruction effort Gotō Shinpei declared, "The most important issue in national politics at the moment is to promote the opportunity of reviving the country . . . and renewing people's minds is our top priority."[39] His handwritten verse for the Kōdansha volume (fig. 5.18) expressed his faith that some good could come out of the misfortune of the quake on the road to reconstruction: "While for a thousand years the volcano has smoldered without cease, at times like this, life's tragedies and blessings become inextricably connected."[40] This personal statement was followed by a terse four-character couplet handwritten by the commander of martial law, Fukuda Masatarō, that read simply: "National solidarity" (*kyōkoku itchi*).

The two imperial edicts issued two weeks and two months after the quake, respectively, translated these sentiments into powerful moral proclamations. They were printed in national newspapers along with annotated commentary for the general public and then reprinted in other government publications. The second, the Imperial Edict Enjoining Sincere and Strenuous Life (Kokumin seishin no sakkō ni kansuru shōsho), issued 11 November 1923, made the moral connection even more explicit:

> In recent years, much progress has been made in science and human wisdom. At the same time frivolous and extravagant habits have set in, and even rash and extreme tendencies are not unknown. If these habits and tendencies are not checked now, the future of the country, we fear, is dark, the disaster which befell the Japanese nation being very severe. It may not

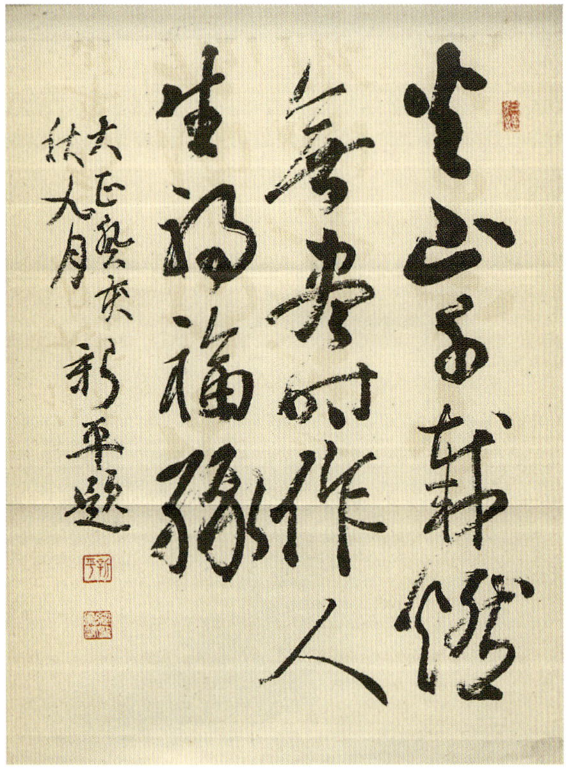

5.18 Gotō Shinpei, "While for a thousand years the volcano has smoldered without cease, at times like this, life's tragedies and blessings become inextricably connected," handwritten verse. Dai Nihon Yūbenkai Kōdansha, *Taishō daishinsai daikasai* (The Great Taishō Earthquake and Conflagration) (Tokyo: Dai Nihon Yūbenkai Kōdansha, 1923).

> be possible to hope for the restoration of national culture and prosperity unless the determined will of the whole nation is aroused. . . . Our subjects must faithfully observe the honourable tradition of national education and must endeavour to elevate the twin cardinal virtues of knowledge and morals. They must discard frivolous habits and lead a sincere and strong life, avoiding rashness and extremities.[41]

This need to protect the nation from "frivolous and extravagant habits" would become the justification for a series of "peace preservation" laws (*chian iji hō*) in the decade after the quake, particularly those aimed at suppressing the leftist political activity that had expanded in Japan after the Russian Revolution, which was considered a threat to the national polity (*kokutai*).[42]

Various moral pundits identified different sources of excess. Some pinpointed these bad tendencies in the vanity, ostentation, and liberated behavior of women. For example, in a 1923 article about traditional earthquake lore, Tōhoku Imperial University seismologist and chair of geophysics Kusakabe Shirōta (1875–1924) noted that while men are associated with the positive geomantic entity *yang*/sky or heaven (*yō* in Japanese), women are associated with the negative *yin*/land (*in*). Thus, in the human realm, the shaking of earthquakes was connected to women—specifically, to their unrest or restlessness.[43] Rakuten's cover image (fig. 5.19) for the 14 October issue of *Jiji manga* echoed this denunciation of women. Bearing the caption "vices rectified by the earthquake," it features the imposing figure of an irate catfish-man ripping off the jacket of a startled woman; on the lining of her jacket is the four-character expression "vanity and ostentation" (*kyoei kyoshoku*). With his other hand, the catfish thrusts a more subdued garment toward her that reads "substance and moral and physical fortitude" (*shitsujitsu gōken*). The woman's kimono, not coincidentally, has a pattern of hearts cascading over the fabric, likely a reference to the perceived excessive tendencies of the popular cultural trend known as Taishō romanticism, or more specifically, an indictment of the potentially disruptive new phenomenon of "love marriages" (*renai kekkon*) that women's rights advocates were promoting to replace the more established practice of arranged marriages (*o-miai kekkon*).

Kusakabe completes his discussion of women and quakes with a chilling anecdote that reveals the depth of animosity toward the new liberated activities of modern women:

> I have heard about the following story. Once there was a well-known "new woman" (*atarashii onna*) who traveled abroad to the West by steamship via the Indian Ocean; on the ship, she spoke loudly about the equality and rights of both sexes. She made a fuss and got into heated arguments; she was considered arrogant and outrageous, and people on the ship could not deal with her. On the ship, there was a guy who was a notorious womanizing playboy. After discussing this topic with the new woman, he invited her to come to his room to continue their chat since they were annoying other passengers. Let's decide whether the two sexes have equal rights there, he said. She accepted. Then the playboy locked the door. In that room, there were only the woman and the man, and even if anyone yelled or cried, there was only the roaring sound of the Indian Ocean to answer. Nobody knows what happened in the room. However, after the incident, whenever the playboy asked her whether she still thought that men and women have the same rights, the new woman left the room like a mouse in front of a cat.[44]

Like the punitive catfish-man in Rakuten's cartoon, the rake's brutal rape to silence the new woman points to the broad retributive social impulses against women's liberation that were raging in the 1920s and were brought to the surface by the temblor.

Western journalists noted the continued Japanese belief in the myth of the catfish as purveyor of divine punishment, but they failed to understand the metaphorical aspects of this imagery and tended to dismiss it as a relic of primitive folkloric belief.[45] Bates

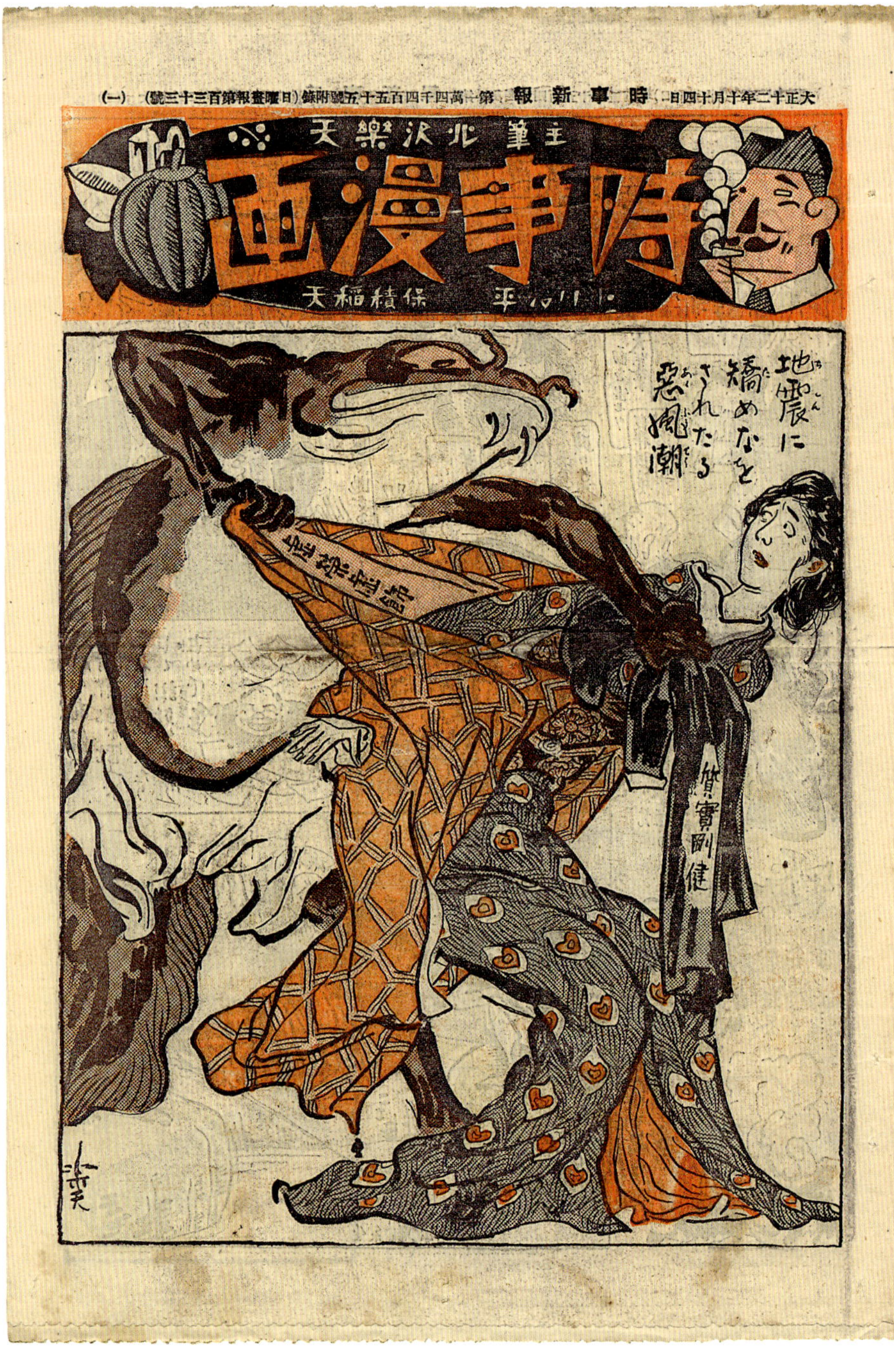

5.19 Kitazawa Rakuten, "Vices rectified by the earthquake" (*Jishin ni tamenaosaretaru akufūchō*), cartoon, cover. *Jiji manga* 133, 14 October 1923. Collection of Ohio State University Billy Ireland Cartoon Library & Museum. By permission of Sangyō Keizai Shinbunsha.

has rightly argued that it is misleading to view the critical deployment of divine punishment as a literal attribution of the disaster to divine power; rather it should be seen as a rhetorical strategy designed to bolster preexisting desires for social reform, by both the right and the left.[46] The government clearly wanted to parlay the disaster into an opportunity for moral and social reform, particularly through the system of national education. Just three months after the quake, the Ministry of Education published a three-volume series titled *Shinsai ni kansuru kyōiku shiryō* (Education Materials Related to the Earthquake) for use in the national education curriculum. Educators saw the earthquake as an opportunity to shore up moral pedagogy, emphasizing loyalty to the emperor, sacrifice, and bravery as the critical triumvirate of values.[47]

The rhetorical linking of divine punishment with natural events in the *tenken* discourse resonated with earlier disasters in Western history, such as the vociferous public calls for repentance by prominent Jesuits like Gabriel Malagrida after the 1755 Lisbon earthquake, but it is in sharp contrast to the standard explanations of the San Francisco quake more than a century and a half later. As San Francisco quake observer Charles Morris noted, there was no lack of moralizing about the ravages of human sin and of frivolity more generally: "When the weakness and the folly and the sin of men bring woe upon other men, there are plenty of texts for the preacher and no scarcity of earnest preachers. But here is a vast and awful catastrophe that befell from an act of Nature apparently no more extraordinary than the shrinkage of hot metal in the process of cooling. . . . No doubt, the process goes on somewhere within the earth almost continuously, and it no more involves the theory of malignant Nature than that of an angry God."[48] But in Japan nature was still invested with great moral authority, and believers in the divine auspices of the quake or their symbolic value predominated.

Rhetorical deployment of divine punishment continued well into reconstruction. Two years after the quake, Rakuten again invoked the specter of the retributive catfish in his print on the 30 August 1925 cover of *Jiji manga* (fig. 5.20) to warn unscrupulous profiteers, idle public officials, and immoral elites of the consequences of their profligate behavior. Rakuten's looming catfish exclaims, "Shall I shake the ground one more time to really wake them up?" The caption reads, "Behold the status quo of the second anniversary of the earthquake. The land readjustment has not been completed yet. Reconstruction materials are depleted and there is no building. The government does not know how to use the damage [donation] funds of 7.2 million yen. Women are dancing on the erupted volcano with their Western dress and short hair (*yōsai danpatsu*)." In a plume of smoke below the catfish, a stylish modern couple dances on the ruins; the woman's parasol reads, "Frivolous" (*keichō fuhaku*). Behind them is a corrupt politician stuffing his pockets with money; the lining of his jacket reads, "The dona-

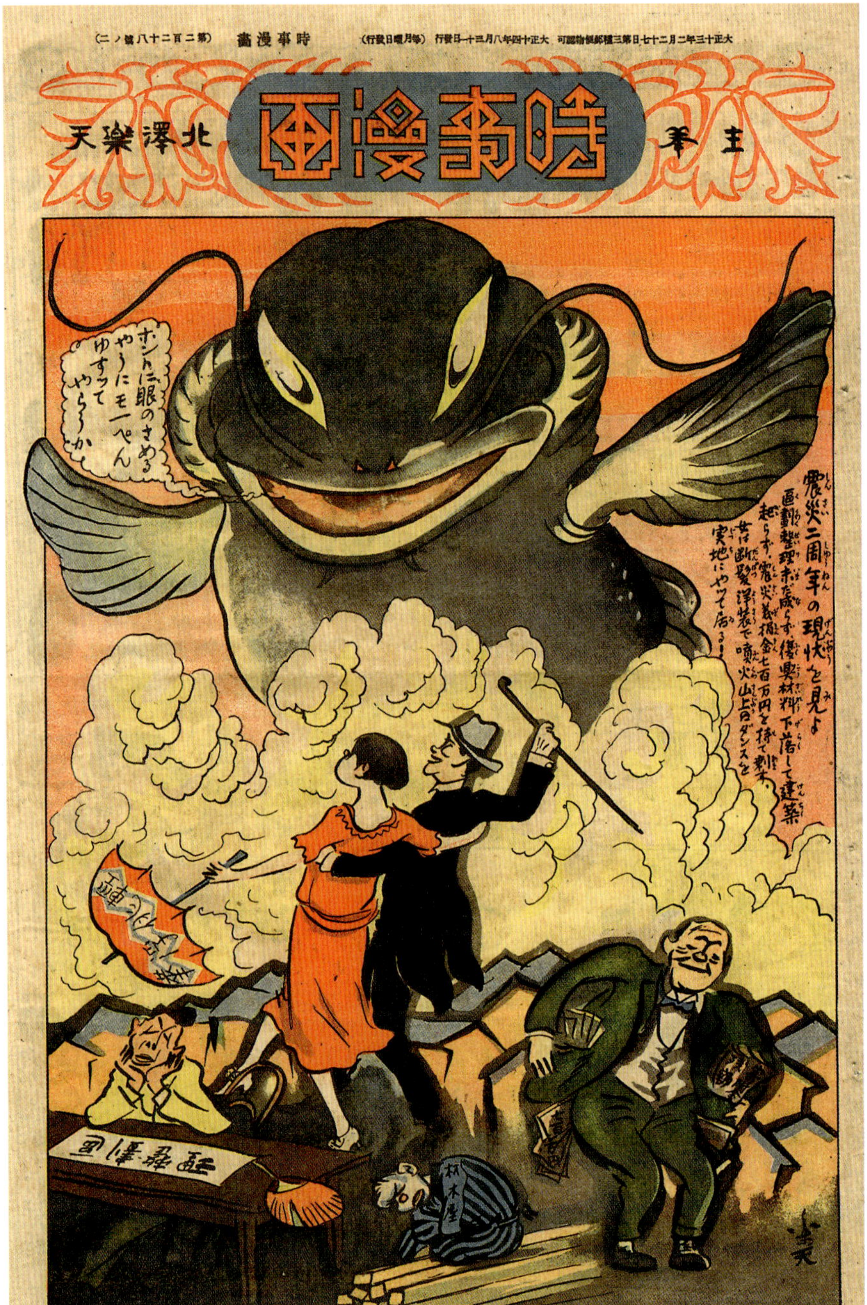

5.20 Kitazawa Rakuten, "Behold the Status Quo of the Second Anniversary of the Earthquake" (*Shinsai nishūnen no genjō o miyo*), cartoon, cover. *Jiji manga* 228, 31 August 1925. Collection of Ohio State University Billy Ireland Cartoon Library & Museum. By permission of Sangyō Keizai Shinbunsha.

tions still remain 7 million." A timber dealer naps on a pile of lumber in the middle of the scene, and a government official in the land readjustment office (*kukaku seiri*) falls asleep at his desk on the lower left.

Men of science, like the famous feuding Japanese seismologists Ōmori Fusakichi and Imamura Akitsune, however, generally followed the model championed by Voltaire after the Lisbon quake, which marked a shift in Western philosophy from clericalism to the secular humanism of the Enlightenment.[49] Voltaire, who took issue with both the optimistic and the pessimistic forms of religious fatalism of the time, penned his now-famous philosophical tract *Candide* in response to this event, eschewing religious fatalism for notions of man's rational free will and achievement of civilizational progress through science and reason.[50] The incessant visual presentation of seismological records in Japan and statements by the scientific community similarly attempted to use reason and logic to explain the geological rather than the moral causes of the quake, thus providing a counterweight to *tenken* discourse in the debates. These efforts were "seismology trying to beat the catfish with earthquake prediction," observes Clancey.[51]

Barracks: Symbol of Altruism and Corrosion

In the wake of the temblor and amid these social fissures and fractures, authorities struggled to reestablish basic services for imperial subjects in the capital, who continued to live as refugees for months. The ongoing refugee saga similarly provided fodder for criticism and comical entertainment as well as serious reflection on the possibility of reordering social structures. As we have seen, humor and even games were not alien to catastrophe, particularly if they could make the discussion of harsh realities more palatable.

Yamamura Kōka designed a board game called the Great Taishō Earthquake *Sugoroku* Game (Taishō Daishinsai Sugoroku) (fig. 5.21) based on the earthquake experience. Played like backgammon, or more commonly like Snakes and Ladders, *sugoroku* dated back to at least the eighth century. It was often a mode of gambling and a forum for intense competition. *Sugoroku* games were based on a dizzying array of themes, from religion and politics to adult topics, and they were produced in large numbers, so the creation of an earthquake game was not as strange as it might seem. Still very popular during the modern period, *sugoroku* games were often inserted as supplements into mass-market magazines. On Kōka's board, players as refugee pawns follow the conventional *sugoroku* directions "go back to the beginning" (*furidashi ni modoru*), "proceed" (*agari*), and "lose a turn" (*ikkai yasumi*). With the roll of the dice, they proceed through the spaces representing the now-familiar experiences and places

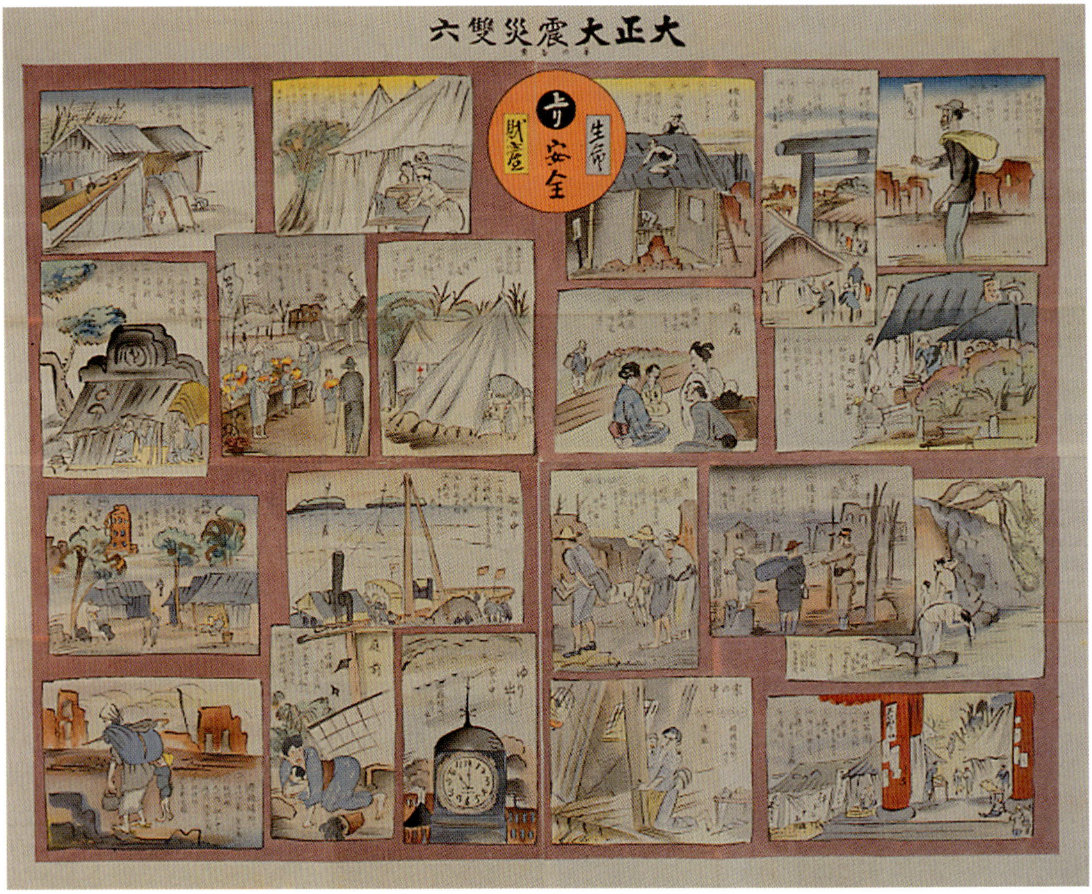

5.21 Yamamura Kōka, Great Taishō Earthquake *Sugoroku* Game (*Taishō Daishinsai Sugoroku*), game board, c. 1923. Collection of Tōkyō-to Edo Tōkyō Hakubutsukan.

of the event, starting with the Taishō temblor itself, symbolized by a clock frozen at 11:58 A.M. and perilous tumbling houses. Then, depending on the roll, the player moves from the space "in the house" (*ie no naka*), where figures are pictured fearfully holding onto swaying supporting posts, and either goes to the refugee camp (*nanminjo yuki*) or is injured (literally, "meets disaster," *sōnan*) and moves to the space for rescue. Spaces feature the burned landscape and its famous destroyed sites (the Twelve Stories, Ueno Park, Shiba Park, etc.), encounters with army troops, care in hospitals, stays in refugee camps, and residence in barrack towns. A space for paying homage at the Honjo Army Clothing Depot is also on the route. The goal (*agari*) is to reach the central red circle, labeled "Property, Life, Safety" (*zaisan, seimei, anzen*), the ultimate objectives of any earthquake survivor. The game could be played or simply appreciated through viewing. It distilled the earthquake into a series of moments and experiences through which everyone had to proceed according to chance. It strategically omitted the experience of outsiders and reinforced the normative view of the typical Japanese refugee as a stoic victim in an altruistic and resilient society, struggling with the collective to survive.

In the real world, before arriving at that coveted final destination of property, life, and safety, many earthquake survivors had to reside in makeshift temporary shelters known as barracks (*barakku*), which figured prominently in the picturing of daily life in the aftermath of disaster. The term *barakku* was used broadly after the quake to refer to a diversity of structures that included ephemeral tentlike shelters and huts of iron sheet metal for refugees and businesses as well as sturdier, sometimes elaborately decorated wooden edifices designed to stand for several years until permanent reconstruction could be completed. Similar makeshift structures had been erected after previous fires and earthquakes, but they were generally referred to as "temporary architecture" (*kari kenchiku*) or "temporary small huts" (*karigoya*), and these terms continued to be used in 1923 as well.[52] *Barracks* was a new term emerging from the 1923 quake and subsequently came into widespread use in Japan. Undoubtedly, the unprecedented number of like structures assembled in long repetitive rows was visually reminiscent of military and workers' barracks, and this association was most likely the first inspiration for the term, but it also came to refer to individual structures, implying their provisional status.

Barracks were concentrated in the capital's low city, in the areas most heavily damaged by the earthquake. One of the largest assemblages of barracks was erected in the grounds around the Imperial Palace (fig. 5.22).[53] The primitive structures became a vivid symbol of basic human existence shorn of the extraneous trappings of modern life, and for some on the political left, they even symbolized the potential for radical social transformation out of the ashes of the calamity. The barracks were also aesthetically objectified, both in the spontaneous artistic movement to decorate them and in artistic representations of them, which included appealing prints and photographs of orderly rows of pristine white tents and affecting depictions of family life in the rickety shacks.

Barrack imagery was particularly touching because of its connotations of noble perseverance as people struggled to regain a sense of normalcy in the extreme postearthquake conditions. Focusing on compassion, it spotlighted the most helpless members of society: children and the elderly. A large-scale oil painting entitled *Refugee Barracks in front of the Imperial Palace* (*Kyūjō mae no nanmin barakku*) (fig. 5.23), created by the academic artist Tokunaga Ryūshū (better known as Tokunaga Hitoomi, 1871–1936) to adorn the walls of the earthquake memorial museum, extols the multigenerational

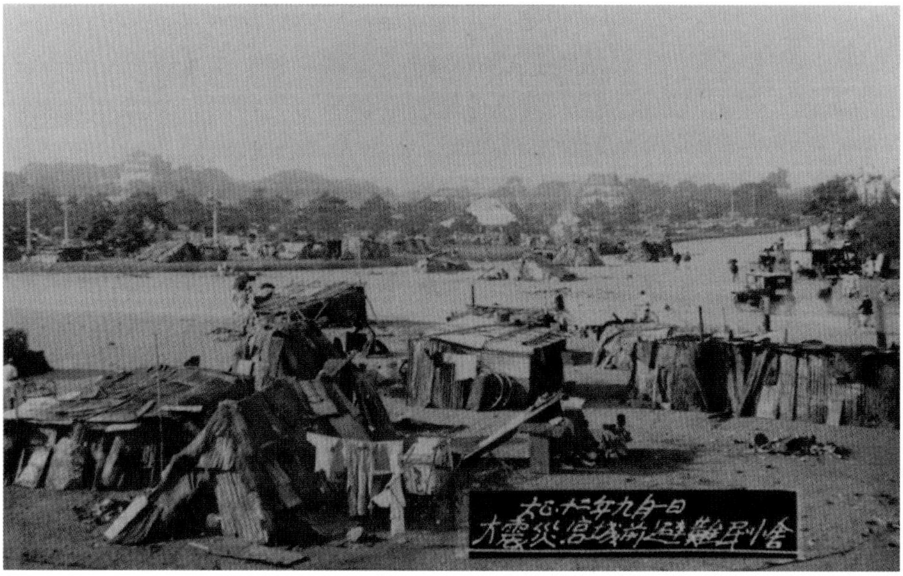

5.22 *Taishō 12, September 1; Earthquake Refugee Huts in front of the Imperial Palace* (*Daishinsai kyūjōmae hinanmin koya*), photograph, 1923. Kon Wajirō Archive, Kōgakuin Daigaku Toshokan.

care among refugees. In it, a daughter tends to her elderly mother in the barracks, and a young mother with her infant strapped to her back plods through the rain to carry water to her family.[54] Souvenir postcards of the barracks similarly carried figures of mothers caring for children, undoubtedly one of the most sympathetic images of refugees in its ability to convey innocence as well as an innate maternal compassion among the populace. One frequently reproduced photograph shows a mother breastfeeding her infant in a makeshift camp that she has erected on the presumably defunct railway tracks (fig. 5.24). In an age when Christian iconography was increasingly familiar to Japanese viewers, this earthly updated reference to the divine Madonna and maternal sacrifice would have been widely recognized.

In actuality, the barrack towns provided only squalid living conditions and offered little shelter from inclement weather. Refugee life consisted of gray, depressing days of interminable waiting in leaking hovels, a reality vividly captured by Yamada Minoru (1889–1925) (fig. 5.25), a painter and well-liked cartoonist who worked for the *Tōkyō asahi shinbun*. Even children's magazines pictured the miserable lives of families in the barracks. "People whose houses have burned to the ground must live sad lives in the little huts when it rains," reads the simple, phonetically rendered caption for Murayama Tomoyoshi's illustration in the October 1923 earthquake special issue of *Kodomo no*

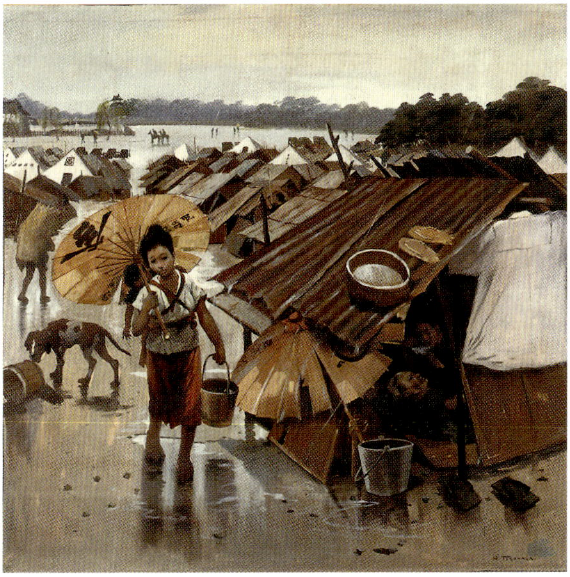

5.23 Tokunaga Ryūshū, *Refugee Barracks in front of the Imperial Palace* (*Kyūjō mae no nanmin barakku*), oil on canvas, c. 1923. Reconstruction Memorial Museum. Collection of Tōkyō-to Irei Kyōkai. Photograph courtesy of Art Restoration Studio 21.

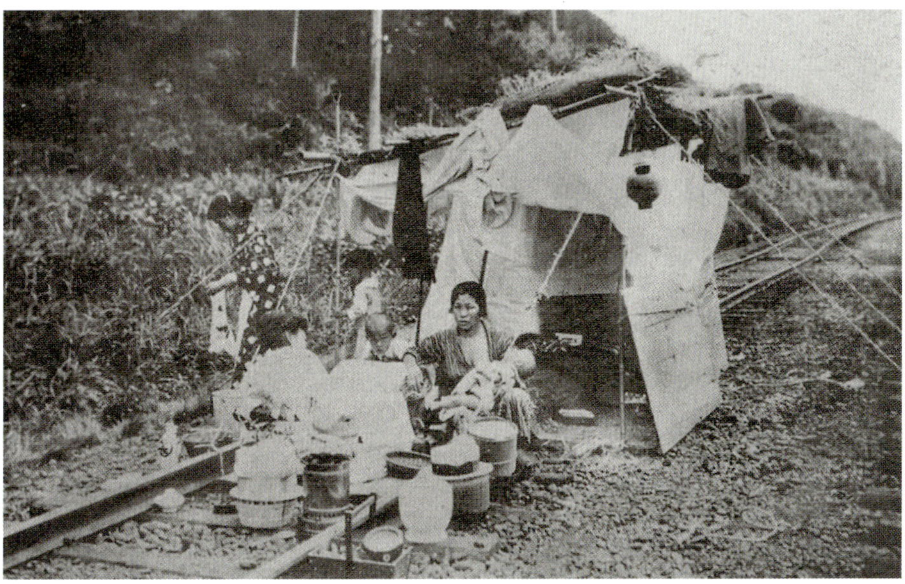

5.24 *Miserable Conditions of Victims of the Great Kantō Earthquake* (*Kantō Daishinsai risaisha no awarena jōtai*), postcard, 1923. Ishii Toshio Collection. By permission of Japan Publishing Copyright Association.

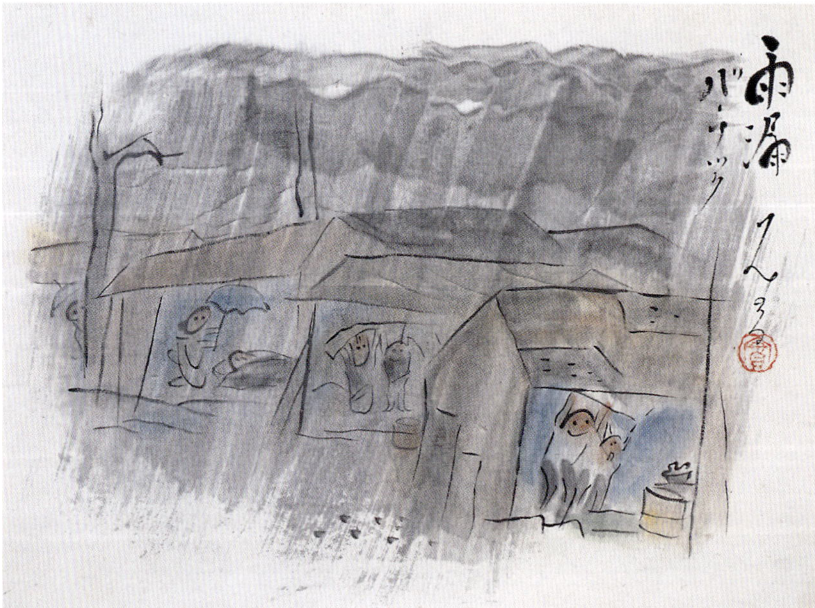

5.25 Yamada Minoru, *Tears Like Rain, Barracks* (*Urui barakku*), *Paintings of the Great Kantō Earthquake* (*Kantō daijishinga*), ink and color on paper, c. 1923. Collection of Tōkyō-to Edo Tōkyō Hakubutsukan.

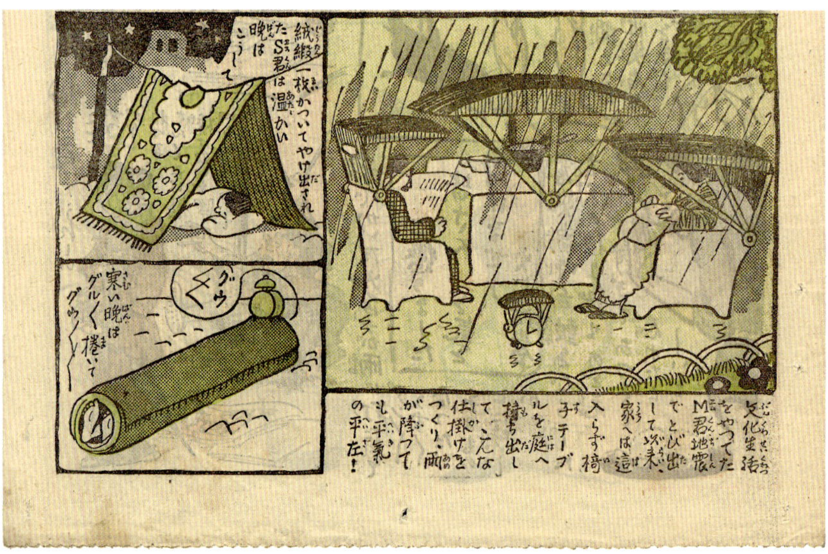

5.26 Arita Shigeru, "Mr. M lived the cultured life . . ." (*Bunka seikatsu o yatteta M-kun*), cartoon. *Jiji manga* 133, 14 October 1923, 3. Collection of Ohio State University Billy Ireland Cartoon Library & Museum. By permission of Sangyō Keizai Shinbunsha.

tomo (Children's Companion). A destitute family of four sits forlornly in its barrack hut as the rain falls; the mother clutches the infant to her for warmth and security as her son stares blankly.[55] There was no way to sugarcoat the predicament of quake refugees. Some viewed these makeshift communities as ghettos for the lower classes. The government expressed concern about the social environment of the barracks, particularly for children, who were living with strangers and might be learning bad habits like lying and stealing.[56] Frequent public statements of concern implicitly impugned the moral character of the classes of refugees who inhabited the barracks. These comments tied into a broader contemporary social discourse on slums among government officials and new urban social scientists, which cast Tokyo or "the city" as "a site of baleful elements portending disease and degeneration if not properly administered."[57]

This negative view of barrack inhabitants was in striking contrast to the imaging of middle-class refugees, whose hardships often centered on their loss of possessions and the disturbance of their formerly comfortable lifestyles. For instance, one cartoon (fig. 5.26) satirizes the condition of those who lived "the cultured life" (*bunka seikatsu*) in new Western-style homes in the trendy upscale urban residential neighborhoods of Tokyo. Describing the figures on the right in their temporarily reconstructed bourgeois lifestyle, the caption states, "Mr. M lived the cultured life. Since he fled from his home because of the earthquake, he never enters into the house, and brought his chairs and tables out to his garden. Thanks to these [awning] contraptions, even rainy days are okay." Mr. M and his wife sit in large, comfortable armchairs reading the newspaper and sewing—their previous existence essentially undisturbed—protected by mounted awnings over each chair, the table, and even the alarm clock. To the left, "Mr. S, who was burned out by the fire, has only a rug. During warm nights, he uses it like this [as a shelter], and on cold nights, he wraps himself up in it [as a blanket] and falls fast asleep."

BARRACK BATHER, DISASTER MUSE

In the barrack towns, life was literally stripped down to nothing, leaving residents perilously shorn of civilized decorum—not to mention their clothing. Kōdansha editors noted with opprobrium that people living on the Imperial Palace grounds would jump over the crumbled stone wall, strip off their clothes, and bathe in the moat. They remarked with consternation that they did so right across from the Tokyo Municipal Police Station and that this behavior would have been unthinkable before the quake.[58] In 1871, the Meiji government issued an ordinance prohibiting public nudity, urination, obscene acts, and the sale of pornographic pictures as part of its campaign for civilization and enlightenment. Naked bodies on public display were deemed incompatible with the civilizing goals of the nation. At the same time, Japanese artists returning from Europe in the late nineteenth century were arguing for the value of the nude as a vital canonical subject in Western high culture, which they had studied extensively in European art ateliers and the École des Beaux-Arts. The scandal created by the exhibition of Kuroda Seiki's nude painting *La Toilette* at a public exhibition in Kyoto in 1895 after his return from Paris is indicative of the multivalent cultural status of the naked body at the turn of the century—simultaneously a "corruptor of manners" (*fūzoku kairan*) and a touchstone of progressive civilization.[59]

A quarter century later, commentators like Miyatake Gaikotsu noted that nudes—a curious marker of modernity—were everywhere: art exhibitions, newspaper advertisements, store displays, and other public venues. So prevalent were these images that "it felt like one's nose was being thrust into them."[60] Still, concerned about the moral degeneracy of public nudity after the quake, Japanese authorities prohibited the sale of images with nude figures along with those of dead bodies, which included the now well-known photograph of naked bathers in the muddy groundwater near the Imperial Palace (fig. 5.27). Such images circulated clandestinely. Not only did they flaunt indecent nudity, but they also suggested that the disaster had caused the complete abandonment of social propriety. Refugees, particularly women, were ignoring formerly sacrosanct spaces such as the grounds of the Imperial Palace and were reportedly tearing off their clothes while standing in line to get into makeshift bathing facilities, emphasizing both their animality and their sexuality.[61] Any concerns about the abandonment of civilized decorum, however, did not stop the purveyors of such images from profiting off voyeurism at the expense of these miserable refugees. Yet there was also an undercurrent of sympathy in many images of naked people bathing in rivers and improvised tubs, such as beer barrels and goldfish basins, in recognition that the bathers were displaying the practical resilience of the Japanese people. Bathing was shown as a pleasurable outlet for releasing the physical and psychological stress of the quake experience. It allowed refugees to reclaim their bodily pleasure after suffering unimaginable physical hardships. Moreover, infectious diseases were a major public health concern after the disaster, so proper hygiene was commendable.

Bathing scenes in mass-media photographs as well as artists' renderings also evoked strong aesthetic appreciation, transforming naked bodies into nudes that drew from classical artistic templates ranging from well-known *ukiyo-e* prints of bathhouses to the impressionist and postimpressionist paintings of women washing, such as the series of bathers by Edgar Degas, which were inspired by Japanese prints. Images of bathers also clearly referred to the modern Japanese print revivalists of the new prints move-

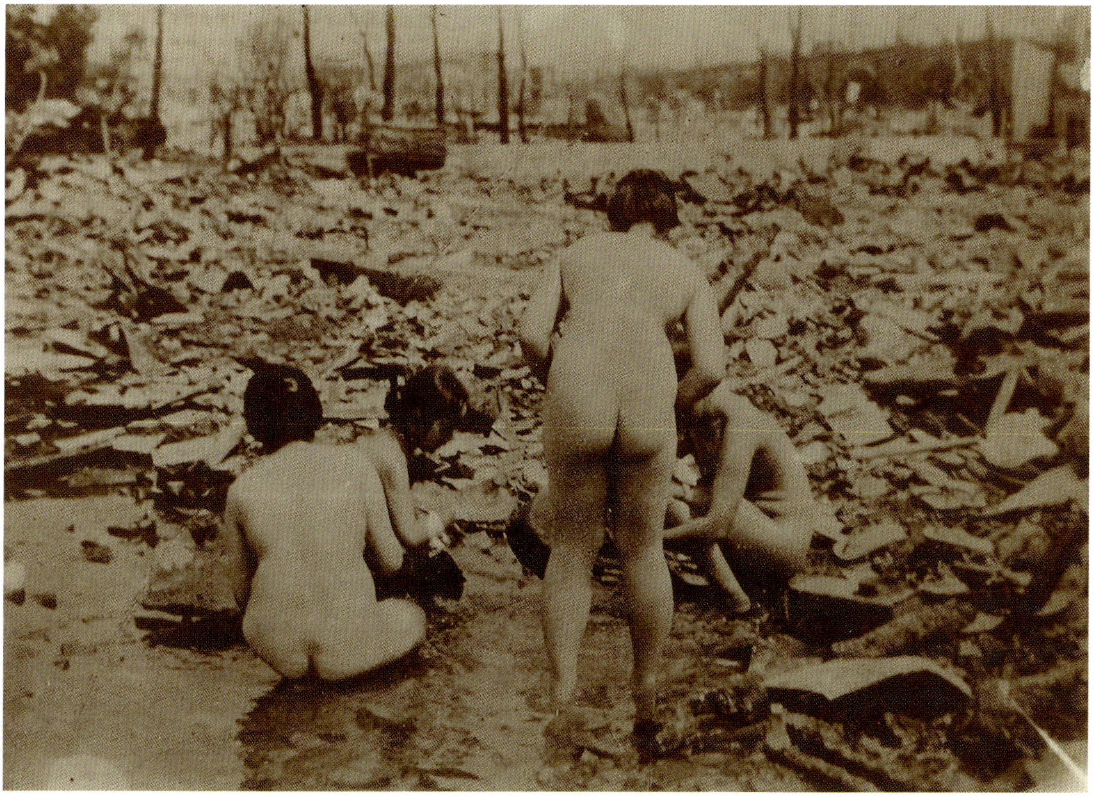

5.27 Artist unknown, women washing in the muddy groundwater near the Imperial Hotel across from the Imperial Palace, photograph, 1923. Courtesy of Sashikata Eisuke.

ment, particularly Itō Shinsui and Hashiguchi Gōyō, who were known for their eroticized images of women bathers in the 1910s and 1920s. In this visual context, the nude female refugee became an artistic muse, and the artists, such as Rakuten's disciple Kōuchi Junichi (1886–1970), lingered over the careful composition and poses of the women, as in Kōuchi's evocative sketch *Roadside Bath* (*Robō suiyoku*) (fig. 5.28). Such images extended the long-standing visual (and erotic) pleasure of washing into the spaces of the disaster.

Kawamura Karyō, responding to Kōka's expressionistic and ebullient sketch of bathers in Shiba Park (fig. 5.29), described the pure blue water and the delightful vision of naked bathers as a return to the idyllic times of ancient folklore or his boyhood memories of island life in Japan's rural periphery. The bathing men and women, returned to the nakedness of their birth, symbolized for Kawamura a return to nature. But he lamented, while the people should find sustenance in the purity of life and the simple

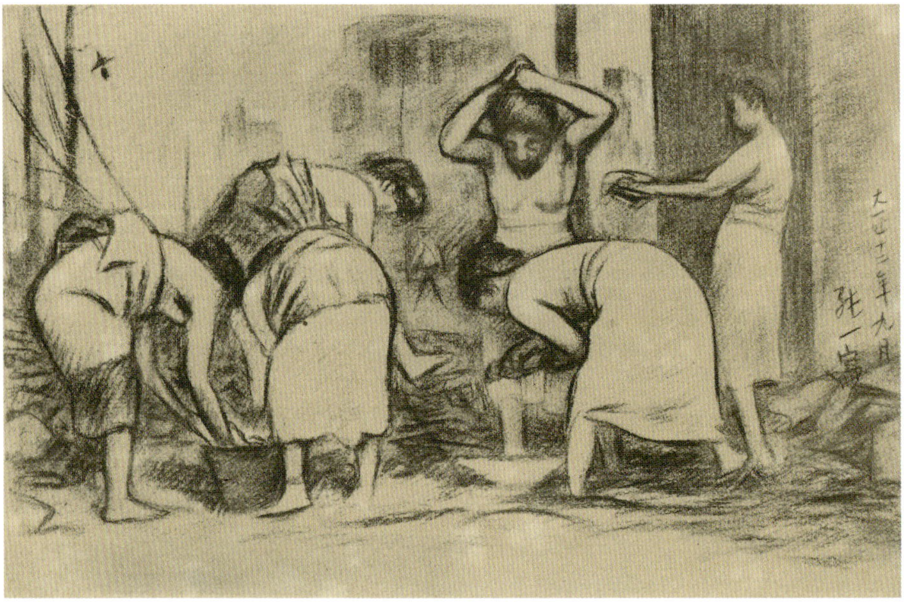

5.28 Kōuchi Junichi, *Roadside Bath* (*Robō suiyoku*), drawing. Nihon Mangakai, *Daishinsai gashū* (Pictorial Collection of the Great Earthquake Disaster) (Tokyo: Kanao Bunendō, 1923).

joys of being, they focused only on their losses and misery. Modern man, like Adam and Eve, did not appreciate the simple pleasures, even when he was given an opportunity to contemplate them.[62]

Even as bathing images conveyed powerful aesthetic sentiments, they undeniably objectified their subjects and had troubling class implications. Degas's bathers were modeled on women from the lower classes, about whom he stated, "Hitherto the nude has always been represented in poses which presuppose an audience, but these women of mine are honest, simple folk, unconcerned by any other interests than those involved in their physical condition. Here is another; she is washing her feet. It is as if you looked through the keyhole."[63] While radically departing from the elegant aesthetics of the classical nude, Degas not only admitted his voyeuristic peeping into the private lives of women but also subordinated them as "simple" others. Japanese bathing scenes in *ukiyo-e* woodblock prints also had a voyeuristic aspect, bordering on softcore pornography. They were one of the print subjects commercially available that frequently featured nude figures, mostly female types of the lower classes, who were sexually available like courtesans, entertainers, and teahouse girls. During periods of strict censorship under the Tokugawa shogunate, bathing images were produced to circumvent restrictions on explicit erotica and categorized as *abuna-e* (dangerous pictures).

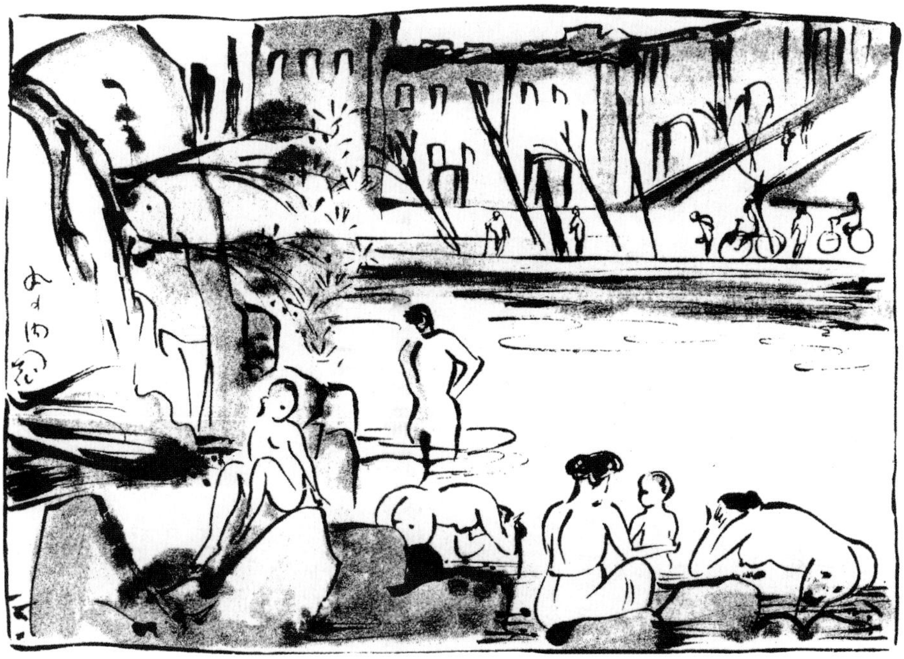

5.29 Yamamura Kōka, *Bathing* (*Suiyoku*), from *Record of Impressions of the Great Earthquake Disaster: Taishō Stirrups of Musashi Province* (*Daishinsai inshōki: Taishō Musashi abumi*), drawing, originally in *Hōchi shinbun*, 30 September–1 November 1923, P.M. ed. Reproduced in Takehisa Yumeji, Kawamura Karyō, and Yamamura Kōka, *Yumeji to Karyō, Kōka no Kantō Daishinsai rupo* (Tokyo, Kress Shuppan, 2003), 101. Courtesy of Kress Shuppan.

Viewing the naked bodies bared by the Kantō quake also evoked memories of Ansei-era prints of frantic male and female bathers running en masse into the street and being thrown to the ground from the force of the temblor.[64] These images produced an eroticized public gaze that resubjugated and resubordinated women of the lower classes as it had with the bodies and corpses of the women of the Yoshiwara licensed prostitution quarter from Ansei through 1923.

BARRACK HUMOR

People tried to speak back to the exploitative circumstances after the quake through humorous jabs that exposed the cramped and unsanitary barracks and the rampant profiteering in the reconstruction effort. These conditions provided ample material for the sardonic humor of Japanese cartoonists who feasted on the ironies and absurdities of everyday life. Okamoto Ippei, renowned satirist and founder of Japan's first associa-

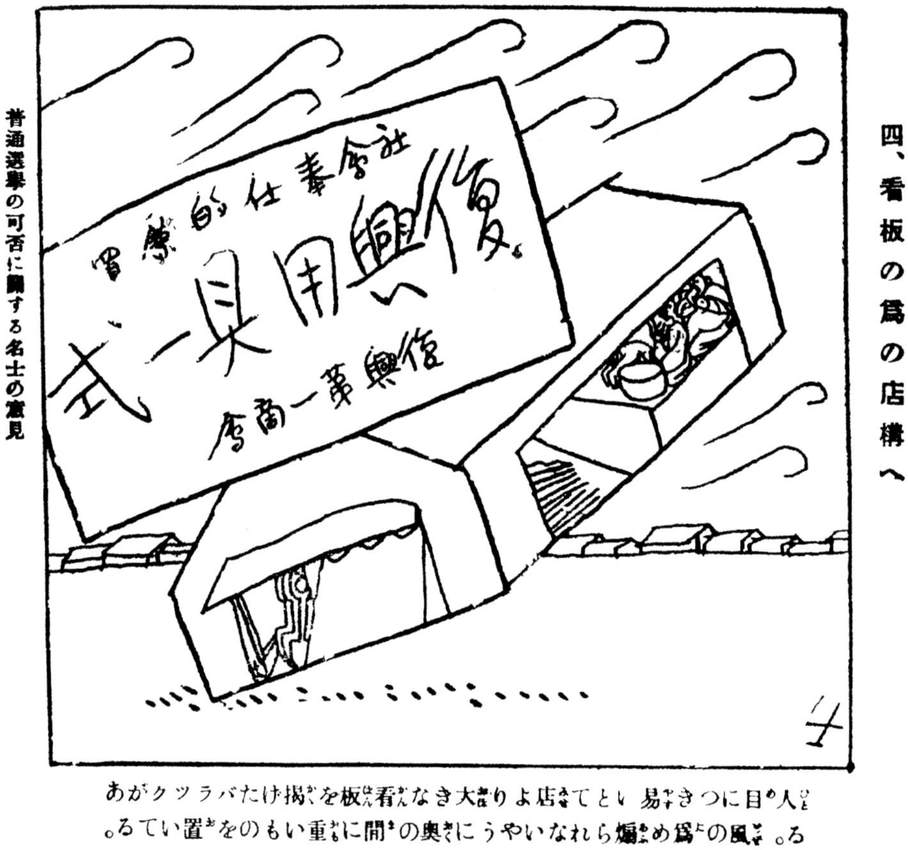

5.30 Okamoto Ippei, *Store Built for the Sake of Its Signboard* (*Kanban no tame no misegamae*), cartoon. Okamoto Ippei, "Fukkō ni arawaretaru yūmoa jūdai" (Ten Humorous Topics Appearing during Reconstruction), *Taiyō* 30, no. 1 (1924): 103.

tion of cartoon artists (Nihon Mangakai), drew several series of cartoon sketches for *Taiyō* (the Sun) magazine in which he depicted humorous barrack scenes.[65]

One Ippei cartoon (fig. 5.30) shows a flimsy barrack structure with such a large, heavy signboard that all of the possessions have to be loaded in the back of the building to counterbalance the weight. The shack teetering on one corner is a metaphor for the disproportionate weight placed on profiting from the reconstruction effort. This is elaborated through the visual pun on the expression *kanban daore* ("the signboard falls over"), which commonly meant that something was not as good as it looked. The lines on the signboard read "Bargain Prices as a Social Service" (*Shakaihōshi-teki renbai*), "Complete Set of Reconstruction Tools" (*Fukkō yōgu isshiki*), and "Reconstruction Is Number One Trading Company"(*Fukkō Daiichi Shōkai*), in mockery of the charitable veneer placed on reconstruction efforts that were in fact serving greedy business interests; the cartoon juxtaposes the spiritual aspects of *fukkō* (reconstruction) with the profit-seeking imagery of the *shōkai* or trading company. The name of the company provides the final coup de grâce in Ippei's social satire: *Fukkō Daiichi Shōkai* would have sounded particularly absurd to contemporary readers, with its play on the term *daiichi* (number one), which appeared both in the government reconstruction slogan *fukkō daiichi* (Reconstruction Is Number One) and in the standard local company name Daiichi Shōkai (Number One Trading Company).[66]

Profiteering and greed were continual topics of criticism, as they had been in the aftermath of the Ansei quake. One man's misfortune was often another's good fortune. In the cartoon "High Profit Earthquake," an ecstatic businessman throws his hands and papers up in joy over his financial bonanza from the earthquake: "Other traders in the same business are either bankrupted or burned out, so I monopolize orders." A figure to his left runs off with a briefcase full of loot, exclaiming, "My single workers are burned and hospitalized, so I don't have to pay boarding-house rent for them. More than that, I get an allowance as a special victim of the quake!" Another smiling figure touts the benefits for debtors, crowing, "The deed for my debts was destroyed in the fire," as a small winged "deed" (*shōmon*) flies away nearby.[67] Not only were financial records widely destroyed in the quake, but there was also an imperial ordinance suspending the collection of debts. Self-interest, however, went on uninterrupted.

The catfish continued to caution against self-interested behavior many years into reconstruction. Rakuten drew another sinister, leering catfish figure with wild whiskers hovering menacingly over the city and wearing a hat labeled "Three Whole Years after the Earthquake" (fig. 5.31). The catfish addresses the viewer directly: "Look at the land readjustment that never moves forward even an inch. And look at the corruption of the plan itself. Unless government officials stop committing injustice and people give up being greedy, the real construction will never be realized not in three years not even in a hundred years!" At the bottom of the image, a man wearing a hat labeled "Tokyo City" breaks a branch trying to lever a seated man off the log of the "Land Readjustment Project." The log sitter's worker's cap is labeled "Citizens" (*shimin*), and he holds a box of "Greed" (*yokushin*).

Rakuten derided the process relentlessly, in another cartoon picturing an oversize figure of a corpulent government bureaucrat (perhaps a likeness of Suzuki Kisaburō, the home minister) in a large red chair looming over a warren of pathetic wood shanties in the capital (fig. 5.32). The caption to the right begins, "Officials Honored, People Despised" (*kanson minpi*), invoking a common Meiji-era expression derived from Chinese that criticized the imbalance between the government authorities' reverence

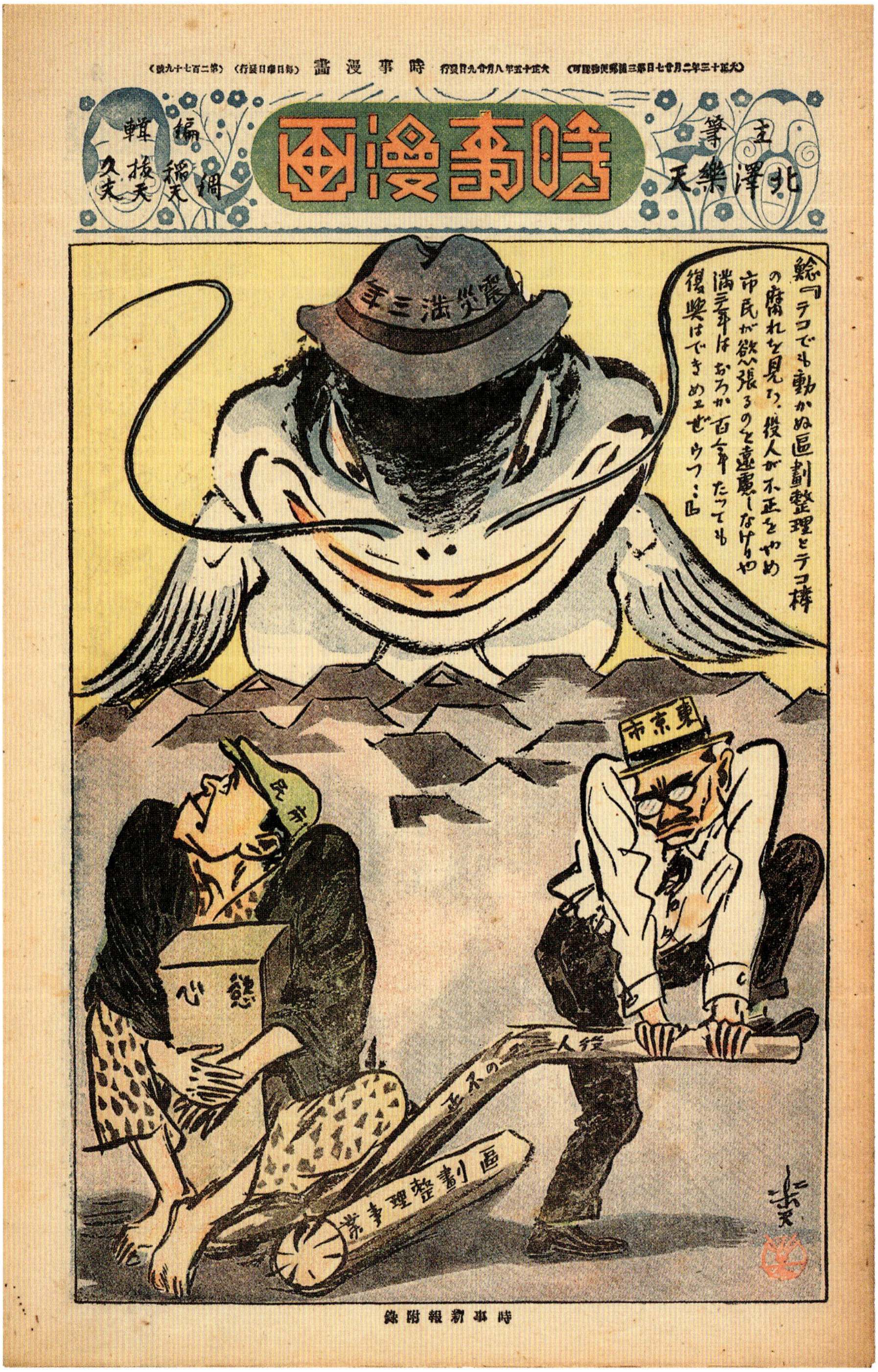

5.31 Kitazawa Rakuten, *Three Whole Years after the Earthquake* (*Shinsai man sannen*), cartoon, cover. *Jiji manga* 279, 29 August 1926. Collection of Ohio State University Billy Ireland Cartoon Library & Museum. By permission of Sangyō Keizai Shinbunsha.

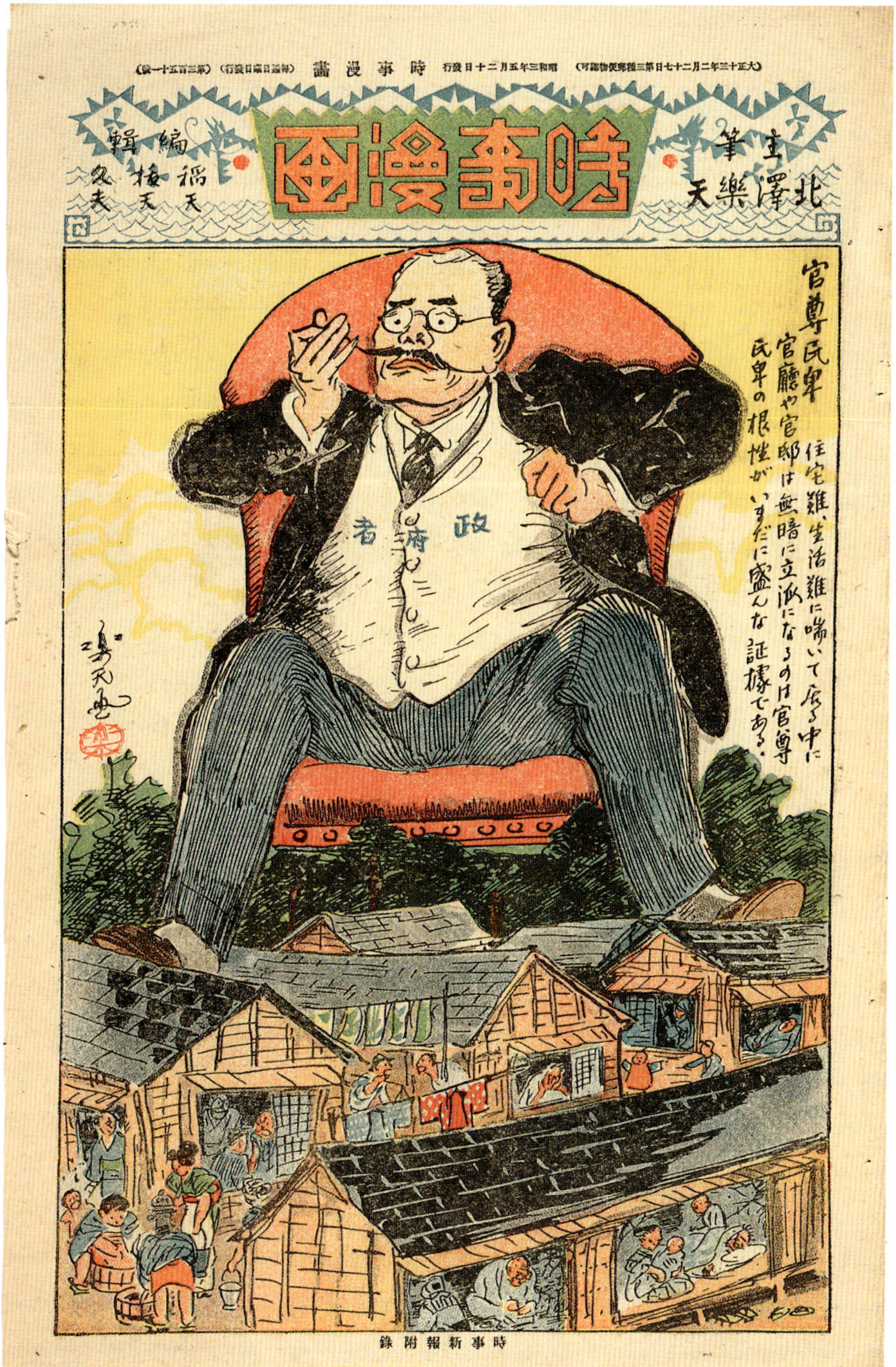

5.32 Kitazawa Rakuten, *Officials Honored, People Despised* (*Kanson minpi*), cartoon, cover. *Jiji manga* 351, 20 May 1928. Collection of Ohio State University Billy Ireland Cartoon Library & Museum. By permission of Sangyō Keizai Shinbunsha.

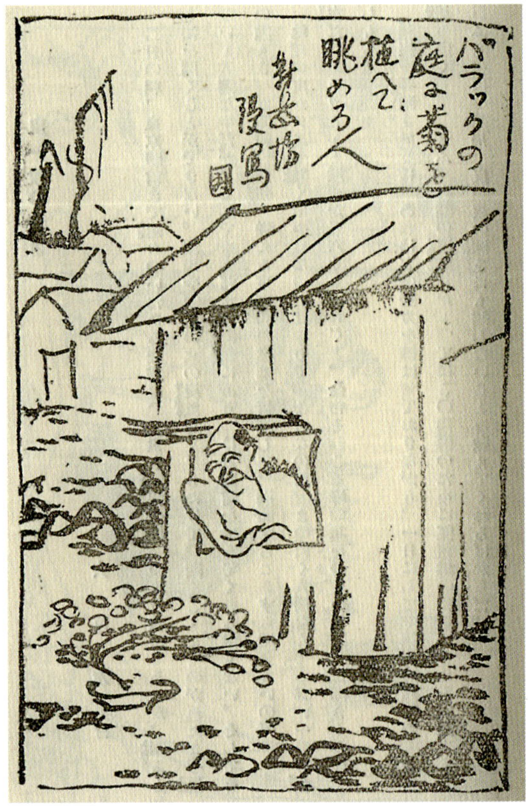

5.33 Shimizu Taigakubō, *Person Gazing at a Chrysanthemum* (*Kiku o nagameru hito*), drawing. Reimeisha Henshūbu, *Shinsai gafu: Gaka no me* (Earthquake Sketchbook: Eye of the Artist) (Tokyo: Reimeisha, 1923), 55. Private collection.

for officials and the upper classes and their disdain for average people. The text continues, "In spite of people's suffering from housing shortages and poor living conditions, the government buildings and residences are strangely getting better and better. This is proof that the traditional mind-set 'officials honored, people despised' survives." While refugees continued to live in fetid barracks, a luxurious building for the prime minister was inaugurated on 3 May 1928 at a cost of over 2 million yen. In addition, the government constructed a new central police station, a new Tokyo municipal government office, a new building for the Ministry of Education, and new residences for several cabinet ministers.[68] The cartoon graphically conveys that the mustache-twirling government fat cats had gorged themselves on the disaster-relief funds they had siphoned off, while leaving the puny inferior residents in the barracks to wait for assistance at their feet.

The residents of the barracks had no choice but to wait. In a spoof by former *Yorozu chōhō* political cartoonist Shimizu Taigakubō (1883–1970) (fig. 5.33), a man stares out the small window of his shack at the "garden" he is growing, which consists of a single potted chrysanthemum sitting on the barren earth. The image is ambiguous; it could be read as a message of hope, fertility, and rebirth, as the flower blooms even in the scorched earth—the proverbial new shoot. However, the pitiful drooping trees in the background convey disappointment rather than hope. It is not coincidental that the flower, the chrysanthemum, is the symbol of the imperial household; hence, the man looks wishfully to the emperor for hope—or perhaps looks out dejectedly, thinking of the promise the emperor has not fulfilled.[69]

BARRACK DECORATION: SYMBOLIZING SOCIAL TRANSFORMATION

Despite the wretched reality of the barracks, conceptually the structures became the inspiration for building a new, freer society among left-leaning intellectuals, including several artists' groups. For them, the symbol of social transformation was not the heaven-sent catfish but the lived-in reality of the barracks. Hoping to take advantage of the opportunity the earthquake presented to reshape the capital, artists thrust themselves and their work into the public eye through highly visible barrack-decoration projects covered in the popular news media.[70] The news stories portrayed the artists as heroes, actively engaged in their decoration labor. A simple cartoon that appeared in the evening edition of the *Tōkyō nichinichi shinbun* on 29 October 1923—titled "The Power of Art" and signed simply "Rakan"—read, "Barracks in the burned expanse are bleak, but art has tremendous power. However, by doing this, artist groups will also be revived."[71]

With the rallying cry "From the atelier to the streets!" (*atorie kara gairo e*), artists declared, "The first step toward reconstruction was to relieve the damaged spirit [of the city and the people in the city] through art."[72] The most active of these groups was the Barrack Decoration Company (Barakku Sōshokusha), an enterprise started by the Waseda University architecture professor Kon Wajirō (1887–1973) and a group of graduates of the design department of the Tokyo School of Fine Arts called the Forefront Company (Sentōsha), and artists from Action (Akushon), a splinter group of the well-established modernist association the Nika-kai.[73] Active from September 1923 until the summer of the following year, the Barrack Decoration Company issued a public statement on 2 October: "We have become the avant-garde of the imperial reconstruction. In an effort to create unconventional beautiful buildings, we have taken to working in the streets. We believe that Tokyo in the age of barracks has afforded a good opportunity to experiment with our art."[74]

Kon and other company members saw barracks decoration not only as a service to society and a means to achieve the "spiritual renovation" (*seishin fukkō*) of Tokyo but also as the first step toward revivifying the arts. As art redesigned daily life, daily life would "revive the arts" (*geijutsu fukkō*).[75] The common slogan "Reconstruction of the imperial capital" (*teito fukkō*) implied physical and spiritual revival. A variety of intellectuals saw this as a pivotal moment, an opportunity for change analogous to the profound social and cultural reordering brought on by the tumultuous historic events of World War I and the Russian Revolution. The barracks were emblematic of this moment of potential change.

The Japanese press and other print media were greatly taken with the image of artists working in the streets. Newspapers reported that after just a week the company already had several projects.[76] Publicity of artists' magnanimous assistance in the reconstruction effort fit well within the popular literary category of *bidan* (stories of noble deeds), which typically featured an altruistic individual and communal outpourings of support during and after the disaster.[77] The heroization of the artist is typified by the *Asahi Graph*'s inclusion of a photograph of three Barrack Decoration Company members on a ladder painting the front of the Tōjō bookstore barracks in a photo collage of the bustling activity of reconstruction work titled "Glimpses of Reviving Tokyo!"[78]

The company advertised for "young" clients who were willing to take a chance with experimental projects. The artists offered their services at cost, without commission, and announced that they were willing to decorate all kinds of structures, including stores, factories, restaurants, cafés, residences, and storage sheds. They saw barracks as monumental canvases on which they could express themselves freely.[79] For the bookstore, the company produced a seemingly random assortment of playful and expressionistic motifs accentuating the architectural members of the building with paintings scattered across the exterior. Architectural historian Fujimori Terunobu has argued that the barracks offered a new generation of architects and designers the opportunity to revel in and enjoy design—an indulgence the previous generation would not countenance—producing a great sense of liberation after the earthquake.[80] This contributed to a major shift in architectural practice in the postquake period away from stalwart, institutional structures toward more individualized, expressive forms with playful façade and interior ornamentation.

Describing the Barrack Decoration Company's work, member Asuka Tetsuo stated, "Our job is really like an oasis in the middle of the desert because we have to bring each beautiful thing to life on the rough, burned earth." He was one of several members who saw the company's mission in largely aesthetic and spiritual terms. For Asuka, beauty (*bi*) was an indispensable part of life and reflective of one's cultivation (*kyōyō*). He viewed the barracks as repulsive and pathetically devoid of "civilization," and this conviction compelled him to intervene in order to restore a bit of humanity and culture to the barren landscape. While acknowledging that the group's efforts were grossly insufficient to the task, Asuka expressed the hope that he and the other company members could perhaps help remedy at least some of the spiritual despair among the disaster refugees.[81]

Kon and painter Kambara Tai, in contrast, were inspired by the austerity of the barracks to rhapsodize about the beauty of the "simple life" (*sobokuna seikatsu*).[82] Kon found profound spiritual meaning in the stripped-down state of the barracks, idealizing the simplicity of poverty and affirming the sublimity of a subsistence-level existence. Unlike Asuka, he saw great beauty in the crude environment of the barracks, which he associated with the dignity of rural poverty. In this spirit, he called for a return to basics and reminded his readers to differentiate between the necessities of existence and the material desires of modern life.[83] Kon's conception seems particularly indebted to the views of major European design theorists like William Morris and Henry van de Velde, who were active in the arts and crafts movement and endeavored to simulate the mental state and rustic surroundings of country life in their designed environments. His invocation of the simple life in responding to the barracks was also rooted in his idealization of communal living, which accorded with the socialist political values infused into simple-life philosophy.

Kon emphasized the artisanal character of the decoration work, implying an honorable moral position in the transformation of the artist into a laborer. The manual labor of painting barracks temporarily recast artist-intellectuals as members of the working class, an identity they tried to express through their informal work outfits consisting of cut-off pants and gaiters. While sheepishly admitting that he felt more like an actor playing the role of a laborer than an actual laborer, he still expressed great delight in getting the chance to experience this blue-collar lifestyle. He even joined a labor union of craftsmen to improve his understanding of the profession.[84]

The closest the members of the Barrack Decoration Company came to designing a structure that might realize their proletarian ideals was their final project, the Tokyo Imperial University Settlement (Teidai Setsurumento), which was planned around October 1923 and completed in June 1924 after three months of construction. Unlike previous commissions, the settlement building was both designed and decorated by the company. It was a simple two-story structure with minimal decoration on the exterior except for red triangular bay windows protruding three-dimensionally from the façade. The building had ten rooms, each approximately 13.2 square meters (4 *tsubo*) in area and each containing a bed, a cupboard, and a bookcase.[85] Inspired by the project name, residents, predominantly students from Tokyo Imperial University, came to refer to themselves as "settlers." According to Naitō Sadako, a reporter for *Fujin no tomo* (Women's Companion) magazine, after the earthquake the settlers came to feel strongly that they had a social responsibility to share their knowledge with the workers alongside whom they had struggled in the reconstruction effort, since workers were still excluded from the benefits of higher education under the current system. At the time of Naitō's visit, they were planning to open a school in the settlement, and about eighty people were already attending classes there.[86]

Mary Beard, an American historian residing in Japan, noted the prominent role the working class took in disaster relief in her *Shūkan asahi* (Weekly Asahi) article entitled "The Supreme Test of the Japanese Nation." She wrote,

> On all sides we hear tales of the daring, resourcefulness, and unselfishness that the laboring class exhibited when the crash came and the fire spread. Certainly they have earned by their sacrifice at this time the most sincere consideration and the best possible treatment in the plans for the city that is to be. It sometimes takes a catastrophe on an immense scale to teach us true values. In normal times we are so apt to be engrossed in our personal enterprises and immediate circle of acquaintances that we forget or ignore the problems presented to our fellow citizens whose fate is really so closely bound up with our own.[87]

The sudden catapulting of the working class onto center stage in the popular press and the ostensible momentary leveling of Japanese society in the ashes of the earthquake inspired many idealistic expressions of a classless brotherhood and further fueled the mythicization of the noble manual laborer. Yet, troubled by the disturbing view of the masses as prone to lawlessness, racism, and self-interest that was conveyed in widely circulated images of the vigilantes, left-leaning intellectuals struggled to reconcile their heroic notions with the visible evidence.

Company artists performed the noble labor of barrack decoration rapidly, using whatever materials were available. Detailed accounts of a well-publicized company project, the Kaishin restaurant in Hibiya, provide an invaluable window into the group's working process.[88] The job was estimated to cost around eighty yen for materials (about eight months' rent for a detached house in 1924) and eventually took three days and ten people to complete.[89] Despite the poetic rhapsodizing about the simple life, this was clearly not a bare-bones endeavor and was more indicative of a middle-class sensibility. Yoshida Kenkichi was responsible for the building design and the overall decorative program. He produced numerous still lifes of foodstuffs for the interior sliding doors and wainscoting, painting dried sardines (*mezashi*), turnips (*kabu*), sweet potatoes (*satsumaimo*), and other items that he hoped would conjure up associa-

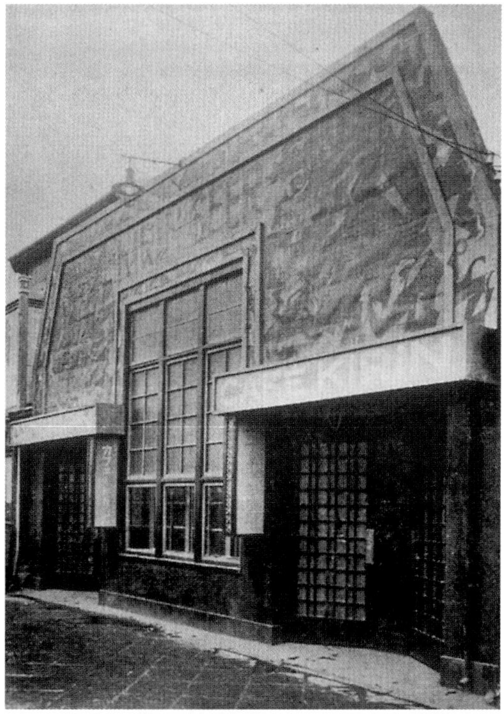

5.34 Barrack Decoration Company, exterior of Café Kirin on the Ginza, photograph, 1923. Kon Wajirō Archive, Kōgakuin Daigaku Toshokan.

tions with pleasant aromas. He also designed the cheerful menus. Kon was in charge of painting small floral patterns on the interior pillars. The exterior of the structure was painted in green with an ivy motif winding exuberantly over the surface. The building was capped off by a tall sign-flag poking out into the sky among the cluster of barracks.[90]

One of the company's grandest projects was the Café Kirin in Ginza, a structure designed by the Takenaka Construction Company (Takenaka Kōmuten) for Kirin Beer (fig. 5.34). Photographs and verbal descriptions reveal a stylish design with the front doors pushed off to the sides and a large series of windows inserted in the center of the boldly colored façade. The lower portion of the façade was painted an earth-black hue, but the areas around the windowpanes were rendered in a mixture of colors that became increasingly intense as one looked up the building, eventually climaxing in what one reviewer termed a "rococo-style" signboard emblazoned with the words *Kirin Beer* and *Café Kirin* in romanized capital letters.[91] The sign displayed two crouching *kirin* figures in profile glaring mischievously from their corner perches, bestial renditions of the eponymous mythological Chinese animal that served as Kirin's emblem—incidentally the same figures that stood at the ends of Nihonbashi Bridge.

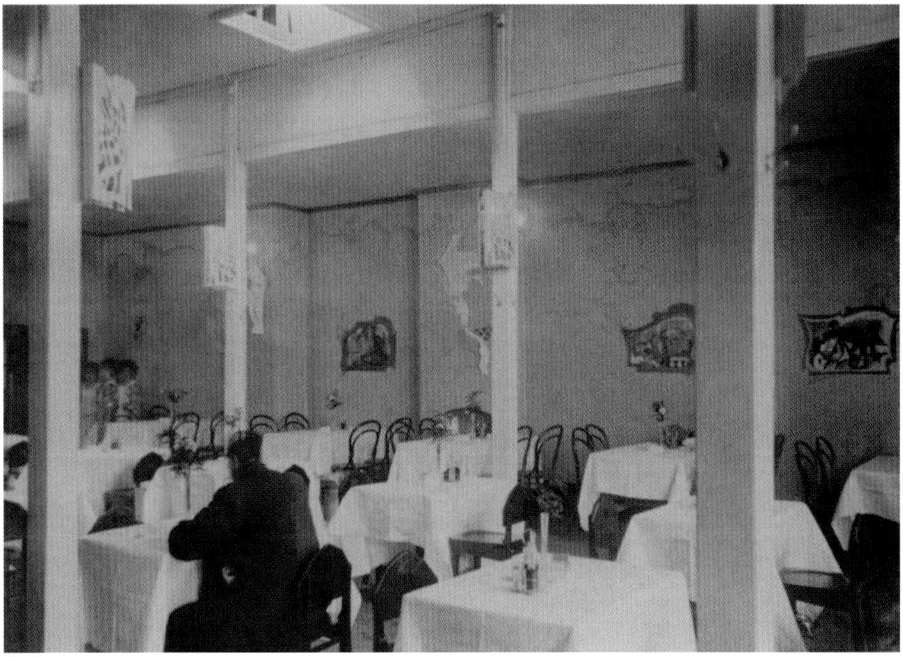

5.35 Barrack Decoration Company, interior of Café Kirin, photograph, 1923. Kon Wajirō Archive, Kōgakuin Daigaku Toshokan.

In general, the café interior (fig. 5.35) exuded a gay and cosmopolitan ambience meant to provide earthquake survivors a welcome temporary refuge from the grim and laborious task of reconstruction.[92] The interior decorations consisted of wall paintings and sculptural reliefs. An undulating line defined a lyrical, pastoral frieze around the top of the room, dividing the wall into different tones. Periodically, along the wall, a bold abstract design would appear, extending down to the height of the café tables, almost as if the frieze were being unzipped and a barrage of abstracted foliate and geometrical forms were falling out of the wall. At the bottom of this motif was a small, sculpted nude figure protruding from the wall.

Irregularly shaped oblong cartouches were placed intermittently along the walls, slightly above the eye level of viewers seated at the tables. Some of these displayed figurative landscape scenes, while others were bold expressionistic abstract compositions composed of dynamically pulsating lines and shapes. These vignettes read like paintings hanging on the wall or like windows peering out onto a dreamscape—a world of distraction away from the reality of the earthquake. Running down the middle of the establishment from front to back were two rows of columns on which were mounted ornate, perforated box-shaped light fixtures designed by Tōyama Shizuo that guided the viewer's eye toward a large painting on the wall behind the bar. The painted scene, which displayed seemingly classical figures under an arch in an abstract landscape with a fountain, contained sculptural elements in low relief.[93]

By December 1923, a debate had begun in the popular press over the aesthetic value of barrack decoration done largely by nonarchitects.[94] Supporters hailed the "beautification" and "artification" of the city during a time of intense hardship, but detractors flatly rejected these solutions as structurally impractical and overly concerned with subjective expression. For example, Endō Arata, a protégé of Frank Lloyd Wright and one of the most active barrack architects, having designed at least eight structures, was not amenable to the idea and roundly criticized Kon for the meaninglessness of the Barrack Decoration Company's work.[95] Architects from the prominent modernist group the Bunriha Kenchikukai (Secessionist Architecture Association) echoed his complaints. In his response, Kon affirmed that the company's playful, effervescent aesthetic was a legitimate response to the liberated space of the barracks.[96] The company was joined in its activities and defense of the barrack decoration project by the avant-garde group Mavo, a diverse collection of Western-style artists led by the rebellious Murayama Tomoyoshi and including Yanase Masamu. Even harsher criticism was leveled at Mavo's work. Vividly demonstrating the moral and territorial tug-of-war under way in the culture, Tokyo Imperial University architecture professor Kishida Hideto lambasted Mavo's designs as "temporary gratification" (*shunkan kōfun*) architecture and called instead for a "forceful primitivism" (*chikaratsuyoki genshisei*) that he deemed more appropriate to the extreme circumstances. He concluded, "Cursing polish and running away to intense curved lines and toy-store architecture is the same as being fed up with the princess and running to a prostitute."[97] In every sphere, the earthquake amplified previous concerns that modernity itself was a cultural crisis for Japan. For some, this was expressed in the generalized rhetoric of divine punishment; others directed more focused vitriol at the seemingly incomprehensible abstraction, uninhibited subjectivity, and rampant individualism of modernism.

While horrified by the extent of earthquake fatalities and the brutal murder of leftist colleagues, Mavo responded to the destruction of Tokyo and part of its institutional infrastructure with creative fervor, mounting traveling street exhibitions of members' artwork and designing architectural signboards (*kanban kenchiku*) and barracks across the city. Like Kon, the group's members reveled in the provisional, impermanent, and liberatory nature of the barracks. Moreover, the chaos, randomness, shock, violence, and fragmentation of the quake experience resonated with some of the fundamental aesthetic (or anti-aesthetic) principles of avant-garde art in the interwar period, particularly futurism, dadaism, and expressionism. And in an ironic convergence, like

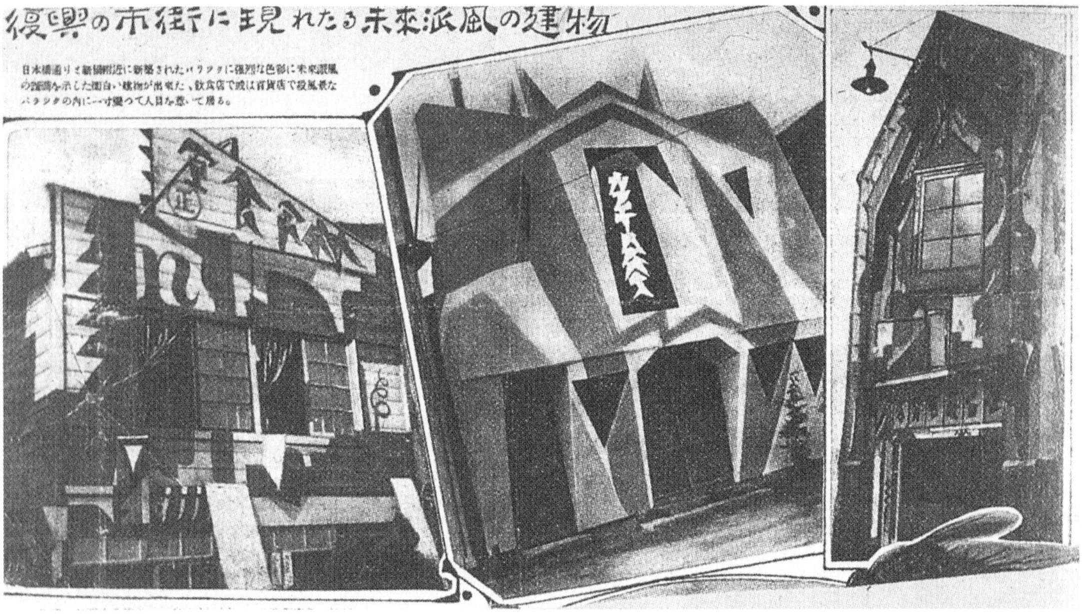

5.36 Mavo, Hayashiya Restaurant (Hayashiya Shokudō) *(far left)* and other modernist barrack decoration projects, "Futurist-Style Buildings Appearing on the City Streets of the Reconstruction" (*Fukkō shigai ni arawaretaru miraihafū no tatemono*), newspaper clipping featuring photo collage, c. 1923–early 1924. Murayama Tomoyoshi Scrapbook (original source unknown). Courtesy of Murayama Harue.

the conservative purveyors of the divine-punishment discourse, these Japanese artists similarly viewed the quake as an opportunity to combat immorality, corruption, and iniquity, but for them, these injustices were the product of capitalist exploitation of the working class and the social suppression of individualism. Thus, for Mavo and many leftist-oriented intellectuals, barracks were a true revelation of the spirit of the proletariat and embodied the possibility of leveling social relations. Many of Mavo's barrack-decoration projects, however, like those of the Barrack Decoration Company, were commercial merchant establishments such as the Hayashiya Restaurant pictured in a dramatic newspaper photo collage (on the far left in fig. 5.36) under the headline "Futurist-Style Buildings Appearing on the City Streets of the Reconstruction," and while they may have provided valuable relief from the harrowing conditions of the postquake cityscape, it is unclear how they helped relieve class conflict.[98]

Despite the dynamic flurry of creative activity in the wake of the disaster, the period of barrack construction began to come to an end about five months after the quake in early 1924, when the Japanese state and Tokyo municipal government began planning for permanent reconstruction of the city. Demand for barrack decoration decreased accordingly. While ephemeral, the barrack-decoration projects had a lasting impact on the expansion of architectural practice in the postquake period and are preserved in photographic compilations and in the exhibits at the earthquake memorial museum, where tattered scrapbooks display newspaper clippings of these fanciful incursions into the cityscape—and Café Kirin sits prominently at the top of the page.

The three powerful emblems of the 1923 earthquake—refugees, catfish, and barracks—illuminate the vigorous public debate that developed in Japan as people sought to claim the meaning of the disaster for their own purposes. Was the postearthquake community altruistic or corrosive? Representations of heroic survivors, loyal imperial subjects, compassionate family members, magnanimous humanitarians, sympathetic artists, advocates of social unity, and champions of the working class offered myriad positive images of an altruistic Japan. At the same time, lawless vigilantes, bigoted colonialists, misogynists, corrupt profiteers, political hacks, and menacing modernists revealed the deeply corrosive nature of the community. Interpretations differed widely according to a person's politics, class, gender, and ethnicity. These differences gradually became more muted, however, as the discourse of reconstruction ultimately yielded a retrospective official master narrative of the quake that then passed into collective memory.

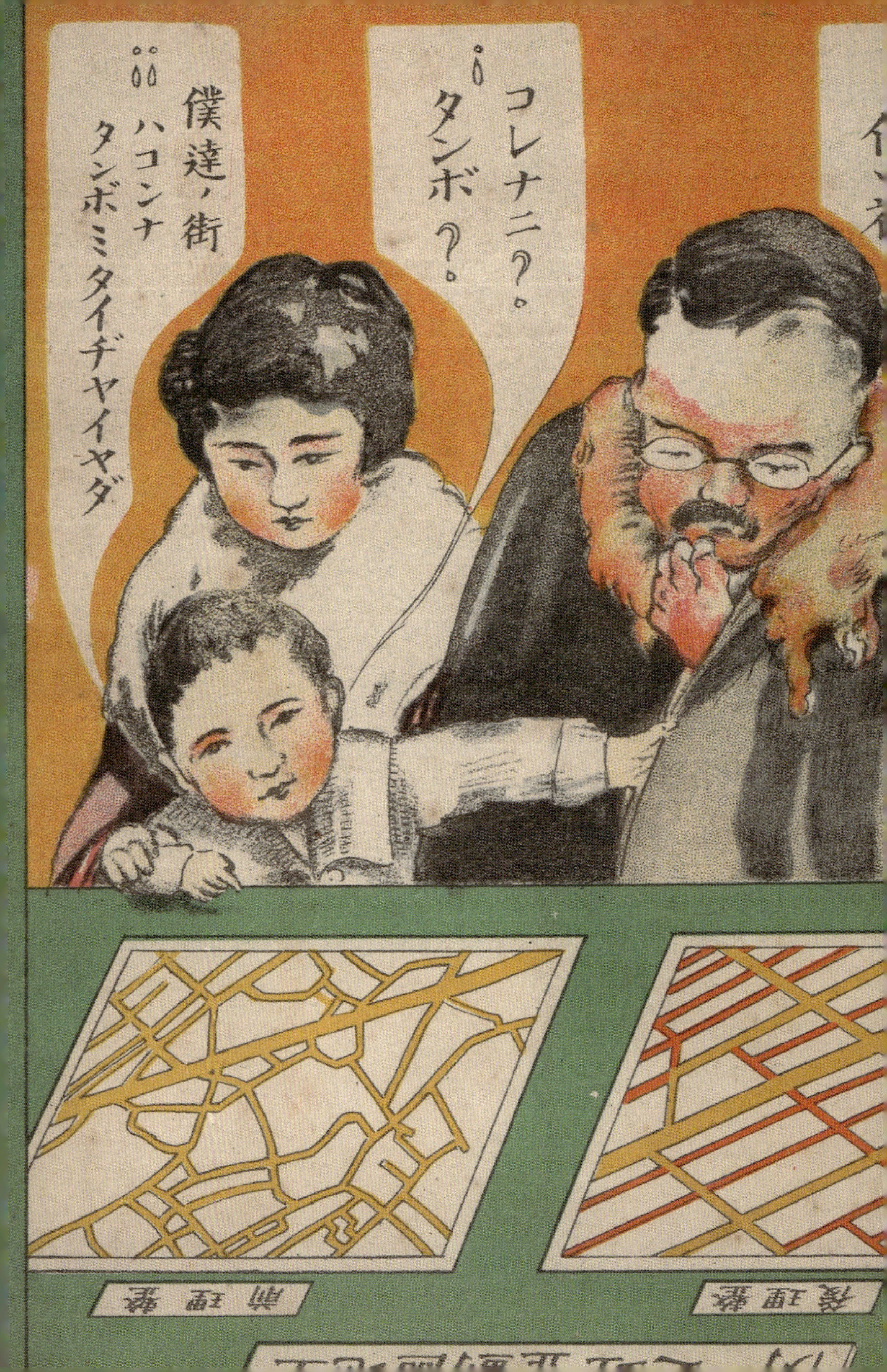
コレナニ？。タンボ？。
僕達ノ街ハコンナタンボミタイヂヤイヤダ……
整理前
整理後

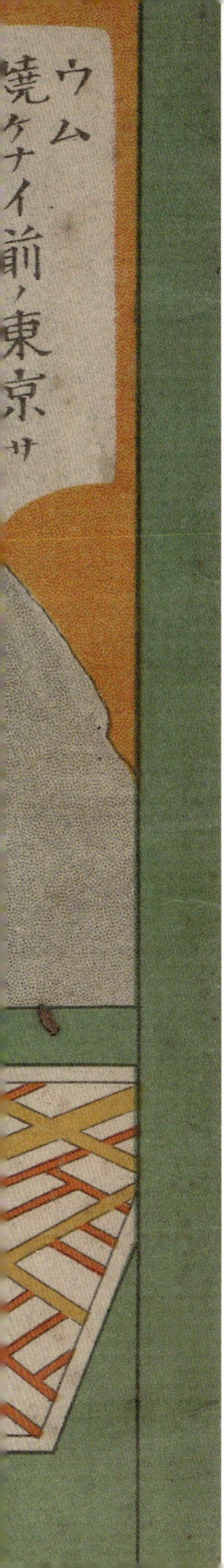

6

RECONSTRUCTION'S VISUAL RHETORIC

Though fire destroyed the houses
The spirits of the sons of Edo will never dim, just watch us
Ara-ma, oya-ma
Barracks shacks
Standing in a row
At night we lie down to sleep and gaze at the moon
eezo eezo
The Imperial Capital rebuilds
eezo eezo
As the wife says to her husband
Be strong!
Ara-ma, oya-ma
Haven't even the Imagawa cakes been renamed Reconstruction cakes?
Be strong!
eezo eezo
The Imperial Capital rebuilds
eezo eezo
Children born in the midst of the tumult
Are named Shintarō
Ara-ma, oya-ma
There's Shinji and Shinsaku, Shinko and Fukuko
When these kids are grown up
The earthquake will be the seed of their story
eezo eezo
The Imperial Capital rebuilds
eezo eezo

To the upbeat rhythms of a lively march, residents of the imperial capital hummed "The Reconstruction Song" (*Fukkō bushi*) as they began the process of recovery and rebuilding. The tune expressed their resilient spirit for reviving the capital, a process they had undertaken on many previous occasions, although never on such an enormous scale. There was a saying among merchants in Edo that if a business could not get up and running again within three days, it would not survive. Such was the rapidity of reconstruction. Some long, wooden single-story row houses in back alleys known as *nagaya* burned so often that they were humorously referred to as *yakeya* (burn houses). Yet their rental was so lucrative and their rebuilding so quick and cheap that *yakeya* were profitable even if they burned down regularly.[1] That towns and cities would be rebuilt was a given in Japan, and in most previous instances, the task was considered a local responsibility and private individuals were left to reconstruct their own property. The shogunal government was mainly responsible for reconstructing official buildings and for providing social and economic assistance such as extra food rations, provisional debt relief, and temporary housing. But the unprecedented scale of the earthquake destruction and the fact that Tokyo was now the national capital of the burgeoning Japanese empire made reconstruction a negotiated public-private process of great interest to both the government and its citizens, one that was deeply implicated in contemporary social and political issues.

As the authors in *The Resilient City* ask, "What does each particular process of recovery reveal about the balance of power in the society seeking to rebuild? Whose vision for the future gets built, and why?"[2] In 1923 Japan, the conflicting interests of bureaucrats, the prime minister, the nationally elected parliament, individual property owners, and other earthquake victims, combined with severe budgetary constraints, preempted implementation of the most radical proposals for Tokyo's urban reorganization and resulted in a severely truncated vision. While the main architectural and urban outcomes of the reconstruction process have been well studied, how the negotiated process of reconstruction was visualized has not.[3] Thus, this chapter considers the potent visuality of Tokyo's reconstruction process, as the official visual language of high modernist, rationalized urban design came into conflict with other modes of seeing the city, and with the divergent concerns of its constituents. Some social critics expressed deep mistrust of the government plans and stakeholders' self-interest by visualizing the abstractions and perceived absurdities of the reconstruction process. They particularly targeted Home Minister Gotō Shinpei as the embodiment of government overreach and partisan politics. Influential artists, architects, and designers sought to modify the Reconstruction Bureau's (Fukkōin) rationalized approach to urban planning on the grounds that it did not consider the aesthetic aspects of urban space and the city's nature as a built environment. As the capital went through the classic stages after a major catastrophe—rescue, recovery, rebuilding, and remembrance—it became clear, however, that the overriding productivist ethos of reconstruction was generating an authoritative "progressive narrative" of the disaster—a forceful reiteration of modernity's logic of creative destruction—aimed at legitimating urban development and renewal.[4] Eventually, this became the metanarrative of the event, enfolding individual interests and sacrifice with remembrance of the quake itself, but only after a vigorously contested process took place in which the residents of the capital were both audience and patrons in the sphere of visual culture. By examining the dynamic evolution of this process in this chapter and the next, I aim to illuminate the complexity of reconstruction's visual rhetoric to show its integral connection with the social demands of remembrance.[5]

Reconstruction's Rhetoric of Urban Renewal

Despite the immense trauma of the catastrophe, many government bureaucrats treated it as a fortuitous urban planning opportunity to rationalize and modernize the capital. However, their unremittingly optimistic focus on long-term national goals often came at the expense of the individual rights and concerns of the suffering populace. When Gotō Shinpei wired his mentor, American public administration expert Charles Beard, for advice about how to deal with the catastrophe, Beard famously wired back, "Lay out new streets, forbid building within street lines, unify railway stations."[6] Beard's oversimplified response is indicative of the pragmatic urban planning perspective of Gotō and his technocratic followers, who viewed reconstruction as a scientific, depersonalized process of rationalization that could be articulated and made understandable to the population in clinical maps and statistical charts to pave the way for state interventions in the social life of the city.[7] Adherents of this "high modernist" scientific and technological progressivism believed that reconstruction required a forward-looking approach that relegated the disaster and Tokyo's urban history to the past. Yet social demands for remembrance necessitated the development of a framework within which the visceral and traumatic losses of the event could be acknowledged. In this narrative, Gotō and the Japanese leadership were visionaries that would benevolently guide their country back from the brink of apocalypse to the status of respected world power within a decade. As historian Yoshimi Shunya has noted, the earthquake and reconstruction enabled the spectacle of the new metropolis of modern Tokyo to emerge, with its lively cafés, department stores, tall buildings, impressive railway transportation hubs, and major boulevards crowded with automobiles—producing a symbolic shift in urban social consciousness and cultural topography from the bawdy entertainments of Asakusa to the chic leisure economy of Ginza.[8] However, while we must acknowledge the tremendous achievements of reconstruction and some of its underlying social welfare benefits, we must look more closely at the rhetoric and reality of the reconstruction process.

The sense of opportunity for revamping the capital was clearly expressed in many of the retrospective accounts of the reconstruction project, which showed little fondness for the historical capital and its Edo roots. When describing Tokyo's roadways, for example, one official chronicle noted,

> Roads: The roads of Tokyo before the earthquake were mostly relics of Yedo days. The roads plan of Yedo days was drawn mainly with the idea of being prepared for street fighting, and was full of bends and windings with the Yedo castle (the present Imperial Palace Precincts) for the center. In consequence, the streets of Tokyo abounded too much in inflexions and crookedness at the cost of systematization. Besides, they were generally narrow, and become fearfully muddy in the rain; so they were exceedingly unsatisfactory as thoroughfares for a great metropolis. The City of Tokyo had long since taken the matter into consideration, and been making efforts to effect improvement of roads and streets from as far back as 1880. Roads that offered free traffic for automobiles were yet very few in number, innumerable narrow pathways running between them, that could not but occasion great concern from the point of view of maintaining peace, health and sanitation and safe traffic.
>
> A revolutionary improvement of roads and streets was a matter of absolute necessity. But improvement could not be forced drastically on busy business centers all at once, owing to old usages and individual economic considerations. But the great earthquake did its part in burning down more than half of the busiest center in downtown Tokyo, and unexpectedly gave the people an opportunity to carry out their long-standing desire thoroughly to improve the roads and streets by reconstructing and widening them.[9]

Government statements such as this did not display even a tinge of nostalgia or lament for the loss of the "crooked" roads of Edo and their attendant urban cultural fabric. In fact, many public commentaries maligned Tokyo for its obsolete infrastructure, random organization, inadequate housing for the urban working class, poor hygiene conditions, and even bureaucratic corruption.[10] As mayor of the city from December 1920 to April 1923, Gotō had already tried to implement a grand vision of urban planning that included radical land readjustment for thousands of city residents to accommodate new main boulevards, more public green spaces, and more logical neighborhood dispositions.[11] Historical continuity with the shogunal capital of the past was not important to him. Gotō sought to create a majestic modern metropolis whose broad boulevards would provide easy access for domestic and international commerce.[12] The budget for his initial plan was estimated at a whopping 800 million yen, at a time when the entire gross domestic product of Japan was only 1.3 billion yen. "Mayor Gotō's 800 Million Yen Plan," as it was called, was presented in the scientific language of modernity: clinical bar graphs, pie charts, and large-scale maps (fig. 6.1).[13] This language presupposed the progressive value of such radical incursions into the urban landscape for the improvement of trade and, ostensibly, the betterment of society. It was, in effect, an affirmative declaration of modernity's logic of creative destruction. And it was a language that the state actively deployed after the earthquake in the discourse of reconstruction.

At the time, people sarcastically referred to the plan as "Gotō's Big Talk" (*Gotō no daifuroshiki*) because of its wildly extravagant and all-encompassing scale, and its exorbitant budget and unpopularity prevented its passage.[14] When the earthquake hit, however, it was a blessing in disguise—or so it seemed to Gotō and urban planning reformers. Reportedly saying that the earthquake created an "ideal opportunity to make a perfect new city," Gotō presented a celebratory attitude that was in stark contrast to widespread public sentiments of bereavement.[15] Reconstruction optimism was subtly parodied in "A Reconstruction Ditty" (*Fukkō kouta*) by unknown *enka* composer Beniya Shunshō, using the melancholic musical melody of the 1921 prequake hit "A Boatman's Ditty" (*Sendō kouta*) by Noguchi Ujō (lyrics) and Nakayama Shinpei (music). "A Boatman's Ditty" was infamous for its extreme expression of despair and pessimism, a mood so dark, in fact, that its "sinful" attitude was even widely blamed for bringing on the divine punishment of the earthquake.[16] Replacing the original lyrics "I am withered grass on the river bank. You, the same. We'll never flower in this world," "A Reconstruction Ditty" began:

> All around me the earth is scorched.
> Yet the vendors are selling five-*sen* dumpling soup [*suiton*].
> I am a man, and I will never die.
> I will never be brought to ruin.
> Even though our house has burned down,
> our Tokyo spirit did not die.
> Look at us![17]

The gloomy undertone of the ditty's minor pentatonic melody stands in ironic contrast to the hopeful lyrics. The irony is underscored by the cheerful cover illustration on the sheet music (fig. 6.2), which shows a woman dressed up in festival attire displaying on her folding fan and jacket lapel the homonyms *fukkō,* which can mean either "reconstruction" or "felicity and good luck." The personal pathos of the music is on a different

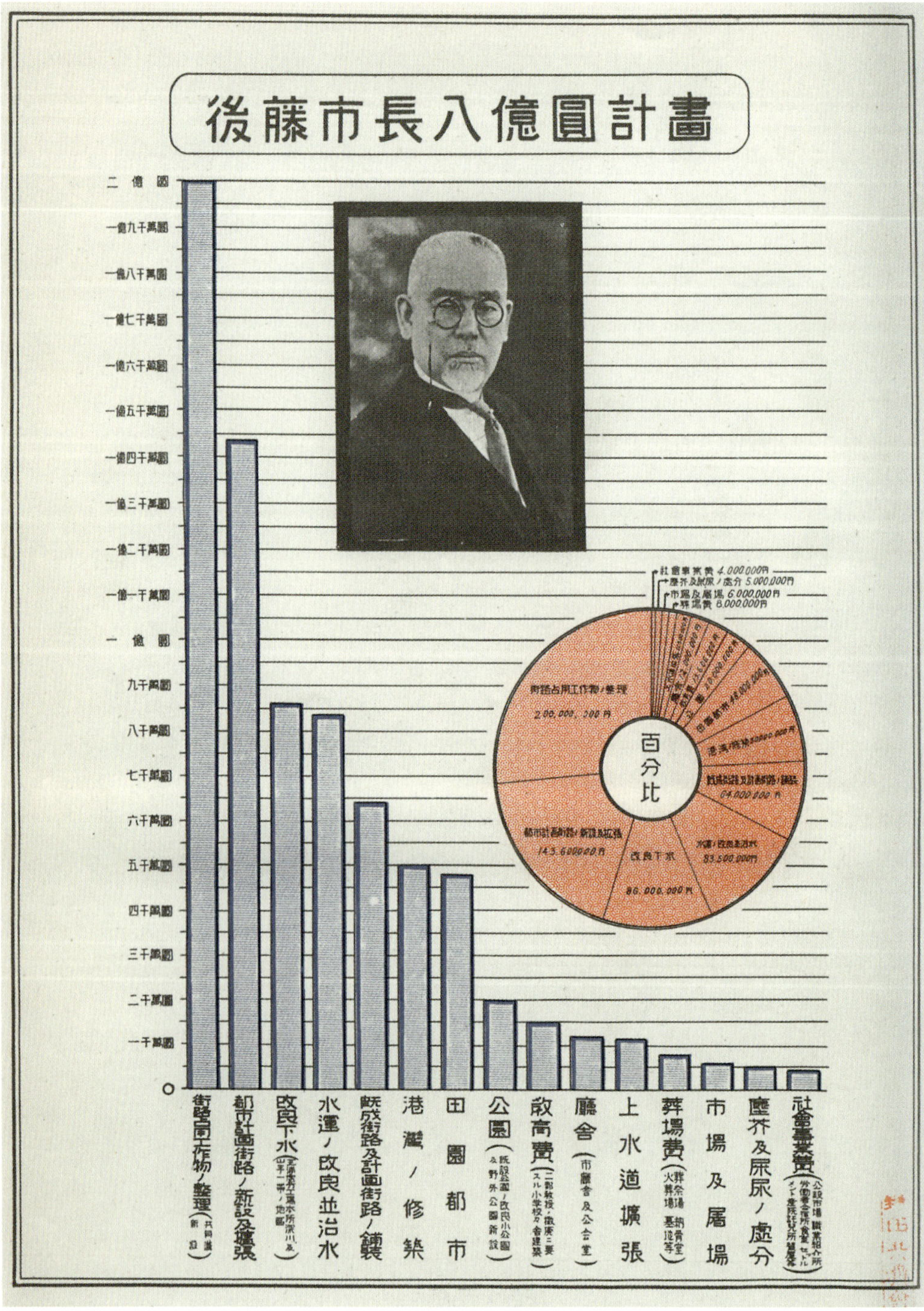

6.1 Mayor Gotō's 800 Million Yen Plan, chart. Uchiyama Zenzaburō, ed., *Teito fukkō jigyō taikan* (General Survey of the Imperial Capital Reconstruction Project) (Tokyo: Nihon Tōkei Fukyūkai, 1930), 1:ch9, 14.

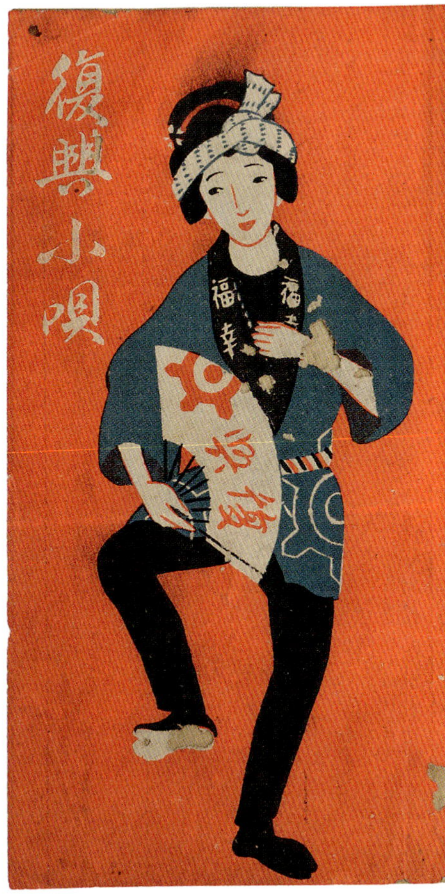

6.2 Artist unknown, "Reconstruction Ditty" (*Fukkō kouta*), cover, sheet music, woodblock print, $6\frac{1}{2} \times 3\frac{3}{8}$ in. From the set *Ten Taisho-Era Songbooks,* Beniya Shunshō (composer) (Tokyo: Shunkōdō, January 1924). Honolulu Academy of Arts: Gift of Patricia Salmon, 1994 (25371.10).

emotional register than the affirmative national symbolism of the cherry blossom in the figure's hair and the display on her fan and jacket of the wheel-shaped crest of the city of Tokyo (adopted in 1889), which represented the sun with six radiating rays to signify a thriving Tokyo as the center of Japan. While a chorus of voices extolled reconstruction as providential for the capital, this cheery overlay of national resilience could not drown out the public's ongoing laments of loss, despair, and uncertainty.

Reconstruction's Critics

Appointed chair of the Reconstruction Bureau, Gotō was the impresario of reconstruction planning. He took on so many different roles within the government and the reconstruction process that social critics soon took notice. A *Jiji manga* cartoon (fig. 6.3) by satirist Ogawa Jihei shows Gotō about to exit a building and staring at

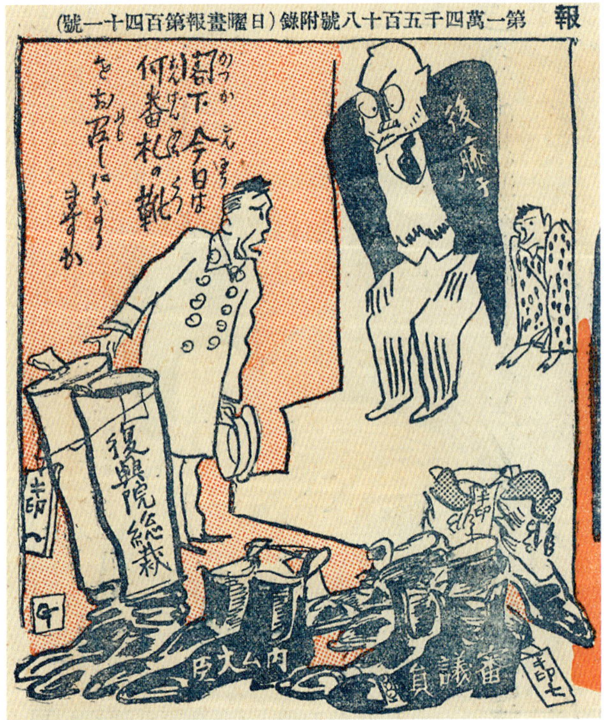

6.3 Ogawa Jihei, "Your Excellency, which shoes will you be wearing today?" (*Kakka kyō wa nanbansatsu no kutsu o omeshi ni narimasu ka*), cartoon. *Jiji manga* 141, 16 December 1923, 4. Collection of Ohio State University Billy Ireland Cartoon Library & Museum. By permission of Sangyō Keizai Shinbunsha.

the shoes in the entryway (*genkan*), the place where people commonly left their footwear in Japan. A doorman presents him with an array of shoes labeled "President of the Reconstruction Bureau," "Member of the Deliberative Council," and "Home Minister," with the tall boots for the Reconstruction Bureau conspicuously larger than the others. The doorman asks, "Your Excellency, which shoes will you be wearing today?" Gotō's visage—along with Yamamoto's—became inextricably linked with the government reconstruction process, and his distinctive pince-nez, somewhat sinister furrowed eyebrows, and well-manicured mustache and beard were often exaggerated for satirical effect. Visual satirists took on the mantle of the vox populi to express widespread apprehension about the government's future plans.

In the Imperial Edict on Reconstruction (*Teito fukkō ni kansuru shōsho*), the imperial household charged the prime minister and parliament with overseeing the reconstruction, which included convening an advisory board of eminent statesmen and public figures called the Imperial Capital Reconstruction Deliberative Council (Teito Fukkō Shingikai) to keep Gotō and his urban planners in check.[18] The necessity of this oversight soon became obvious when Gotō submitted his initial recommendation that the government buy all of the city's land, even in the Yamanote (high-city) areas less affected by the quake, to facilitate complete erasure and rational reorganization. In this scheme, the city would sell back standardized parcels to the people, although not necessarily in the areas of their original property. Not only was this fiscally impossible, it was predicated upon an extreme undervaluation of property that would have cheated landowners out of the full worth of their legitimate assets. After hearing the outpouring of public protests, the prime minister and parliament strenuously rejected this proposal as antithetical to the nation's core value of respect for private property.

Rakuten and his colleagues voiced concern about the closed-door nature of these discussions, which gave the people no access to deliberations that profoundly affected their future. His cover illustration for the 28 October 1923 issue of *Jiji manga* (fig. 6.4) visualized this lack of transparency and the exquisite hierarchy of bureaucrats and influential elites, whom he shows seated at long tables organized in a clear pecking order inside and outside the temporary halls of power. The cynical upper caption reads, "The Reconstruction Bureau seen through a hole in [the fence along] the moat by the people of Tokyo." The people, including a figure representing the Municipal Assembly, hover at the bottom, outside the fence, only able to peep at the inner circles of power through the broken boards. Shooed away from the fence, one man comments to another, "It's an absolute contradiction that these people who love the hierarchical order so much had a popular election," pointing to the seemingly antithetical autocratic tendencies of public figures who had created the new system of democratic elections. Closest to the fence in the outer garden are the seats for the Imperial Capital Reconstruction Consultative Committee (Teito Fukkō Hyōgikai), composed of urban planners, engineers, and architects, and the Imperial Capital Reconstruction Councilors Committee (Teito Fukkō Sanyokai), which included advisors from government and academia, clearly indicating their position on the outermost ring of power.[19] Slightly closer to the inner hallowed halls but still outside in the garden sits the table for the Imperial Capital Reconstruction Bureau Board of Directors (Teito Fukkōin Rijikai). And visible only in silhouette inside the building through the windows is the Imperial Capital Reconstruction Deliberative Council, which convenes in a room down the corridor from the shuttered special room off in the distance for the president (*sōsaidono*), Gotō himself. Such cartoons played on the well-publicized news photographs of Gotō seated outdoors in consultation with the earthquake cabinet, which seemed to suggest that the people had access to the political machinery of power, though in actuality they did not.

In his *Reconstruction Theater* (*Fukkōza*) (fig. 6.5), depicted as a mock kabuki play, Ogawa also lampooned the critical debates emerging about reconstruction. In the image, a circle of critics jeer at Gotō and Yamamoto, clearly identifiable in the center, as they pound *mochi* rice cakes (*mochitsuki*). A common visual metaphor for prepa-

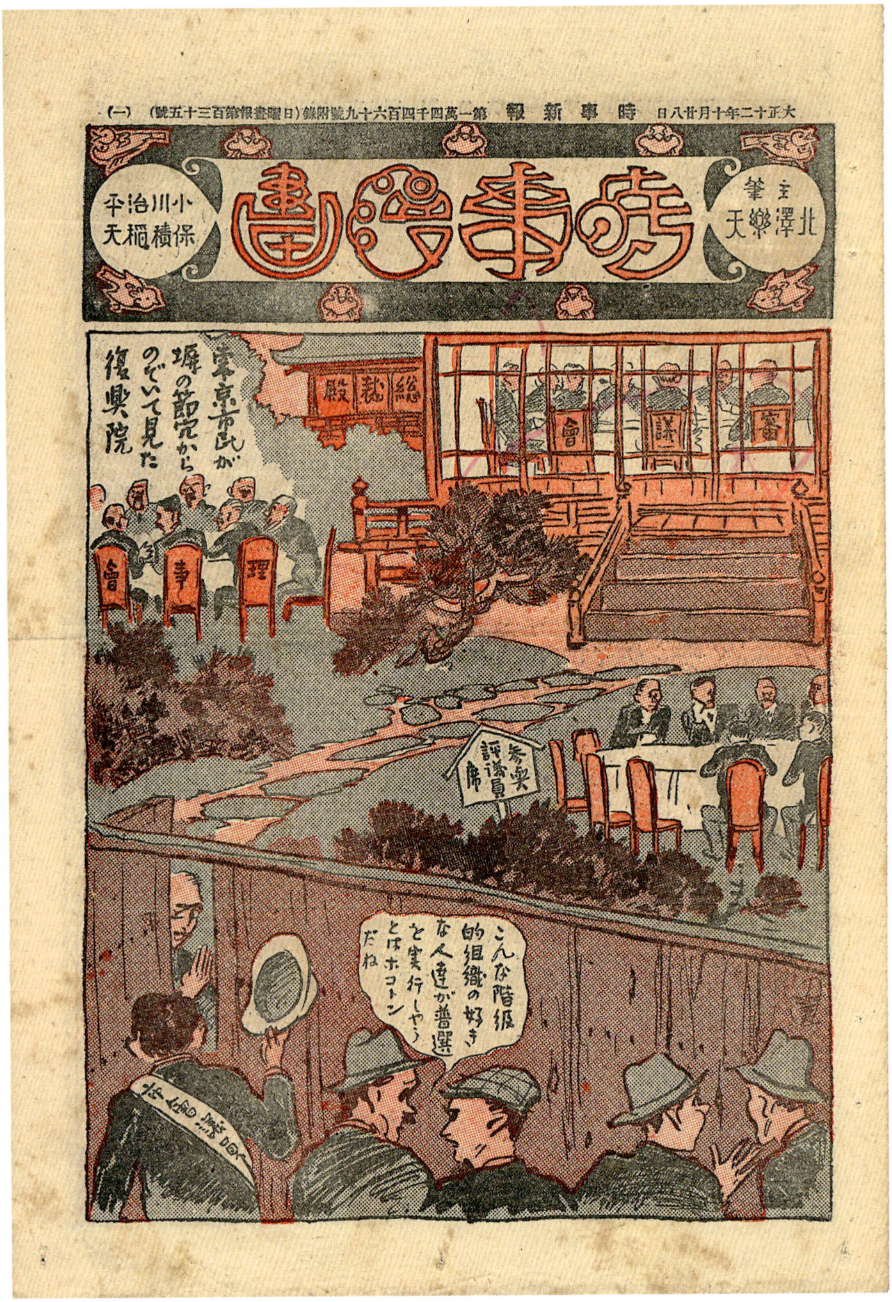

6.4 Kitazawa Rakuten, "The Reconstruction Bureau Seen through a Hole in [the Fence along] the Moat by the People of Tokyo" (*Tōkyō shimin ga hori no fushiana kara nozoitemita fukkōin*), cartoon, cover. *Jiji manga* 135, 28 October 1923. Collection of Ohio State University Billy Ireland Cartoon Library & Museum. By permission of Sangyō Keizai Shinbunsha.

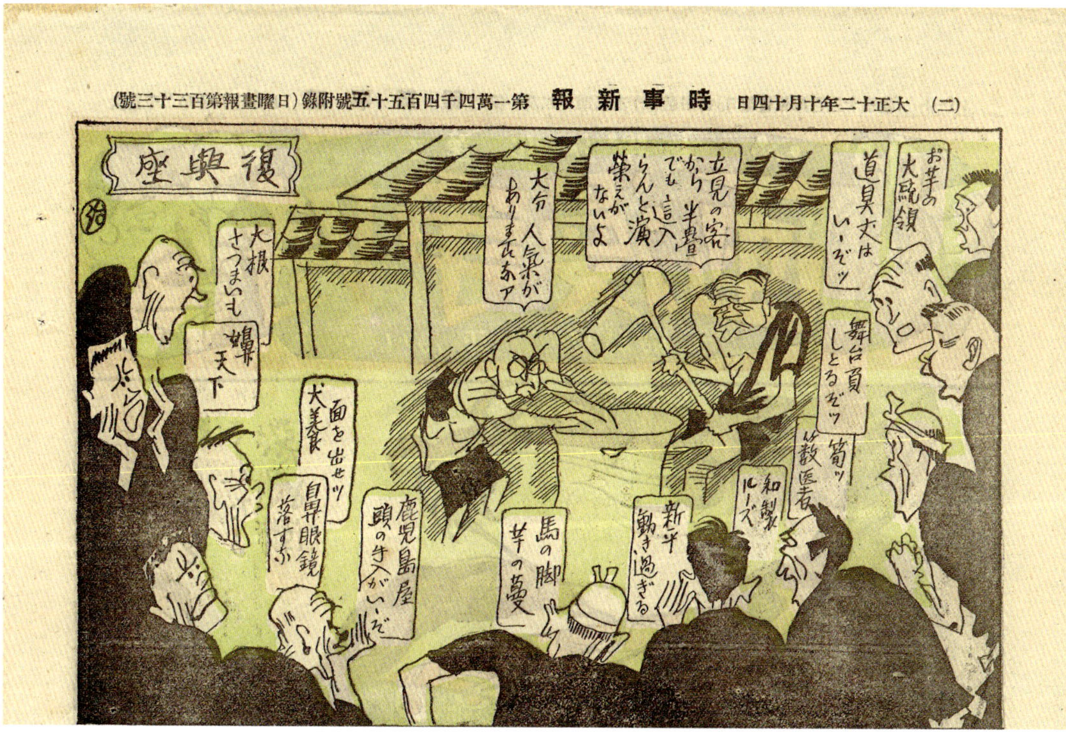

6.5 Ogawa Jihei, *Reconstruction Theater* (*Fukkōza*), cartoon. *Jiji manga* 133, 14 October 1923, 2. Collection of Ohio State University Billy Ireland Cartoon Library & Museum. By permission of Sangyō Keizai Shinbunsha.

ration, *mochi* pounding here represents the reconstruction process. Yamamoto comments, "If the people standing around watching don't even jeer at our performance, it is not a success." Gotō replies, "It's not very popular, is it?" The hilarity of seeing these typically dignified bureaucratic figures in plebeian work clothes engaged in hard labor was obvious. In addition, a group of men, some elites and others blue-collar workers, stand around commenting on the pair the way that kabuki viewers regularly yelled out to actors during performances. From right to left, they call out, "Dud Prime Minister," "The only good thing is your tools," "You are a terrible actor, an impostor!" "Japanese Roosevelt," "Shinpei goes too far," "Horse legs, inexperienced actor," "Kagoshima-ya [Yamamoto], like your hairstyle," "Inukai [Tsuyoshi, minister of education], show your face," "Don't drop your glasses," "Petticoat government," "Lousy actor! Sweet potato [Yamamoto]!" In the reconstruction process, everyone was a critic.[20]

Parodying the self-interested input about urban planning from various parties, Ogawa also offered the *Athletes' Plans for Reconstruction* (fig. 6.6). The absurd propositions for urban reconstruction based on an athlete's perspective obliquely critiqued the similarly self-interested and sometimes impractical suggestions of stakeholders in the city. One sportsman in the scene suggests, "Let's connect the Sumida and Rokugo [Tamagawa] rivers. I want to enter a boat race there through the moat in spring beneath the green pine trees." Another says, "Construct a spiral road that can also be used as a marathon track." The runner is shown on a large spiral road going inward toward the center without getting anywhere. "Construct a grand athletic field at the center of the park, if necessary; it can certainly be used as an evacuation area for refugees," chirps another. The incongruous suggestions continue: construct pools to gather rainwater for use in emergencies and also heat them for use as public baths; construct golf courses around the city that could double as airports; have companies build unisex tennis courts that could be used as evacuation areas. Clearly, one person's rationality was another's absurdity.

One pithy Ogawa image captures Gotō's delicate balancing act in moving reconstruction forward and appeas-

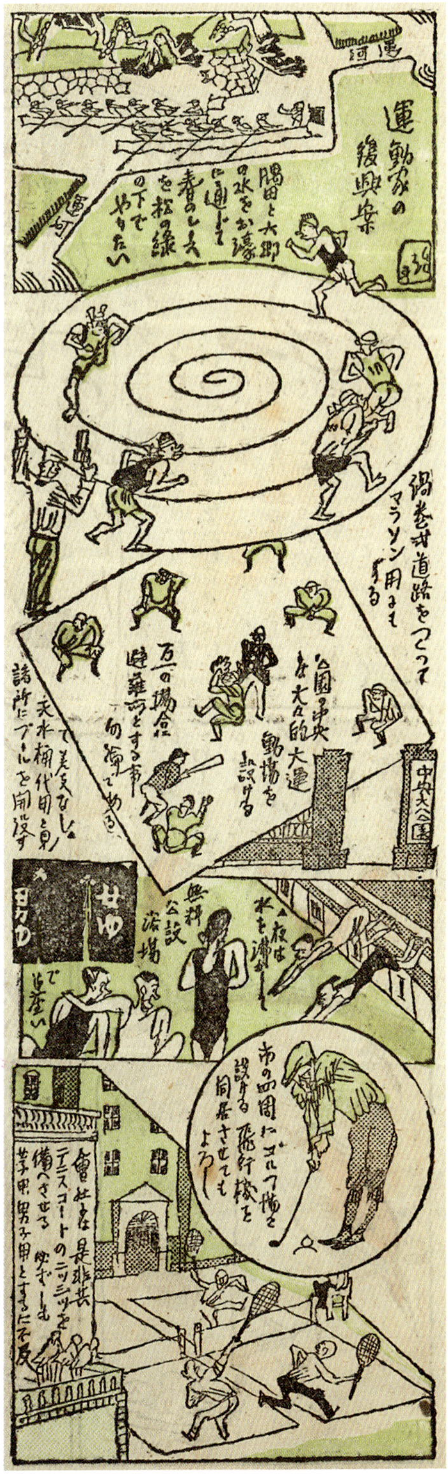

6.6 Ogawa Jihei, *Athletes' Plans for Reconstruction* (*Undōka no fukkōan*), cartoon. *Jiji manga* 133, 14 October 1923, 7. Collection of Ohio State University Billy Ireland Cartoon Library & Museum. By permission of Sangyō Keizai Shinbunsha.

ing the diverse interests of stakeholders. The drawing shows him doing a handstand on top of an architect's compass, balancing precariously like the traditional ladder-top acrobats often pictured in Japanese woodblock prints (fig. 6.7). "Home Minister Gotō standing on the big stage of the Imperial Capital Reconstruction," reads the caption.

Among all of the clamoring interest groups, Gotō was most beholden to the Tokyo municipality. His first concern was to assert Tokyo's continuation as the seat of government and to have his plans affirmed as a "national" effort to ensure state support; otherwise, his reconstruction project could have been treated as a regional endeavor, which had been the case in all previous major disasters. The new national status of the capital changed the political lay of the land, but Tokyo city still had to argue for its viability. After this simple yet crucial reaffirmation was accomplished, Gotō's reconstruction bureau and the cabinet submitted their plans to parliament in December 1923, a scant four months after the quake.

By that time, the Deliberative Council had already significantly reduced the budget from 3 billion yen to less than 597 million yen. The parliament engaged in another round of intense debates about the proposed plan, eventually removing an additional 130 million yen from the budget, leaving 468 million yen.[21]

The treacherous discussion and brutal fiscal cuts were viscerally represented in Rakuten's cover illustration *The Imperial Reconstruction Plan on the Chopping Board* (*Sojō no teito fukkōan*) (fig. 6.8), which depicted Gotō's plan as a turkey about to be slaughtered on the bloody cutting board of "the extraordinary session

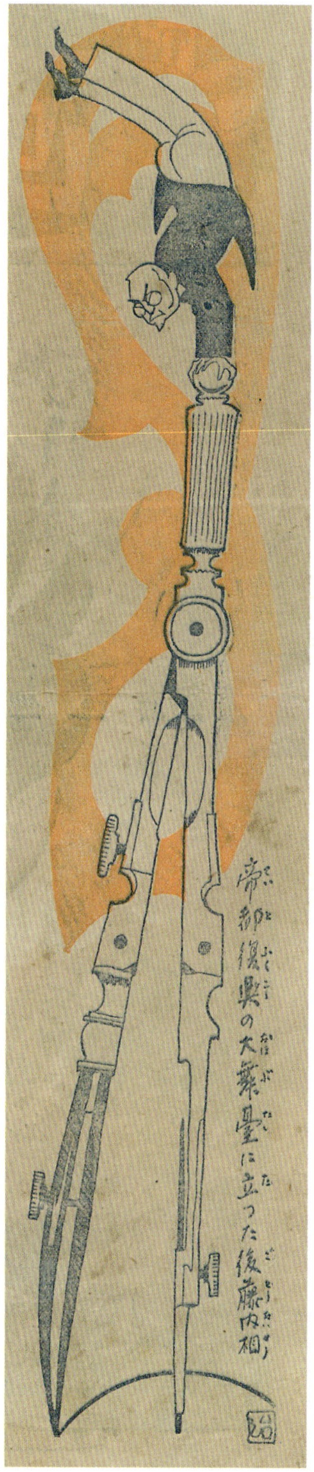

6.7 Ogawa Jihei, *Home Minister Gotō Standing on the Big Stage of the Imperial Capital Reconstruction* (*Teito fukkō no ōbutai ni tatta Gotō naishō*), cartoon. *Jiji manga* 132, 7 October 1923, 2. By permission of Sangyō Keizai Shinbunsha.

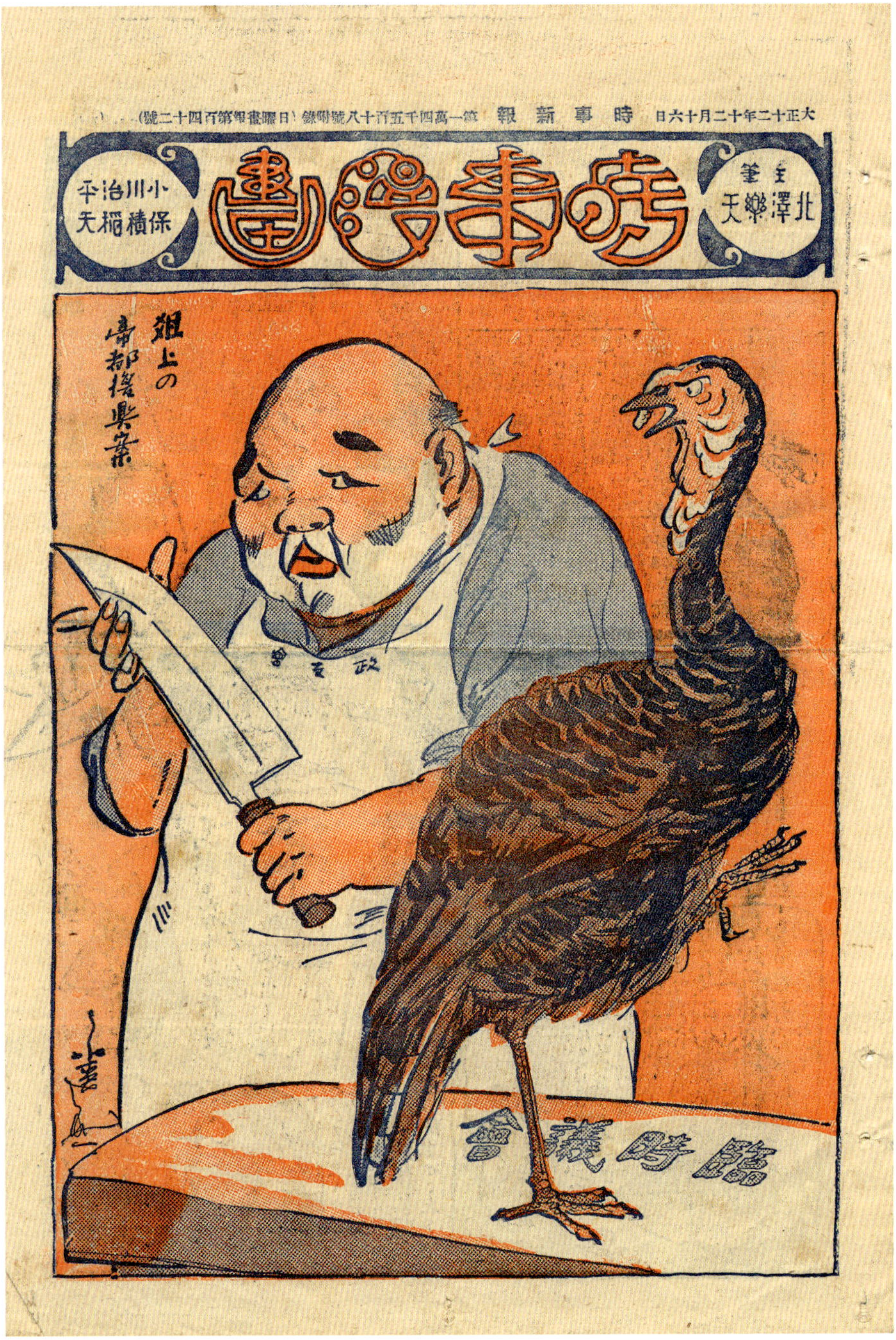

6.8 Kitazawa Rakuten, *The Imperial Reconstruction Plan on the Chopping Board* (*Sojō no teito fukkōan*), cartoon, cover. *Jiji manga* 141, 16 December 1923. Collection of Ohio State University Billy Ireland Cartoon Library & Museum. By permission of Sangyō Keizai Shinbunsha.

of parliament" (*rinji gikai*).[22] Testing his sharpened knife, the butcher, with an apron labeled "Seiyūkai," is a thinly veiled portrait of Seiyūkai party president and former finance minister in the first Yamamoto administration, Takahashi Korekiyo (1854–1936). The Seiyūkai was one of the most powerful political parties of the prewar period until the early 1920s and monopolized the office of the prime minister and the seats of parliament. In principle, parliamentarians opposed Gotō's heavy-handed approach, and the Seiyūkai used the earthquake in its own political maneuverings to leverage power.

The tensions within the government were fierce. The Ministry of Finance was, in Sorensen's words, "the nemesis of the development of the [urban] planning system" and had routinely opposed acceptance of any central government support for city planning, a position that elicited strong recriminations from the Home Ministry in 1922 as it was trying to implement newly enacted laws.[23] One foreboding cartoon by Ogawa shows Gotō in silhouette speaking furtively to Minister of Education Inukai Tsuyoshi, labeled with the first character of his last name and easily recognizable by his distinctive protruding ears. The double-entendre text, pivoting on the near homophones *naishō* (home minister) and *naisho* (secret), says it all: "The home minister's conversation (speaking secretly). This seems to be a consultation that will bring down the political parties one of these days."[24] Gotō's plan exhibited paternalism in its purest form: ignore the people for the sake of the people and the supposed greater collective good. But many critics were uncomfortable with the imposition of eminent domain at the expense of landowners. They were also acutely aware of the immense outcry against land readjustment or relocation by the general populace, given that many people were preemptively squatting on their properties. Ultimately, the reality of limited monetary resources and intense opposition sharply curtailed the plans.

The architect of the earthquake memorial hall and museum, Itō Chūta (1867–1954), characterized this eviscerating process in a caricature depicting Gotō's budget as a deflating balloon (fig. 6.9). As the numbers of the budget and the air in the balloon shrink from 3 billion to 2 billion to 1 billion to 500 million, the figure of Gotō becomes similarly deflated, ending up pale and emaciated.[25] Kabashima Katsuichi (1888–1965), a popular illustrator who worked for *Shōnen kurabu* (Boys' Club) magazine and was a staff illustrator for the *Asahi shinbun* newspaper company, put an even more cynical spin on the issue in an *Asahi Graph* cartoon (fig. 6.10) showing Gotō as Santa Claus promising to bring a big bag of presents to the waiting arms of Tokyo's fat cats, who then turn their backs on him when he cannot deliver, sneering, "What's this gift of only 700 million yen from the big talk 4 billion yen reconstruction plan?" Titled in English *Much Ado about (Nearly) Nothing,* the cartoon is also labeled in

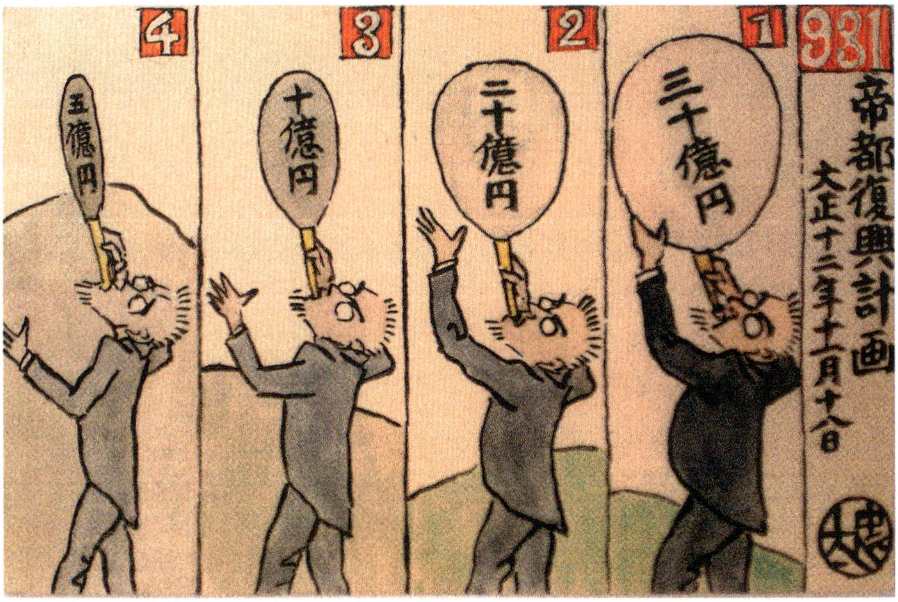

6.9 Itō Chūta, *Plans for the Imperial Capital Reconstruction* (*Teito fukkō keikaku*), original drawing, c. 1923. By permission of Nihon Kenchiku Gakkai Kenchiku Hakubutsukan.

Japanese with the well-known proverb "The big mountains roar but have brought forth only a mouse" (*daizan meidō shite nezumi ippiki*), which is based on the Latin saying *parturient montes, nascetur ridiculus mus* (essentially, "all that work and nothing to show for it").[26]

Hein has convincingly argued that while urban disasters can bring about an opportunity for changes in the built environment, they do not necessarily induce innovation, and rebuilding is often predicated on preexisting plans. "Disasters alone do not revolutionize planning," she concludes. In her opinion, and in contrast to the more definitive historical account of the success of postquake reconstruction by prominent architectural and urban planning historian Koshizawa Akira, the reconstruction plan for Tokyo, as radically invasive as it might have seemed, did not go far enough in considering the future growth of the city. It was "neither progressive nor responsive to the needs of the expanding metropolis; it was even more conservative and pragmatic than plans discussed before the disaster. It ignored the developing suburbs and did not provide guidelines for urban extension, although many people rebuilt their livelihoods in the suburbs after the earthquake."[27] However, she and other scholars do not interrogate the impact of the rationalized, scientific state language of urban planning that was linked to large-scale social engineering.[28]

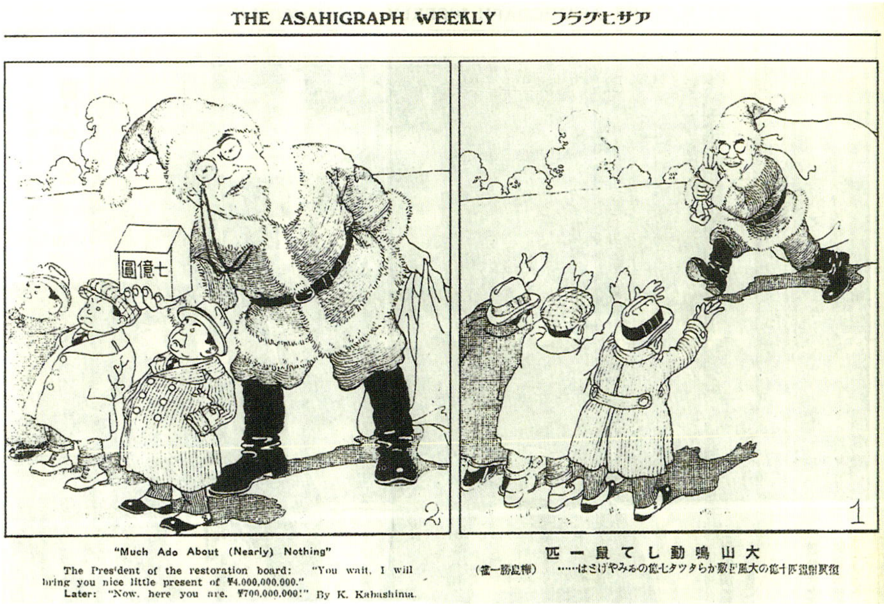

6.10 Kabashima Katsuichi, *Much Ado about (Nearly) Nothing.* "What's this gift of only 700 million yen from the big talk 4 billion yen reconstruction plan?" Cartoon. *Asahi Graph,* 5 December 1923.

In the end, Gotō's final scheme, which stayed in effect until 1930, resulted in approximately 8,895 acres of land-readjustment projects, construction of 157 miles of standardized roads, and the creation of fifty-five parks for recreation and disaster refuge.[29] Many of the parks were adjacent to newly constructed earthquake-resistant primary schools built to provide Tokyo's children with healthy green spaces as well as safe learning environments (figs. 6.11 and 6.12). Government reconstruction concentrated on the central metropolitan area, significantly expanding main trunk roads and auxiliary thoroughfares into the city. It also included an important program to replace wooden bridges, many dating from Edo times, with 298 modern steel-suspension and concrete structures, a plan that acknowledged bridges' critical role as transportation and evacuation conduits across the many waterways of the city, as well as the fire-retardant benefits of steel.[30] In addition to creating the Central Wholesale Market at Tsukiji, the Reconstruction Bureau established subsidiary food markets near railway lines to systematize food distribution. The reconstruction plan called for critical repairs of harbors, rivers, and canals along with their drainage systems, as well as major infrastructure improvements to the city's water supply and its sewage system. The city also created "fire prevention zones," in which the use of less combustible construction materials was encouraged, along with the removal of flammable industrial production away

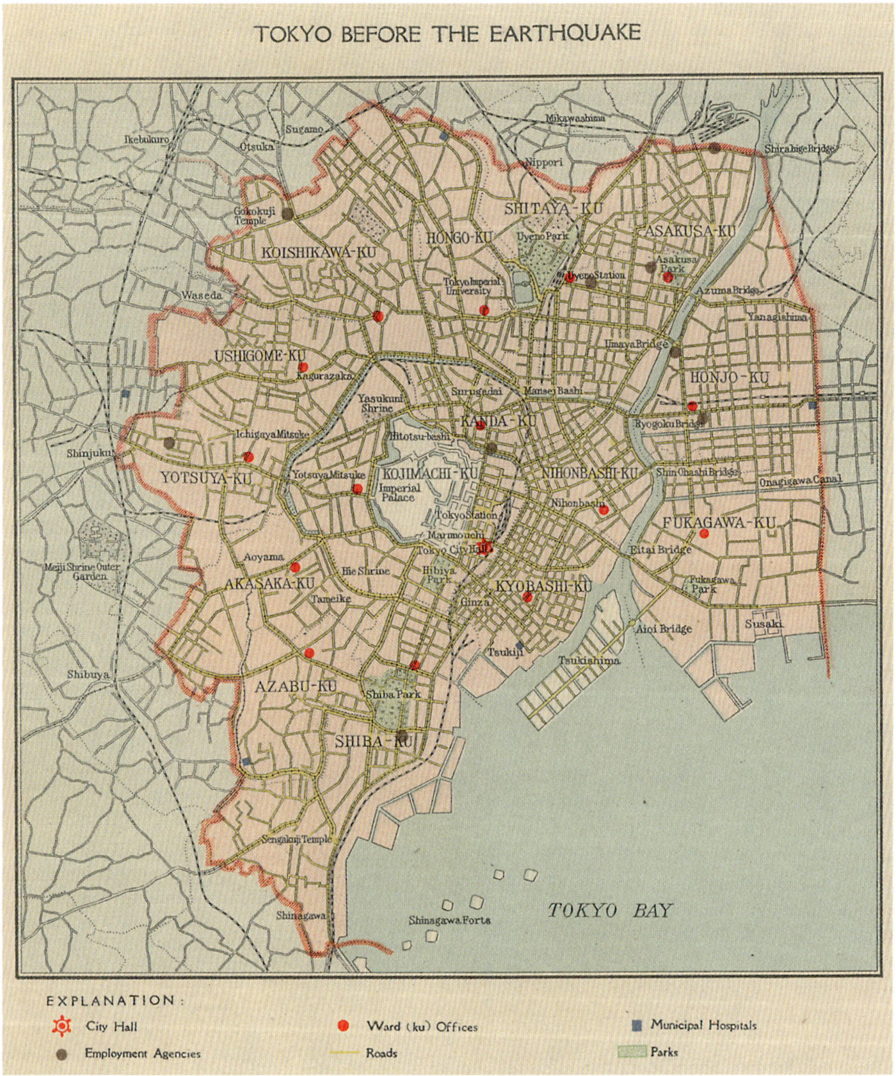

6.11 Tokyo before the Earthquake, map. Tokyo Municipal Office, *The Reconstruction of Tokyo* (Tokyo: Tokyo Municipal Office, 1933), 5.

from residential neighborhoods, where it had been one of the major incendiary triggers responsible for the incineration of the low city.[31]

Even in its reduced form, Gotō's plan was a monumental undertaking. People fretted over how it would be completed. Rakuten offered a satiric view of Tokyo twenty years in the future, picturing a tall bronze statue of Gotō in a triumphant pose confidently presenting his plans and budget for reconstruction (fig. 6.13). Gotō is dressed

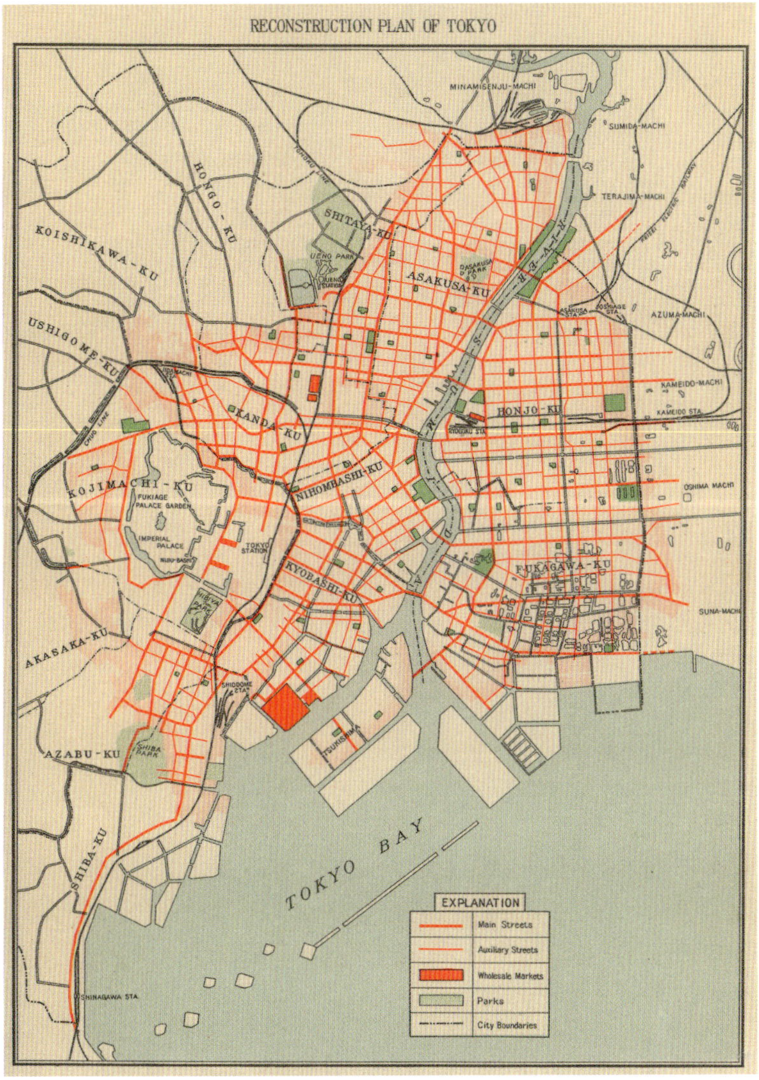

6.12 Reconstruction Plan of Tokyo, map. Tokyo Municipal Office, *The Reconstruction of Tokyo* (Tokyo: Tokyo Municipal Office, 1933), 103.

like a war hero in a uniform adorned with medals and accolades. His statue, set in a large park to herald the distinguished service of the director of the imperial capital's reconstruction, is the only project that's been completed. With the city having run out of funds, its buildings, construction, and streets remain incomplete; half-erected monumental structures encased in scaffolding are visible in the background, impassable roads are seen in the foreground, and refugees are still displaced, their walking

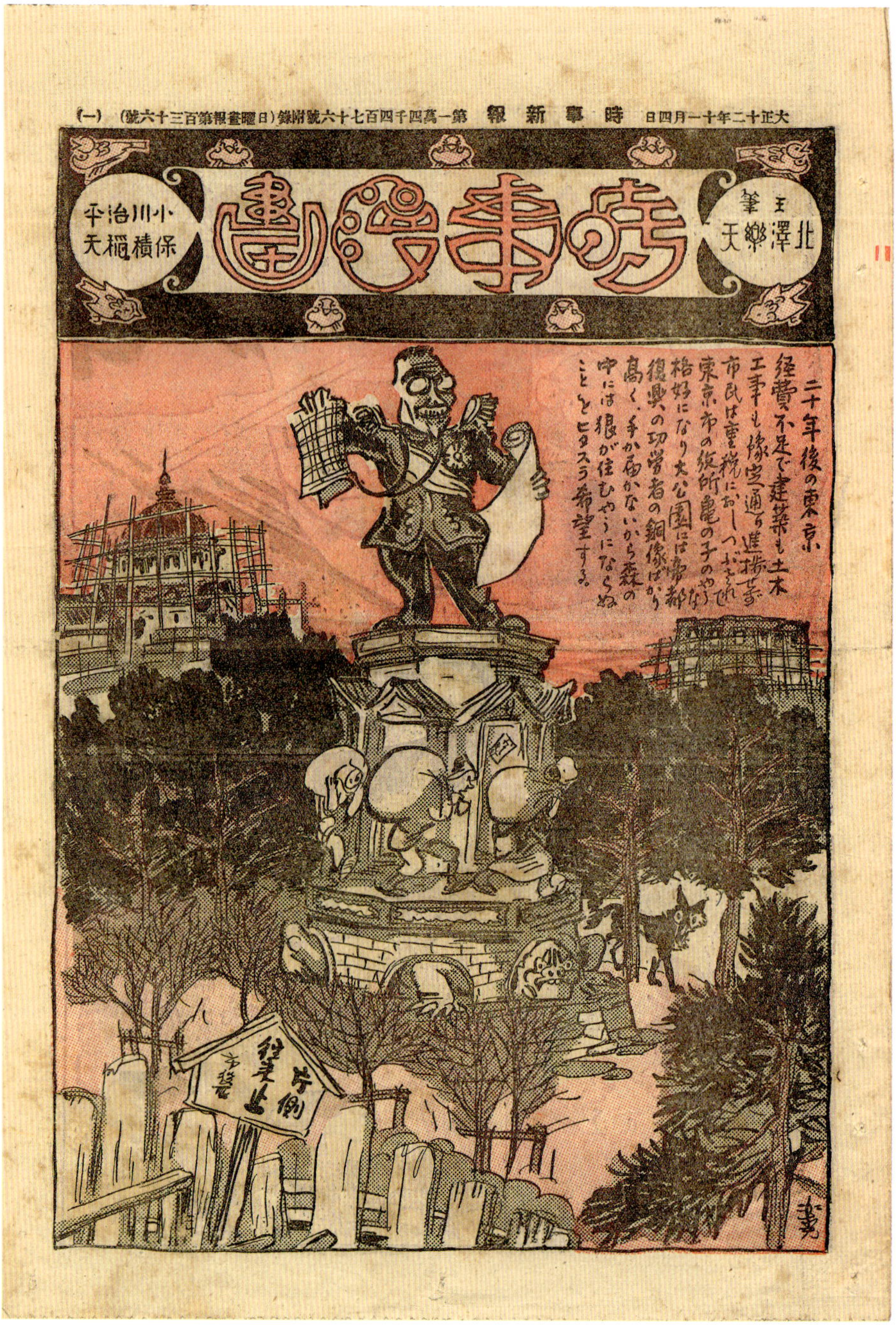

6.13 Kitazawa Rakuten, *Tokyo in Twenty Years* (*Nijū nengo no Tōkyō*), cartoon, cover. *Jiji manga* 136, 4 November 1923. Collection of Ohio State University Billy Ireland Cartoon Library & Museum. By permission of Sangyō Keizai Shinbunsha.

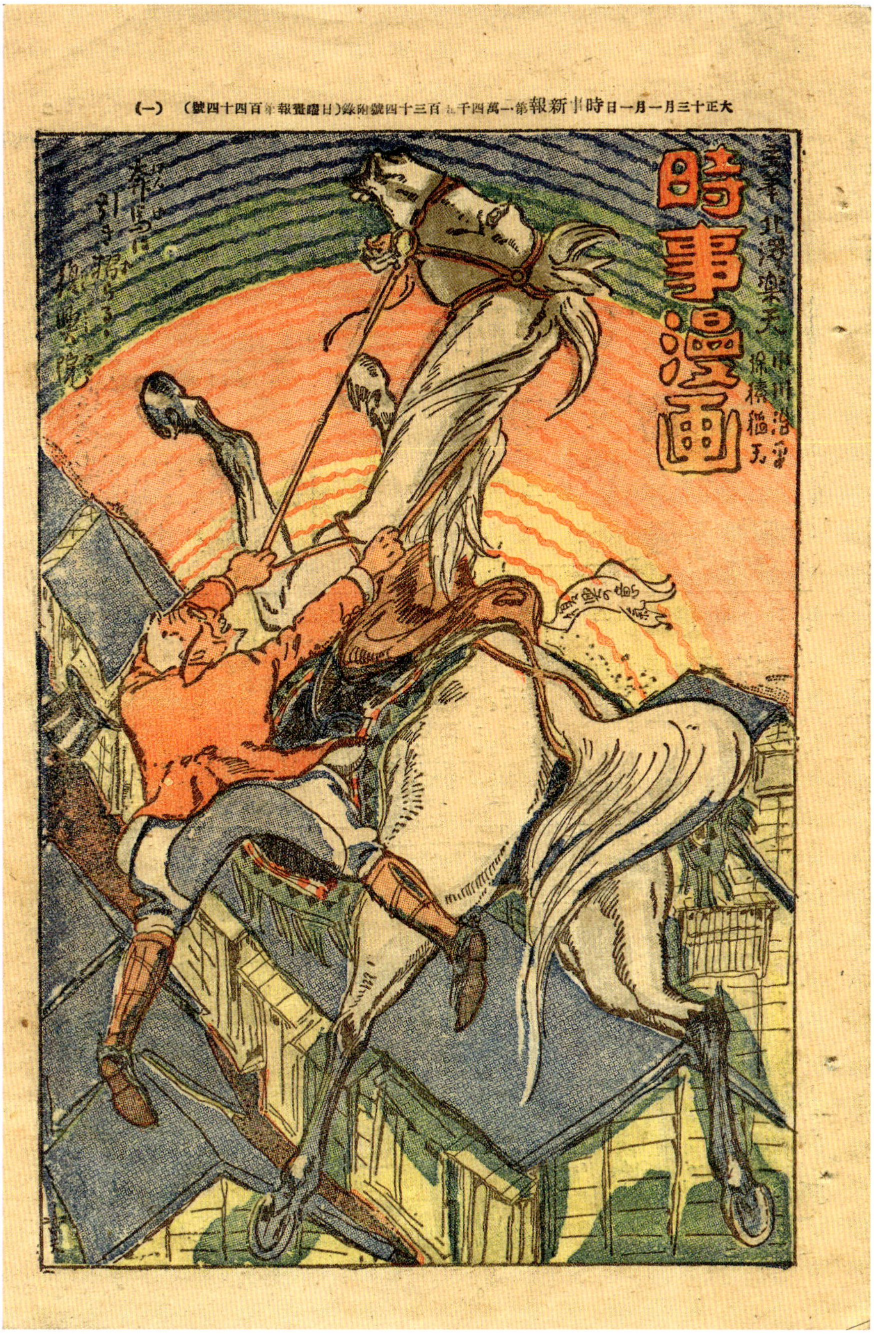

6.14 Kitazawa Rakuten, *The Galloping Horse Dragging the Reconstruction Bureau* (*Honba ni hikizuraru Fukkōin*), cartoon, cover. *Jiji manga* 143, 1 January 1924. Collection of Ohio State University Billy Ireland Cartoon Library & Museum. By permission of Sangyō Keizai Shinbunsha.

figures, burdened with belongings, adorning the base of the statue. The caption concludes, "Because no one is able to care for the statue and it is in the middle of the woods, we fervently hope that wolves don't start to live in it." The large green spaces for which Gotō had fought so vociferously are here a lonely wilderness with feral animals in the middle of an unfinished city. Grand visions would mean nothing in the long run if practical economic limitations were not heeded.

To the average imperial subject, the Gotō Reconstruction Bureau seemed out of control.[32] Under the title *The Galloping Horse Dragging the Reconstruction Bureau (Honba ni hikizuraru Fukkōin)*, Gotō is shown on the cover of *Jiji manga* in riding gear barely holding on to the reins of the runaway horse of the reconstruction process (fig. 6.14). He careens off the saddle as the horse gallops over the barracks-ridden cityscape below. Shock waves radiate outward. Such criticisms were not uncommon in postdisaster reconstruction around the world; Chicago officials were similarly castigated by the press for their "reckless go-aheaditiveness" in rebuilding after the city's Great Fire in 1871.[33] The intense anxiety about this unrestrained process was visible and palpable; it was clear that the city would not be rebuilt as it had been.

The Rationalist Response

To ease people's anxiety, the mayor of Tokyo, Nagata Hidejirō (1876–1943), issued special pamphlets detailing the reconstruction plans, and the city produced a series of graphs, pie charts, diagrams, and large-scale maps to explain the process in scientific terms.[34] These explanatory props buttressed the top-down directives circulated through the mass media and public exhibitions and were subsequently published and elaborated upon in a comprehensive set of volumes titled *Teito fukkō jigyō taikan* (General Survey of the Imperial Capital Reconstruction Project, 1930), overseen by the Tokyo Institute for Municipal Research (Tōkyō Shisei Chōsakai, hereafter TIMR, founded by Gotō) and edited by the Japanese Society for the Dissemination of Statistics (Nihon Tōkei Fukyūkai). Examples of diagrammatic images were eventually permanently exhibited in the earthquake memorial museum, tellingly named the Reconstruction Memorial Museum (Fukkō Kinenkan).[35] This was a clear continuation of Gotō's prequake visual rhetoric in which he used the data-driven, scientific language of the state to try to coerce popular opinion. However curtailed, the plan took the city layout one step closer to his goal of legibility for effective state governance.

The cover of the pamphlet *Kukaku seiri no hayawakari: Teito fukkō no kiso* (Quick Guide to the Land Boundary Readjustment: Basics of the Imperial Capital Reconstruction, 1924) (fig. 6.15), edited by the Tokyo municipal government and distributed

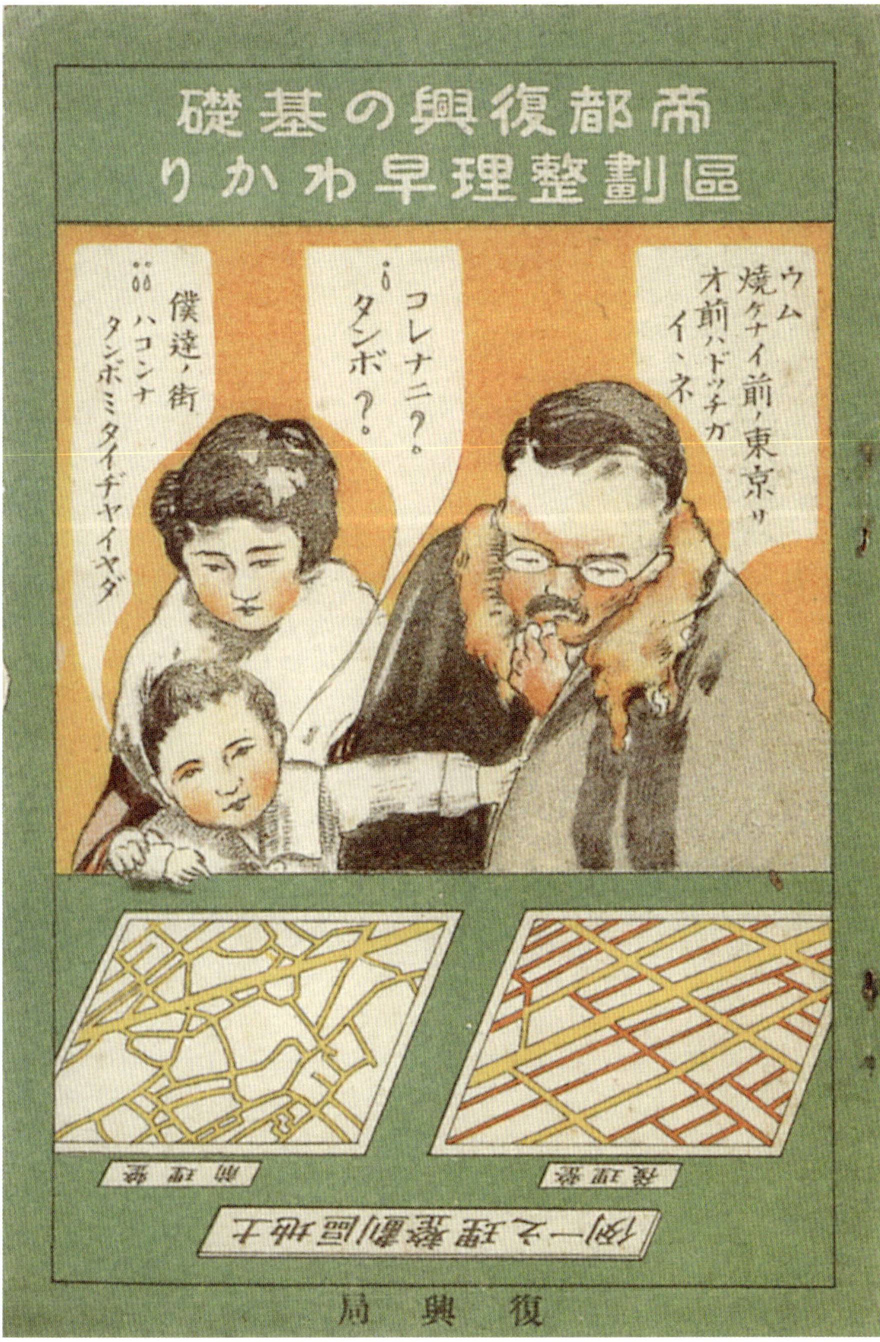

6.15 Artist unknown, *Quick Guide to the Land Boundary Readjustment: Basics of the Imperial Capital Reconstruction* (*Kukaku seiri no hayawakari: Teito fukkō no kiso*), cover, 1924. Edited by the Tokyo Municipal Government and distributed by the Reconstruction Agency. Tanaka Masaru Collection.

by the Reconstruction Agency (Fukkō Kyoku), an extraministerial board of the Home Ministry that took over supervising reconstruction work after the Reconstruction Bureau was dissolved on 25 February 1924, shows a middle-class family perusing the new information about reconstruction, which is labeled and displayed as if in an exhibition before them.[36] The husband, wife, and small child gaze intently at two maps, one a tangled city plan in yellow consisting of irregular roadways and uneven land parcels and the other an ordered grid in bright red, composed of regular land parcels and enlarged yellow transecting arterial roadways—clearly, before-and-after images of Tokyo. Tugging on his father's coat, the small child exclaims, "What are these? Rice fields?!" To which his father replies, "Uh-huh, unburned Tokyo before [the quake]. Which do you think is better?" The boy answers, "I don't like that our city looks this much like rice fields!" This setup makes clear that the correct answer is the rationalized reconstruction plan. This notion is reinforced in the rest of the pamphlet by comparisons between pleasing diagrams of the ordered structures of the reconstruction plan with the "disorganized" plan of the unadjusted layout, the assumption being that regularity and grids are good. The visual language of modernization and civilization is used here to shame Tokyo residents into looking beyond their personal interests to consider the larger importance of the capital's image in the world theater of nations. How could Japan's imperial capital afford to remain a warren of crooked streets and rice paddies? Reconstruction now meant not only rebuilding but rebuilding to the vision of the state. As James Scott persuasively contends in *Seeing Like a State,* the abstraction of cityscapes and their conversion into statistical data are effective means of creating legibility and transparency for state governance, but this controlling synoptic vision can reduce complex urban society into two-dimensional schemata that eventually remake the reality they purport to depict.[37] The postquake emergency situation of the Japanese capital was fertile ground for implementation of this kind of authoritarian visualization, and the technocratic reconstruction agencies used the ends to justify the means, giving little consideration to the disproportionate impact on certain populations within the city and the immense cultural losses that would result. Yet while a prostrate civil society may often lack the capacity to resist such plans, as Scott notes, a determined Tokyo populace and a divided Japanese central government did not fall into this trap. While they grudgingly cooperated with the reduced government plans, people did not abandon their own interests—a fact confirmed by the many social satires depicting the intransigent citizenry stymieing the government's program of land readjustment.

This resistance did not stop the government from polemically championing its cause. In fact, public intransigence and lingering discontent may have been the rea-

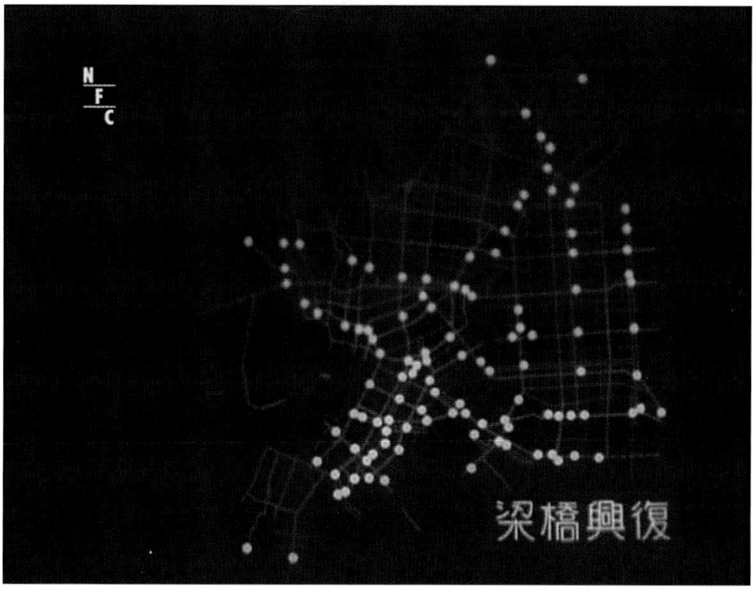

6.16 A constellation of "reconstruction bridges," screen shot. *Reconstruction of the Imperial Capital* (*Teito fukkō*), 1930. Oda Hamatarō, director; Fukkō Kyoku, producer. By permission of Tōkyō Kokuritsu Kindai Bijutsukan Film Center.

son that the authorities continued their efforts to elucidate the plans throughout the decade. As it had during the quake, the government used a diverse range of media to communicate its plans for reconstruction to the public. In the film *Reconstruction of the Imperial Capital* (*Teito fukkō*), a retrospective look at the disaster through reconstruction, the aesthetics of artistic modernism were harnessed for the media apparatus of the reconstruction agencies. This film shows strong affinities with the modernist cinematic style of landmark European works like *Berlin: Symphony of a Great City* (*Berlin: Die Sinfonie der Großstadt*, 1927) by German director Walter Ruttmann, commercially released to critical acclaim in Japan in 1928 as *Berurin: daitokai kokyōgaku*. For example, a sparkling constellation of lights flickering over the gridded map of the city labeled "reconstruction bridges" (*fukkō kyōryō*) (fig. 6.16) fades into a glittering montage of steel and concrete bridges to present a magical world of modernity (fig. 6.17) that would have appealed to any card-carrying member of the avant-garde.[38]

Another segment visualizes land readjustment through abstract pictorialization of irregular land parcels as boxes that shimmy and nestle together into a more uniform grid pattern to make a happy, stable cityscape (fig. 6.18). Applying the modernist aesthetics of a Dutch De Stijl neoplasticist style such as that in paintings by Piet

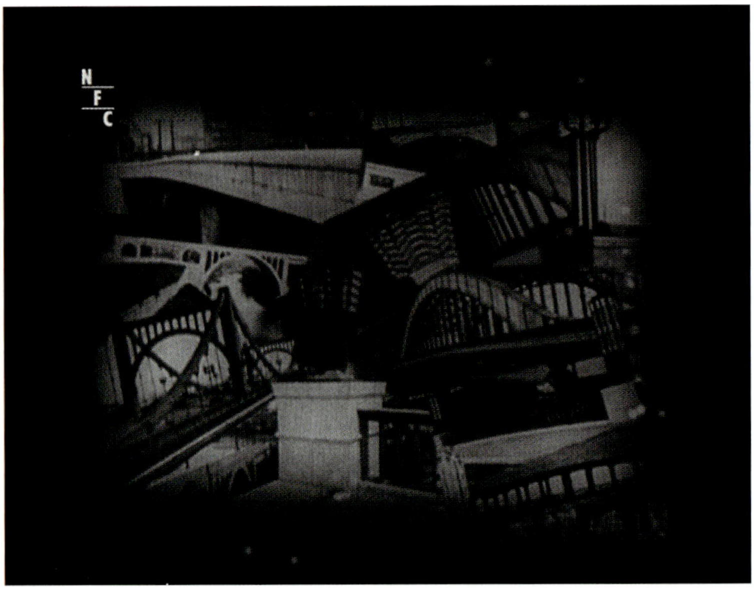

6.17 A montage of "reconstruction bridges," screen shot. *Reconstruction of the Imperial Capital* (*Teito fukkō*), 1930. Oda Hamatarō, director; Fukkō Kyoku, producer. By permission of Tōkyō Kokuritsu Kindai Bijutsukan Film Center.

Mondrian or Theo van Doesburg, which were inspired by the modern metropolis, this conceptual abstraction writ large is superimposed onto the living city of Tokyo, a movement from painting to architecture that the De Stijl group strongly advocated.[39] However, in the Gotō or Nagata urban planning ideology of high modernism, gridded cities were modern because of their stability and legibility and their amenability to the mechanisms of industrial capital, not because of their cultural *élan vital.* The condensed city in the film is compressed into a square outline of regularity, resulting in a truncated version of the more totalizing plans for utopian urban space that went awry with planned cities like Chandigarh and Brasília. In his publications throughout the 1920s and 1930s, such as the seminal *Toward an Architecture* (*Vers une architecture,* 1923), renowned modernist architect and urban planner Le Corbusier posited the rectification and salvation of the world through his geometrical cities, replacing the mess of disorderly human civilization with the crisp, clean order of grids.[40]

In its efforts to implement plans for such a new urban order, the Tokyo city government reminded citizens in its explanatory pamphlet on land readjustment that the rebuilding of the capital was absolutely not a "restoration" (*fukkyū*) of the old, dangerous, and disorderly city that existed before the quake, but rather a "reform" (*fukkō*)

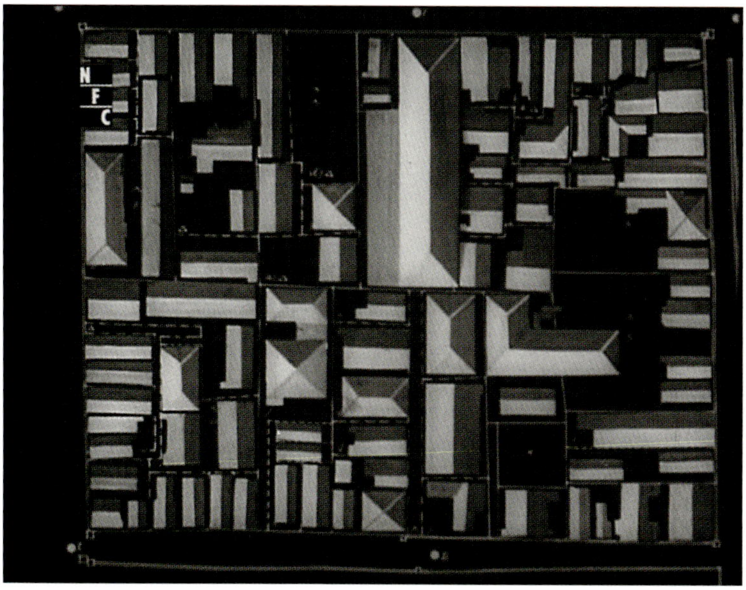

6.18 Land readjustment, screen shot. *Reconstruction of the Imperial Capital* (*Teito fukkō*), 1930. Oda Hamatarō, director; Fukkō Kyoku, producer. By permission of Tōkyō Kokuritsu Kindai Bijutsukan Film Center.

to produce a new metropolis that would be a safer and more hygienic, comfortable, accessible, efficient, and profitable new setting for the social activities of the citizenry. Although this undertaking would be difficult, the pamphlet noted, "turning misfortune into fortune" (*wazawai o tenjite saiwai to nasu*) could be achieved only by constructing orderly, systematized (*chitsujo seizen*) roads and residential land plots. Each resident of the city would need to make a little sacrifice. San Francisco had done it; Tokyo could do it. And any citizens now squatting on their land who might be thinking of resisting this plan were reminded that in times of emergency, the country and the city had the right to impose land readjustment by force.[41]

The Reconstruction Agency's film underscored the expectation of voluntary compliance by showing various citizens dutifully cooperating with the process of land readjustment, described in the intertitles as moving along in an "orderly" fashion. Cooperation was a prerequisite for realizing the abstract grid, and orderly people make orderly cities. In one scene, a handsome, well-dressed mother and daughter receive an official notification to relocate and respectfully comply. This is followed by displays of the yeoman engineering efforts to move structures into the grid and pictures of the family's hygienic daily life in the state-constructed barrack town where they will live

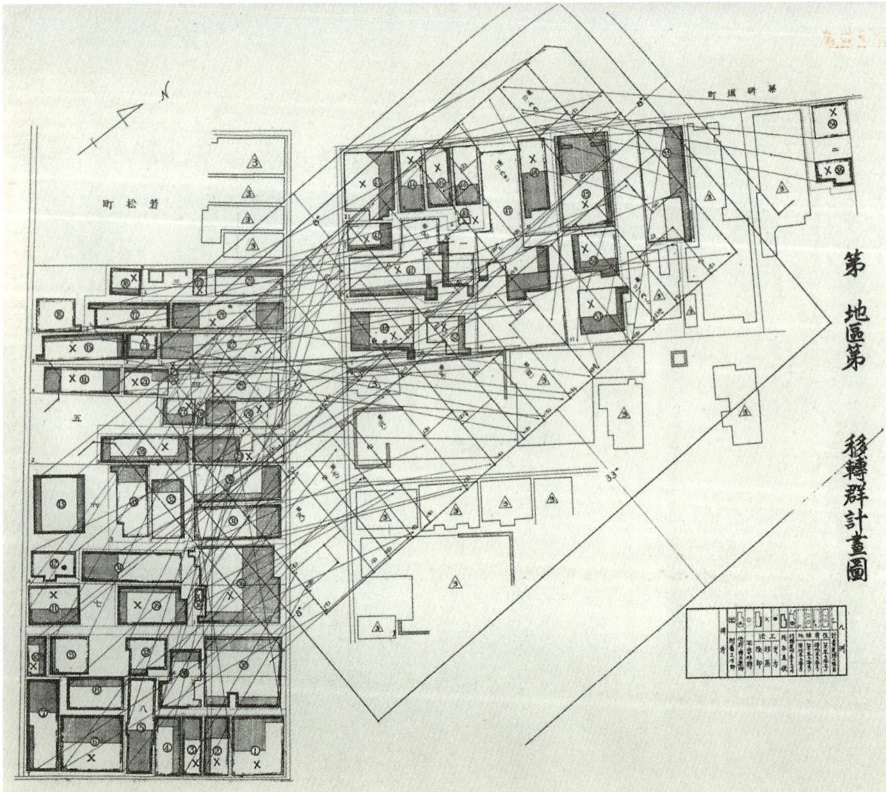

6.19 Chart of Plans for Groups of Moved Residences (*Itengun keikakuzu*), Tokyo Land Readjustment. Uchiyama Zenzaburō, ed., *Teito fukkō jigyō taikan* (General Survey of the Imperial Capital Reconstruction Project) (Tokyo: Nihon Tōkei Fukyūkai, 1930), 1:sec10, 13–14.

temporarily. In the narrative structure of the film, the earthquake disaster is represented as the darkness of night and reconstruction as the dawn of a new day, when people will return to work and economic productivity in their new gleaming metropolis.

Some of the clearest surviving visual exemplars of the scientific language of reconstruction are the collected images in the *General Survey of the Imperial Capital Reconstruction Project.* This retrospective account of the "public-private collaborative success" of reconstruction illustrates the work of land readjustment through annual color-coded maps charting relocations between 1924 and 1928. It concludes with a complicated diagram showing where each property parcel has been moved, which would probably not have been easily legible to the average viewer (fig. 6.19).[42]

The volume also uses scientific visual rhetoric to speak on a national and international register, displaying statistics in a world context, with a bar chart topped by

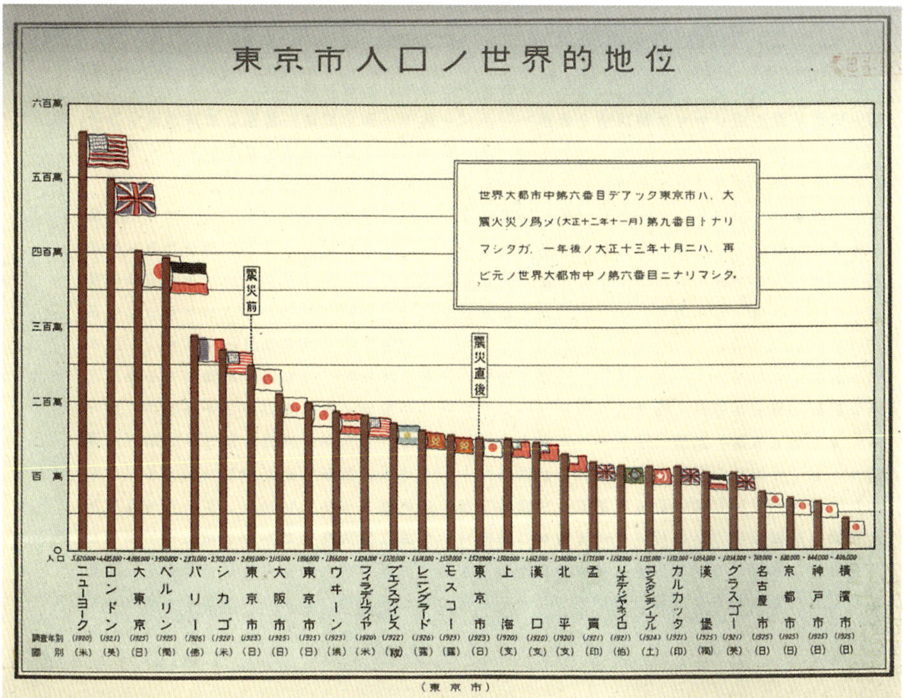

6.20 The World Rank of Tokyo City's Population (*Tōkyōshi jinkō no sekaiteki chii*), chart. Uchiyama Zenzaburō, ed., *Teito fukkō jigyō taikan* (General Survey of the Imperial Capital Reconstruction Project) (Tokyo: Nihon Tōkei Fukyūkai, 1930), 1:sec4, 3.

national flags that compares the populations of cities across the world, showing Tokyo as the sixth most populous city (nearly 2.5 million people) before the quake, coming after major metropolises like New York and London (fig. 6.20). The chart is so confusing, however, that annotation is inserted to explain its significance, proving that seemingly scientific diagrams do not necessarily enhance legibility. The text explains that Tokyo, while dipping to ninth place right after the quake, was back up to sixth, with 1.5 million people, just one year later in 1924.[43] The message: the city had bounced back with a vengeance.

Bar charts were often paired with pie charts that reinforced the relative proportions of the statistical data. In a diagram of economic losses by municipality, for example, the overwhelming losses of Tokyo compared to those of Yokohama or other areas is underscored by the brightly colored, large section of pie occupied by the city.[44] Some charts used Japanese coins or the Tokyo city wheel emblem as ornamental markers to symbolize the financial capital or the metropolis.[45] This schematic language could

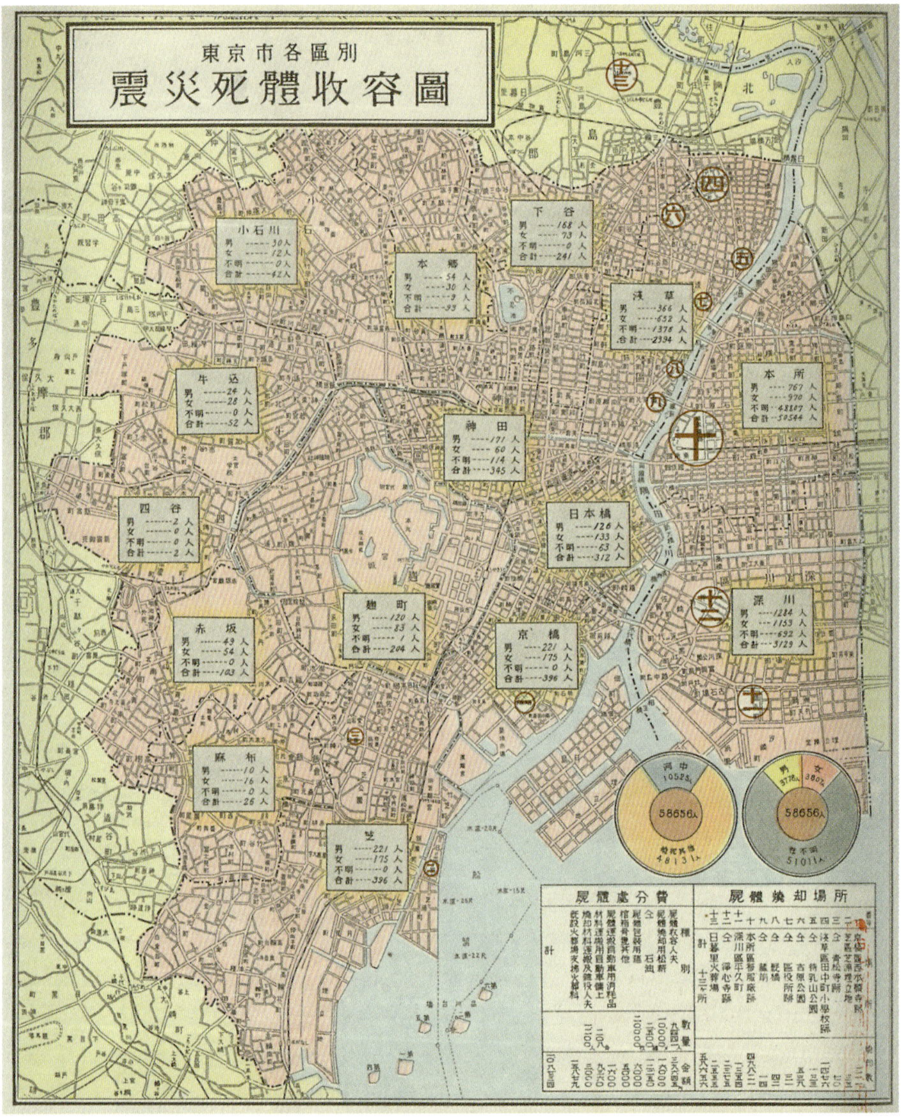

6.21 Chart of Earthquake Corpse Distribution (*Shinsai shitai shūyōzu*). Uchiyama Zenzaburō, ed., *Teito fukkō jigyō taikan* (General Survey of the Imperial Capital Reconstruction Project) (Tokyo: Nihon Tōkei Fukyūkai, 1930), 1:sec4, 27.

both reveal and conceal. For example, a map displaying ward casualties throughout the city (fig. 6.21) uses a similar graphic for every area, thereby obscuring the massive relative losses in Honjo, estimated at 50,544 people and far above the 3,129 deaths in the second-place neighboring area of Fukagawa.[46] The wards are equalized in the visual rhetoric, which subordinates the suffering of the low city to the larger metropolitan trauma—a strategy that is not surprising in a publication edited by the research arm of the Tokyo municipal government. While it may seem that anything can be reduced to statistical information and rendered in diagrammatic form, method and metrics in disaster assessment can be very deceptive.[47]

Disaster's Exhibitionary Logic

Media historians and sociologists such as Yoshimi Shunya, Yamamoto Tadahito, Satō Miya, and Takano Hiroyasu have explored the use of this polemical visual rhetoric in the exhibitionary field, bringing to light much valuable information on the cultural mediation of reconstruction and the means by which it transformed the popular social consciousness of the city.[48] As Satō describes, the financial news organization Chūgai Shōgyō Shinpōsha (now the preeminent Nihon Keizai Shinbunsha), for example, sponsored the widely attended Exhibition of the Actual Conditions of the Reconstruction of the Imperial Capital (Teito Fukkō no Jitsujō Tenrankai) at its newly rebuilt Tokyo headquarters in mid-April 1924. The exhibit laid out the city's reconstruction process in extensive displays of maps and charts, featuring at center stage six large official diagrams of land readjustment. It was one of the earliest exhibitions to visualize the broad-based status of recovery after the quake with detailed, clear information about the government's plan. A variety of state agencies such as the Home Ministry's Reconstruction Agency and the Urban Planning Division participated along with private entities from finance, insurance, and retail; altogether, 550 images and objects sprawled over five exhibit spaces. While capitalizing on the public entertainment value of images showing the spectacular rapidity of reconstruction, the organizers also implicitly questioned the process, telling viewers that the exhibition would expose the true process, now happening before their very eyes.

A highly orchestrated, popular "media event" with displays and films to entertain the whole family, the show was well advertised in posters throughout the city and heavily promoted in the company's *Chūgai shōgyō shinpō* newspaper, purportedly drawing close to 179,000 visitors. Photographs of viewers at the exhibition and stories reporting their positive comments not only attested to the significance of this visual mode of communication, but also touted the company's central position in the Japanese news media during a period when Tokyo-based newspapers were greatly weakened in their competition with Osaka papers. One photo showed the imperial family attending the show (fig. 6.22), followed by bureaucrats and prominent businessmen, aiming to give the event an official imprimatur. A story about Mayor Nagata's visit to the invitational preview quoted his admiring words for the impressive maps of land readjustment and

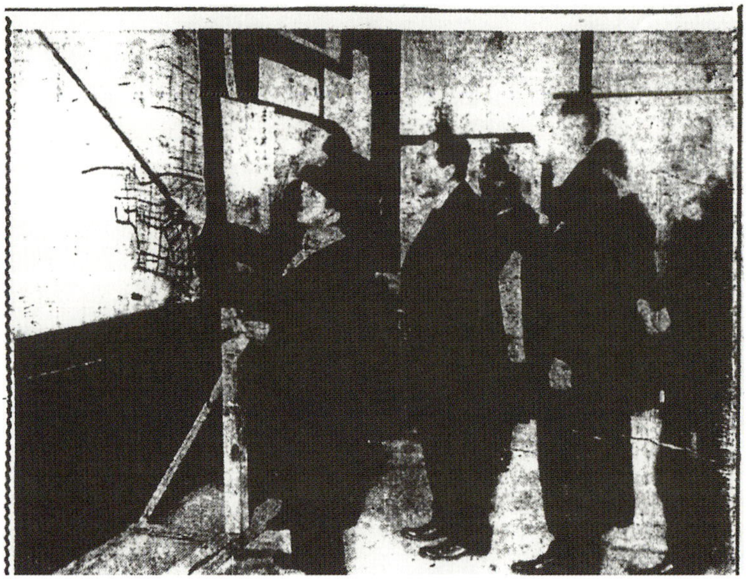

6.22 Prince Kuninomiya Kuniyoshi (with pointing stick) looking at a map of Tokyo and listening to an explanation of the plans for the reconstruction of the imperial capital by the *Chūgai shōgyō shinpō* newspaper's senior managing director, Yanada, at the Exhibition of the Actual Conditions of the Reconstruction of the Imperial Capital (Teito Fukkō no Jitsujō Tenrankai), photograph. *Chūgai shōgyō shinpō,* 26 April 1924, 7. By permission of Nihon Keizai Shinbunsha.

his recommendation of them to all viewers as the easiest means of appreciating the extraordinary project the city was undertaking. The newspaper noted that visitors from every class, not only white-collar workers, students, and members of the intelligentsia, flocked to see the straightforward explanations. It also recorded the active attendance of workers, senior citizens, and women.[49] This emphasis on spectatorial witnessing implied that visitors saw and were convinced of the plan's excellence; it was a crucial state strategy for generating a broader consensus among the populace, even if viewing did not mean concord.

The Tokyo metropolitan government sponsored two major exhibitions of its own, one in 1924 at the beginning of the process and one in 1929 toward the end. The Earthquake Reconstruction Exhibition (Shinsai Fukkō Tenrankai) was held at two venues in Ueno—the Jichi Kaikan and an exposition hall near the Kangetsukyō Bridge by Shinobazu Pond—throughout September 1924. The former featured objects memorializing the disaster: materials from earthquake victims, documentary photographs and printed images, scholarly studies of the destruction, models, and plans. The latter featured diagrams and materials related to reconstruction: urban planning, improvements in citizens' daily life, construction practices, and new industrial development. The first exhibition inaugurated the sensational practice of exhibiting earthquake debris, objects from daily life that had endured the disaster and offered visible evidence of the physical trauma it had wrought. City officials advertised in the press for public contributions to the exhibit. Many of the items were painstakingly collected from private individuals over several months.[50] Vividly described in the *Yomiuri shinbun* as "the enormously popular Earthquake Exhibition where the display space seemed about to fracture" and in the *Tōkyō asahi shinbun* as the exhibit of "blood-soaked clothes and trees torn up by the cyclone," the show was said to include thousands of "hideous" burned objects retrieved from the rubble of the city after the quake, "evoking memories that brought people to tears."[51] Wildly exceeding the organizers' expectations, "people drawn to the frightening spectacle of death" came in droves, marking the next phase of the disaster's dark tourism. Three thousand people were admitted per hour, and about thirty thousand people visited the first day. The galleries became so overcrowded that glass cases were smashed and precious documentary objects disappeared in the melee.[52] The visceral aesthetic of trauma communicated by the objects elicited a strong emotional response from the viewing public, with many of the visitors themselves refugees who had shared this horrific experience and were seeking psychological compensation for their losses. The auratic nature of the objects incited an emotional riot, transforming the crowd from the bereaved masses into an incendiary mob.

This sensational exhibitionary rhetoric of physical trauma and relic veneration was powerfully continued in the quake's official memorial and taken up later in the memorialization of other disasters, such as the wartime atomic bombings. In the early postquake period, such visceral displays coexisted in a tentative balance with the urban-renewal rhetoric of reconstruction, even as government officials tried mightily to harness the powerful emotions of public trauma and empathy in the cause of the future city. Through calculated display of these objects and representations, they importuned the living to honor the sacrifice of the dead by cooperating with the reconstruction effort, going so far as to compare the deaths of earthquake victims to those of soldiers killed for the nation on the battlefields of the Russo-Japanese War. By the end of the decade, this balance would tip, however, as the national and municipal governments felt a need to enfold mourning and memory within the progressive narrative of a forward-looking reconstruction.

Around 1929, there was a clear shift toward consolidating remembrance as authorities started to bring down the curtain on reconstruction. The 1929 Imperial Capital Reconstruction Exhibition (Teito Fukkō Tenrankai), held from 19 to 30 October at the TIMR headquarters in Hibiya, was a sign of this shift and stands as a visual testament to the official vision of reconstruction.[53] The exhibition displayed over seventy thousand items, including "hideous photographs of the broken Twelve Stories," along with screenings of films about the reconstruction in the basement. Three hundred of these objects, primarily diagrams, were chosen for inclusion in the two-volume *General Survey of the Imperial Capital Reconstruction Project*. Over 110,000 viewers attended, and the exhibition was extended for several days to accommodate the crush of visitors.[54] Spanning four floors, the exhibition devoted spaces to all of the Kantō region's major reconstruction stakeholders: Tokyo city, the Reconstruction Agency, Yokohama city, Tokyo prefecture, the Imperial Household Ministry, the army and navy, the Tokyo Municipal Architectural Agency, the Ministry of Communications, the Ministry of Finance, the Ministry of Education, the Home Ministry, the municipal police and fire departments, the Tokyo Imperial University architecture and seismology division, the Waseda University architecture division, Tokyo Industrial University, the Tokyo Museum, the Tokyo Subway Company, the Dōjunkai housing association, and a host of private corporate associations and industries representing electricity, communications, libraries, oil, mass media, and railroads. Tokyo city's exhibit, the largest in the show by far, featured extensive schematic plans and posters along with documentation of the region's history of earthquakes, notably including Ansei-era maps and catfish representations. This space exhibited Gotō's prequake 800 million yen project, the reconstruction authority's comprehensive postquake plans, and the final reduced reconstruction scheme. This public display of the radical downsizing prompts one to wonder if it was designed to quell protests and not merely to explain the logic of reconstruction planning.

A small section of the exhibition was sponsored by the Tokyo Association for the Earthquake Memorial Project (Tōkyō Shinsai Kinen Jigyō Kyōkai), the committee formed in August 1924 to oversee the creation of a memorial.[55] The section displayed a few artistic renderings and models of the commissioned design that was near completion, as well as some of the commemorative objects and oil paintings that would be housed there. The Tokyo Museum exhibited other earthquake-related objects and documentary photographs that would eventually become part of the permanent display at the memorial. The exhibition's overwhelmingly official vision enfolded mourning and memory within the moralistic, progressive narrative of a nearly completed reconstruction.

In the interim, other civic groups voiced their opinions about how to redesign the city, organizing independent exhibitions to offer suggestions for enhancements to the plans for rebuilding the metropolis. One of the most prominent was the Exhibition of

Plans for the Reconstruction of the Imperial Capital (Teito Fukkō Sōan Tenrankai), a juried show sponsored by the Citizens' Art Association (Kokumin Bijutsu Kyōkai) held from 13 to 28 April 1924 at the Takenodai Chinretsukan in Ueno.[56] As one of the group members, Western-style painter Kobayashi Mango, said, "People won't be satisfied with lining up antiseismic and fire-retardant buildings; the beautification of the city will not be accomplished by simply widening the streets; beyond necessity, there are still many more essential things." Exhibition jury member and famed sculptor Asakura Fumio echoed these sentiments, challenging artists to grasp the opportunity of the earthquake to beautify the city and go beyond mere practicality. This meant providing aesthetic "spiritual reconstruction" (*seishinteki fukkō*) along with material reconstruction.[57] Like artists and architects in other countries grappling with disaster, they felt entitled to a special jurisdiction over the visual interpretation and commemoration of the quake.

The Citizens' Art Association exhibition was the first reconstruction show to focus on the social and aesthetic function of the city and the built environment. It was open to submissions of any architectural project that would enhance the beautiful sites of the city and contribute to the success of the effort to rebuild the imperial capital: large-scale urban plans, single architectural models or drawings, or any kind of interior decoration. Submissions could also include designs for streets, public squares, canals, bridges, gardens, commemorative sculpture and towers, fountains, graves, window decorations, signboards, posters, wall paintings, wall reliefs, paintings, and furniture.[58] Many prominent young modernist architects participated, including Takizawa Mayumi (1896–1983) and Yamaguchi Bunzō (1902–78) from the now-celebrated Bunriha Kenchikukai (Secessionist Architecture Association) and Sōusha Kenchikukai (Creation of the Universe Architectural Society). The architects of this generation had a profound impact on postquake building and the design of public space in Japan, including the innovative plans for modern schools built to new disaster-prevention construction standards. Their emphasis on the plasticity of architecture and its expressive capacities was evident in their myriad designs for commercial barrack structures and other structures across the rebuilt cityscape, and their work contributed to a new civic style.[59] And, as architectural historian Ken Oshima has compellingly argued in his analysis of the controversial reinforced-concrete arch design for the postquake construction of Hijiribashi (Saints' Bridge) in Ochanomizu by Bunriha leading member Yamada Mamoru (1894–1966), the group's expressive, artistic approach to architecture "challenge[d] the utilitarian notion of a modern bridge as an earthquake and fireproof structure spanning and providing passage over an obstacle." In this project undertaken for the reconstruction authority's bridge-building division, Yamada transcended the

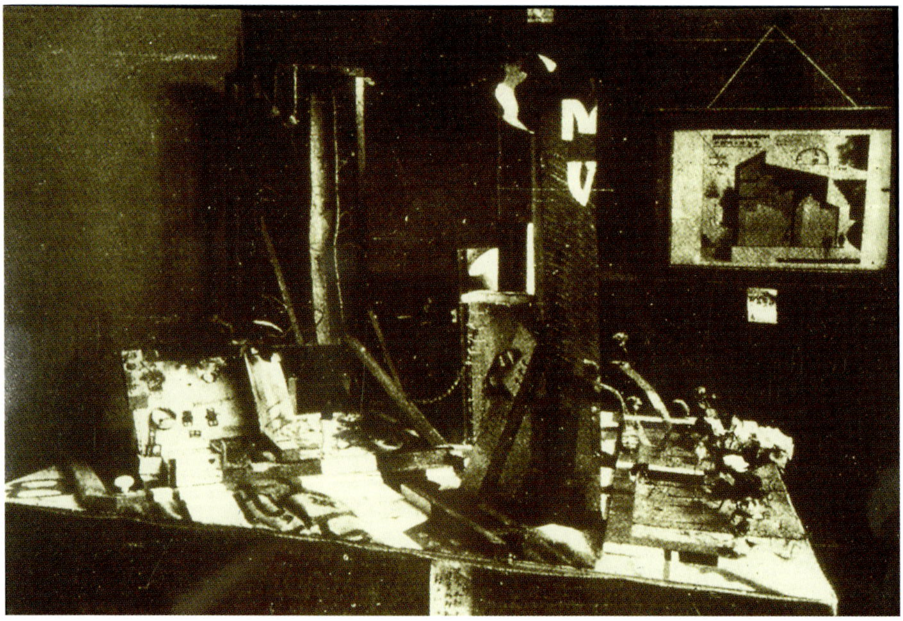

6.23 Murayama Tomoyoshi, *Architectural Idea for Mavo Headquarters* (*Mavo honbu no kenchikuteki rinen*), model, mixed media. Exhibited at the Exhibition of Plans for the Reconstruction of the Imperial Capital (Teito Fukkō Sōan Tenrankai), April 1924. Courtesy of Murayama Harue.

structural requirements initially laid out by division engineers with his streamlined "abstract composition" of line and plan, aesthetically reenvisioning Hijiribashi "as an urban centerpiece of Tokyo's new modern transportation and green space network."[60]

The Barrack Decoration Company architect and outspoken cultural critic Kon Wajirō played a major role in orchestrating the conceptual parameters of the Citizens' Art Association reconstruction show to maximize its impact on future design practice. Kon insisted that the exhibition set aside discrete spaces for individual groups to accentuate their distinct theoretical approaches; focus on generating visionary plans rather than polished, buildable architectural designs; and invite participation by nonspecialists.[61] Consequently, the show presented diverse visions of reconstruction. According to association president Chūjō Seiichirō, the exhibition was a "symphony of volume" (*yōseki no shinfonī*), and future shows would add a "harmony of color" (*shikisai no hāmonī*).[62]

Kon's approach opened the door to the participation of artistic groups like Mavo, and this avant-garde association was granted a sizable space in the show. Critics disapproved of the flamboyantly anarchic design aesthetic in the members' extensive contributions. Extending their barrack designs into reconstruction, the Mavo artists were avowedly more concerned with expressing the violent ruptures of the postquake

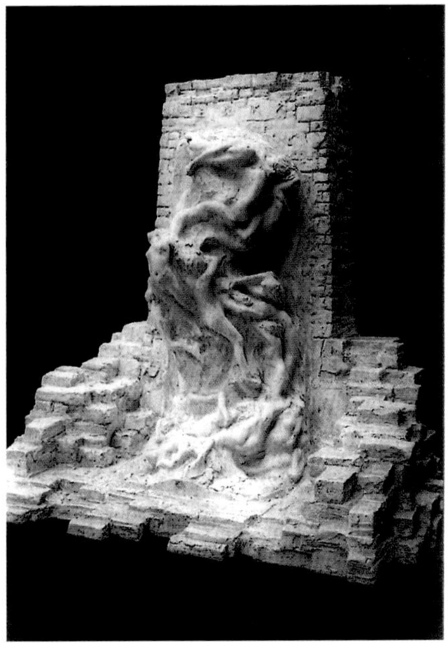

6.24 Hinako Jitsuzō, *Monument to the Incineration of Edo-Tokyo Culture* (*Edo Tōkyō bunka enjō hi*), model. Exhibited at the Exhibition of Plans for the Reconstruction of the Imperial Capital (Teito Fukkō Sōan Tenrankai), April 1924. Hirota Toshikazu, *Shōwa shoki chōkoku no kisai: Hinago Jitsuzō no sekai* (The World of Hinago Jitsuzō: The Remarkable Talent of an Early Shōwa-Period Sculptor) (Kyoto: Shibunkaku Shuppan, 2008), 34. Courtesy of Shibunkaku Shuppan and Hirota Toshikazu.

moment than with providing plans for orderly urban reconstruction. Next to the state's visual barrage of scientific urban planning documentation, Murayama Tomoyoshi's *Architectural Idea for Mavo Headquarters* (*Mavo honbu no kenchikuteki rinen*) (fig. 6.23), with its seemingly random bricolage of material fragments formed into a kind of urban planning model, can only be seen as a deeply ironic parody.[63] Mavo's other projects were an array of playfully misshapen edifices in colorful hues that called for a new urbanistic agenda based on the liberation of expression rather than the rationalization of space. Their severest critic was coexhibitor Tokyo Imperial University architecture professor Kishida Hideto (1899–1966), associated with the Ratō group, whose review of the show hailed the efforts of his fellow young architects to give birth to a new form of architecture but derided those (unnamed) individuals who focused merely on expressive lines without concern for the actual massing of buildings, which, in his words, "left much to be desired." He even compared the design progeny of this drunken expressionistic revelry with "the idiot offspring of the alcoholic."[64]

The winners of the exhibition prizes were Nishimura Yoshitoki (1886–1961), head of the architecture division of Daiichi Ginkō bank, for his *Map of Tokyo's Reconstruction Plans* (*Tōkyō fukkō keikakuzu*); celebrated academic salon sculptor Hinako Jitsuzō (1893–1945), for his commanding sculptural relief *Monument to the Incineration of Edo-Tokyo Culture* (*Edo Tōkyō bunka enjō hi*), and little-known designer Okamoto Kaoru (dates unknown) for his project *Cooperative Housing* (*Kyōdō jūtaku*).[65] Okamoto's proj-

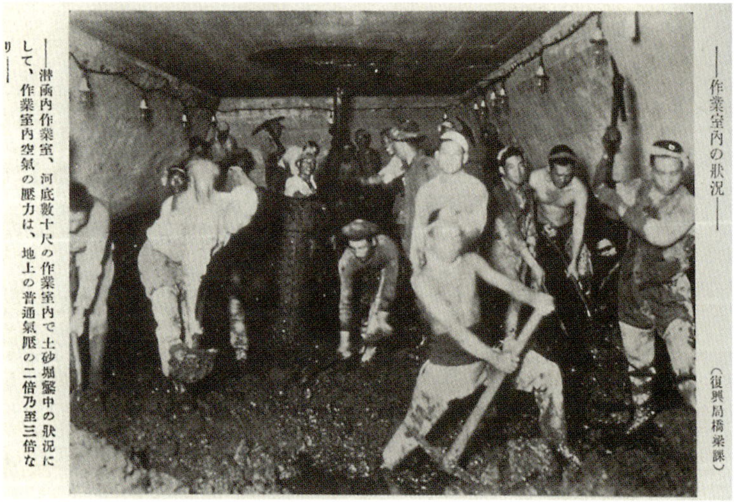

6.25 *Situation Inside the Work Room (Reconstruction Agency Bridge Division)* (*Sagyōshitsunai no jōkyō [Fukkōin kyōryōka]*), photograph. Uchiyama Zenzaburō, ed., *Teito fukkō jigyō taikan* (General Survey of the Imperial Capital Reconstruction Project) (Tokyo: Nihon Tōkei Fukyūkai, 1930), 2:sec13, 11.

ect does not survive, and all that remains of Nishimura's plan is a poor, largely illegible illustration.[66] Never realized, Hinako's design (fig. 6.24) survives only in photographs, but these images convey the work's effective physical embodiment of a fearsome conflagration that consumes the landscape and its tormented inhabitants, their nude bodies writhing upward in a cyclone of fire set against the backdrop of ruins; the work is a powerful cenotaph to the actual and metaphorical incineration of Tokyo's culture.

In the meantime, rebuilding continued, using the hard labor of the working class. Photographs of laborers in dangerous and filthy conditions, crowded into small work areas as they excavated the subterranean spaces for the city's new infrastructure of bridges and subway tunnels, graced the front pages of the news media as a testament to the city's grit and determination to rebuild (fig. 6.25).

A small cartoon titled *Smoke of the Reconstruction* (*Fukkō no kemuri*) (fig. 6.26) provides quite another view, speaking volumes about the inequitable social sacrifices among the city's residents. The silhouette of a laborer pushing a cart gradually amalgamates into the smoke from a factory, conveying that the reconstruction process is literally immolating his body for the interests of the nation and industrial capital. In the final frame, the cart is fully transformed into a factory building from which the smoke,

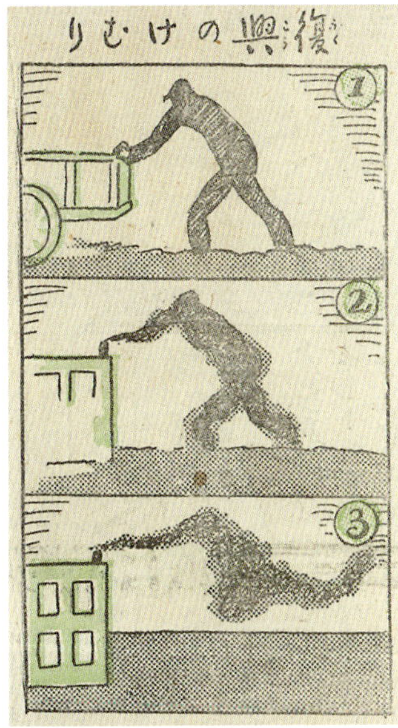

6.26 Artist unknown, *Smoke of the Reconstruction* (*Fukkō no kemuri*), cartoon. *Jiji manga* 141, 16 December 1923, 7. Collection of Ohio State University Billy Ireland Cartoon Library & Museum. By permission of Sangyō Keizai Shinbunsha.

curiously similar to the shape of Japan, emanates: the government and business were in cahoots. This played on the familiar images of incinerated bodies that were still raw in people's minds after the deadly conflagration and the cremation of victims' remains soon after the earthquake. Both disaster and reconstruction victimized the body of the proletariat, which was sacrificed to the machinery of capitalism.

But this was a minority viewpoint that was largely obscured by the authoritative progressive narrative of the disaster experience that repressed disproportionate suffering and sacrifices in favor of a unifying message. Affirming Tokyo's resilience in the face of catastrophe, the scientific and data-driven visual rhetoric of reconstruction promoted a productivist view of urban development and renewal. The logic of the plans for the capital was to make them more legible and amenable to incursions of industrial capital. State bureaucrats, particularly Gotō and other technocratic high modernist sympathizers, assumed that a more orderly and efficient city was a better city and that such a city needed to be created at any expense. However, the degree to which they had to alter their plans indicates that this discourse had limited appeal to many in the Tokyo populace, who resisted on multiple fronts. Moreover, Gotō's instrumentalist approach did not value the city as a historical, social, or aesthetic entity, rejecting a component that others thought of great importance for building a truly equitable and cultured new metropolis. The tension between these two views is also evident in the politics of remembrance that infiltrated the reconstruction effort from the minute the fires were extinguished.

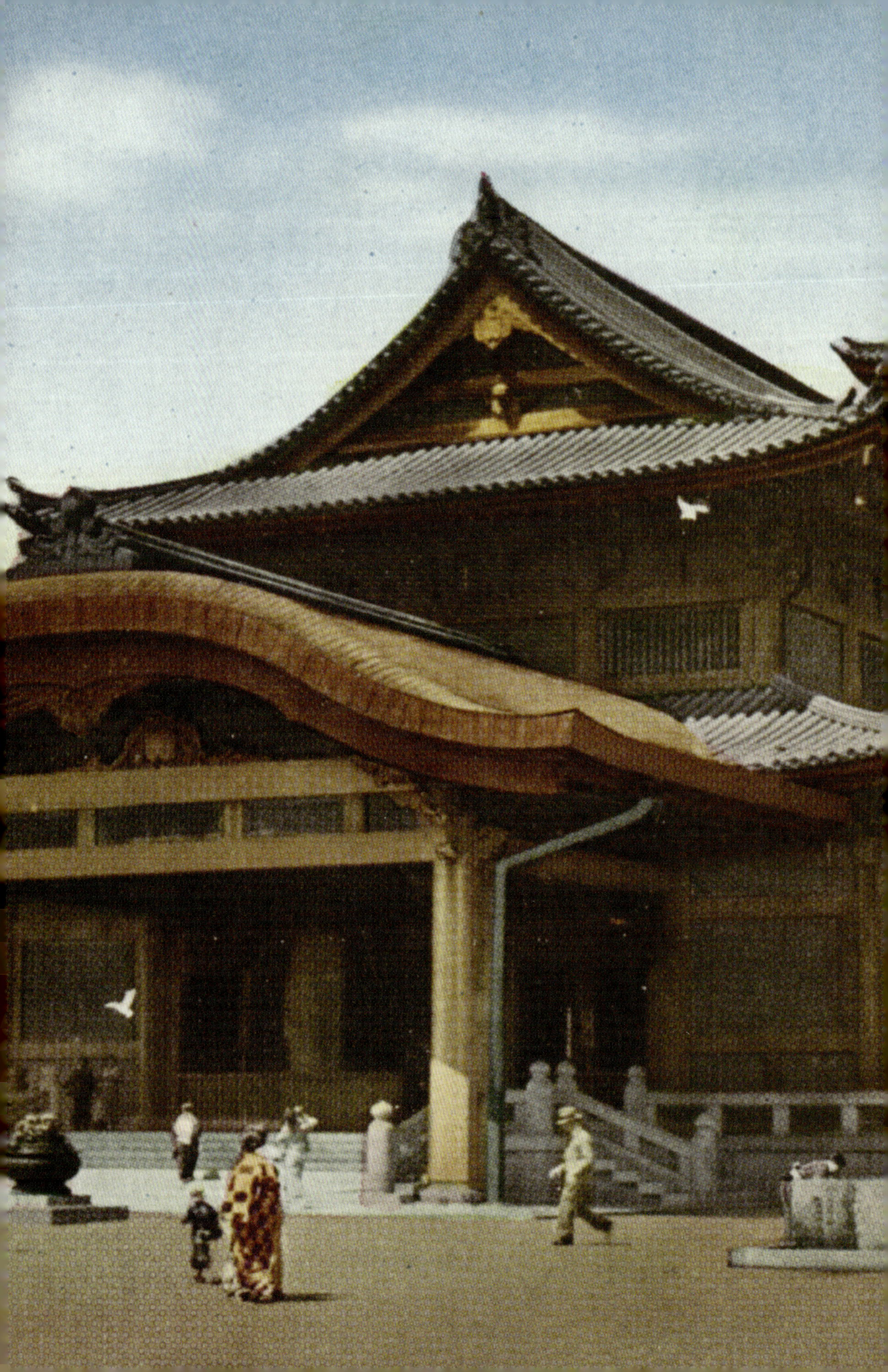

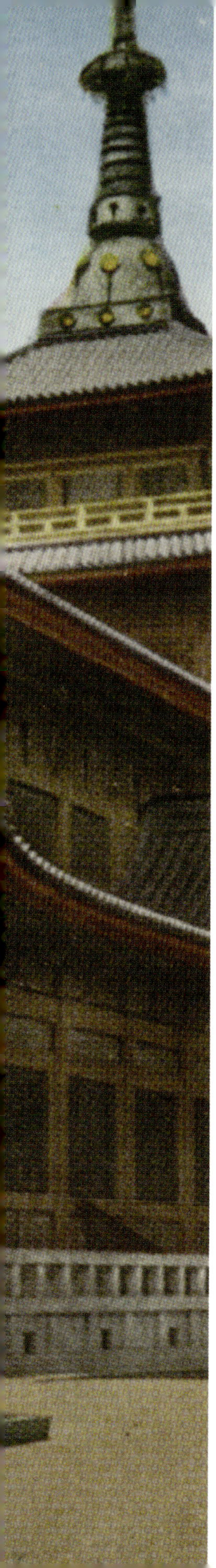

7

REMEMBRANCE

WITH TENS OF THOUSANDS OF PEOPLE DEAD in the streets, mourning and memory could not wait for reconstruction. Remembrance was an intrinsic part of recovery from the Great Kantō Earthquake, intimately linked to the visual rhetoric of reconstruction. It was also an evolving process. Close examination of this process, which culminated in the celebration of the completion of Tokyo's reconstruction in March 1930 and the inauguration of the official earthquake memorial complex erected on the site of the Former Army Clothing Depot in Honjo, clearly reveals two inherent cultural tensions: one between sacred and historical remembrance, and the other between memorialization of the past and celebration of the future.

"Traditionally, memorials organize emotions and attitudes in time and give them a public dimension by rooting them in space. They act as temporal signposts duly recording the traces of tradition onto the built environment and offering themselves as sites for the worship of the past," notes Andrew Shanken.[1] In the earthquake memorial complex, a large landscaped park with a hall for religious rites and a museum dedicated to reconstruction (added in 1931), the Tokyo municipal government's reconstruction rhetoric sought to relegate the disaster to the past by encompassing the tragedy within a larger progressive narrative of urban renewal and national resilience: even memory was modernized. Bereaved families and the religious community, however, made sure that memories of their loved ones were not swept away in modernity's erasures; they foregrounded the primary sacred function of the memorial to honor and appease the spirits of the dead—a critical form of insurance for the future prosperity of the living. The subsequent construction of the separate Reconstruction Memorial Museum with an architectural style distinct from that of the Earthquake Memorial Hall indicates the city's attempts to deemphasize the site as an "altar to tradition" and make it a secular, historical space for narrating the nation. As a forward-looking, didactic vision of modern Tokyo, filled with scientific illustrations and models of the successful reconstruction project, the museum gestured to the active role of the living, even as it lingered among the dead.

Burying the Dead

At the time of the disaster, survivors placed flowers and burned incense at makeshift memorial sites as a way of transforming horror and grief into active communal mourning. In accordance with Buddhist practices, a large-scale memorial service for the dead was held on the third day after the quake, amid massive piles of bones as high as ten feet. Mass cremation and religious ceremonies were held on-site in Honjo at the Former Army Clothing Depot because of the logistical impossibility of transferring all the remains to a crematorium.

These spiritual rites were immortalized in films, photojournalism, commemorative postcards, watercolors, and a large oil painting by Tokunaga Ryūshū that hangs in the Earthquake Memorial Hall (Shinsai Kinendō) adjacent to the museum. Photographs include many images of heaps of "bleached white bones" (*hakkotsu*) (fig. 7.1), purportedly all thirty-five thousand victims, often pictured with mourners and Buddhist monks performing funerary rites before their interment. Other images (fig. 7.2) feature large ceramic urns of victims' bones labeled by city ward—the vigorous individual inhabitants of the city reduced to tubs of anonymous remains. These bones were later reinterred in the foundation of the ossuary tower at the memorial hall.

Films panned from pile to pile, zooming in for excruciating detail of the tragic remains of tens of thousands of people. In the background, the earth still smolders. People wait in line to pay their respects and listen to the priests intoning prayers amid a cornucopia of flowers and dense clusters of memorial wooden slats (*tōba*) inscribed with the Buddhist names of the dead (*kaimyō*), which would usually be placed next to the household altar and grave after the ceremony. Covering their faces, they give the viewer a sense of the acrid and choking odors.[2]

Relic veneration was central to the development of Buddhist mortuary practice. The cremation of the body of the historical Buddha Shakyamuni served as a didactic performance of the main doctrinal precept of impermanence. It also generated powerful, enduring physical relics for enshrinement and veneration that transcended his-

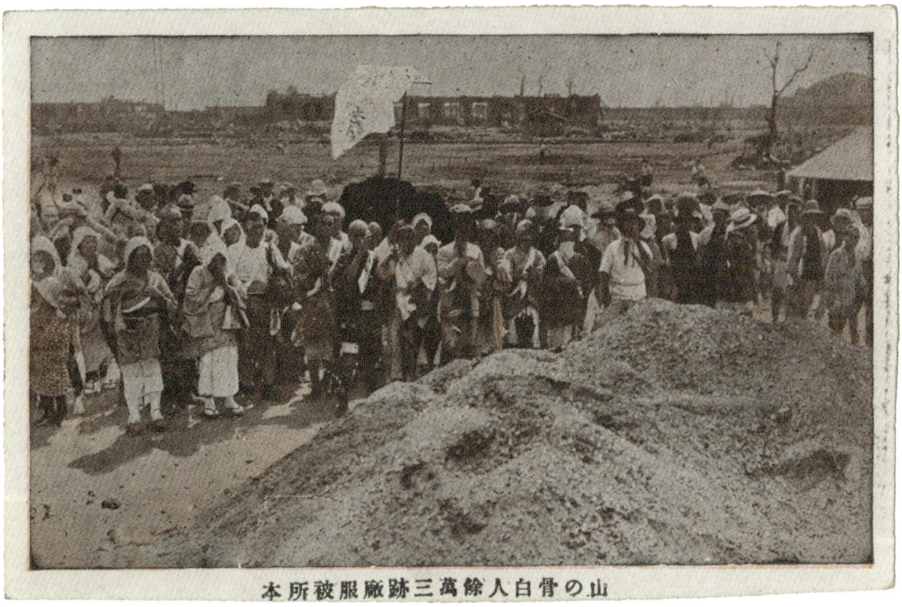

7.1 *Mountain of White Bones of the More Than Thirty Thousand from the Former Honjo Clothing Depot* (*Honjo Hifukushōato sanman yojin hakkotsu no yama*), postcard, 1923. Collection of the author.

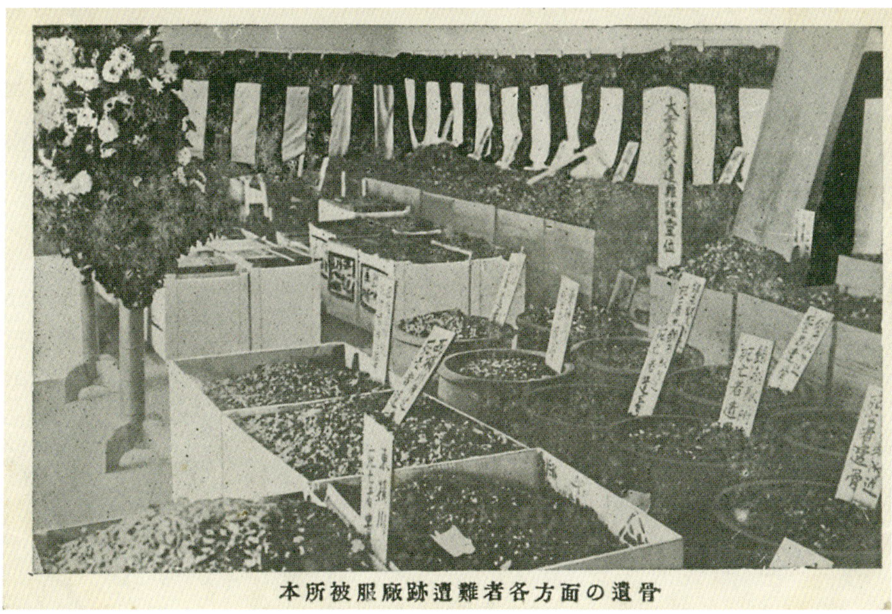

7.2 *Remains of the Victims of the Former Honjo Clothing Depot in Every Direction* (*Hifukushō sōnansha kaku hōmen no ikotsu*), postcard, 1923. Collection of the author.

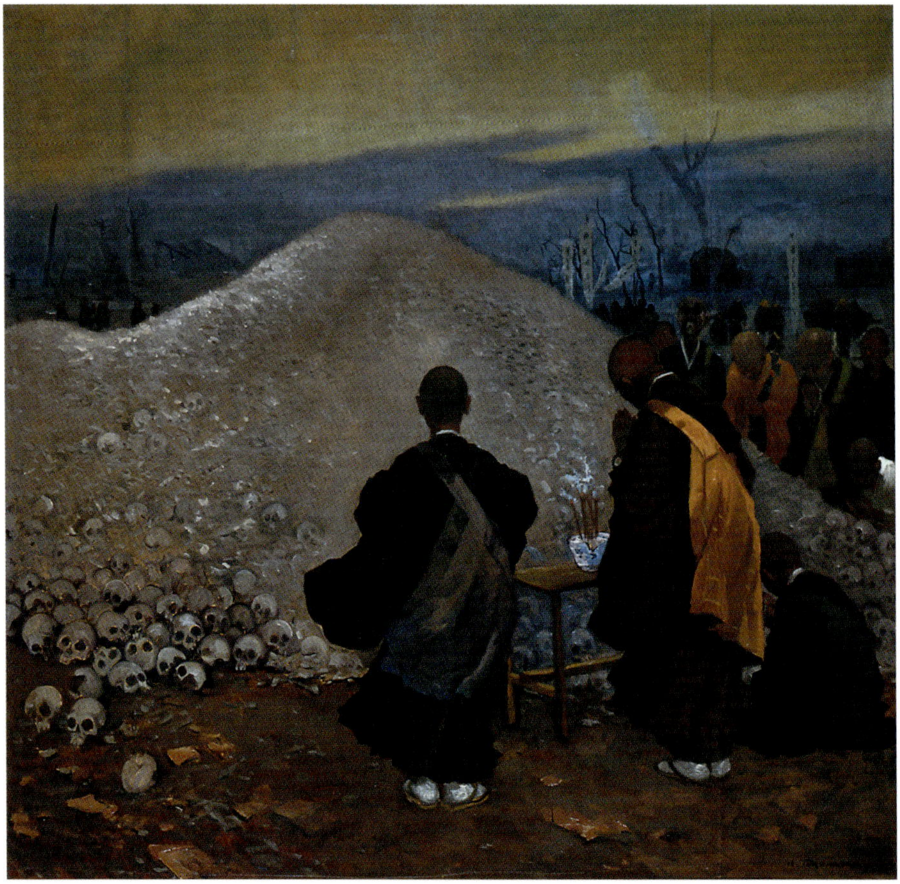

7.3 Tokunaga Ryūshū, *Site of the Former Clothing Depot* (*Hifukushōato*), oil on canvas, c. 1923. Earthquake Memorial Hall. Collection of Tōkyō-to Irei Kyōkai. Photograph courtesy of Art Restoration Studio 21.

tory. Relic veneration gradually extended to other human remains, giving rise to an increasing number of burials and interments, which in turn led to the development of independent mortuary temples (*bodaisho*) for the preservation and veneration of bones and cremated remains.[3] In an oil painting produced the year of the quake, Tokunaga Ryūshū vividly depicted the overwhelming piles of bones, showing them towering over the priests and threatening to engulf them in collective suffering (fig. 7.3). Unlike the random and undifferentiated piles in photographs, the individual skulls at the base of the mound in the painting stare out at the viewer, seeming to speak from beyond the grave about human frailty and mortality. By pacifying the souls of the dead, the prayers provide the living both solace and protection against the awful specter of retribution from beyond the grave.

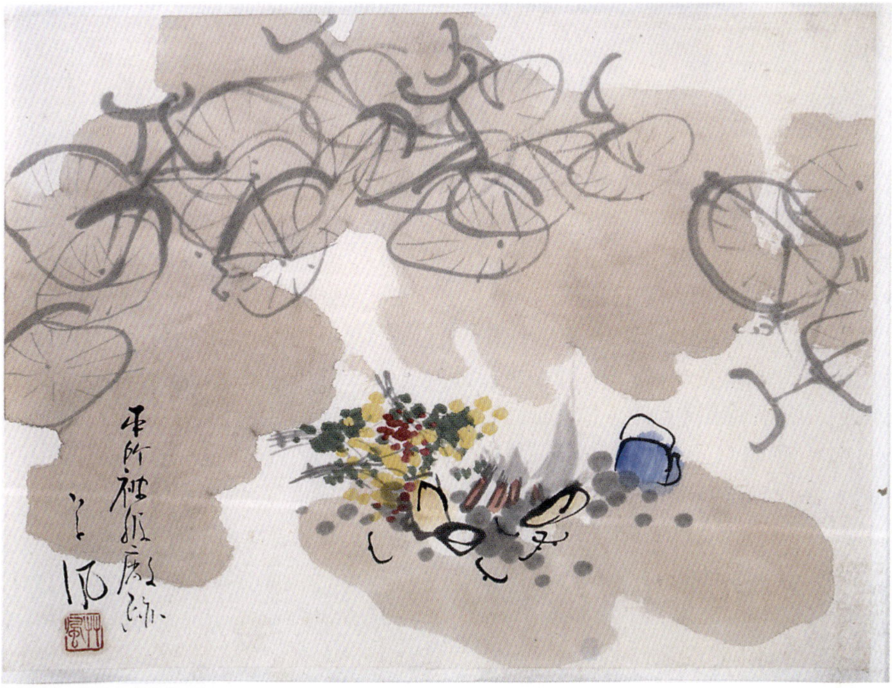

7.4 Nagano Sōfu, *Former Honjo Clothing Depot* (*Honjo Hifukushōato*). *Paintings of the Great Kantō Earthquake* (*Kantō daijishinga*), ink and color on paper, c. 1923. Collection of Tōkyō-to Edo Tōkyō Hakubutsukan.

Remembering the Dead, Forgetting the Past

Many painters conveyed the human horror and sense of profound loss of the quake experience through symbolic objects of daily life that bore the physical marks of trauma. An ink-and-color image by Nagano Sōfu (1885–1949) (fig. 7.4) captures a semicircle of charred bicycles, their riders incinerated into invisibility.[4] Mournful and misshapen, the wheels and frames of the abandoned bicycles lie around a still-smoking pile of household and personal items. As stand-ins for the people who rode them, the bicycles convey lamentations to the once-bustling city of riders, now silent and empty. Nagano's intentionally elegiac image, evacuated of people, stands in stark contrast to the lively, populated scenes of postquake street life represented in film. The practice he depicts of creating makeshift memorials on roadsides has become common around the world, often marking the tragic deaths of friends and family killed in traffic accidents. It also relates to more recent improvised memorials, such as the memorial fence erected in Oklahoma City after the 1995 bombing or even the tributes left at the gates of Kensington Palace after the death of Princess Diana. In the moments of tragedy before permanent monuments are erected, transitory memorials serve as spatial markers of loss and aids to the grieving process.

Soon after the quake, family members and officials heatedly debated whether the victims should be placed in a cemetery or a public monument. Religious observances for the dead (*kuyō*) at temples and gravesites was the traditional form of commemoration, but it was jointly decided that a collective tragedy of this magnitude warranted a public monument to house the proper civic funerary rites, and this became a primary objective of the Earthquake Memorial Hall.[5] The Tokyo municipal government, however, always intended for the memorial to function as a sacred and historical site of remembrance, from the moment it commissioned a public competition to select a design for the memorial, which was originally to be one structure containing both a religious hall and display spaces for objects narrating the earthquake disaster and reconstruction.

Everyone agreed that there was no more fitting symbolic site for the monument than that of the Honjo Former Army Clothing Depot, which was already located near a site of historical remembrance of past urban disasters and untimely deaths, Ekōin Temple. Prior to the quake, some of the land (5,890 *tsubo,* or 4.8 acres) had been combined with donated property from the estate of Yasuda Zenjirō and cleared to become a public park (this clearing is what drew the ill-fated quake refugees there in the first place), ultimately named Yokoamichō Park after the street it faced.

A temporary wooden memorial ossuary, the Cinerarium or Reliquarium (Nōkotsudō), with a traditional hip-and-gable tile roof was erected on the site in October to house and protect the bones of the dead as they waited for reinterment in the permanent memorial. This was the site of the important Buddhist memorial service (*tsuitōkai*) on the forty-ninth day after cremation (this day is particularly important in Buddhist belief because it marks the end of the intermediate stage from this life to the next). Over a hundred thousand mourners attended, many of them grieving relatives, including the imperial family. The Tokyo municipal social education division, not a Buddhist religious institution, convened the ceremony, asserting the city's responsibility for administering the site. In the following months, thousands more came from around the nation to visit the temporary memorial.[6] Hiratsuka Unichi captured this modest wooden structure in a woodblock print in 1925 (fig. 7.5). Standing among the rows of barracks, at the end of a row of fluttering banners with Buddhist prayers, the lonely structure seems to be a sad testament to the thousands of souls within. Like his image of the Nikoraidō, Unichi's cinerarium is epitaphic, conveying the building's symbolic solemnity.

Because reconstruction was an extended process, remembrance had to be phased.

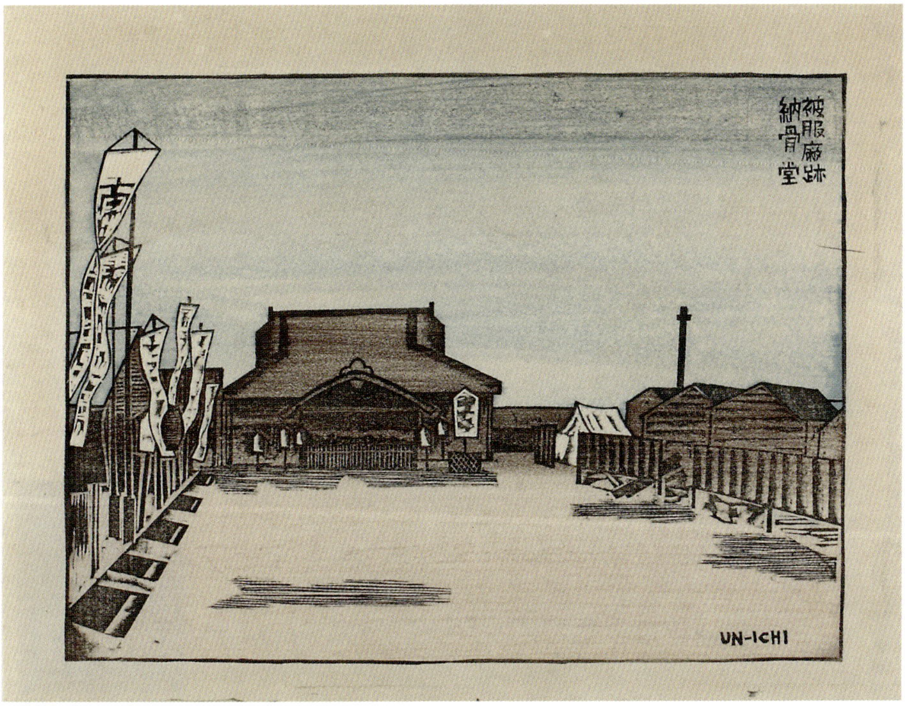

7.5 Hiratsuka Unichi, *Cinerarium* (*Nōkotsudō*), from the series *Scenes after the Tokyo Earthquake* (*Tōkyō Shinsai ato fūkei*), woodblock print, 1925. Collection of Machida Shiritsu Kokusai Hanga Bijutsukan.

Memorial steles were erected in locations that had suffered damage throughout the Kantō region, some dedicated to the tragic dead and others commemorating the rebuilding of destroyed temples, shrines, roads, and bridges.[7] Yearly anniversaries became symbolic, at least for the first few years. Then as the final phase of the process drew to a close, there was a crescendo of "reconstruction commemoration" (*fukkō kinen*) events leading up to the "completion" celebration. The one-year anniversary of the quake marked the beginning of an important stage in the strategic rewriting of the disaster narrative, during which time the government actively tried to purge the record of residual toxic elements, such as the officially instigated massacre of Koreans, that had been suppressed at the time of the disaster, only to see questions about these disturbing facts resurge several months after, which spurred it to actively suppress them again and continue doing so through the war years. A month before the anniversary, the *Asahi shinbun* began collecting stories of "hidden heroes" (*kakureta kōrōsha*) whose meritorious deeds had not been made public, including those who had bravely protected Koreans. These stories were published as a regular column in the newspaper.[8]

The Tokyo metropolitan government sponsored a collection of nearly a hundred personal stories titled *Taishō Shinsai biseki* (Excellent Deeds of the Taishō Earthquake) to help codify the progressive narrative.[9] It is no surprise that this official commemoration was invested in positive and heroic memories of the event and attempted to erase counternarratives of dissension, inequity, or massacre. The resulting story, Mai Denawa notes, emphasized that "the Japanese people behaved courageously and admirably through acts of self-sacrifice, and the Japanese government successfully supported its citizens in time of emergency." As she has convincingly shown, the government was concerned to distance itself from the murderous rampage of the vigilantes and to erase the connection between official actions and the Korean massacre. This distancing included downplaying the explicit order sent by Tokyo Chief of Police Gotō Fumio to every prefectural governor to "take firm measures in dealing with the activities of Koreans," which legitimated the vigilante violence to come and the widespread participation of police and soldiers in the killings.[10] On the first anniversary in 1924, a number of Kantō-area cities even erected steles to memorialize "Korean victims" of the quake, although their laments over the tragic fate of these unwitting "victims" (*giseisha*) never specified who had perpetrated the atrocities, instead passively incorporating Korean murder victims into the larger category of unfortunate casualties of the disaster.[11] Remembering was the cornerstone of restoration; forgetting was critical to remembering.

The progressive narrative used the rhetoric of renewal to put the quake in the past.[12] In its first-anniversary special edition, *Jiji shinpō* heralded the progress of reconstruction by showing expansive aerial photographs of a densely rebuilt Tokyo cityscape with orderly gridded barracks, lively streets, revived commerce, and a dynamic citizenry (fig. 7.6). These images were juxtaposed with photographs showing the previous glories of the city and the burned wasteland of the immediate postquake condition to emphasize the progressive recovery and the desire to supersede the former glory of the capital.

The forgetting aspect of remembering was mischievously evoked in Miyatake Gaikotsu's experimental word poem "Composition as a Commemoration of the Great Earthquake," printed on the cover of the journal he edited, *Hentai chishiki* (Abnormal Knowledge), on the first anniversary of the quake (fig. 7.7). The poem is labeled "Box with Overturned Printed Text Scattered Here and There" (*Tenpuku no katsuji o hiroi-yoseta hako)* in a five-seven-five–syllable *waka* format, and it emulates (perhaps, parodies) the avant-gardist visual literary experimentations of groups like Mavo. The visual chaos and overturned elements parallel the state of the capital after the quake—a condition that persisted a year later. Differently sized text floating in a conglomeration of print reads, "today's remembrance of last year," "Korean invasion," "Prime

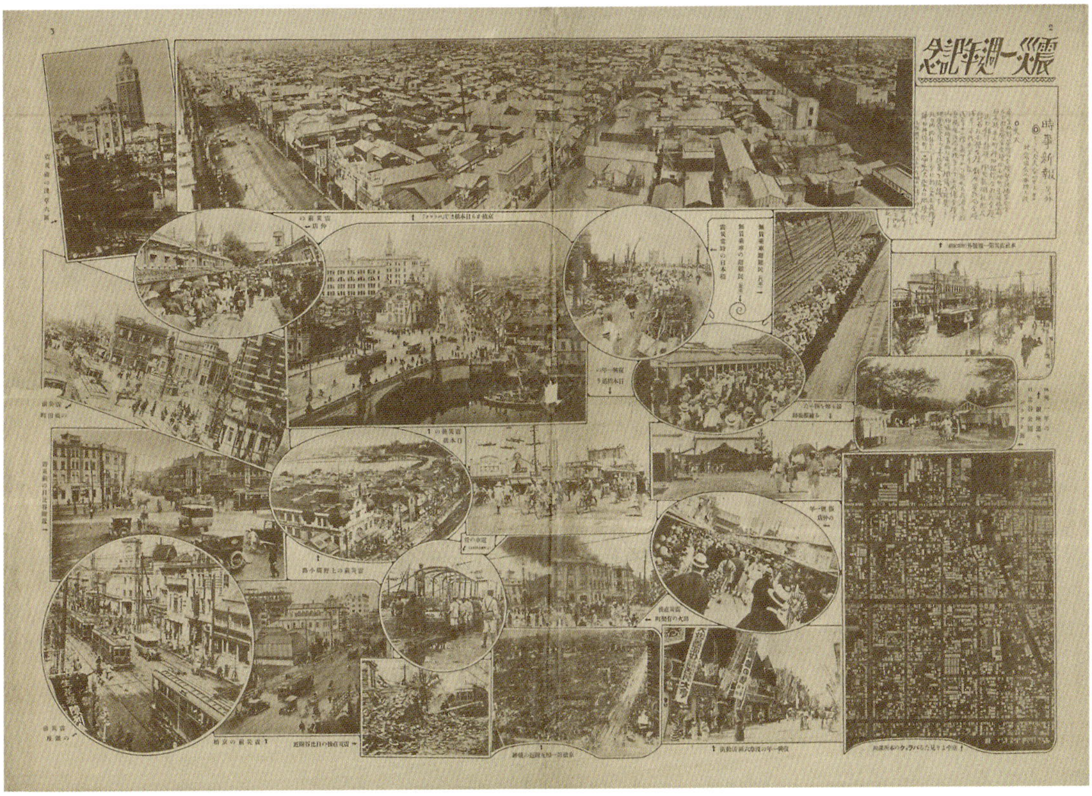

7.6 "One-Year Anniversary of the Earthquake" (*Shinsai isshūnen kinen*), special issue, *Jiji shinpō*, 1 September 1924, 2–3. Collection of the author. By permission of Sangyō Keizai Shinbunsha.

Minister Yamamoto assassination," "police station burned to the ground," "trams and steam trains not running," "water service stopped," "Honjo Clothing Depot," "100,000 people burned to death," "poisoned well water." As in a word-jumble puzzle, phrases emerge from the mishmash of characters to produce a meaningful narrative that commemorates the discordant elements of the past year: the unfounded rumors of Korean sabotage through poisoned well water, the mass death of thousands, the complete and continued disruption of vital services, and the resignation of Prime Minister Yamamoto and his entire cabinet in January 1924 over the infamous failed attempt by student and communist sympathizer Nanba Daisuke, son of a member of parliament, to assassinate Prince Regent Hirohito on 27 December 1923, popularly known as the Toranomon Incident. Nanba was subsequently executed for his crime. The poem made clear that the first year of reconstruction had not been such a sunny affair. Signs of corrosion and dissension were everywhere.

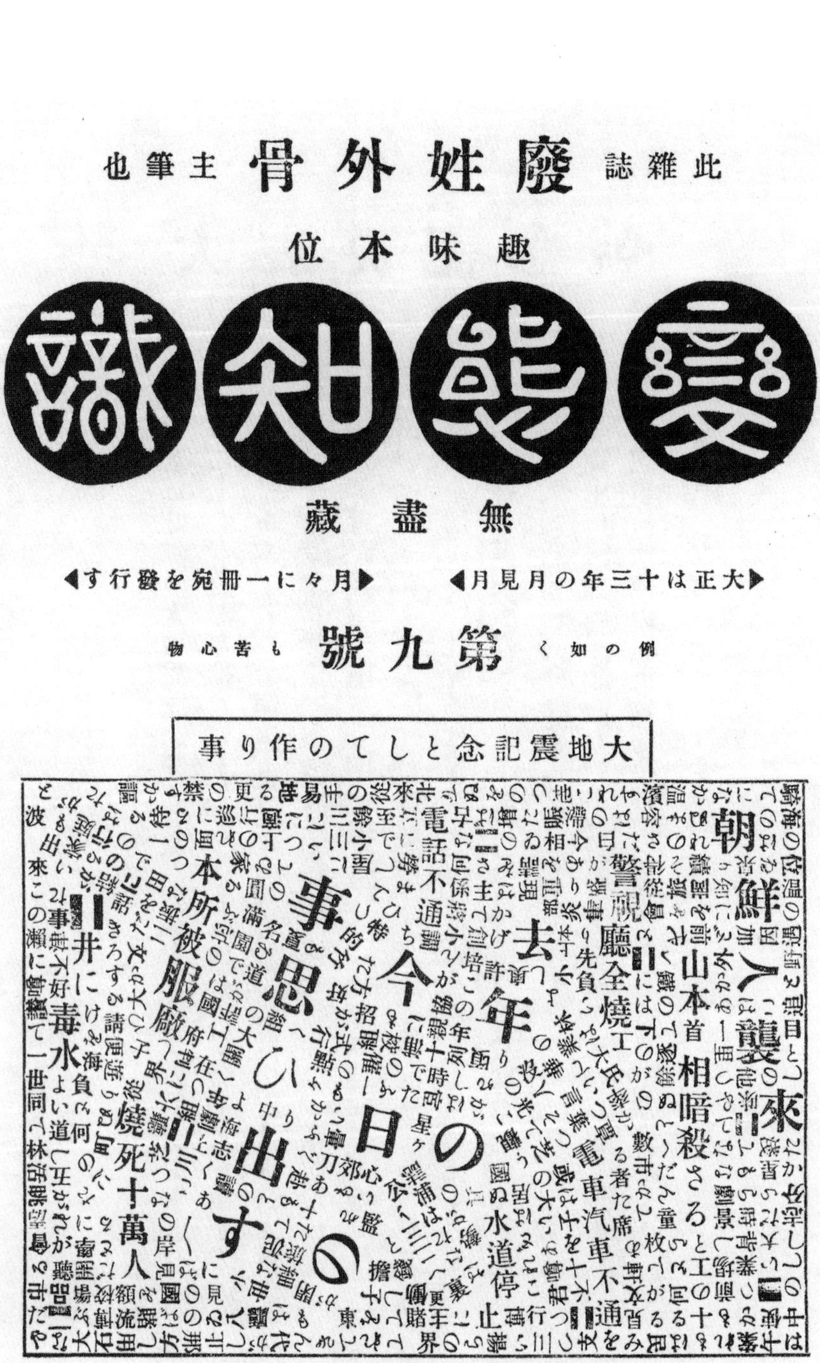

7.7 Miyatake Gaikotsu, "Composition as a Commemoration of the Great Earthquake" (*Daijishin kinen to shite no tsukurikoto*), word poem, cover. *Hentai chishiki* (Abnormal Knowledge) 9, September 1924. Reproduced in *Miyatake Gaikotsu chosakushū* (Collected Works of Miyatake Gaikotsu), ed. Tanizawa Eiichi and Yoshino Takao (Tokyo: Kawade Shobō Shinsha, 1986), 6:407. Courtesy of Kawade Shobō Shinsha.

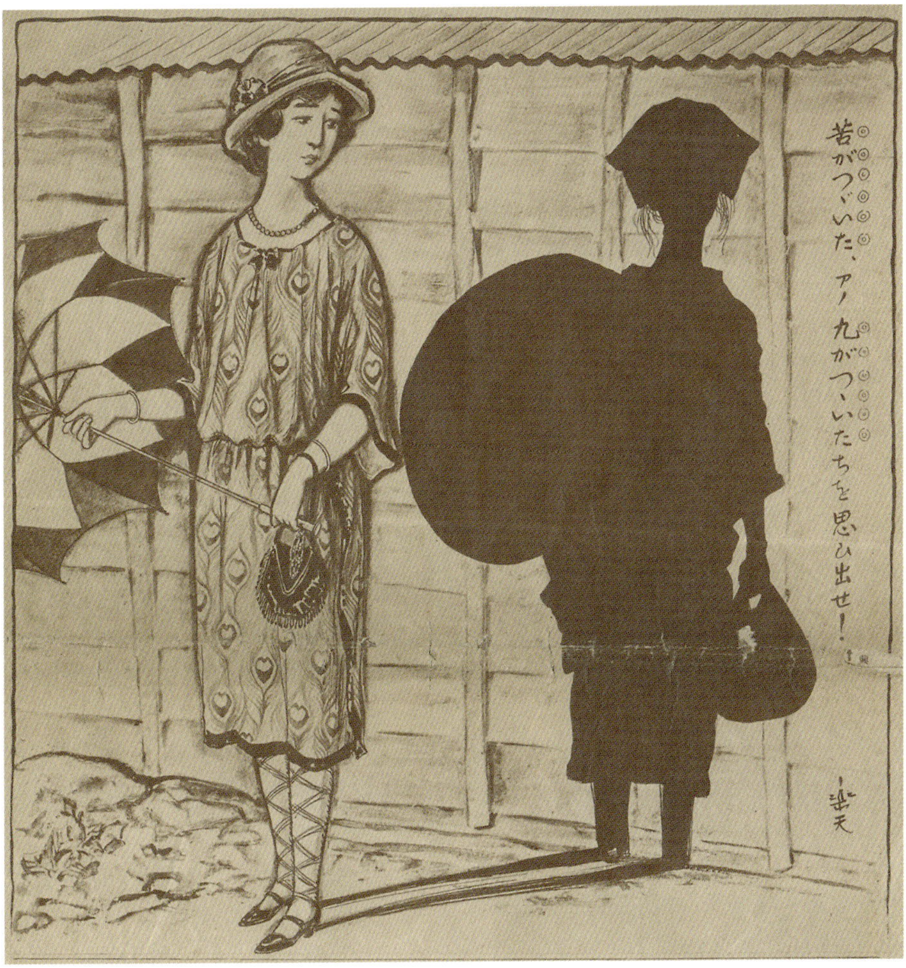

7.8 Kitazawa Rakuten, *The Suffering Continued, Remember That September 1!* (*Ku ga tsuzuita, ano kugatsu tsuitachi o omoidase!*). "One-Year Anniversary of the Earthquake" (*Shinsai isshūnen kinen*), special issue, *Jiji shinpō*, 1 September 1924, 4. Collection of the author. By permission of Sangyō Keizai Shinbunsha.

In marking this one-year milestone, Rakuten's popular caricatures reminded people that while life had returned to normal for some, many still struggled. In his illustration in *Jiji shinpō*'s anniversary issue, a stylish modern girl with cloche hat, flapper dress, decorative stockings, and parasol casts a shadow on the clapboard wall of the barracks behind her (fig. 7.8). Her shadow reveals the outline of a disheveled refugee in tattered clothing forlornly carrying her belongings. Within everyone, this image suggests, was the shadow of the trauma and the memory of being reduced to nothing. By tapping into readers' latent collective identities as refugees, Rakuten seems to be cautioning them to remember those still suffering.

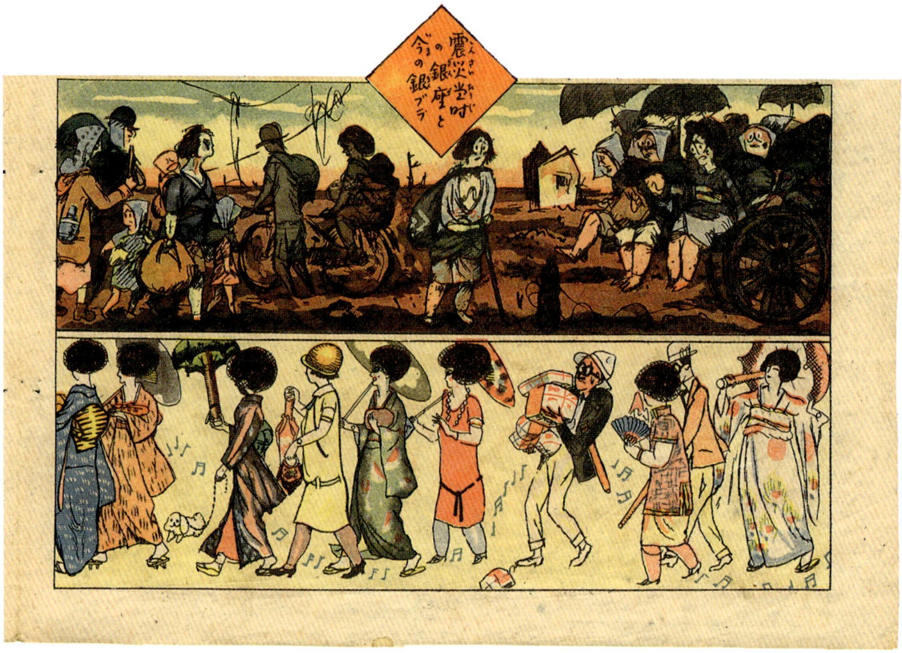

7.9 Kawamori Hisao, *Ginza Immediately after the Earthquake and Strolling through Ginza Now* (*Shinsai chokugo Ginza to ima no ginbura*), cartoon. *Jiji manga* 228, 31 August 1925, 4. Collection of Ohio State University Billy Ireland Cartoon Library & Museum. By permission of Sangyō Keizai Shinbunsha.

Along with Rakuten, some critics also wondered if reconstruction meant simply returning to the materialist and consumerist practices before the quake without having learned any moral lessons. In a *Jiji manga* cartoon in August 1925, Kawamori Hisao (1898–1968) juxtaposed the "Ginza stroll" (*ginbura*) right after the quake, when bereft refugees made their way through a blighted landscape of loss and sorrow, with a later stroll through the reconstructed Ginza district (fig. 7.9). In the depiction of the new Ginza, stylishly dressed women in kimono and Western-style dresses promenade along the boulevard as their male companions follow behind laden with packages. The cartoon returns to the social criticism of the liberated woman that was voiced immediately after the quake. Was reconstruction destined to become a world of henpecked and emasculated men led by frivolous women in gaudy attire inappropriately sashaying in public to the floating tunes of jazz? As a harbinger of future divine punishment, Rakuten's catfish continued to remind readers of the potential hazards of their profligate, selfish ways, warning against focusing on personal gratification and political self-interest that would morally and logistically hamper meaningful reconstruction down the line. Dancing on the volcano might bring the catfish back.

By the second anniversary of the quake, only a few pages in the *Asahi Graph* were devoted to reconstruction, and these continued the tenor of progressive rebuilding

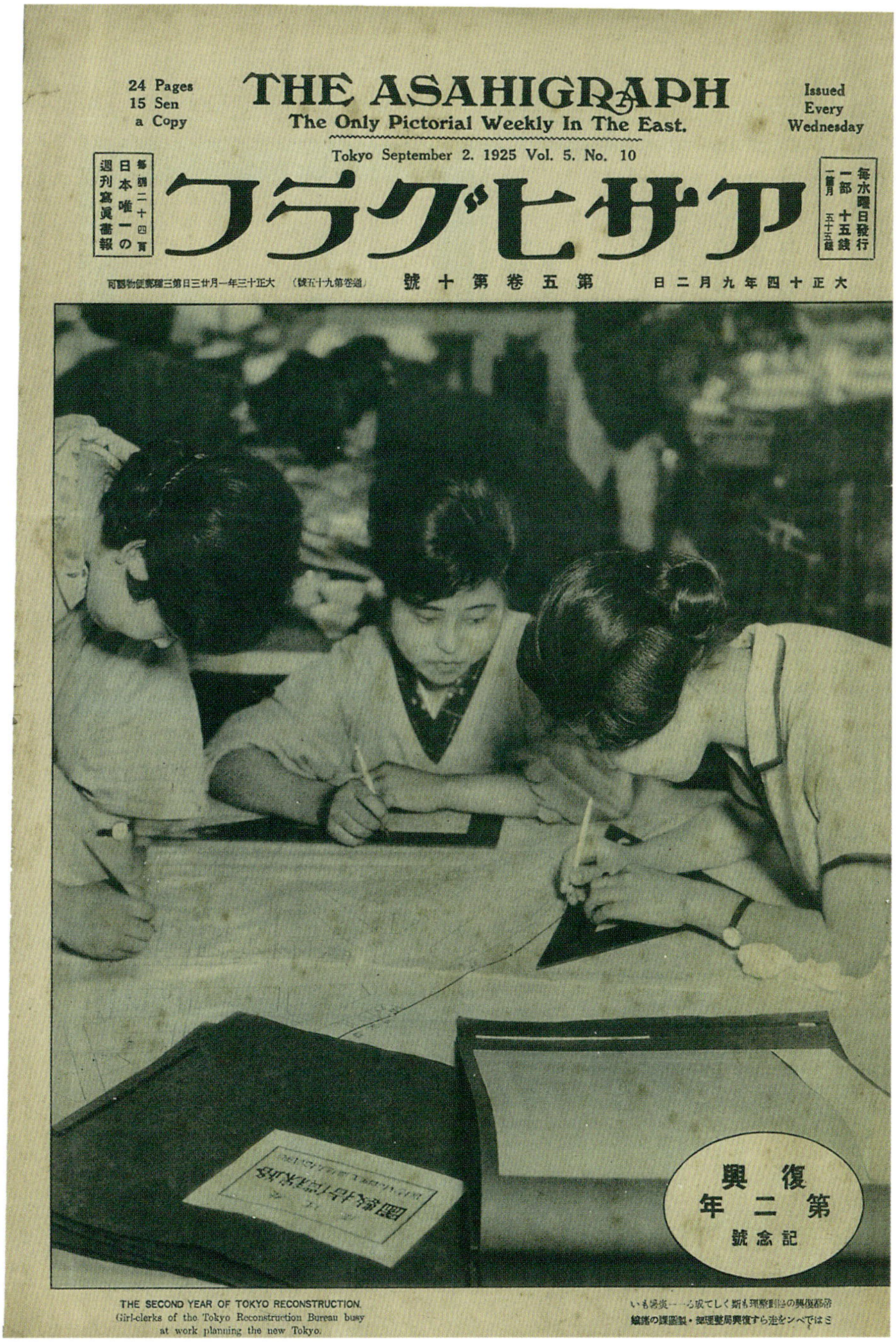

7.10 "Girl clerks" engaged in reconstruction planning, photograph, cover. "Commemorative Issue for the Second Anniversary of Reconstruction" (*Fukkō daininen kinengō*), special issue, *Asahi Graph* 5, no. 10 (2 September 1925).

of the previous year.[13] The cover (fig. 7.10) displays a curious photograph of three modern young women—"girl clerks"—engaged in earthquake reconstruction planning. Leaning over a table with detailed maps of land readjustment, the young women engage in a new technical job, combining the scientific language of modernity and the professionalization of women—two things that were troubling to critics, albeit sometimes for opposite reasons. Even those who supported modernization and rationalized urban planning did not necessarily want liberalization for women. Others championed increased freedoms and rights for women but questioned the state's remapping of the cityscape for social management. This photograph demonstrated that both were moving forward as the forces unleashed by modernity continued undaunted—perhaps even assisted—by the quake.

Constructing Remembrance

The official effort to erect a memorial to the quake got under way around May 1924. As with all previous official endeavors, the government formed a committee to administer the construction of a memorial, convening it in August and assigning it a budget of 50,000 yen. The Tokyo Association for the Earthquake Memorial Project (Tōkyō Shinsai Kinen Jigyō Kyōkai) included Gotō Shinpei, business entrepreneur Shibusawa Eiichi, director of the TIMR Sakutani Nozomu, and mayor of Tokyo Nagata Hidejirō. In addition to providing government and imperial financial assistance, the authorities decided to conduct a public fund-raising campaign to amass the 1 million yen they estimated would be necessary to finance the memorial.[14]

At the time the committee was formed, a design was publicly issued by Tokyo city for a traditional Nara-period-style historical memorial building with an octagonal hipped roof, constructed of steel-reinforced concrete (fig. 7.11). The building was to contain a large circular memorial hall that could accommodate a thousand people seated facing a front altar, and the entire space was to be ringed with a wide corridor. The design called for enshrining the spirits of the deceased in a space behind the altar and for locating the cinerarium for their remains in a subterranean crypt directly below.[15] This plan was abandoned, however, in response to widespread demand for a public design competition for the memorial. The competition jury comprised celebrated architects who were well ensconced in Japan's higher institutions of architecture and engineering—Satō Kōichi, Sano Toshikata, Itō Chūta, and Tsukamoto Yasushi—as well as important academic and civic leaders like the president of the Tokyo School of Fine Arts, Masaki Naohiko; the director of the Tokyo municipal parks division, landscape architect Inoshita Kiyoshi; and the deputy mayor of Tokyo,

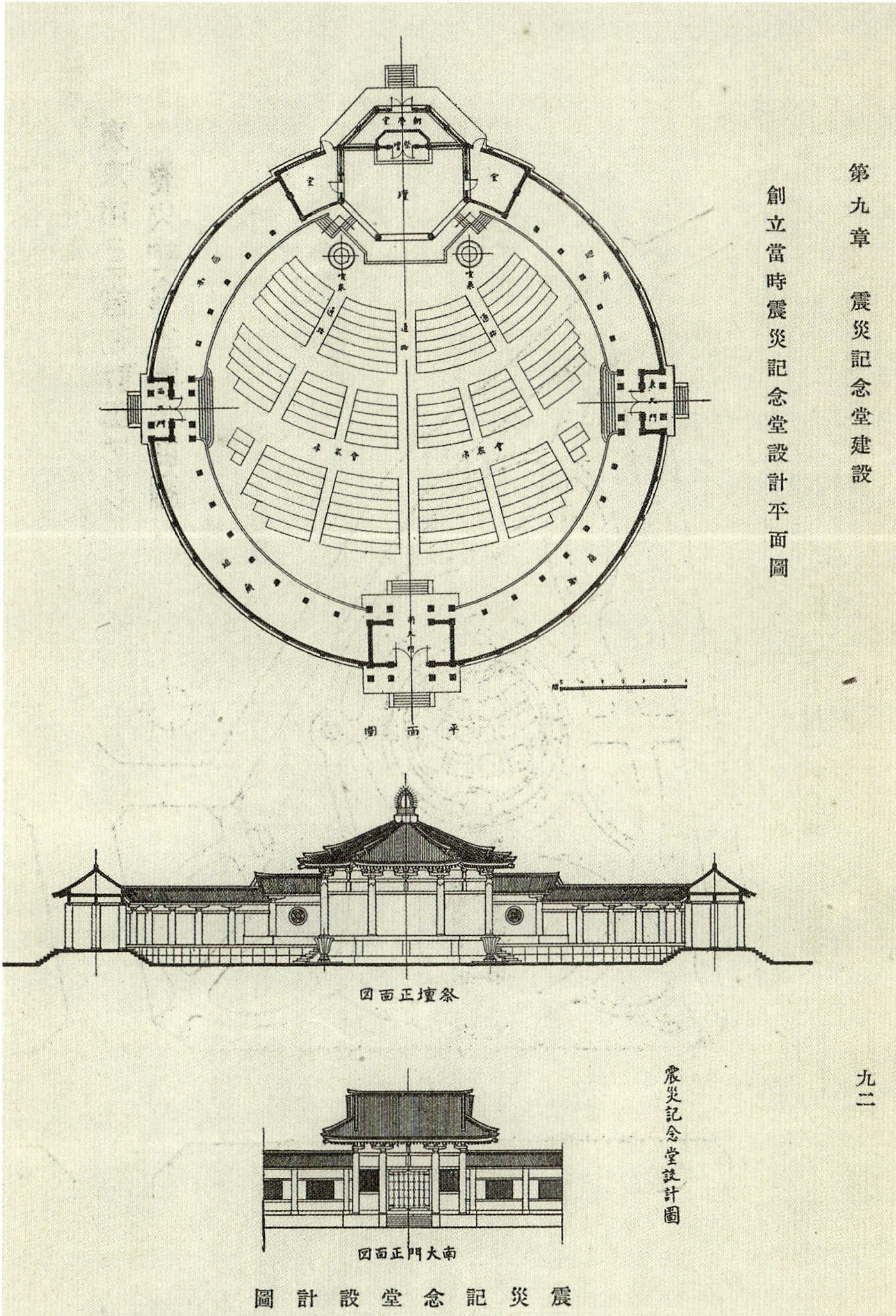

7.11 Artist unknown, Nara-period-style memorial building, original plan and elevations, 1924. Tōkyō Shinsai Kinen Jigyō Kyōkai Seisan Jimusho, *Hifukushōato* (Site of the Former Army Clothing Depot) (Tokyo: Tōkyō Shinsai Jigyō Kyōkai Seisan Jimusho, 1932), 92.

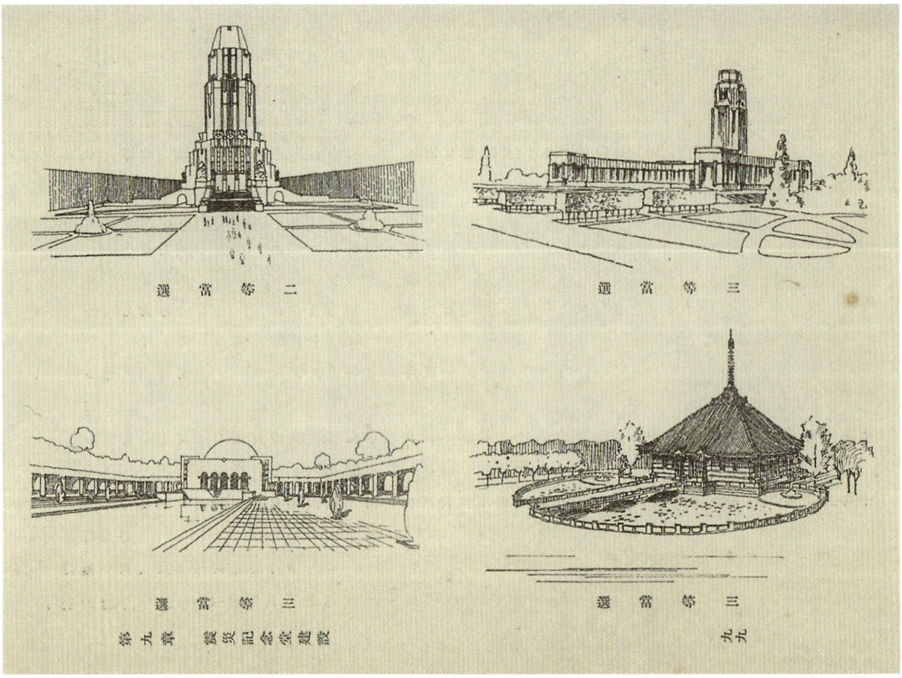

7.12 Sketches by the second- and third-prize winners in the design competition for the Earthquake Memorial Hall, 1925. Tōkyō Shinsai Kinen Jigyō Kyōkai Seisan Jimusho, *Hifukushōato* (Site of the Former Clothing Depot) (Tokyo: Tōkyō Shinsai Jigyō Kyōkai Seisan Jimusho, 1932), 99.

Okada Tadahiko, who served as director of the Tokyo Association for the Earthquake Memorial Project. The rules for the competition stipulated that the building, with a budget of 65,000 yen, be fire retardant and antiseismic in construction. It also needed to harmonize with its park context. Cash prizes of 3,000, 2,000, and 1,000 yen—considerable sums at the time—would go to the top three selections. After two months, the committee had received 220 entries from architects, ranging in style from neotraditional structures with hip-and-gable roofs to neoclassical Western-style edifices with tall projecting vertical towers (fig. 7.12).[16] The majority of design submissions presented a strong geometrical base structure enclosing a main sanctuary hall above a subterranean crypt for the interment of the deceased. Some designs were ziggurat-like towers with stepped massing that framed dramatic central entries to the main chamber and crypt. Many also featured projecting vertical elements such as pagoda finials or columnar towers. The designs were clearly inspired by both the Asian Buddhist architectural mortuary traditions of the pagoda (stupa in Sanskrit) tower, a temple reliquary monument, and the Western classical architectural rhetoric of mausoleums that had developed in a long genealogy from the originary Mausoleum at Halicarnassus (also known as the Tomb of Mausolus, from which the term for stately tombs *mausoleum* originates). The Mausoleum at Halicarnassus, one of the Seven Wonders of the World, had a raised square base, colonnaded sanctuary, and pyramidal stepped roof structure. Modern updated interpretations dotted the world landscape at the time of the competition. Some entries drew on the common memorial architectural forms of vertical obelisks, columns, and arches.

Contemporary architectural obsession with soaring towers, particularly among modernist designers, also influenced these designs. There were numerous prominent examples of towers erected abroad and in Japan during this period, ranging from the widely publicized submissions for the 1922 Chicago Tribune competition in the United States to the heralded Machinery Hall by Japanese architect Horiguchi Sutemi at the 1922 Peace Memorial Exposition in Ueno. Many of the architecture submissions to the 1924 Citizens' Art Association exhibition of reconstruction plans also featured tall towers—notably Kishida Hideto's striking *Model of a Tower for the Memorial Services for Victims* (*Giseisha kuyōtō mokei*), which was a tall, conical structure, and Hinako Jitsuzō's *Tower of Death* (*Shi no tō*) (fig. 7.13), which had a stolid ziggurat-shaped tower with grand sweeping stairways on either side that presumably led to the raised crypt chamber.[17] Some of these competition designs had an afterlife in the wartime period, when numerous monuments to the sacred war dead (*chūreitō*) were built in Japan and on the continent. Hinako would later realize a related design in 1940 in his war memorial in Miyazaki, Kyūshu, which was inscribed on the front with the imperialist expansionist slogan "all the world under one roof" (*hakkō ichiu*).

The judges for the earthquake memorial awarded first place to the design by esteemed young modern architect Maeda Kenjirō (1892–1975), who had trained at the Tokyo School of Fine Arts; worked for the Ministry of Communications; joined Daiichi Ginkō, where he helped with the construction of Nishimura Yoshitoki's design for the bank's headquarters in 1919; and designed stylish commercial establishments for the Shiseido cosmetics company in Ginza in the late 1920s. Maeda won so many competitions during this period that he was known as "competition Maeda," but few of his winning designs were actually realized.[18] His design for the memorial evinced a strong stylistic vocabulary of stripped-down neoclassicism with art deco decorative embellishments.[19] It was a single-story steel-reinforced semicircular building with a projecting columnar tower at the center (fig. 7.14). The recessed colonnaded portico was simplified into a square opening with two columns. The first floor featured a large memorial sanctuary hall in the front that could accommodate three hundred people facing the main altar at the end of the room. Behind it was the base of the tower, which would be

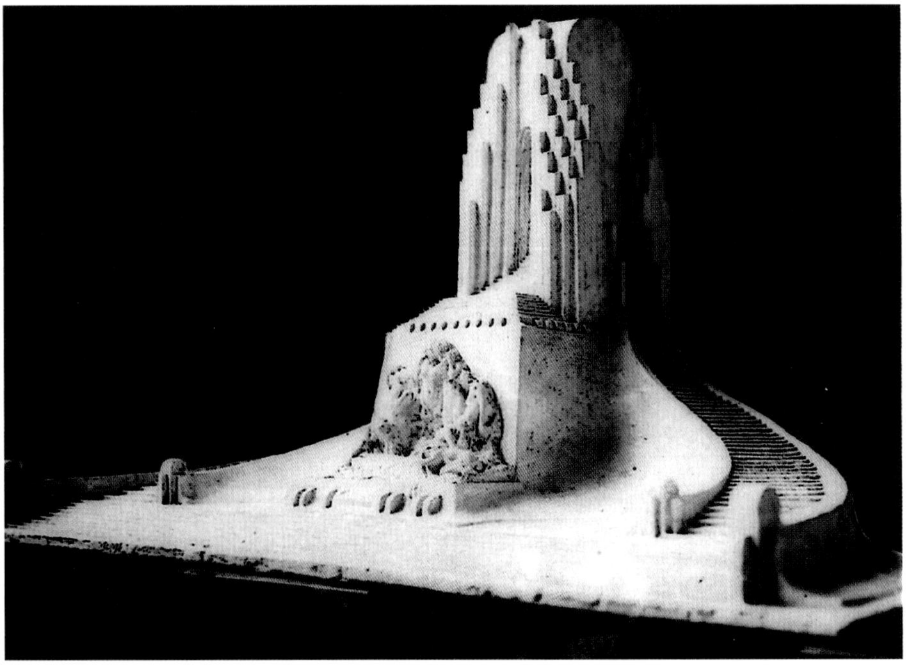

7.13 Hinako Jitsuzō, *Tower of Death* (*Shi no tō*), model, 1924. Hirota Toshikazu, *Shōwa shoki chōkoku no kisai: Hinago Jitsuzō no sekai* (The World of Hinago Jitsuzō: The Remarkable Talent of an Early Shōwa-Period Sculptor) (Kyoto: Shibunkaku Shuppan, 2008), 34. Courtesy of Shibunkaku Shuppan and Hirota Toshikazu.

a focus of the monument, with a circular sacred space that housed a white marble cenotaph embodying the spirits of the deceased (*reitai*) in the center. The cenotaph would be surrounded by eleven large black marble pillars supporting a stained-glass covering that would reflect light onto the space. Directly under this was the subterranean sepulcher, a cinerarium for the cremated remains of earthquake victims. The tower, projecting nearly 175 feet above the ground, was divided into small display and storage spaces, and the back rooms on either side of the tower on the first floor were spacious exhibition rooms for earthquake-related documents and objects. The entire structure was bordered by water: curved ponds in the back and decorative reflecting pools with fountains in the front.[20] The design and renderings for the six noteworthy submissions were publicly exhibited at the Jichi Kaikan in Ueno for three days.[21]

Maeda's design relied on the engineering and construction expertise of Tokyo Imperial University professors Sano Toshikata (also known as Sano Riki, 1880–1956) and Itō Chūta. Having witnessed the effects of the San Francisco earthquake firsthand, Sano made an important contribution to the 1917 *Earthquake Disaster Prevention*

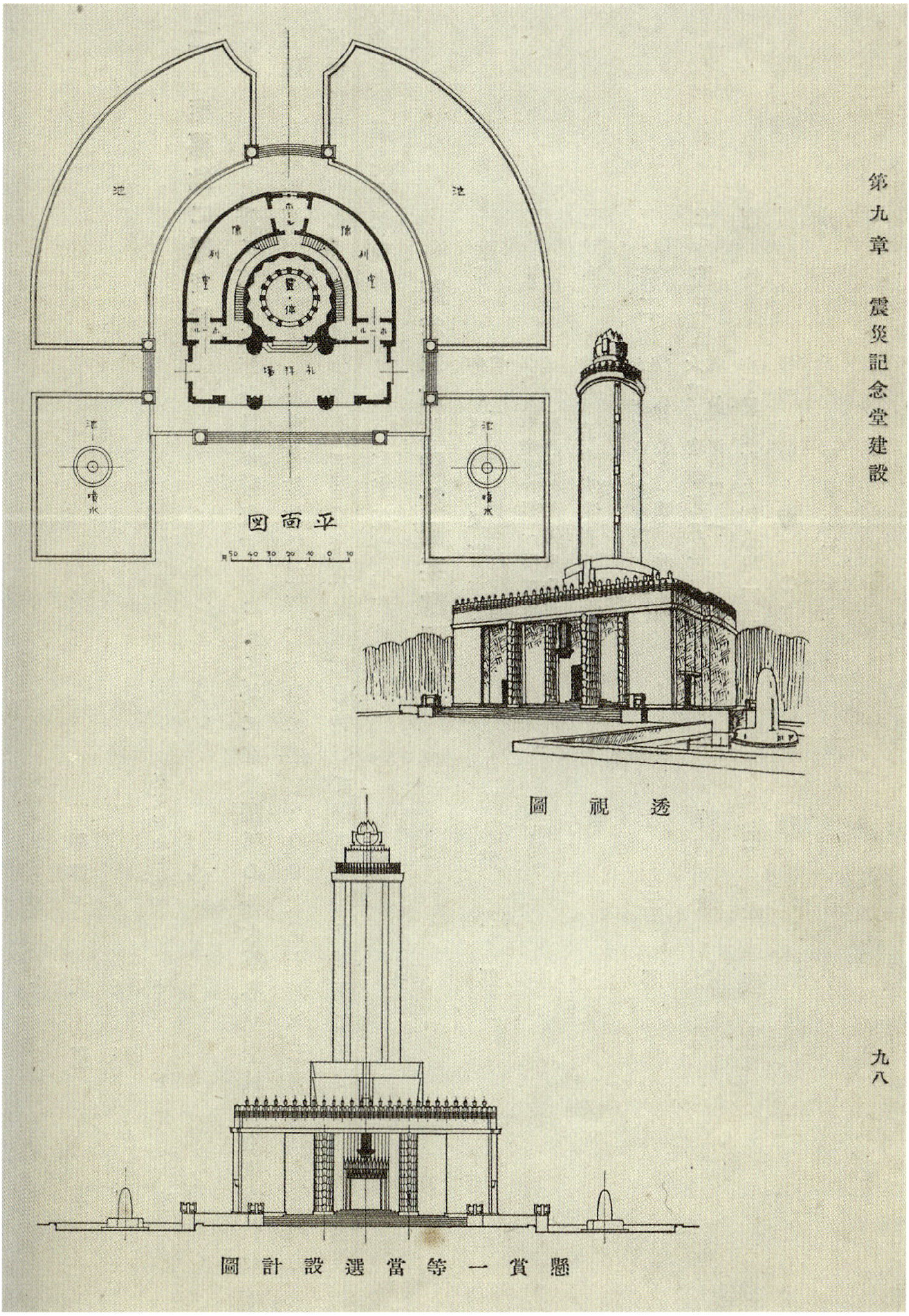

7.14 Maeda Kenjirō, *Earthquake Memorial Hall Building* (*Shinsai Kinendō kensetsu*), plan and elevations, 1925. Tōkyō Shinsai Kinen Jigyō Kyōkai Seisan Jimusho, *Hifukushōato* (Site of the Former Clothing Depot) (Tokyo: Tōkyō Shinsai Jigyō Kyōkai Seisan Jimusho, 1932), 98.

Study Association Report on earthquake-resistant construction (*kaoku taishi kōzōron*) and devoted his career to advocating for the safety benefits of steel-reinforced-concrete construction.[22] He was a member of the board of the Reconstruction Bureau, where he assisted with the work and land readjustment planning. In 1924, he became director of the Tokyo Municipal Architectural Agency and oversaw the design of many steel-reinforced-concrete elementary schools, devoting his professional efforts to making the city less vulnerable to fire disasters and providing greater protection for schoolchildren, more than five thousand of whom had died in the 1923 quake.[23]

By mid-August 1926, only 300,000 yen of the projected 1 billion yen project budget had been raised. Fund-raising was stepped up among the Buddhist community, which the government asked to raise another 300,000 yen.[24] Newspapers reported that thirteen hundred Buddhist monks from prominent Tokyo temples had marched through the city asking for donations, telling people, "We can't just sit here in silence watching."[25] Efforts were also broadened to include other civic groups and religions. The Earthquake Collaborative Fund-Raising Association (Shinsai Kyōdō Kikinkai), established in late August 1926, included youth groups, women's groups, and Christian groups. Its slogan was "Let's put money aside for our unforeseen disaster" (*otagai no fuji no saigai sonaemashō*). Although Buddhism was without question the funerary religion of Japan, the involvement of the two other major religious groups, Shinto and Christianity, heralded the ecumenical religiosity of the memorial.[26]

At this time, a model was commissioned to stoke interest in contributions, but it had a decidedly opposite effect. Upon seeing the design in three dimensions, the city's residents—particularly those from the Honjo and Fukagawa low-city wards hit hardest by the quake—raised a public outcry. They objected to the design rhetoric of the tower, which newspapers described as resembling the turret on an imperial palace (*gosho no yagura*) and which residents said looked too much like the Tower of Pisa.[27] Inappropriately "compromising on the distinct ethnic characteristics of spiritual culture," argued the Honjo Ward Association in its formal statement to the city, would fundamentally undermine the memorial's religious and educational objectives. As a sacred burial site, the design first had to satisfy the spiritual longings of the citizens. In sympathy with the general citizenry, the Allied Buddhist Association (Bukkyō Rengōkai) had already submitted a letter of formal opposition to the design at the end of April 1926, explaining the clergy's concurrence that the architectural rhetoric of the memorial was too modern and too Western to express the profoundly national and spiritual nature of the quake tragedy. These and other opposition groups petitioned to change the design. The committee and mayor acceded to these demands, and Maeda's design was scrapped.[28]

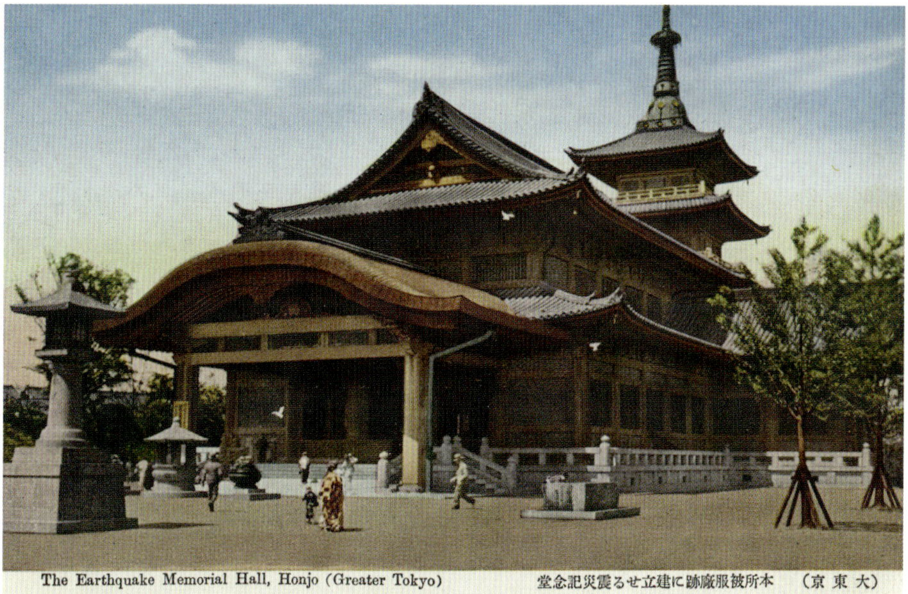

7.15 *The Earthquake Memorial Hall, Honjo (Greater Tokyo)* (*Dai Tōkyō Honjo Hifukushōato ni konryū seru Shinsai Kinendō*), postcard, c. 1930. Collection of the author.

Jury member and well-known historicist architect Itō Chūta, professor at Tokyo Imperial University, was asked to ameliorate the situation, and he devised a pan-Asian design that was promoted as purely Japanese when it in fact exuded an overarching visual "Asianness" and drew from Japanese, Chinese, Indian, and even Christian architectural idioms (fig. 7.15). A prolific architectural theorist with broad interests and influences, Itō had designed a Gothic cathedral for his graduation project, and his doctoral thesis had been on the famed Horyūji Temple in the ancient imperial capital of Nara. He is renowned as the first architectural historian in Japan, and he devoted extensive time to antiquarian activities, studying and writing about ancient monuments across Asia. Itō had the uncanny ability to design buildings that seemed both traditional and modern, supreme examples of his notion of East-West eclecticism that could still express "national taste." After designing the memorial hall, he would go on to design many buildings in variations of this pan-Asian style, most prominently Tsukiji Honganji Temple (1934) in Tokyo for the True Pure Land Sect of Buddhism, as well as a series of monuments to the war dead (*chūreitō*) for the same sect on the Asian continent as the Japanese empire expanded during wartime.[29]

Construction finally began in November 1927. In consultation with engineering expert Sano Toshikata, Itō's Earthquake Memorial Hall was built out of steel-

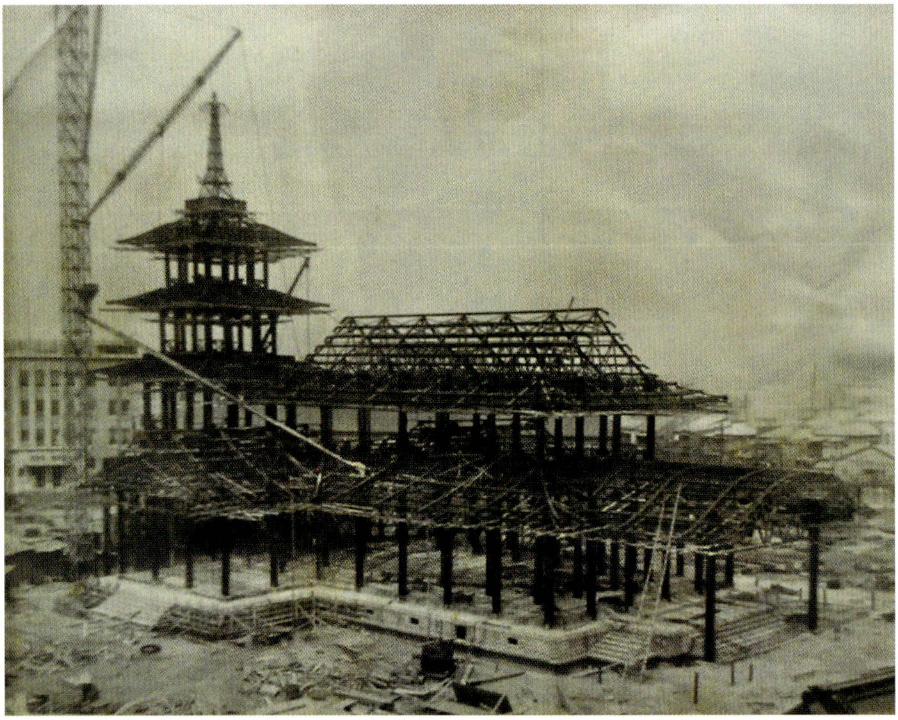

7.16 *Earthquake Memorial Hall* (*Shinsai Kinendō*), steel-reinforced-concrete construction, photograph, c. 1929. Reconstruction Memorial Museum. Collection of Tōkyō-to Irei Kyōkai.

reinforced concrete, its progress extensively documented in photographs (fig. 7.16). Construction was difficult because of the poor condition of the earth foundation at the site. In light of the strong advocacy for this as an antiseismic, fire-retardant mode of construction, the building materials were themselves symbolic of Japan's modern response to the disaster and its adaptable resiliency.

The basilica-like cruciform layout is reminiscent of Christian cathedrals with a nave, transept, and apse (fig. 7.17). But at the apse end stands a large stupa, or pagoda-like tower, with a vaguely Indian-style finial (fig. 7.18). The building in general emulated ancient wooden Japanese Buddhist temples, which Itō had studied closely and helped reconstruct in Kyoto and Nara with tiled hip-and-gable roofs. In front, a prominent portico with a Tokugawa-style cusped gable projects outward to cover the dramatic entryway.

The building (fig. 7.19) enshrines the bones of fifty-eight thousand victims in the stone base of the tower, functioning as a reliquarium, or perhaps more aptly, a martyrium, if we consider the patriotic overlay of the victims' sacrifices for the nation

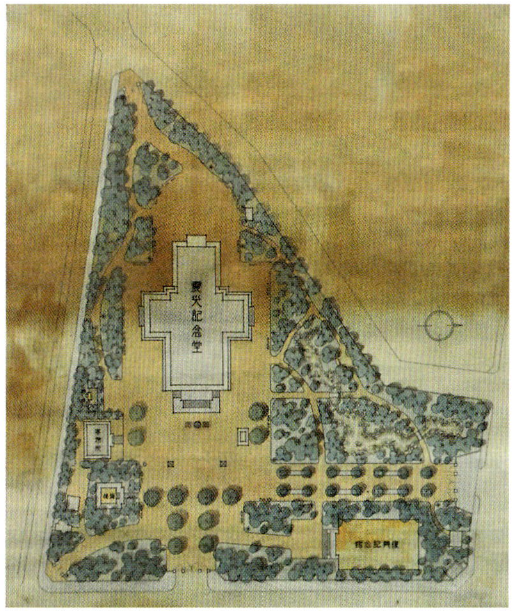

7.17 *Earthquake Memorial Hall Park Completed Plan* (*Shinsai Kinendō teien shūkō heimenzu*), cruciform layout of the hall, c. 1930. Reconstruction Memorial Museum. Collection of Tōkyō-to Irei Kyōkai.

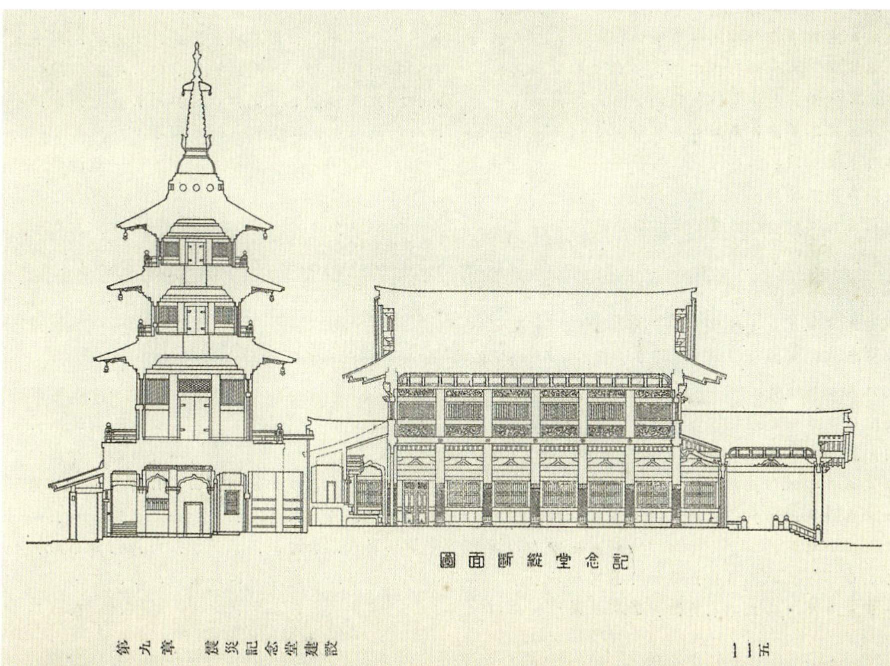

7.18 Itō Chūta, *Earthquake Memorial Hall* (*Shinsai Kinendō*), section, c. 1926–27. Tōkyō Shinsai Kinen Jigyō Kyōkai Seisan Jimusho, *Hifukushōato* (Site of the Former Clothing Depot) (Tokyo: Tōkyō Shinsai Jigyō Kyōkai Seisan Jimusho, 1932), 115.

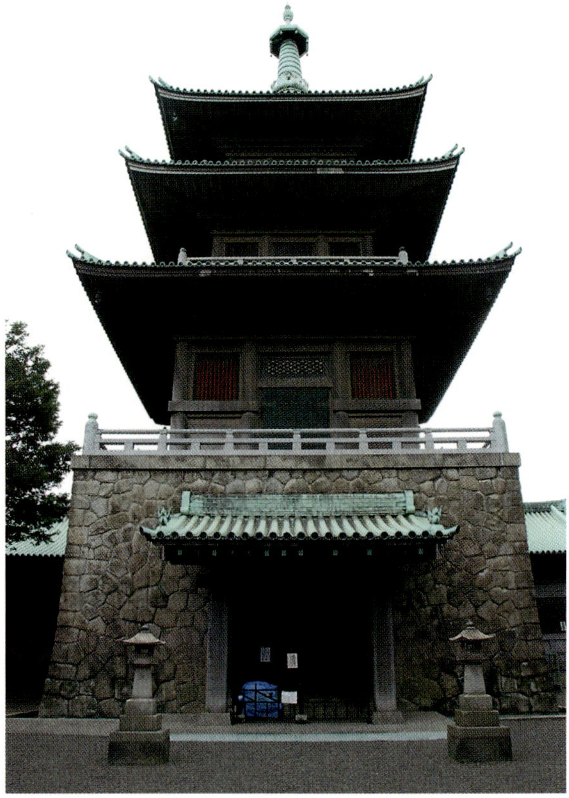

7.19 Itō Chūta, ossuary (pagoda), Earthquake Memorial Hall (Shinsai Kinendō), completed in 1930, photograph by Kita Takaomi, 2011. By permission of Tōkyō-to Irei Kyōkai.

and official references to them as "heroic spirits" (*eirei*).[30] Like the eternal flame at the Tomb of the Unknown Soldier at Arlington National Cemetery in Washington, DC, the incense reportedly has burned continuously in front of the altar since four days after the disaster. Sites that have human remains embedded in them are often seen as "open graves" and described as "sacred ground."[31] Memorial services for the dead (*irei hōyō*) are still held at the memorial hall annually on the anniversary of the quake.[32] In Buddhist belief, those who have their lives interrupted for an aberrant reason are considered to have had "unnatural deaths" (*higō no shi*) and are seen as disaccording with karma. The annual memorial service provided a social mechanism by which Tokyo's citizens could express and ease their discomfort over the exceptional spiritual status of those whose lives had been violently interrupted.[33] As mandated in the original plans for the memorial, a door to the cinerarium is opened each year on 1 September for the entire month so that people can view the bones.[34] In the years following the quake, many visitors came to the annual service and throughout the year to visit the memorial as a popular sightseeing destination. It was soon incorporated into the new famous sites of the imperial capital (*teito shin-meisho*) and immortalized in widely circulated photographic postcards (see fig. 7.15) and popular woodblock print series, such as Koizumi Kishio's (1893–1945) *One Hundred Pictures of Great Tokyo in the Shōwa Era* (*Shōwa dai Tōkyō hyakuzue,* 1928–40), an homage to the reborn metropolis.[35]

The interior of the main worship hall was as impressive and eclectic in design as the exterior, combining two Buddhist altars (one for general worship and the other enshrining the spirits of the dead) with a Western-style devotional space containing pews on a tile floor (fig. 7.20). Elegant electric chandeliers and sconces together with clerestory windows illuminated the large central space, designed to hold eight hundred people; Itō's signature attention to details is evident throughout, from the raised arabesque flourishes on the massive bronze front doors (fig. 7.21) to the elaborated mythical beasts holding the light fixtures. The expressive decorative creatures on the exterior, featured in the popular press, were described as talismans to keep the evil spirits away.[36] The space also serves a historical pedagogical function, for it is ringed above the sidewall transoms by Tokunaga Ryūshū's large-scale oil paintings depicting spectacular scenes of the disaster, and on a lower tier, a series of documentary photographs are on display.

Outside the building, rows of trees line the walkways leading visitors to the main entryway and the north entrance. In the large open plaza in front of the main entryway stand two tall sepulchers for burning incense, offering flowers, lighting votive candles, and washing hands before entering the sacred space. A large cadre of garden designers and park officials consulted on the project. Around the hall, they created a picturesque, Japanese-style landscaped garden with many varieties of elegant ornamental trees and plants, decorative rock groupings, small ponds, stone lanterns, a bridge, and benches for resting; the garden, which was electrically illuminated for nighttime viewing, was also designated a neighborhood evacuation site for future quakes.[37] Along one of the main walkways visitors encountered a host of disturbing burned and deformed memorial objects from daily life that offered foci for contemplation and remembrance of the tragic dead, as well as ominous civic morality lessons—a so-called silent warning (*fugen no keikoku*)—about the transience of the material world and the importance of fire prevention.[38] Many artifacts large and small were carefully collected and preserved after the quake with the explicit purpose of being exhibited at this site. The city also issued numerous formal requests for object donations to the memorial.[39] Various groups donated commemorative items for the garden, such as the inscribed bronze bell and belfry given in sympathy from the Chinese people to memorialize their dead countrymen.[40] For example, the teachers and classmates of the more than five thousand schoolchildren who perished in the quake sponsored a bronze sculpture in their

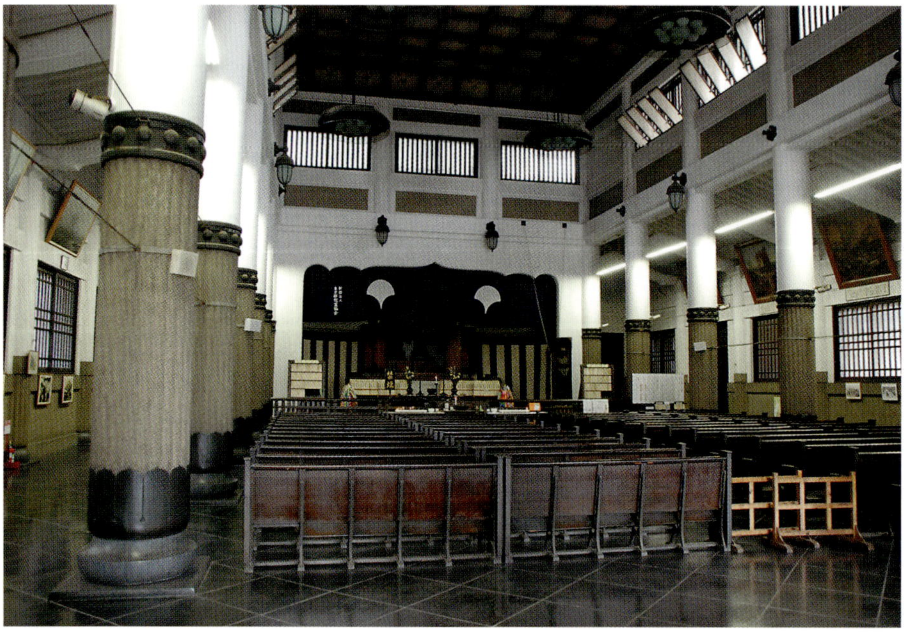

7.20 Itō Chūta, Western-style devotional space in Earthquake Memorial Hall (Shinsai Kinendō), photograph by Kita Takaomi, 2011. By permission of Tōkyō-to Irei Kyōkai.

7.21 Itō Chūta, detail of main entrance doors, Earthquake Memorial Hall (Shinsai Kinendō), photograph by Kita Takaomi, 2011. By permission of Tōkyō-to Irei Kyōkai.

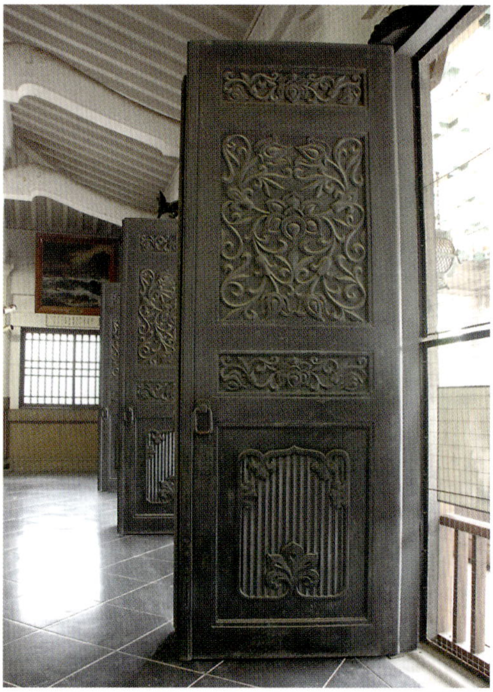

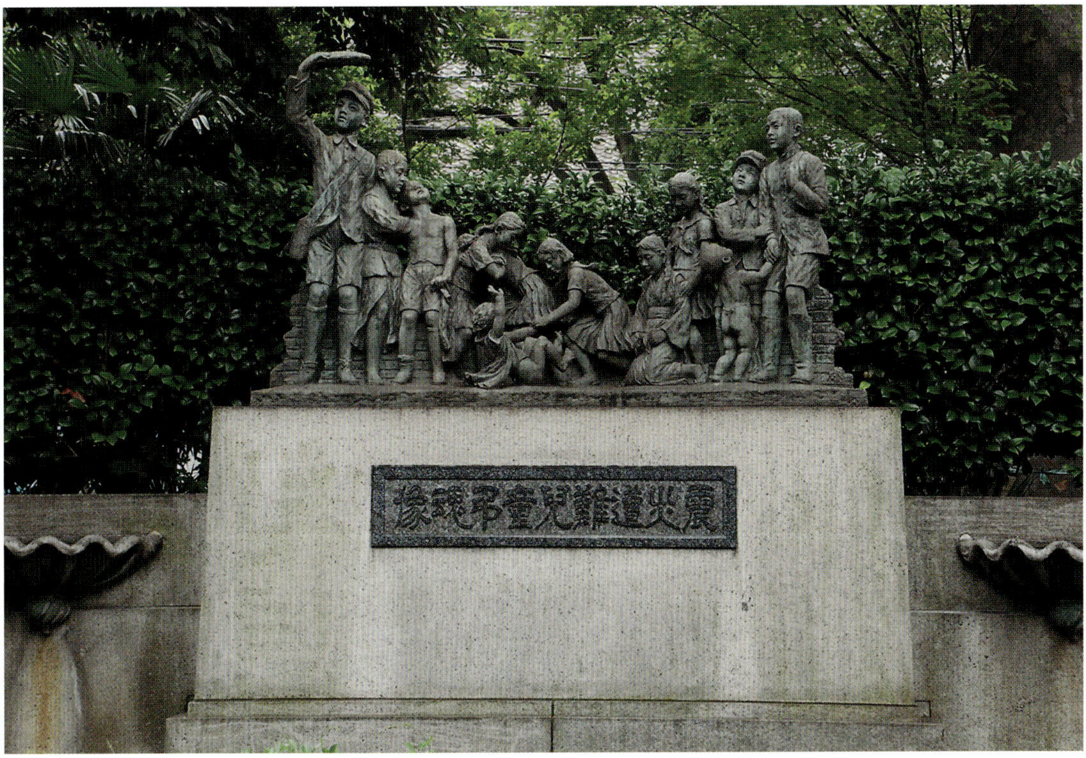

7.22 Ogura Uichirō, *Statue to Console the Spirits of Children Who Were Casualties of the Earthquake* (*Shinsai sōnan jidō chōkonzō*), Yokoamichō Park Earthquake Memorial, completed c. 1932, photograph by Kita Takaomi, 2011. By permission of Tōkyō-to Irei Kyōkai.

memory by prominent academic sculptor Ogura Uichirō (1881–1962) (fig. 7.22). Ogura's highly melodramatic figurative ensemble, set on a tall stone base, features a group of imperiled schoolchildren clustered together, hugging one another, and collapsing to the ground in fear and sorrow.[41] The monument resonated with the harrowing artwork of surviving schoolchildren that was exhibited in the adjacent reconstruction museum. Planners designed this garden so that visitors would encounter the individual memorials and deformed objects of the quake during their strolls. The silent warnings emit a cacophony of voices, all clamoring to be heard and remembered.

After completion of the memorial hall's steel frame, the ridgepole-raising ceremony was held in June 1929 for imperial inspection, and seven hundred illustrious visitors attended. This ceremony served as a public announcement of the imminent completion of the memorial, which government bureaucrats used to mark the official end of reconstruction.[42]

Enfolding Disaster into Reconstruction

Officials created the holiday Tokyo City Reconstruction Commemoration Day, first celebrated on 26 March 1930, which thereafter became the official anniversary of reconstruction.[43] This day was differentiated from the actual anniversary of the quake, still honored as a day of mourning on 1 September, because of a prohibition on rejoicing on days of mourning: funerary and festive remembrance had to be separated. In preparation for the momentous new holiday, the buzzword of 1930 became "reconstruction commemoration" (*fukkō kinen*), and the phrase was used in everything from sports events and flower displays to advertising for retail stores. The multiple-day "reconstruction festival" (*fukkōsai*) began on 24 March with a citywide imperial procession leading to the palace. The procession route was designed to show the emperor (and the citizens) all the new thoroughfares by physically tracing the reconstruction projects through the city. Thousands of Tokyoites watched as the imperial motorcade passed. Temporary triumphal arches were erected at major gates, intersections, and bridges around Tokyo; the arch at Babasakimon street, illuminated with 376 electric lights, was captured in its brilliant glory in the *Yomiuri shinbun*. Seven decorated trams (*hanadensha*) with electric lights traveled along the reconstructed tracks, playing a continuous concert of music for three days leading up to the festival. Some trams featured dioramas of the reconstructed metropolitan skyline with its monumental skyscrapers. One float (fig. 7.23) showed the new metropolis emerging from the heavenly cave (*amanoiwato*) of the Sun Goddess (Amaterasu), where, according to ancient legend, this ancestral deity of the imperial family had retreated with the light of the world after being angered by her trickster brother (Susanoo) until she was coaxed back out by an enormous celebration thrown by the gods, bathing the world in light again. Reporters wrote that the majestic light show transformed the city into a "sea of illumination" (*hikari no umi*).[44] Merging ancient mythology and modern-day symbolism, the rhetoric of reconstruction conveyed the direct link between the divine providence of the city's rebirth and the beneficence of imperial rule.[45]

After his procession, the emperor visited exhibits displaying Tokyo's triumphant reconstruction at several sites throughout the city that he had also publicly inspected shortly after the quake. This full-circle return to formerly devastated sites provided symbolic closure on the disaster and reconstruction. His tour included a stop at the Earthquake Memorial Hall in Honjo, where construction was nearly complete. Greeted by the mayor and clergy members of multiple Shinto and Buddhist sects, the emperor viewed an array of disaster objects that focused on the tragedy of the disaster. He symbolically ended his visit to the memorial hall at exactly 11:58 A.M., when every-

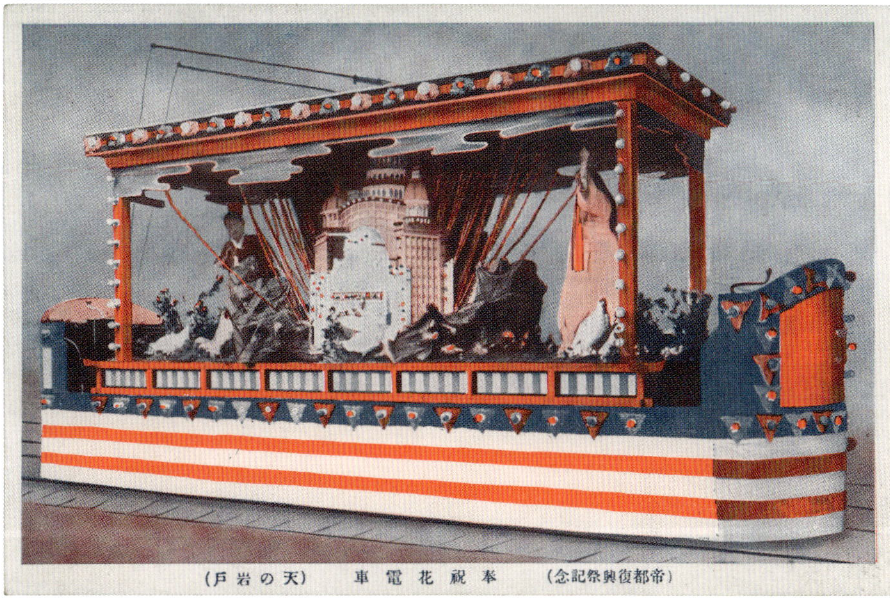

7.23 Reconstruction Festival (Fukkōsai), 24–26 March 1930, postcard, one of seven decorated trams (*hanadensha*) with electric lights that traveled around the city, this one representing the new metropolis emerging from the heavenly cave (*amanoiwato*) of the Sun Goddess (Amaterasu). Collection of the author.

one in the group observed a moment of silent prayer to mark the time of the quake and to remember the dead.[46]

Sacred and Historical Remembrance

While it seemed to many that the work of reconstruction and remembrance was complete, others in the municipal government had gradually realized after the city's enormous exhibition of 1929 that the number of objects narrating the disaster and recovery would far exceed the available space in the memorial hall, and its status as a sacred site would allow only limited public access to important historical artifacts. In addition, heated controversy about the architectural aesthetics of Maeda's original design for this Japanese memorial worship space, which some critics considered inappropriate for a sacred space, reinforced the perception that the two functions of sacred and historical remembrance should be separated. The centrality of religious memorial services during this period of intense westernization is a testament to the enduring power of Buddhist belief in the modern period, particularly for mortuary purposes. In mid-1930, the city announced plans to erect a reconstruction memorial museum (*fukkō kinenkan*) specifi-

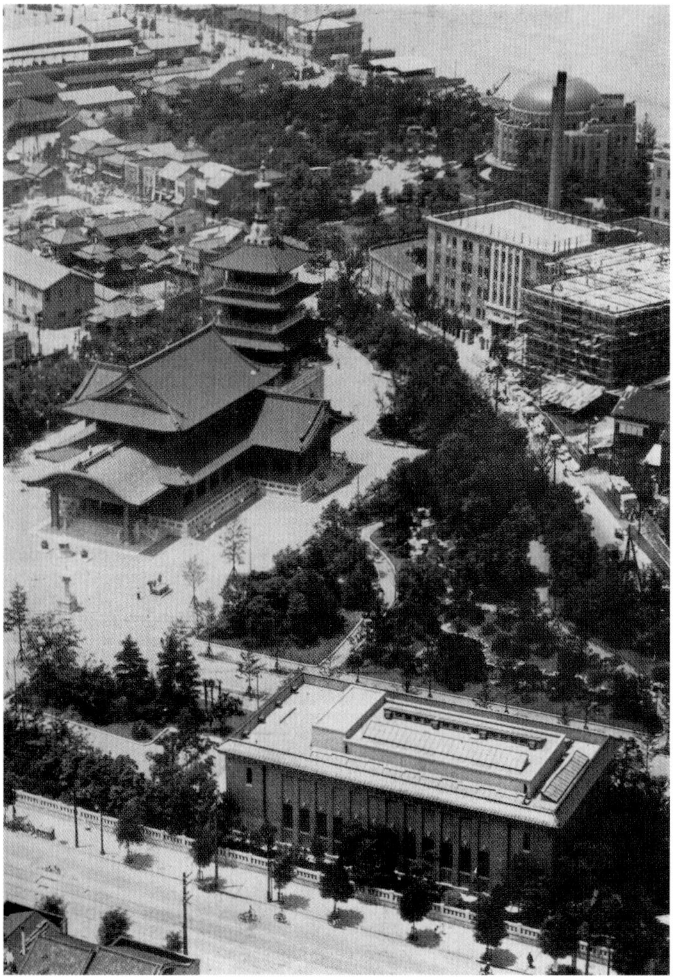

7.24 Aerial view, Yokoamichō Park Earthquake Memorial Complex, photograph, c. 1931. Tokyo Municipal Office, *The Reconstruction of Tokyo* (Tokyo: Tokyo Municipal Office, 1933), 267.

cally to narrate the historical aspects of the disaster within the recovery framework of reconstruction. The project's slogan was "From the earthquake disaster to reconstruction!" (*shinsai yori fukkō e*). The allocated budget for the expedited construction was 200,000 yen. The museum was finished the following year, in August 1931, ostensibly completing the overall memorial complex (fig. 7.24).[47]

The tensions between memorializing death in the disaster and celebrating rebirth in reconstruction were acute. Yet these two functions overlapped in critical ways that make strict division of them impossible. By housing critical mortuary practices, the memorial hall ensured the proper salvific rebirth of the possibly malevolent spirits of the dead, thus contributing to the protection of the survivors and the regenerating society. These religious practices were also part of the turn toward the moral retrenchment of "spiritual reconstruction" advocated in the divine punishment discourse. One might argue that in its ring of paintings and photographs of the disaster, the hall incorporated such representations not as indexical historical documents but as modern substitutes for the admonition imagery of hell paintings common in earlier times. "In pacifying the souls of the unfortunate victims and eternally commemorating the disaster," the hall was, in the words of Mayor Nagata, to be "an institution of silent enlightenment" (*fugen no kyōka shisetsu*).[48] In contrast, the reconstruction museum, as the name attests, was dedicated to narrating a story of recovery. It was decidedly not silent. The objects within it were designed to tell the "before" story of reconstruction as well as celebrate the "after."

And not surprisingly, the "after" story was expressed in the bright, optimistic, rationalized scientific language of the state: maps, statistical charts, and extensive models of urban reconstruction (fig. 7.25). Telling this story was to serve as a form of social education (*shakai kyōka*), a warning (*keikoku*) about the dangers of earthquakes, to prevent such a monumental tragedy from ever befalling the people again and to triumph over disaster through knowledge.[49] The tropes of altruism, imperial benevolence, collective solidarity, and resilience were all strongly conveyed through the displayed objects. Yet to say that these objects were entirely subordinated to this progressive discourse would be to miss the subtle multivalence of material culture. The emotional pathos and physical trauma still echo loudly for modern-day visitors to the museum, for the objects resist reduction to a story of triumph through their metonymical representations of the body in pain and their invocation of the specter of the dead.

Memorial spaces not only serve as spiritual sanctuaries but also provide, in Timothy Brown's words, a kind of "pedagogy of trauma" that goes beyond mere description and narration of the event. In such venues, salvaged artifacts and exhibit spaces are mobilized to bear witness to the event in a visceral and palpable manner that cannot be represented solely by photographs and other documentary materials.[50] While many contemporary memorials are designed as "sites of conscience on the civic landscape," this complex is striking for the complete absence of any objects or visual records that might reveal the blatant social injustices exposed or precipitated by the quake—a void that effectively institutionalizes social amnesia.[51] Thus, while memorial monuments serve as "intensification devices" that deepen emotional resonance, they can also nullify the viewer's guilt and responsibility through a willful partial remembrance.[52]

The memorial hall and museum required not only separate structures, but also radi-

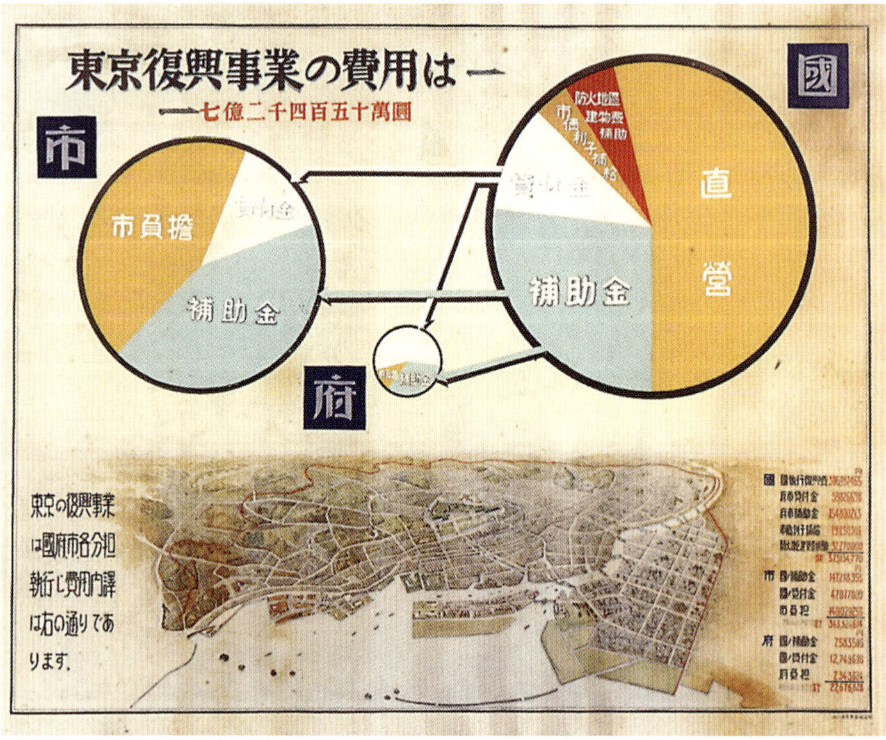

7.25 Tokyo's Reconstruction Work Expenditures Are 724,500,000 Yen (*Tōkyō fukkō jigyō no hiyō wa*), budget allocation divided by city, prefecture, and national government (pie charts from left to right), c. 1929–30. Reconstruction Memorial Museum. Collection of Tōkyō-to Irei Kyōkai.

cally different architectural languages—the former speaking of local traditions and the latter gesturing to a more international rhetoric of museums and other institutional exhibitionary venues as "objective" narrative spaces for preserving history. Thus, the museum's rectangular two-story building (fig. 7.26), thought to have been designed by engineer Hagiwara Kōichi together with Itō Chūta, possibly under the supervision of Sano, was notably more modern and Western in appearance than the memorial hall and clearly drew on the aesthetics and decorative brickwork of contemporary architects such as Frank Lloyd Wright. The style is similar to that of Wright's Imperial Hotel in Tokyo, which famously withstood the quake. The museum building featured a simple façade design in the popular scratch-tile cladding with a prominent series of mullions surrounding the front entrance and windows. The four central mullions were capped by Itō's signature mythical beasts in terra-cotta. The Asian-style cement-tile roof still lent the structure a whiff of localism.

Documentary evidence shows that the displays in this building have changed con-

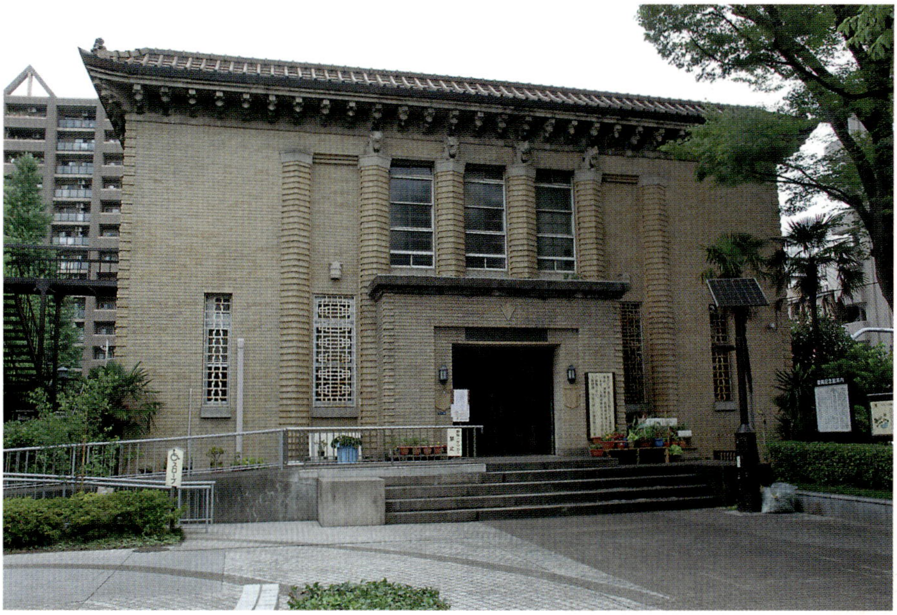

7.26 Hagiwara Kōichi and Itō Chūta, exterior, Reconstruction Memorial Museum (Fukkō Kinenkan), completed August 1931, photograph by Kita Takaomi, 2011. By permission of Tōkyō-to Irei Kyōkai.

siderably over time, particularly since the museum was reorganized in 1951 to include commemoration of the wartime firebombings of Tokyo that occurred in March 1945, which created new relationships of memory and memorialization.[53] As a legally incorporated foundation that received donated funds from the national citizenry, the memorial organizing association felt obliged to publish a report of its activities, and the result was an exquisitely documented account of more than four hundred pages that elucidated all aspects of the memorial complex project, including the state of the facilities in 1931.[54] Although the space has been altered and the objects presumably rotated and reinstalled on several occasions, surviving written records and documentary photographs indicate that the objects currently on display are from the period, if somewhat reorganized. It is clear that the first floor originally displayed objects related to the earthquake experience and the second floor, now also home to materials commemorating the Great Tokyo Air Raids of the Asia-Pacific War, presented illustrative material from the reconstruction exhibitions discussed in the previous chapter. The spatial flow of the museum recapitulated the official narrative showing the nation moving from tragedy to triumph.[55]

On the first floor, rows of scorched and deformed material remnants of the earthquake donated by survivors were exhibited and venerated amid representations of the

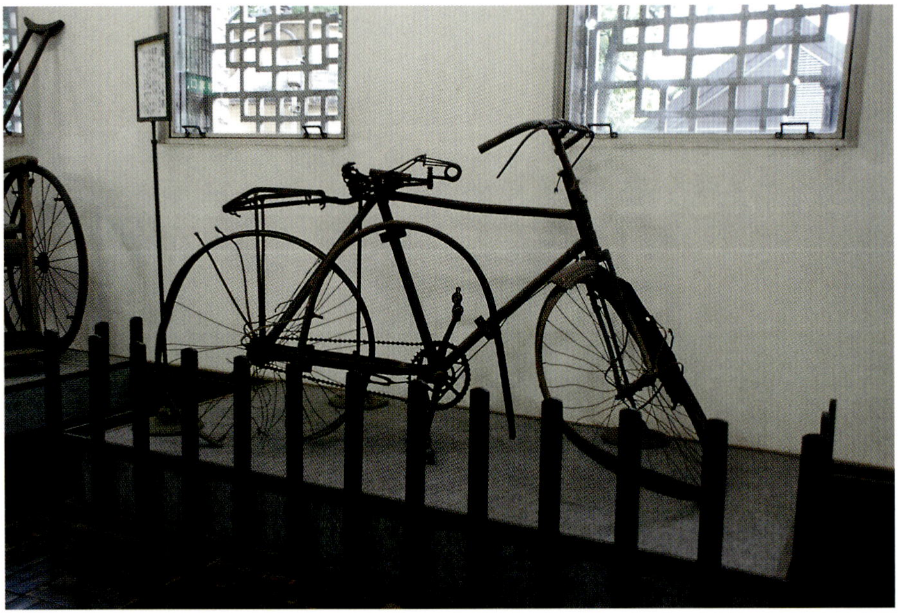

7.27 Bicycle display, Reconstruction Memorial Museum, photograph by Kita Takaomi, 2011. Collection of Tōkyō-to Irei Kyōkai.

event in a range of media. The physical disfigurement of everyday objects came to stand in for the somatic experience of the wounded populace. This staging of an emotional encounter with the debris of the earthquake evoked the sensorium as a means of remembrance.

Burned and mangled bicycles (fig. 7.27) gesture to absent riders and remind the viewer of the sentimental makeshift memorials residents had created in the street. The shock of the blasts is captured by a diorama displaying a piece of sheet metal that had been thrown with such intensity that its twisted body became lodged in the branches of a tree, part of which was also transferred to the gallery (fig. 7.28). The diorama is authenticated by the photograph of the real-world occurrence on the wall behind (fig. 7.29). By laying claim to the truth-telling power of photography, the display invokes the powerful discourse of "the real" through both the material object and its representation.

The destructive power of the quake is expressed in diverse objects: photographs of clocks eternally stopped at 11:58 or the grotesquely contorted bodies of melted bottles displayed in vitrines (fig. 7.30). The observer feels shock blasts that are so powerful that they literally stop time and heat so intense that it melts glass. The iconic image of the stopped clock was widely circulated through a range of media, particularly in commercial and official memorial medallions. People could buy earthquake memorial

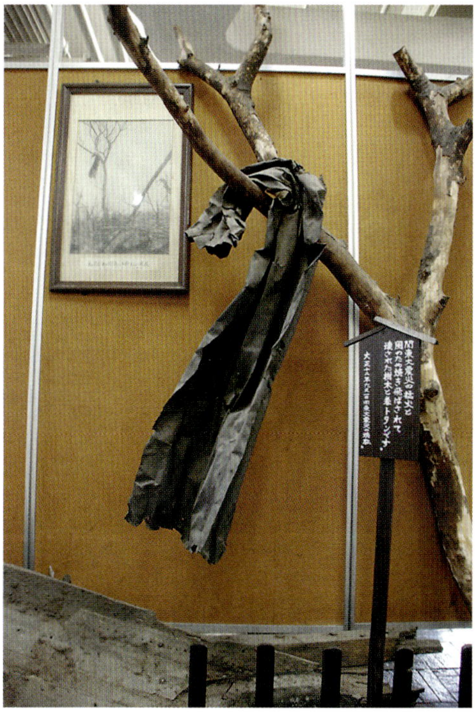

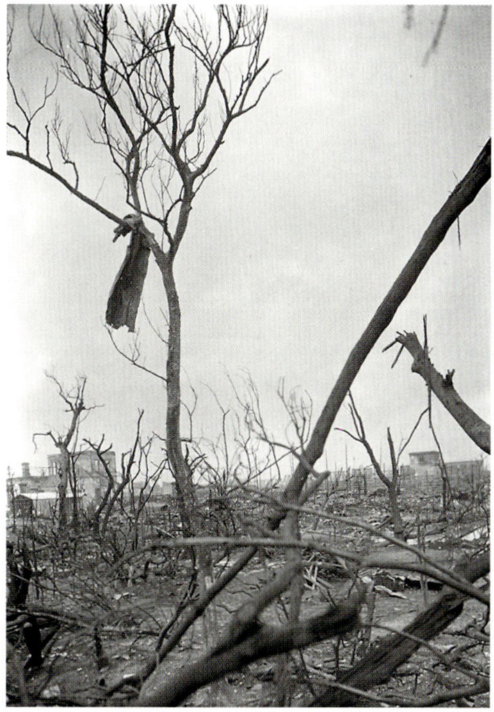

7.28 Diorama with bent sheet metal and tree (photograph of same scene mounted on partition behind), Reconstruction Memorial Museum, photograph by Kita Takaomi, 2011. Collection of Tōkyō-to Irei Kyōkai.

7.29 Artist unknown, bent sheet metal in tree, photograph, c. 1923. Collection of Kagaku Hakubutsukan.

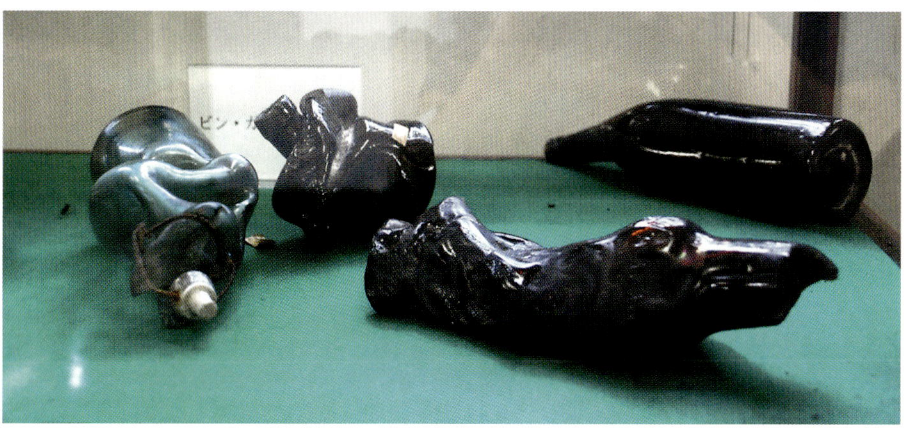

7.30 Melted bottles, Reconstruction Memorial Museum, photograph by Kita Takaomi, 2011. Collection of Tōkyō-to Irei Kyōkai.

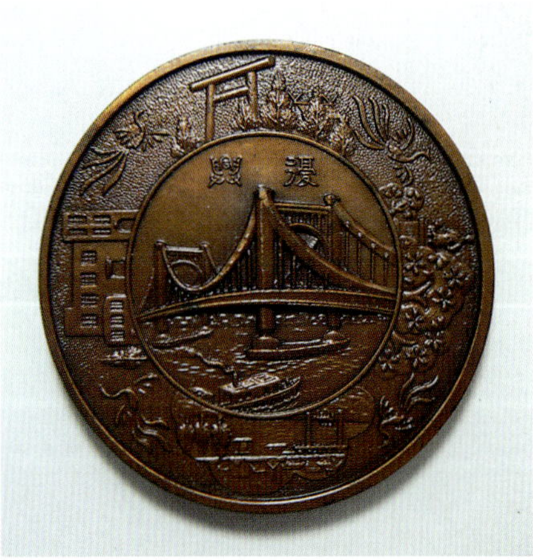

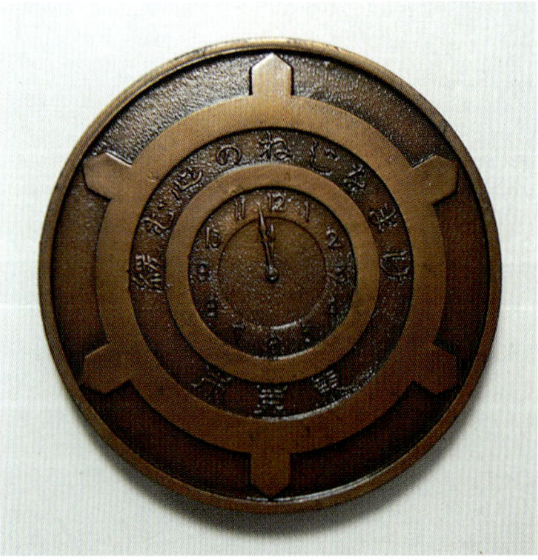

7.31 Earthquake Memorial Medallion (front and back) issued by the Tokyo metropolitan government, 1930. Photographs courtesy of Richard Catalano.

clock medals from the Muramatsu Clock Store or collect the 1930 anniversary medallion (fig. 7.31) that the city of Tokyo issued, which bore the triumphant image of the new steel suspension Kiyosubashi Bridge on the front and the iconic frozen clock on the back.[56] On the medallion, the clock was set within the wheel-shaped emblem of the city encircled by the popular city reconstruction slogan "Rouse your weakened spirit" (*yurumu kokoro no neji o make*) that enjoined the citizenry to wind their internal clocks and start again.[57]

Envisioning a Conclusion

On the second floor (fig. 7.32) of the Reconstruction Memorial Museum now hangs a striking large-scale oil painting titled *Commemoration of the Great Earthquake* (*Taishin kinen,* 1931) (fig. 7.33) by modernist painter and writer Arishima Ikuma (1882–1974), brother of the novelist and infamous suicide victim Arishima Takeo.[58] It currently hangs in a cluttered space filled with model reconstructions of the city, amid a series of small-scale paintings of devastated Tokyo sites, diagonally across from images of the prince regent and empress surveying their traumatized imperial subjects, and across from aerial renderings of the Kantō region in flames.[59] Like the collection of objects with which it is displayed in the museum, Arishima's painted montage narrates the total earthquake experience, taking a retrospective view. It builds a cumulative summary of the quake from images of its iconic moments and sites, condensing the disaster through visual means into a moment of misery, memorial, and mission.

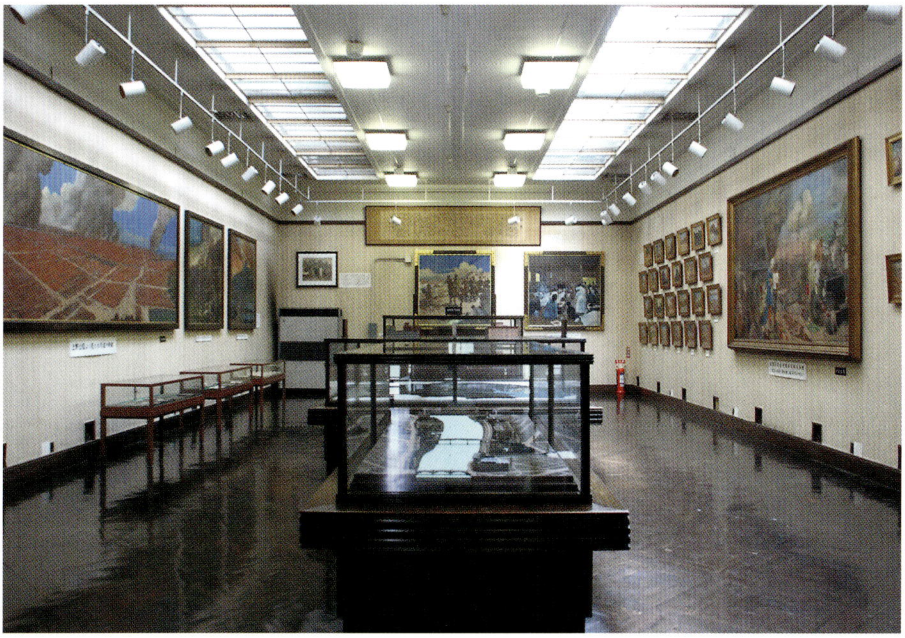

7.32 Exhibition space, second floor, Reconstruction Memorial Museum (Arishima Ikuma's *Commemoration of the Great Earthquake* on far right), photograph by Kita Takaomi, 2011. By permission of Tōkyō-to Irei Kyōkai.

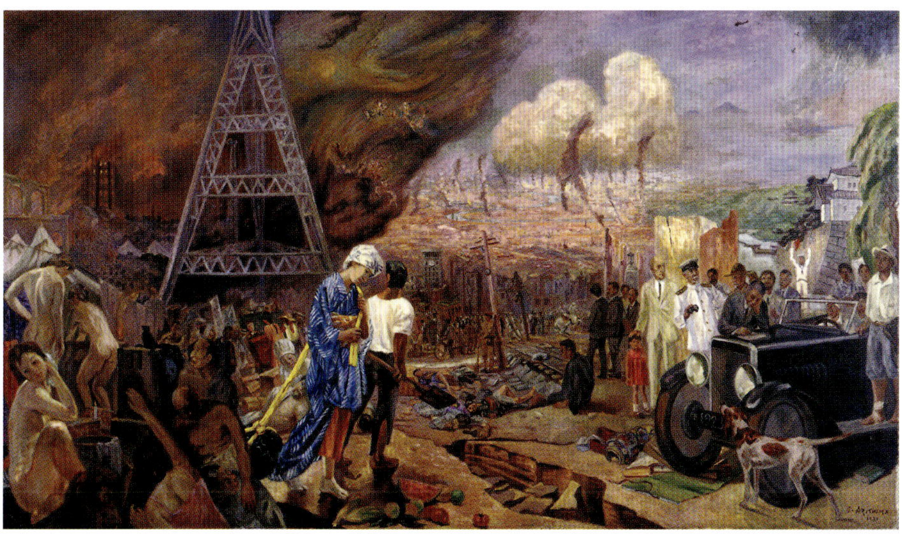

7.33 Arishima Ikuma, *Commemoration of the Great Earthquake* (*Taishin kinen*), oil on canvas, 1931. Reconstruction Memorial Museum. Collection of Tōkyō-to Irei Kyōkai. Photograph courtesy of Art Restoration Studio 21.

The elaborate composition, evocative of Christian images of resurrection, reads from left to right, sweeping from nightfall to daybreak. Although the first tremor occurred near noon, in Arishima's imagined conception the night becomes a metaphor for the darkness of the experience and morning symbolizes the onset of reconstruction. The scene is situated near the Imperial Palace. On the distant left, the decapitated Twelve Stories stands aflame next to savage firestorms whipping overhead that carry off people and souls to heaven, an image reminiscent of Hinako Jitsuzō's memorial sculpture design. In the immediate foreground, naked women stand or crouch in front of barracks awkwardly trying to wash themselves in buckets of water—the artistic muses of the postquake moment.

The composition of figures around the buckets in the left-hand corner evokes classic scenes of tragedy and death by romantic painters such as *Massacre at Chios* (1824) by Eugène Delacroix. Men, women, and children mill around in the middle ground of Arishima's image, salvaging their possessions or tending to the injured or dead; others just sit in disbelief. Two figures stand out in the center: a barefoot man carrying a shovel, his back to the viewer, and a woman in a vibrant blue kimono, identified as Yanagihara Byakuren, famed poet, celebrity beauty, and wife of a wealthy businessman she detested.[60] Yanagihara was known for the scandalous 1921 "Byakuren Affair," when she ran away with her younger lover, Miyazaki Ryūsuke, a leading member of the socialist political group the Shinjinkai (the New Man Society), and publicly declared her advocacy of love marriages. She was divorced by her husband, ostracized by her noble family, and left penniless until she was eventually able to marry Miyazaki. Both a compelling tragic heroine who followed her heart and a symbol of the romantic radicalism of Taishō individualism—radicalism that had driven the artist's brother to suicide—the pitiful Yanagihara in the painting forlornly drags her meager possessions across the canvas as she guides the eye to the new day where a host of officials stand surveying the damage.

On the right side of the painting stands Prime Minister Yamamoto in his white naval uniform holding binoculars. He looks toward the figure bent in concentration over a notebook on the hood of a car bearing the license plate "1923," who has been identified as either renowned journalist Tokutomi Sohō (1863–1957) or shipping magnate and minister of railways Uchida Nobuya (1880–1971). Behind them stands Home Minister Gotō (by 1931 already deceased), in his familiar Boy Scout leader uniform, and Mayor Nagata.[61] To their left in a gray suit is the Belgian ambassador, Albert de Bassompierre, who is patting the head of Arishima's niece Kyōko, who wears a bright red dress. Through the figure of Bassompierre, Arishima highlights Belgium's deep empathic mourning for Japan and the country's extraordinary assistance in the earthquake relief efforts, which included a prominent 1924 art exhibition of works donated by Belgian artists that raised over 22,000 yen for relief aid.[62] Arishima's painting is full of prominent artistic and literary figures, including his brother, White Birch Group author Satomi Ton, who stands to the far right in a panama hat.[63] Even the artist himself is present, represented by the heroic-looking figure on a rearing white horse in the distant background.[64] In the reconstructed cityscape of the Tokyo metropolis, Arishima's vanguard modernists could be the heroes of a new postquake era, having assimilated the radical ruptures of modernity and sutured them together to form a coherent narrative of disaster's devastation and reconstruction's renewal. But while Arishima's image culminates in reconstruction, his scene is fraught with mixed messages, such as the inclusion of murdered anarchist Ōsugi Sakae, standing directly behind the figure leaning over the car, his eyes portentously raised to the sky. Ōsugi stands in the painting as a reminder of the human and moral cost of the disaster.[65]

Even though we can never be present in history, through the prism of this visual continuum, temporally collapsed, we are left with a historical understanding of the Great Kantō Earthquake. Out of the ashes emerged a powerful visual response to the experience that expressed the horror and spectacle of disaster, the sublime nature of ruins, the altruism of national solidarity, and the latent violence in corrosive communities. Remembrance ultimately enfolded unspeakable tragedy into fervor for the possibilities of reconstruction, but this was only after years of contentious civic discourse in which a range of stakeholders sought to claim the meaning of the disaster.

With a living city that constantly renewed itself, reconstruction would not end unless someone declared it officially over. And that is what the city did, confident that the national narration of the Kantō earthquake told a triumphant story of Tokyo's miraculous resurrection from the ruins:

> Now, we are drawing near to the end of the self-imposed work of chronicling all matters bearing on the immediate consequences of the Earthquake, the plans and projects for combating the disastrous effects inseparable from such natural calamity, and then, finally, the actual execution of these projects and schemes for Reconstruction of the Capital; a remarkable trilogy of drama on a grand national scale, in which one who has the eye to see, the mind to think and the heart to feel, cannot fail to perceive a clear revelation of the spirit of heroism, perseverance and resolution, admittedly the priceless legacy of this our Japanese people.[66]

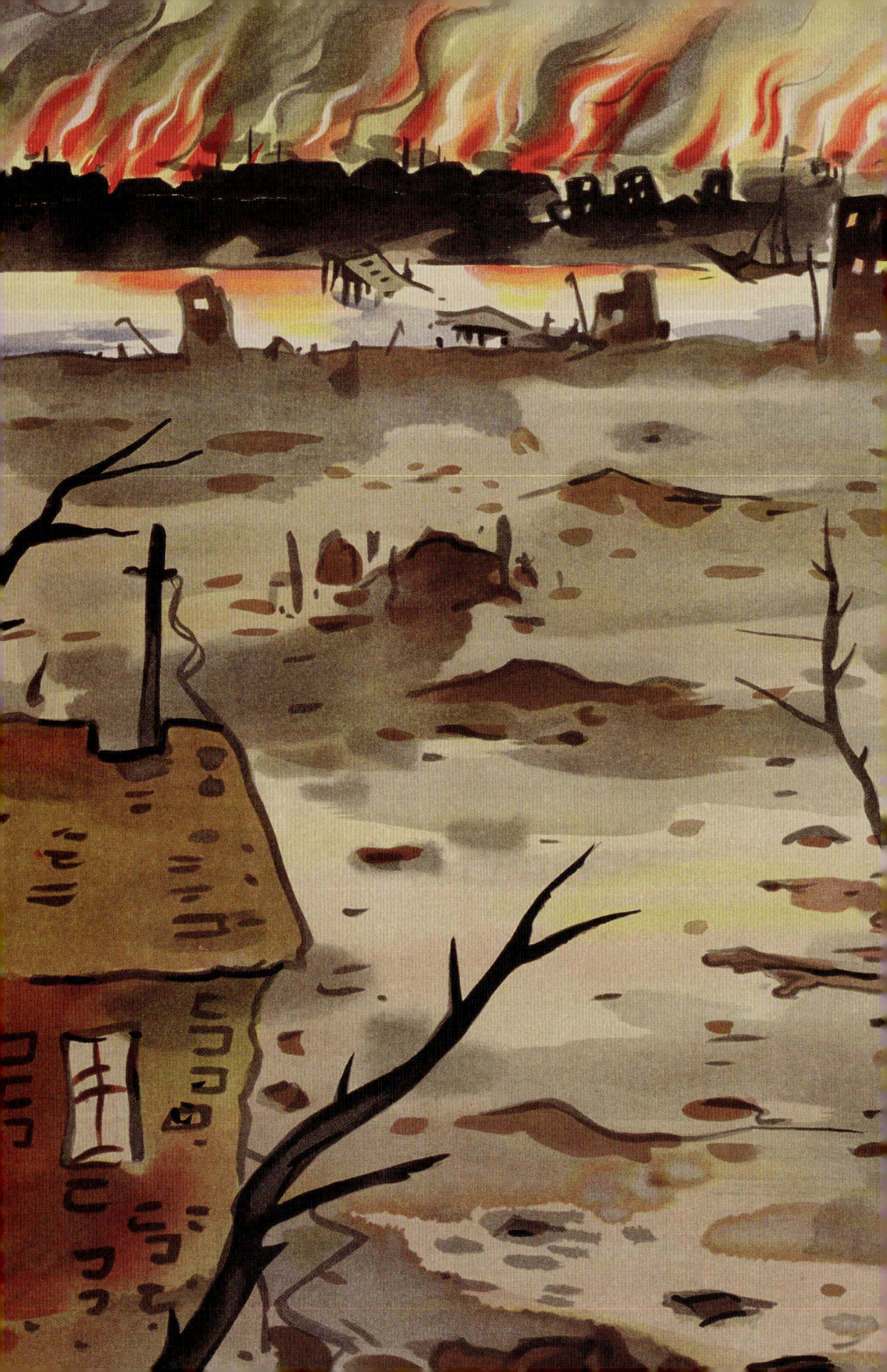

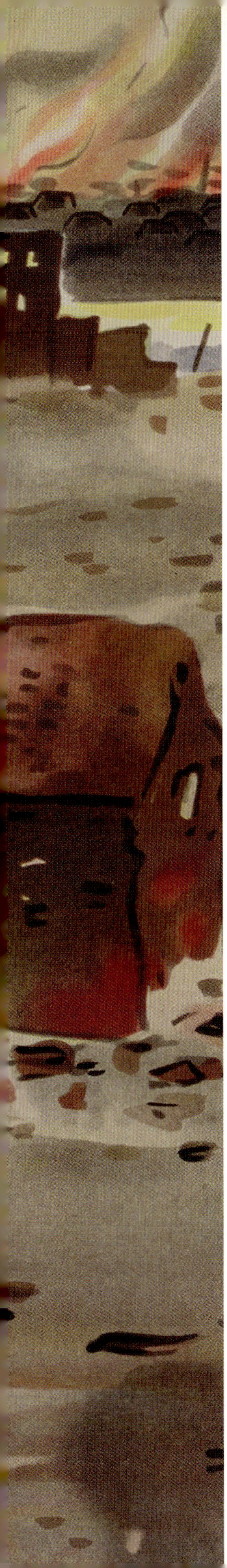

8

EPILOGUE

Afterlives

> The ruins that formed my childhood environment were produced by acts of sudden destruction, unlike those of Greece and Egypt, which had long been in a ruinous state. Wandering among them instilled in me an awareness of the phenomenon of obliteration, rather than a sense of the transience of things.
>
> ISOZAKI ARATA

In the years since the Great Kantō Earthquake, Japan has continued to experience traumatic disasters—the ultimate being the man-made cataclysms of the Asia-Pacific War that reduced the nation's capital and other major cities to ruins once again. Ruminating on the devastated landscape of fire-bombed Tokyo (fig. 8.1) and the postapocalyptic scenes of Hiroshima and Nagasaki flattened by atomic bombs, world-renowned architect Isozaki Arata remembers the profound psychological imprinting of the image of ruins (*haikyo*). Marked by the burned wasteland and evocative ruins of the devastated cityscape, spectacular plumes of smoke, charred human remains, and the wounded personal objects of daily life made into venerable relics, the visual lexicon of disaster emerging from the earthquake continued to have a powerful afterlife in the postwar period. And in an odd twist of fate, the earthquake memorial itself became the site of remembrance for the Great Tokyo Air Raids (Tōkyō Daikūshū) of March 1945. This uneasy cohabitation reveals how the lives of

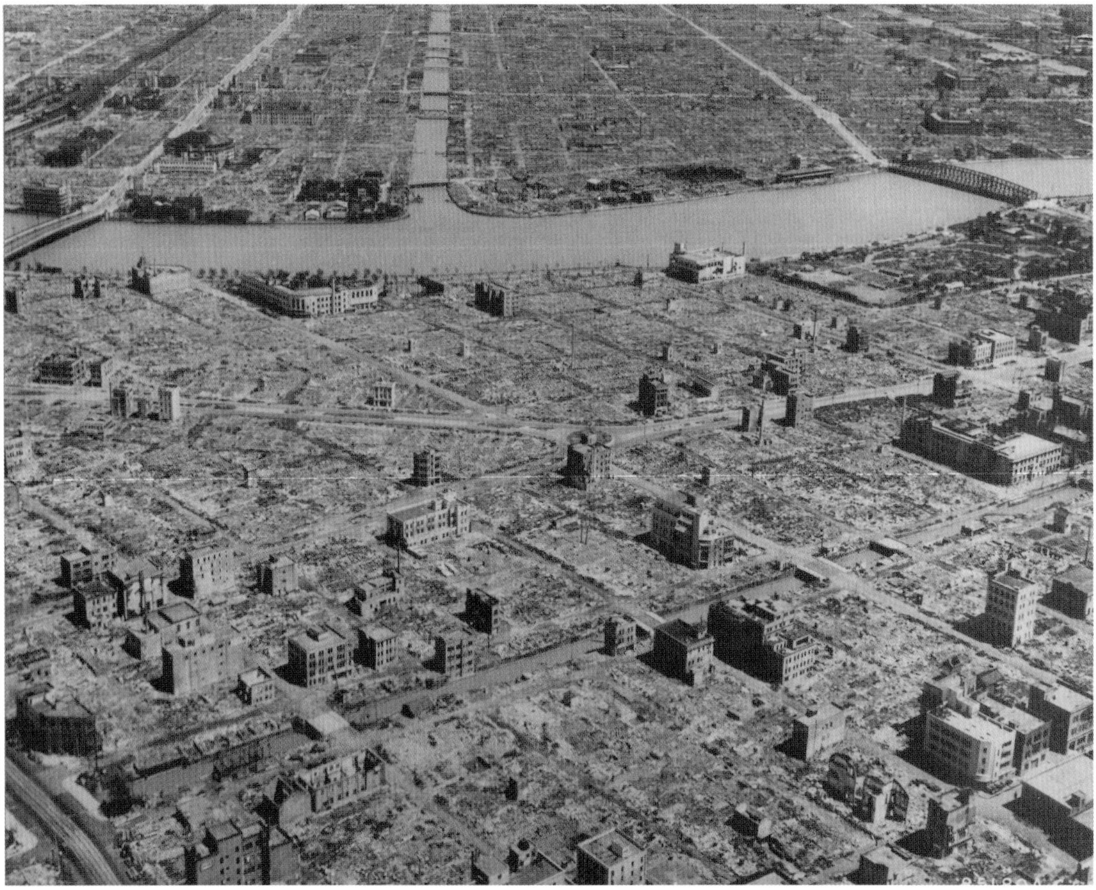

8.1 U.S. Department of Defense, aerial view of fire-bombed Tokyo (looking toward the Sumida River and the Honjo-Ryōgoku area), photograph, 1945. Photograph courtesy of Japan Air Raids.org.

public monuments continue to change over time and how influenced they are by political expedience and the vicissitudes of history.

On 9 and 10 March 1945, in a tragic parallel with the events of 1923, more than half of Tokyo was again incinerated and over one hundred thousand of its residents killed by incendiary bombs dropped by American B-29 bombers (fig. 8.2). Once again the low-city areas—Honjo, Fukagawa, Asakusa—which were still centers of industrial production, specifically military production in wartime, were the hardest hit. Symbolic Ginza, the lively center of the reconstructed modern metropolis, was also in ruins. This was the most destructive urban incendiary bombing that the world had ever seen. An uncanny reminder of the dark days of September 1923, the macabre scenes of mass

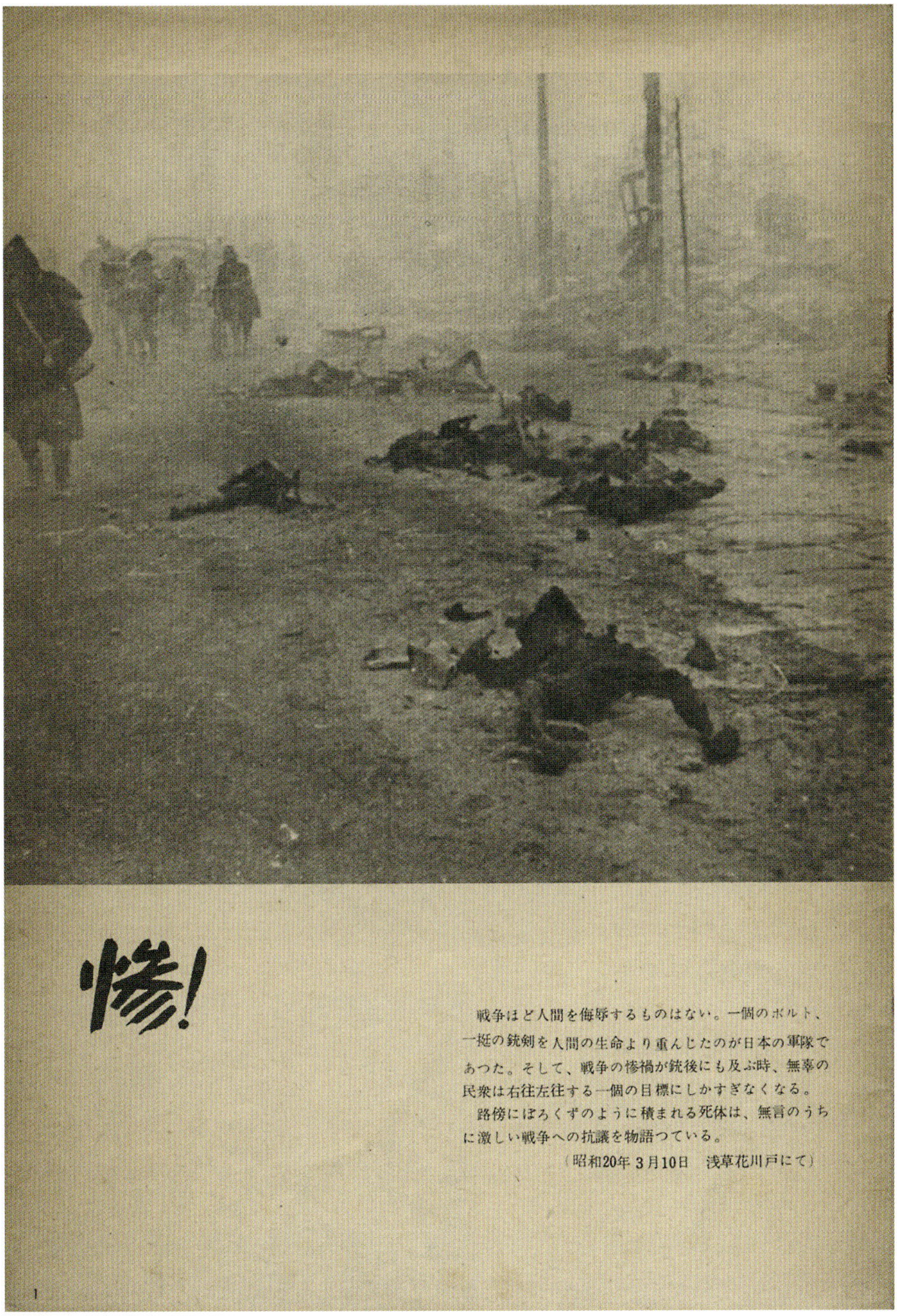

慘!

戦争ほど人間を侮辱するものはない。一個のボルト、一挺の銃剣を人間の生命より重んじたのが日本の軍隊であつた。そして、戦争の惨禍が銃後にも及ぶ時、無辜の民衆は右往左往する一個の目標にしかすぎなくなる。

路傍にぼろくずのように積まれる死体は、無言のうちに激しい戦争への抗議を物語つている。

（昭和20年 3月10日　浅草花川戸にて）

1

8.2 *Tragedy!* (*San!*), scene from Asakusa Hanakawado, 10 March 1945. "Fire Bombing," special issue, *Ondori tsūshin,* no. 1896 (July 1953).

incineration showed the streets of Tokyo again filled with corpses charred beyond recognition. The iconic aerial visualization of the expanse of the burned wasteland (*yakenohara*) and ruins (*haikyo*) in the Honjo-Ryōgoku area of the city confirms the existence of a spatial and topographical trope of tragedy extending back to 1923 and before.[1]

The threat of incendiary bombs became linked to the memory of the Kantō temblor when in 1932 the quake anniversary also became the day for the Tokyo Defense Brigade (Tōkyō Rengō Bōgodan) to lead the city in civil air defense drills.[2] A strong visual association between the catastrophic urban devastation of the earthquake and the potential destruction from aerial bombing was created in the civil-air-defense posters promoting the National Defense Act of 1937, the year the Second Sino-Japanese War (1937–45) began.

Under the cautionary title *Incendiary Bomb Threat* (*Shōidan no kyōi*) (fig. 8.3), a poster warns about the exponentially greater danger of enemy bombing: "[In] the Great Kantō Earthquake, fire originated from a mere *one hundred places,* the imperial capital became a *burned wasteland.* With only one of the enemy's planes, an incendiary bomb, *five thousand places.*"[3] The image of Tokyo ablaze again was used to evoke horror in the minds of city residents in the hope of steeling their determination to prepare for civil defense. But unlike the quake, the bombing was not a random natural event. The destruction was the result of a strategic wartime offensive against the home front of an aggressor nation that was similarly conducting aerial bombings on the Asian continent. Thus, the surface connections between the urban devastation and civilian casualties caused by the two events mask disconcerting differences. They also reveal a host of disturbing convergences in the politics of victimhood—the assignment of guilt and innocence—that are inherent in the process of remembering—and forgetting—history.

At the time of the air raids, Japan was under siege and losing the war. The entire nation was experiencing extreme privation, and Tokyo city was not equipped to handle the disposal of tens of thousands of bodies. Out of expediency and desperation, authorities had most victims interred right away in mass graves in public parks, often to the great distress of their bereaved families. When Japan finally surrendered and the Allied Occupation (1945–52) began, the families soon began to lobby for proper interment and remembrance of their loved ones as gallant victims of the war. As Cary Karacas has well described, addressing their demands posed considerable difficulties for Occupation authorities because of the Occupation censorship of discussion of the American role or responsibilities in the wartime bombings and the liminal status of civilian victims of the war, who were refused burial in the monuments to the sacred war dead, such as

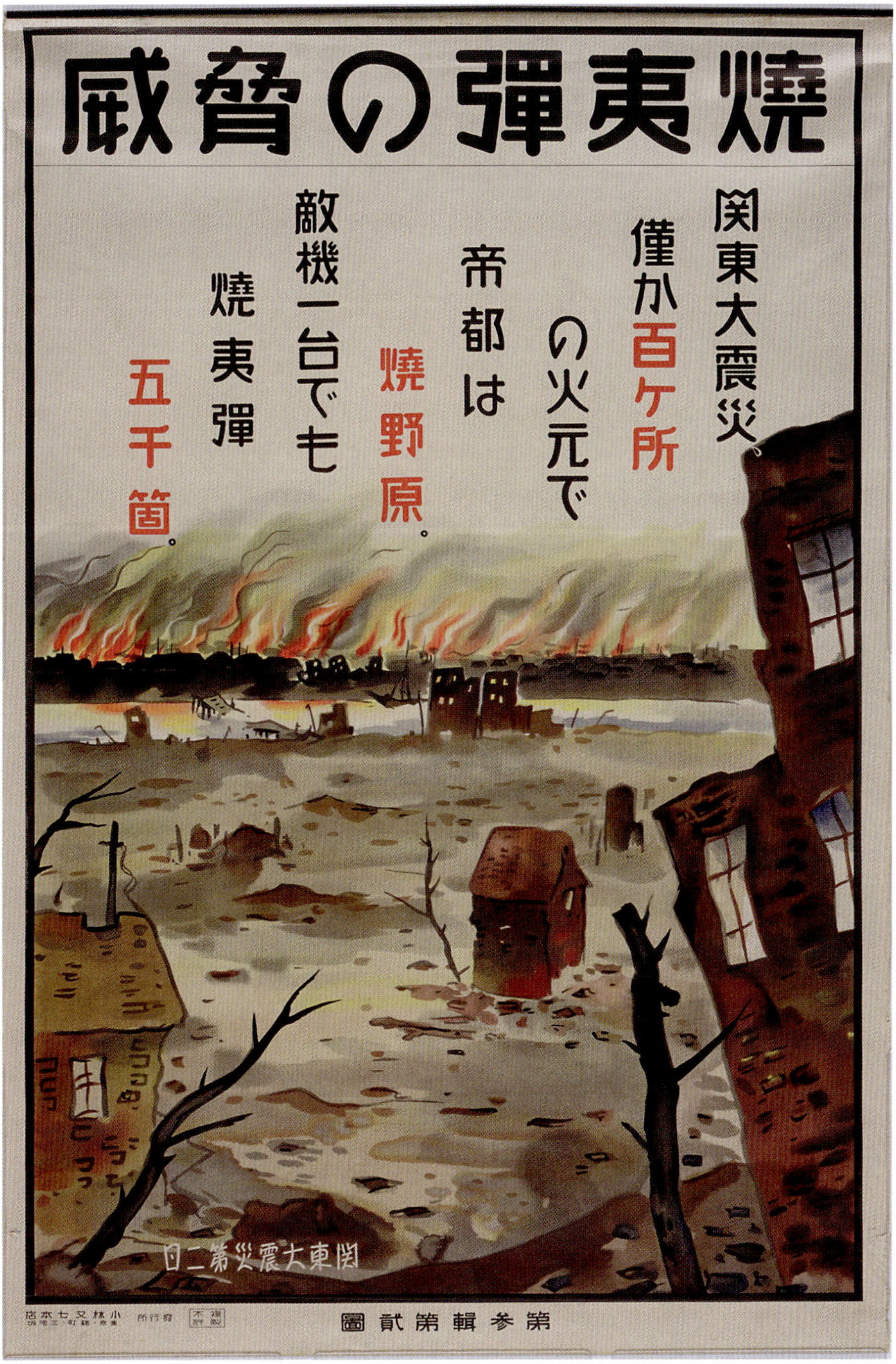

8.3 Artist unknown, *Incendiary Bomb Threat* (*Shōidan no kyōi*), poster, 1937. Collection of Kokuritsu Kōbunshokan.

Yasukuni Shrine, despite the fact that many had died as loyal imperial subjects fulfilling vital wartime home-front activities.[4] Building on the work of Yamamoto Tadahito, Karacas has closely studied the emergence of a grassroots movement among family members to create a memorial to their loved ones, which he describes as an attempt to make Tokyo "a less forgetful city," but he also identifies broader, and deeply troubling, amnesic impulses in the Japanese populace akin to those brilliantly interpreted by Lisa Yoneyama in relation to Hiroshima.[5] Enraged families maintained the untenable position that their deceased relatives were loyal subjects and innocent victims of a duplicitous militarist regime that forced them into mobilizing for war. They demanded a sanitized form of remembrance that indicted Japan's military elite and willfully forgot the decades of citizens' complicities in Japan's imperialist expansionism and wartime aggression. They also, according to Yamamoto, sublimated the tensions of exhibiting and memorializing war while honoring the sacred war dead within monuments dedicated to peace.[6] The painful debate that later emerged about whether German citizens were "willing executioners" under the Nazi regime has still not been meaningfully extended to Japan's regime in the Axis alliance. Instead, as Yoneyama insightfully argues, in the postwar years, the people killed by the atomic bombs were placed within "the global narrative of the universal history of humanity" and situated in a discourse focused on "a national victimology and phantasm of innocence."[7]

Exemplified by the aerial photographs of Tokyo in ruins after the firebombings, the revival of iconic earthquake motifs to memorialize the trauma of the atomic bomb imbricated Hiroshima and Nagasaki in the visual genealogy of Japanese catastrophe regardless of historical context. The spectacular mushroom cloud (fig. 8.4) and the aerial images of the city of Hiroshima engulfed in smoke resonate with the imaging of Kantō. The iconic earthquake motif of the arrested clock, which for Hiroshima stopped at 8:15 A.M. on 6 August 1945, also figures prominently in the wartime visual lexicon, embodied by the personal relics of civilian bomb victims such as Nikawa Kengo's pocket watch (fig. 8.5), donated by his son, Kazuo. The watch, housed in the main building of the Hiroshima Peace Memorial Museum designed by Tange Kenzō and erected after the Occupation in 1955, is accompanied by a personal narrative expressing profound bereavement: "Kengo Nikawa (then 59) was exposed to the bomb crossing the Kan-on Bridge by bike going from his home to his assigned building demolition site in the center of the city. He suffered major burns on his right shoulder, back, and head and took refuge in Kochi-mura Saiki-gun. He died on August 22. Kengo was never without this precious watch given him by his son, Kazuo."[8]

Photographer Domon Ken's (1909–90) famous monograph *Hiroshima* (1958) also features the arrested watch, but the hands are completely obliterated, gesturing to

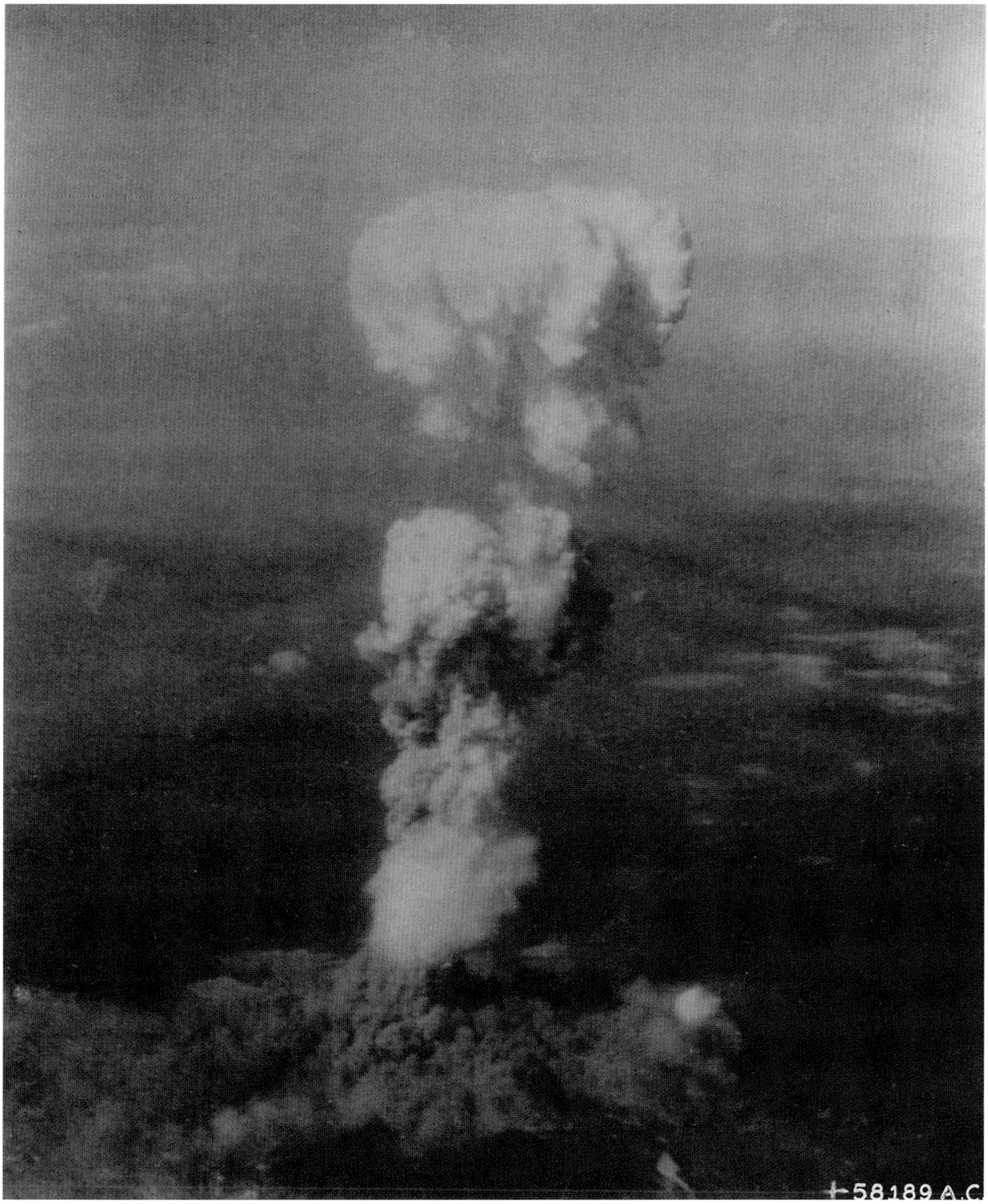

8.4 U.S. Department of Defense, "atomic burst," Hiroshima, Japan, photograph, 6 August 1945. Photograph courtesy of Japan Air Raids.org.

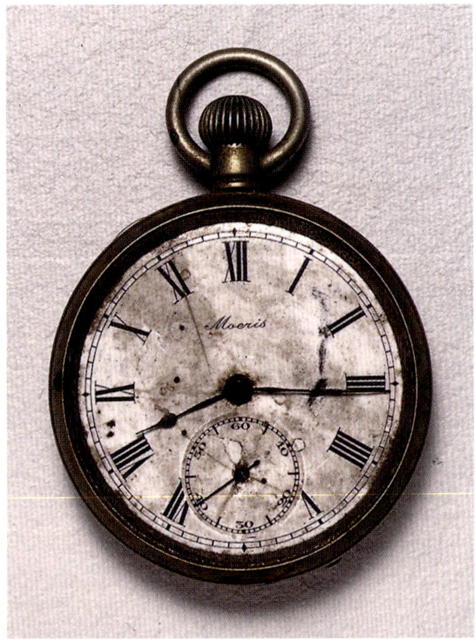

8.5 Pocket watch of Nikawa Kengo, exposed at Kan-on Bridge, 1,600 meters from the hypocenter, photograph. Donated to the Hiroshima Peace Memorial Museum by Nikawa Kazuo. Hiroshima Heiwa Kinen Shiryōkan.

the sense of temporal oblivion after the apocalypse.[9] A few years later, photographer Tōmatsu Shōmei (b. 1930) collaborated with Domon on the 1961 photographic collection *Hiroshima-Nagasaki Document 1961* sponsored by the Japan Council against A and H Bombs (Gensuikyō), which included Tōmatsu's spectral photograph *Time Stopped at 11:02, 1945* (fig. 8.6), showing a small watch floating in a surreal ether stopped at the moment the atomic bomb exploded in Nagasaki on the morning of 9 August 1945.[10]

The Hiroshima memorial displays a panoply of deformed personal objects donated by the bereaved as "material witnesses," focusing particularly on the relics of children—the innocents—such as the scorched lunchbox of Orimen Shigeru (fig. 8.7) donated by his mother, Shigeko, which is displayed with the following heartrending account,

> Shigeru Orimen was a first-year student at Second Hiroshima Prefectural Junior High School. He was exposed to the bomb at his building demolition work site at Nakajima-shin-machi. His mother Shigeko searched for him desperately through the devastated city, but failed to find him. Finally, she got information about him from an acquaintance and, early in the morning of August 9, she found a body with this lunch box clutched to the abdomen. Shigeru had worked diligently in place of his father and brother who were away at the front. He plowed the field and cultivated gardens on the mountain and in a bamboo grove. His lunch that day was made from the first harvest from his new field, which he had brought home so happily. Shigeko's grief deepened when she realized that Shigeru never got a chance to eat the lunch he had been so eagerly anticipating.[11]

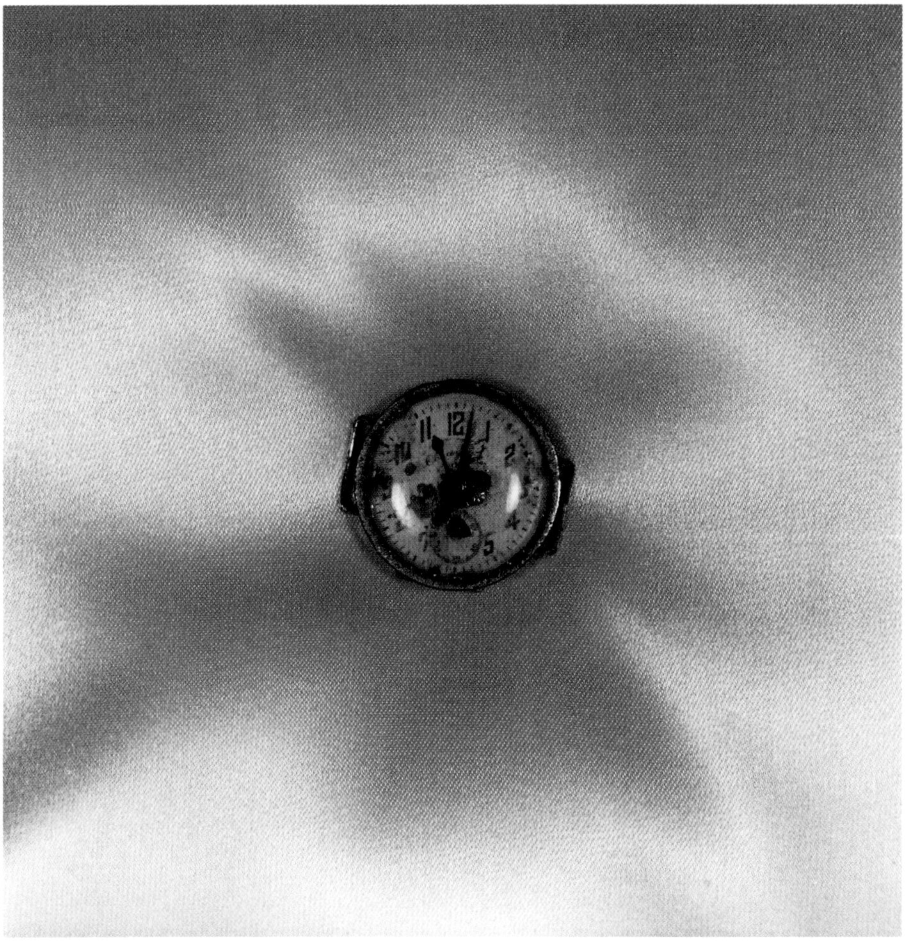

8.6 Tōmatsu Shōmei, *Time Stopped at 11:02, 1945,* Nagasaki, photograph, 1961. Domon Ken and Tōmatsu Shōmei, *Hiroshima-Nagasaki Document 1961* (Tokyo: Japan Council against A and H Bombs, 1961). Printed in 1991, gelatin silver print; 19¹⁵⁄₁₆ × 16 in. (50.64 × 40.64 cm). San Francisco Museum of Modern Art, Accession Committee Fund purchase; © Shomei Tomatsu.

The symbolic center of the Hiroshima Peace Memorial Park, the A-Bomb Dome (Genbaku Dōmu) (fig. 8.8), which stands at the hypocenter of the explosion, is the ruin of the stalwart Western-style building that was once the Hiroshima Prefectural Industrial Promotional Hall. With the steel-frame skeleton of its lofty dome exposed, the monument recalls the many ruins of magnificent destruction immortalized in visualizations of the earthquake, particularly the iconic Yokohama Specie Bank (see chapter 4). While reproductions can also serve as repositories of memory, the ruin itself is left standing in Hiroshima as an eternal material witness to the trauma and lasting

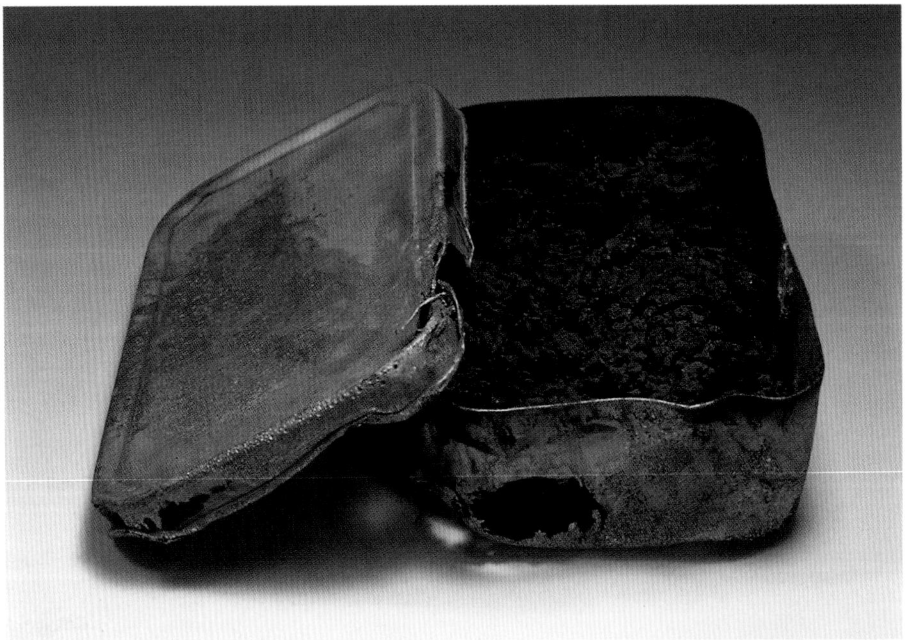

8.7 Lunch box of Orimen Shigeru, exposed at Nakajima-shin-machi (now Nakajima-chō), 600 meters from the hypocenter, photograph. Donated to the Hiroshima Peace Memorial Museum by Orimen Shigeko. By permission of Hiroshima Heiwa Kinen Shiryōkan.

hidden effects of radiation, new to the disasters of the atomic age, forever culturally inscribed on this site. But the causalities of the event remain elusive here. By eliding natural and man-made catastrophes, the enduring visual lexicon of disaster complicates efforts to remember and find meaning in these events.

Within the immediate postwar period's nexus of political censorship and personal and collective amnesia, and after months of debate, Tokyo municipal officials and their Allied Occupation superiors took the most expedient path to accommodating people's desires for a firebombing memorial: they decided to exhume the bodies of the victims and reinter them in the ossuary of the Earthquake Memorial Hall. In 1951, after the gruesome work of exhumation and reinterment was completed, the building was renamed the Tokyo Metropolitan Memorial Hall (Tōkyō-to Ireidō). Authorities also earmarked a section in the adjacent Reconstruction Memorial Museum on the second floor to exhibit materials from the Tokyo air raids. Now two anniversaries are memorialized at the site, and the bones in the stupa are opened to public view in September and March for proper veneration by different constituencies. The elision of the vast causal differences between the natural disaster of the earthquake and the man-made

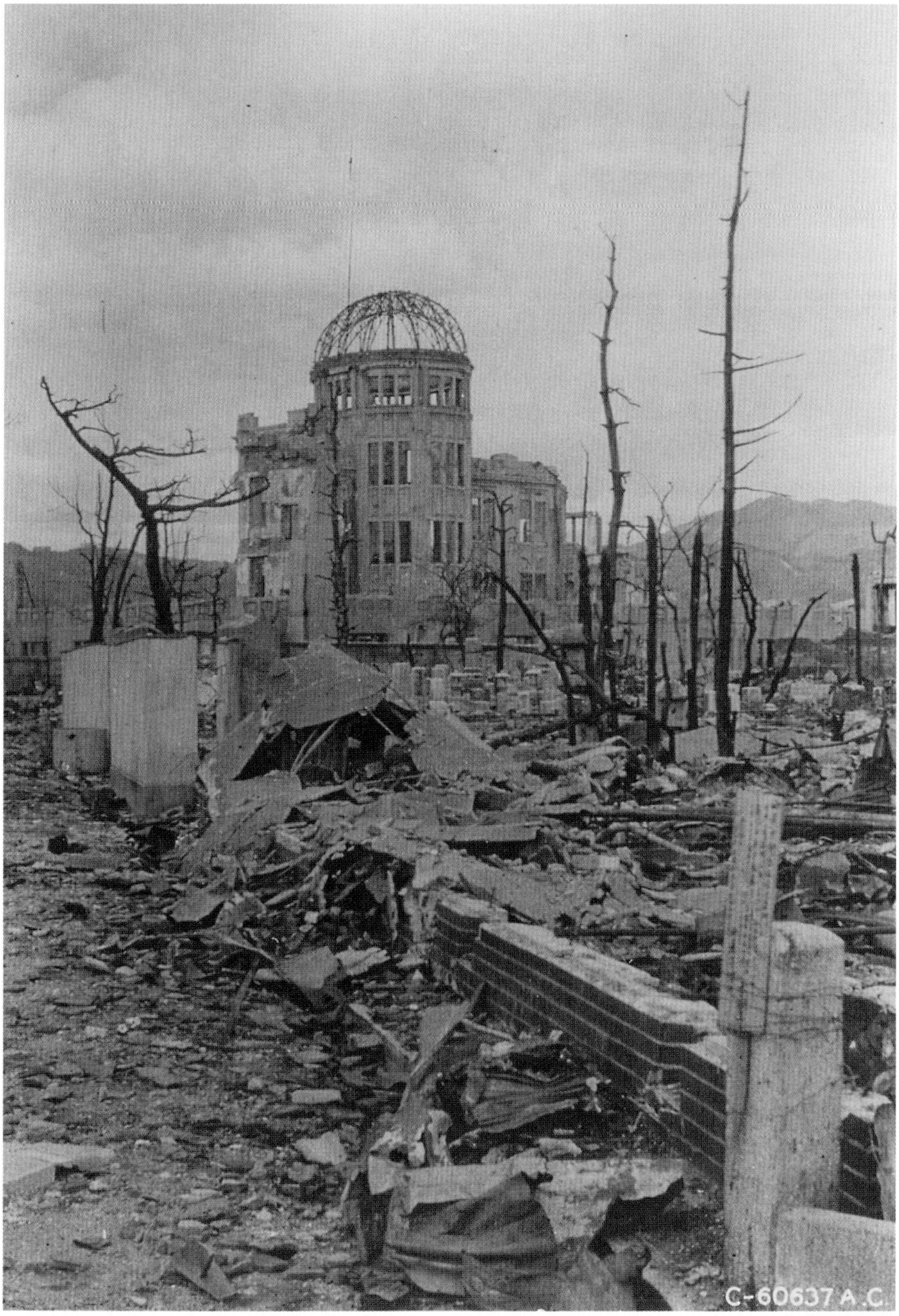

8.8 U.S. Department of Defense, A-Bomb Dome (Genbaku Dōmu), Hiroshima, photograph, c. 1945. Photograph courtesy of Japan Air Raids.org.

catastrophe of the firebombings has rendered this space as a fractured site of remembrance that no longer coheres—giving it a split personality. However, the questions of responsibility that lurk in the shadows of any discussion of the war are just as relevant to the Kantō quake experience, which, as I have demonstrated, was not merely a natural event but also a social disaster that spotlighted the deeply corrosive elements in Japan's national community. Similarly, the complicity of the quake's "innocent" victims and national saviors in the savage retribution visited upon Japan's resident colonial subjects and political dissidents has yet to be fully acknowledged despite repeated calls for a more inclusive remembrance.

Japanese textbooks, well known to have little coverage of the larger imperialist context of Japan's wartime aggression, similarly exclude meaningful discussion of quake atrocities, focusing instead on the master narrative of national solidarity and triumphant reconstruction. While recent textbooks have added brief mentions or sidebars on the massacre, most of these focus on the unfortunate acts of lawless "rogue" vigilantes in a time of extreme chaos or discuss the murders in passive voice to avoid specifying a subject. For the most part, they do not explore the complicity of government officials in perpetrating the atrocities, nor do they examine how the larger context of Japanese imperialism produced the exclusionary ideological structure that enabled the targeting of colonial subjects.[12]

Not until 1973 was a stone stele erected in the memorial complex to commemorate the Korean victims of the earthquake. The explanation for the commission by the Association for Enacting Rites for the Korean Victims of the Great Kantō Earthquake (Kantō Daishinsai Chōsenjin Giseisha Gyōji Jikkō Iinkai) reads, "In the midst of the confusion of the Great Kantō Earthquake that occurred in September 1923, due to regretful mistakes and groundless rumors, nearly 6,000 Koreans had their precious lives stolen from them. Nearing the fiftieth anniversary of the earthquake, we sincerely commemorate the Korean victims. Understanding the truth of this incident so as not to repeat this unfortunate history, we believe will be the foundation for eradicating racial prejudice, respecting human rights, and forging a road to neighborly friendship and peace." The inscription conveys the causes of the atrocities in a depersonalized passive voice that expresses generalized regret but leaves responsibility vague and unarticulated, a common linguistic strategy in Japanese peace memorials commemorating the war. Reinforcing this noncommittal remembrance, the inscription on the stele says, "Never forgetting this history, let us firmly join hands with Koreans living in Japan, and let us establish Japanese-Korean amity and peace in Asia."[13]

In the postwar period, earthquake commemoration was again linked to national emergency management when, in June 1960, the Japanese government designated 1 September as Disaster Prevention Day (Bōsai no Hi) to raise public awareness and to educate people about prevention. Since 1971, national and local governments have teamed up with community organizations on this day each year to sponsor earthquake-preparedness drills across the nation. Usually corresponding with the first day of school in the fall, evacuation drills are often integrated into back-to-school ceremonies. In a return to the wartime posture of civil air defense, sirens are sounded and people don protective headgear while performing the drills.[14] In addition, Japanese schoolchildren must memorize the date of the Great Kantō Earthquake with special mnemonics. As historian Ichinose Toshiya reminisces, in elementary school he learned to memorize the year of the Great Kanto Earthquake—1923—by reciting "*Jishin hito yure kuni sanzan*" (Earthquake one big shake, country is in ruins). *Hito* means one in Japanese. *Kuni,* meaning country, can be separated into the two parts *ku* and *ni,* literally nine and two. The *san* of *sanzan* (in ruins) means three; thus, 1-9-2-3.[15] Children internalize the quake in their bodies by moving through drill procedures and in their minds through a rote exercise without truly understanding the significance of the historical event. Thus, the muted memory of the Great Kantō Earthquake lingers on in the compromised space of the Honjo memorial, in abridged and expurgated textbooks, and in Japan's routine culture of disaster preparedness, without providing the nation the opportunity to learn from the profound social, cultural, and historical lessons of conscience brought to the fore by the event.

NOTES

PREFACE

1. For example, see the images published online by the *Atlantic Monthly:* www.theatlantic.com/infocus/2011/03/japan-earthquake-two-weeks-later/100034/ (accessed 25 March 2011).

2. From the English online edition of *Asahi shinbun*, Oguma Eiji, "POINT OF VIEW/Eiji Oguma: Disaster Highlights Tohoku's Unique Structural Weaknesses," www.asahi.com/english/TKY201106070200.html (accessed 8 June 2011). See also Oguma Eiji, *A Genealogy of "Japanese" Self-Images,* Japanese Society Series (Melbourne: Trans Pacific Press, 2002).

3. For example, NHK (Nippon Hōsō Kyōkai), Japan's national news organization, has posted and archived hundreds of videos of the disaster on its website: www3.nhk.or.jp/news/jishin0311/ (accessed 16 June 2011).

4. The Japan Quake Map, designed by Paul Nicholls of the University of Canterbury's Digital Media Group (Christchurch, New Zealand), presents a time-lapse visualization of the Tōhoku earthquake and its aftershocks, plotting earthquake data from the U.S. Geological Survey on a map using the Google Maps application programming interface. On the quake map, the size of the circle denotes the magnitude (the higher the magnitude, the larger the circle), and the color shows the focal depth. ABC News's "Japan Earthquake: Before and After" online offers a series of aerial satellite maps of the Tōhoku coast over which the viewer can move the cursor to change the landscape view from before the tsunami to after the event. Respectively, www.japanquakemap.com/ (accessed 17 March 2011) and www.abc.net.au/news/events/japan-quake-2011/beforeafter.htm (accessed 13 March 2011).

5. For Sakurai's original video, see www.youtube.com/watch?v=70ZHQ--cK40 (accessed 30 March 2011 and 15 June 2011).

6. See the online version of the North Carolina Research Triangle newspaper the *News & Observer*, www.newsobserver.com/2011/04/12/1123713/a-countrys-strength-and-weakness.html#ixzz1JK44wNb5 (accessed 12 April 2011).

INTRODUCTION

1. W. J. T. Mitchell, *What Do Pictures Want? The Lives and Loves of Images* (Chicago: University of Chicago Press, 2005).

2. I use the term *visuality* here to indicate "the social construction of the visual field, and the visual construction of the social field," reciprocal processes that structure the subjectivity and agency of the viewing public. W. J. T. Mitchell, "Responses to Mieke Bal's 'Visual Essentialism and the Object of Visual Culture' (2003): The Obscure Object of Visual Culture," *Journal of Visual Culture* 2, no. 2 (2003): 252. For an expanded discussion of visuality and visual studies, see Mitchell, "Showing Seeing: A Critique of Visual Culture," *Journal of Visual Culture* 1, no. 2 (August 2002): 165–81.

3. Collective memory tends to reflect master narratives of an event forged by leading social and political groups to become "dominant memory." Daniel J. Sherman, *The Construction of Memory in Interwar France* (Chicago: University of Chicago Press, 1999), 7.

4. Quoted in Edward Thomas Mack, "The Value of Literature: Cultural Authority in Interwar Japan (Ph.D. diss., Harvard University, 2002), 97.

5. The earthquake initially acquired a variety of names, such as Taishō Taishin Kasai (Great Taishō Earthquake and Fire), Taishō Daishinsai (Great Taishō Earthquake), Tōkyō Taishin Kasai (Great Tokyo Earthquake and Fire), until the standard appellation of Kantō Daishinsai emerged about a month after the temblor. Narita Ryūichi has argued that the final naming of the event with the moniker Kantō is indicative of attempts to widen the scope and scale of the catastrophe beyond one municipality and to give it national importance. Narita Ryūichi, "Kantō Daishinsai no meta hisutorii no tame ni: Hōdō, aiwa, bidan" (A Metahistory of the Great Kantō Earthquake: Reportage, Tragic Tales, and Heroic Tales), *Shisō* 866 (August 1996).

6. The epicenter of the earthquake was in Sagami Bay near Oshima Island. Jelle Zeilinga de Boer and Donald Theodore Sanders, *Earthquakes in Human History: The Far-Reaching Effects of Seismic Disruptions* (Princeton, NJ: Princeton University Press, 2005), 170–90.

7. Tokyo Municipal Office, *The Reconstruction of Tokyo* (Tokyo: Tokyo Municipal Office, 1933), 8. One estimate places Japan's gross national product in 1930 at 13.85 billion yen. *Nihon chōki tōkei sōran* (Survey of Japanese Long-Term Statistics) (Tokyo: Nihon Tōkei Kyōkai, 1987–88), 3:350. The yen-dollar exchange rate fluctuated wildly in the 1930s because of the unstable economy, and the GNP also fluctuated during the depression, so it is hard to get a sense of equivalents. Varying methods of conversion can also produce large differences in experts' estimates.

8. Tsuchida Mitsufumi, *Tōkyō kiroku bungaku jiten* (Dictionary of Tokyo Documentary Literature) (Tokyo: Kashiwagi Shobō, 1994), 299–356.

9. The Ansei-era quake that hit Edo city is referred to in Japanese as the Ansei Edo earthquake (*Ansei Edo jishin*), Ansei earthquake (*Ansei jishin*), and sometimes the Great Ansei Earthquake (*Ansei Daijishin*). The Ansei imperial era marked the reign of Emperor Kōmei from 1854 to 1860. Tōkyō-to Edo Tōkyō Hakubutsukan Toshi Rekishi Kenkyūshitsu, *Kantō Daishinsai to Ansei Edo jishin* (The Great Kantō Earthquake and the Ansei Edo Earthquake), Tōkyō-to Edo Tōkyō Hakubutsukan Chōsa Hōkokusho (Tokyo: Tōkyō-to Edo Tōkyō Hakubutsukan: Tōkyō-to Rekishi Bunka Zaidan, 2000), 10:39.

10. Only in the past ten years have scholars in Japan begun to bring together disaster studies with visual and media studies. See, for example, Ozawa Takeshi, ed., *Shashin de miru Kantō*

Daishinsai (The Great Kantō Earthquake Seen through Photographs) (Tokyo: Chikuma Shobō, 2003); Tōkyō Daigaku Daigakuin Jinbun Shakai Kakari Kenkyūka, ed., *Kantō Daishinsai to kiroku eiga: Toshi no shi to saisei* (The Great Kantō Earthquake and Documentary Film: The Death and Regeneration of the City) (Tokyo: Tōkyō Daigaku Daigakuin Jinbun Shakai Kakari Kenkyūka 21 Seiki COE Puroguramu, 2004). The foremost scholars working in this area are Kinoshita Naoyuki, Kitahara Itoko, Narita Ryūichi, Yoshimi Shunya, Satō Kenji, Yamamoto Tadahito, and, more recently, Satō Miya and Takano Hiroyasu. In English, the only overarching cultural account of the earthquake is Edward Seidensticker's well-known two-volume study on the city of Edo/Tokyo, although his sources are predominantly literary and not visual. Edward Seidensticker, *Tokyo Rising: The City since the Great Earthquake* (New York: Knopf, 1990); Seidensticker, *Low City, High City: Tokyo from Edo to the Earthquake* (Cambridge, MA: Harvard University Press, 1991). For scholarship on visual culture, see Gennifer Weisenfeld, "Designing after Disaster: Barrack Decoration and the Great Kantō Earthquake," *Japanese Studies* 18, no. 3 (1998); Weisenfeld, "Omoiegakareru Sanka: Kantō Daishinsaigo no teito o meguru gakatachi" (Imaging Calamity: Artists in the Capital after the Great Kantō Earthquake), in *Mobo moga-ten 1910–1935,* ed. Kamakura Museum of Modern Art (Kamakura: Museum of Modern Art, 1998); Weisenfeld, "Saigai to shikaku hyōshō: Kantō Daishinsai no imējigun o megutte" (Disaster and Vision: On the Visual Representations of the Great Kantō Earthquake), in *Kioku to rekishi: Nihon ni okeru kako no shikakuka o megutte,* ed. Tano Yasunori (Tokyo: Waseda Daigakuku Aizu Yaichi Kinen Hakubutsukan, 2006). For general anecdotal and historical accounts of the quake in English from various individual and official perspectives, see Joseph Dahlmann and Victor Felix Gettelman, *The Great Tokyo Earthquake, September 1, 1923, Experiences and Impressions of an Eye-Witness* (New York: America Press, 1924); Japan Naimushō Shakaikyoku and Fujisawa Morihiko, *The Great Earthquake of 1923 in Japan* (Tokyo: Naimushō Shakaikyoku, 1926); Noel Fairchild Busch, *Two Minutes to Noon* (New York: Simon and Schuster, 1962); Joshua Hammer, *Yokohama Burning: The Deadly 1923 Earthquake and Fire That Helped Forge the Path to World War II* (New York: Free Press, 2006).

11. Sharon H. Nolte, *Liberalism in Modern Japan: Ishibashi Tanzan and His Teachers, 1905–1960* (Berkeley: University of California Press, 1987), 19.

12. Gregory James Kasza, *The State and the Mass Media in Japan, 1918–1945* (Berkeley: University of California Press, 1988), 28–29. Newspaper circulation rates fluctuated significantly during the late Taishō and early Shōwa periods. Some general circulation rates for the major Tokyo daily newspapers around 1927 are approximated as follows: *Tōkyō nichinichi shinbun,* 450,000; *Tōkyō asahi shinbun,* 400,000; *Hōchi shinbun,* 250,000; *Jiji shinpō,* 200,000; *Kokumin shinbun,* 150,000; *Miyako shinbun,* 120,000; *Yomiuri shinbun,* 100,000; *Yorozu chōhō,* 100,000; and *Chūgai shōgyō shinpō,* 100,000.

13. Benjamin as quoted in Eduardo Cadava, "Lapsus Imaginis: The Image in Ruins," *October,* no. 96 (2001): 35.

14. Ariella Azoulay, *The Civil Contract of Photography* (New York: Zone Books, 2008), 85–86.

15. Elizabeth Cowie, "The Spectacle of Actuality," in *Collecting Visible Evidence,* ed. Jane Gaines and Michael Renov (Minneapolis: University of Minnesota Press, 1999), 19.

16. Kasza, *The State and the Mass Media in Japan,* 54–57.

17. Markus Nornes, *Japanese Documentary Film: The Meiji Era through Hiroshima* (Minneapolis: University of Minnesota Press, 2003), 5.

18. By the end of 1923, at least nine fiction films were released that dealt with the earthquake directly; unfortunately, none is extant today. Peter Alexander Bates, "Fractured Communities: Class and Ethnicity in Representations of the Great Kanto Earthquake (Ph.D. diss., University of Michigan, 2006), 56. Documentary footage was later interspliced into fictional films to heighten the sense of realism, a combinatory practice that was already common during the Russo-Japanese War (1904–5), when newsreels were interspliced with dramatic reenactments. Nornes, *Japanese Documentary Film,* 6–10.

19. Denis Wood, John Fels, and John Krygier, *Rethinking the Power of Maps* (New York: Guilford Press, 2010), 51–52.

20. Ibid., 53.

21. Edward R. Tufte, *Visual Explanations: Images and Quantities, Evidence and Narrative* (Cheshire, CT: Graphics Press, 1997), 9.

22. Gennifer Weisenfeld, "The Expanding Arts of the Interwar Period," in *Since Meiji: Perspectives on the Japanese Visual Arts, 1868–2000,* ed. J. Thomas Rimer (Honolulu: University of Hawaii Press, 2011).

23. For example, the print publisher Gahōsha, located in Hongō, issued the series *Collection of Woodblock Prints of the Taishō Earthquake—Thirty-Six Scenes* (*Taishō Shinkasai mokuhangashū zen sanjū rokkei*) in installments of three prints per month for the entire year of 1924. The price for one installment was 1 yen 50 sen, and the price for a yearly subscription was 17 yen. Nishizawa Tekiho et al., *Taishō Shinkasai mokuhangashū zen sanjū rokkei* (Collection of Woodblock Prints of the Taishō Earthquake—Thirty-Six Scenes) (Tokyo: Gahōsha, 1924); Kitahara Itoko, "Egakareta Kantō Daishinsai: Emaki, hanga, sobyō" (The Great Kantō Earthquake Pictured: Illustrated Handscrolls, Prints, and Drawings), *Himoji shiryō kenkyū nenpō,* no. 6 (2010): 25.

24. Marguerite Wells, "Satire and Constraint in Japanese Culture," in *Understanding Humor in Japan,* ed. Jessica Milner Davis, Humor in Life and Letters Series (Detroit: Wayne State University Press, 2006), 194.

25. Ibid., 210–11.

1. EARTHQUAKES IN JAPAN

1. Haruno Ogasawara, "Living with Natural Disasters: Narratives of the Great Kantō and the Great Hanshin Earthquakes" (Ph.D. diss., Northwestern University, 1999), 8–9; Peter Alexander Bates, "Fractured Communities: Class and Ethnicity in Representations of the Great Kanto Earthquake" (Ph.D. diss., University of Michigan, 2006), 2–3.

2. Kitahara Itoko, *Kinsei saigai jōhōron* (Information Theory of Early Modern Disasters) (Tokyo: Hanawa Shobō, 2003); Kitahara, ed., *Nihon saigaishi* (The History of Disasters in Japan) (Tokyo: Yoshikawa Kōbunkan, 2006).

3. Since the destruction caused by earthquakes was largely due to ensuing fires, in Japan earthquakes are often categorized with "fire disasters" (*kasai*), which include nonnatural disasters such as the Great Meireki Fire of 1657 and the subsequent "Flowers of Edo," as people referred to the fires in the shogunal capital.

4. Originally the expression was *nai furu* or *nai yoru* (both meaning "land shakes"), but then *nai* itself came to mean earthquake.

5. Doi Tadao, Morita Takeshi, and Chōnan Minoru, eds., *Hōyaku Nippo jisho* (Japanese Translation of the Japanese-Portuguese Dictionary) (Tokyo: Iwanami Shoten, 1980), 317–18.

6. Ogasawara, "Living with Natural Disasters," 23–26.

7. Ibid.

8. *The Tale of the Heike/Heike monogatari,* trans. Hiroshi Kitagawa and Bruce T. Tsuchida, vol. 1 (Books 1–6) (Tokyo: University of Tokyo Press, 1975; reprint, 1989), 203, 217.

9. Nichiren, *Selected Writings of Nichiren,* trans. Philip Yampolsky and Burton Watson, Translations from the Oriental Classics (New York: Columbia University Press, 1990), 473.

10. Jacqueline Ilyse Stone and Mariko Namba Walter, *Death and the Afterlife in Japanese Buddhism* (Honolulu: University of Hawaii Press, 2008), 4–9.

11. Japanese artists often take "art names" (*gō*), artistic pseudonyms or pen names, by which they are professionally known. Ōkyo is an example of such an art name for the artist born Maruyama Masataka. Many modern artists have continued this practice, although not all have. Throughout this book, if an artist is best known by his art name, that name is used. If an artist does not have an art name, he is first referred to by his surname and given name and subsequently by his surname.

12. Mabuchi Miho, "Maruyama Ōkyo pitsu *Nanpuku Zukan* no fuku no zu ni tsuite" (A Study of the "Fortune" Pictures of the *Nanpuku Zukan* Painted by Maruyama Ōkyo), *Bijutsushi ronsō* 14, no. 3 (1998); Mabuchi, "Maruyama Ōkyo pitsu *Nanpuku Zukan* to *Kannon Kyō Kannon Kyō-e*" (A Study of the *Nanpuku Zukan* Painted by Maruyama Ōkyo and the *Kannon Sutra* and *Kannon Sutra Paintings*), *Bijutsushi ronsō,* no. 19 (2003).

13. Sections of a finished copy of Ōkyo's scroll by nineteenth-century painter Minamoto no Ōko are in the collection of the British Museum. I am grateful to Rosina Buckland for bringing this to my attention. Other copies are in the collection of the Freer and Sackler Galleries.

14. Scholars have different opinions about the identities of the author and illustrators of this work. See Andrew Markus, "*Gesaku* Authors and the Ansei Earthquake of 1855," in *Studies in Modern Japanese Literature: Essays and Translations in Honor of Edwin McClellan,* ed. Edwin McClellan, Dennis C. Washburn, and Alan Tansman (Ann Arbor: Center for Japanese Studies, University of Michigan, 1997); Stephan Köhn, "Between Fiction and Non-Fiction: Documentary Literature in the Late Edo Period," in *Written Texts—Visual Texts: Woodblock-Printed Media in Early Modern Japan,* ed. Susanne Formanek and Sepp Linhart, Hotei Academic European Studies on Japan (Amsterdam: Hotei, 2005).

15. Kinoshita Naoyuki et al., *Nyūsu no tanjō: Kawaraban to shinbun nishikie no jōhō sekai/The Birth of the News: Visual Media in Nineteenth-Century Japan,* Tōkyō Daigaku korekushon (Tokyo: Tōkyō Daigaku Sōgō Kenkyū Hakubutsukan, 1999).

16. I am grateful to John Mertz and Yukiko Kato for their invaluable assistance in translating this passage.

17. Unno Kazutaka, "Maps of Japan Used in Prayer Rites or as Charms," *Imago Mundi,* no. 46 (1994).

18. Gregory Smits, "Shaking Up Japan: Edo Society and the 1855 Catfish Picture Prints," *Journal of Social History* 39, no. 4 (Summer 2006): 1072. Markus, "*Gesaku* Authors and the Ansei Earthquake of 1855," 55.

19. The term *fukkō* signified renewal and revival as well as reconstruction.

20. For more on catfish images, see Cornells Ouwehand, *Namazu-e and Their Themes: An Interpretative Approach to Some Aspects of Japanese Folk Religion* (Leiden: E. J. Brill, 1964); Kitahara Itoko, *Ansei Daijishin to minshū: Jishin no shakaishi* (The Great Ansei Earthquake and the People: A Social History of Earthquakes) (Tokyo: Sanichi Shobō, 1983); Miyata Noboru and Takada Mamoru, *Namazu-e: Shinsai to Nihon bunka* (Catfish Pictures: Earthquakes and Japanese Culture) (Tokyo: Ribun Shuppan, 1995); Shiga Hidemi, "The Catfish Underground: Japan's Earthquake Folklore and Popular Responses to Disaster," *Orientations* 37, no. 3 (2006); Smits, "Shaking Up Japan."

21. Kitahara Itoko, *Ansei Daijishin to minshū,* 162–91. Also discussed in Ogasawara, "Living with Natural Disasters," 43–44.

22. Ouwehand, *Namazu-e and Their Themes.*

23. Sometimes in these prints, the word "earthquake" (*jishin*) is glossed in the phonetic syllabary alongside the image as "catfish" (*namazu*). Miyata and Takada, *Namazu-e,* 106, 110, 264–65, 82–84.

24. Smits, "Shaking Up Japan," 1046–50, 1062–66.

25. Markus, "*Gesaku* Authors and the Ansei Earthquake of 1855," 55.

26. Smits, "Shaking Up Japan," 1052, 1054–55; Miyata and Takada, *Namazu-e,* 106, 266.

27. Miyata and Takada, *Namazu-e,* 235, 268, 324.

28. Köhn, "Between Fiction and Non-Fiction."

29. Kitahara Itoko discusses *Ansei kenmonshi* in *Ansei Daijishin to minshū: Jishin no shakaishi,* 71–142; see also Markus, "*Gesaku* Authors and the Ansei Earthquake of 1855," 69–70.

30. Köhn, "Between Fiction and Non-Fiction," 297, 299.

31. Ibid.

32. Miyata and Takada, *Namazu-e,* 48–51.

33. Gregory K. Clancey, *Earthquake Nation: The Cultural Politics of Japanese Seismicity, 1868–1930* (Berkeley: University of California Press, 2006), 104; Kitahara Itoko, *Bandaisan funka: Saii kara saigai no kagaku e* (Volcanic Eruption of Mount Bandai: From Uncommon Disasters to Disaster Science) (Tokyo: Yoshikawa Kōbunkan, 1998).

34. Kitahara Itoko, "Saigai to shashin media: 1894 Shōnai Jishin no kēsu sutadī" (Disaster and Photography: A Case Study of the 1894 Shōnai Earthquake), *Chōsa kenkyū shiryō,* no. 4 (March 2007).

2. THE MEDIA SCALE OF CATASTROPHE

Epigraph: Ōsaka Mainichi Shinbunsha, *Kantō shinsai gahō* (Kantō Earthquake Pictorial), vol. 1 (Osaka: Ōsaka Mainichi Shinbunsha; Tokyo: Hatsubaijo Tōkyō Nichinichi Shinbunsha, 28 October 1923).

1. Paul Virilio, *Unknown Quantity* (London: Thames & Hudson; Paris: Fondation Cartier pour l'art contemporain, 2003), 63.

2. Susan Sontag, *Regarding the Pain of Others* (New York: Farrar, Straus and Giroux, 2003), 22–23.

3. The earliest information about the disaster was sent by wireless from survivors on ships moored at Yokohama. This information was relayed throughout the country and abroad through the heroic efforts of local telecommunications operators in regional stations throughout Japan, particularly the operator at the Iwaki wireless station in Fukushima. The Kantō earthquake dramatically

demonstrated the importance of wireless for national security, and Japan's experience convinced many governments to fund the technology as a priority mode of telecommunications. Joshua Hammer, *Yokohama Burning: The Deadly 1923 Earthquake and Fire That Helped Forge the Path to World War II* (New York: Free Press, 2006), 178–88.

4. Shinbun Shiryō Raiburarii, *Kantō Daishinsai* (Great Kantō Earthquake), vol. 1 (Tokyo: Ōzorasha, 1992).

5. Because of the difficulties getting access to the devastated Kantō region, aerial reconnaissance was critical in relaying information about the nature and extent of damage to coordinate external relief operations. The Japanese army flew these missions out of its base in Nagoya. J. Charles Schencking, "1923 Tokyo as a Devastated War and Occupation Zone: The Catastrophe One Confronted in Post Earthquake Japan," *Japanese Studies* 29, no. 1 (2009): 117n35.

6. Paul Virilio, *War and Cinema: The Logistics of Perception* (London: Verso, 1989), 1–3, 11.

7. Illah Nourbakhsh et al., "Mapping Disaster Zones," *Nature* 439, no. 7078 (2006).

8. "Daishinsai zenki" (Full Record of the Great Earthquake), special issue, *Asahi Graph,* October 1923; Nihon Shinbun Hakubutsukan, ed., *Daishinsai to hōdō-ten: Kantō daishinsai 80-shūnen kikaku* (Exhibition of the Great Earthquake and the News: Plans for the 80th Anniversary of the Great Kantō Earthquake) (Yokohama-shi: Nihon Shinbun Hakubutsukan, 2003), 29.

9. Ōsaka Asahi Shinbunsha, *Daishinsai shashin gahō* (Photographic Pictorial of the Great Earthquake) (Osaka: Ōsaka Asahi Shinbunsha, 15 September 1923), 1:3.

10. Shirai Shigeru (cameraman), *Kokusen nite fukkō no teito e* (Reconstruction of the Imperial Capital Seen from an Airplane) (Tokyo: Monbushō, 1926). *Kamera to jinsei: Shirai Shigeru kaikoroku* (Camera and Life: The Memoirs of Shirai Shigeru) (Tokyo: Yuni Tsūshinsha, 1983), 51–58.

11. Edo-period earthquake maps were predominantly topographical maps that communicated the location of fires or schematic maps of the layout of the city around the shogunal castle that had arrows indicating the direction of fires, which were called "direction marker maps" (*hōkaku-zuke*). Tōkyō-to Edo Tōkyō Hakubutsukan Toshi Rekishi Kenkyūshitsu, *Kantō Daishinsai to Ansei Edo Jishin* (The Great Kantō Earthquake and the Ansei Edo Earthquake), Tōkyō-to Edo Tōkyō Hakubutsukan Chōsa Hōkokusho, vol. 10 (Tokyo: Tōkyō-to Edo Tōkyō Hakubutsukan: Tōkyō-to Rekishi Bunka Zaidan, 2000), 41, 190; Kinoshita Naoyuki et al., *Nyūsu no tanjō: Kawaraban to shinbun nishikie no jōhō sekai / The Birth of the News, Visual Media in 19th-Century Japan*, Tōkyō Daigaku korekushon (Tokyo: Tōkyō Daigaku Sōgō Kenkyū Hakubutsukan, 1999), 162–63, 78–79; Miyata Noboru and Takada Mamoru, *Namazu-e: Shinsai to Nihon bunka* (Catfish Pictures: Earthquakes and Japanese Culture) (Tokyo: Ribun Shuppan, 1995), 78–79.

12. Kinoshita et al., *Nyūsu no tanjō,* 175–77.

13. Benjamin as quoted and paraphrased in Eduardo Cadava, "Lapsus Imaginis: The Image in Ruins," *October,* no. 96 (2001): 36.

14. Eduardo Cadava, *Words of Light: Theses on the Photography of History* (Princeton, NJ: Princeton University Press, 1997), 23.

15. Gregory Clancey, *Earthquake Nation: The Cultural Politics of Japanese Seismicity, 1868–1930* (Berkeley: University of California Press, 2006), 7, 40.

16. The other major seismograph in Japan, which was at the Central Meteorological Station, was knocked out of commission by the quake.

17. Yamamoto had only recently been appointed to office after the sudden death of Prime Minister Katō Tomosaburō on 24 August 1923. He had not yet chosen his cabinet ministers, so the earthquake in effect caught Japan without a full government in place. Yamamoto acted quickly to make the necessary appointments, however. For a detailed discussion of the dispatch of armed forces into the devastated region, see Schencking, "1923 Tokyo as a Devastated War and Occupation Zone."

18. Janet Borland, "Capitalising on Catastrophe: Reinvigorating the Japanese State with Moral Values through Education Following the 1923 Great Kantō Earthquake," *Modern Asian Studies* 40, no. 4 (2006): 886; Peter Alexander Bates, "Fractured Communities: Class and Ethnicity in Representations of the Great Kanto Earthquake (Ph.D. diss., University of Michigan, 2006), 23–24; Edward Thomas Mack, "The Value of Literature: Cultural Authority in Interwar Japan (Ph.D. diss., Harvard University, 2002), 99.

19. Shinbun Shiryō Raiburarii, *Kantō Daishinsai,* 1–2.

20. Tokyo Municipal Office, *The Reconstruction of Tokyo* (Tokyo: Tokyo Municipal Office, 1933), 37.

21. Jane Gaines, "Political Mimesis," in *Collecting Visible Evidence,* ed. Jane Gaines and Michael Renov (Minneapolis: University of Minnesota Press, 1999), 92. The "Yomiuri Extract" reported that a private citizen, Mishima Akimichi (Shōdō), projected the Mombushō film at his house in Sendagaya for close friends on 5 December 1923. "Yomiuri shō" (Yomiuri Extract), *Yomiuri shinbun,* 5 December 1923, 7.

22. "Dead Are in Heaps in Tokio Streets: Path of Destruction Extends for 50 miles around the Capital—More Earthshocks Sunday," *New York Times,* 3 September 1923.

23. Corpses pictured at the Honjo Clothing Depot, cover, *Hōchi shinbun,* 5 September 1923, A.M. ed., 2. For a brief discussion of earthquake postcards and the depiction of corpses, see Satō Kenji, "Shinsai to shikaku media" (The Earthquake and Visual Media), in *Kantō Daishinsai to kiroku eiga: Toshi no shi to saisei,* ed. Tōkyō Daigaku Daigakuin Jinbun Shakai Kakari Kenkyūka (Tokyo: Tōkyō Daigaku Daigakuin Jinbun Shakai Kakari Kenkyūka 21 Seiki COE Puroguramu, 2004).

24. Mark Driscoll identifies a pervasive necrophilia that emerged in the years after the earthquake in the popular culture trend of "erotic-grotesque-nonsense" (*ero-guro-nansensu*), which he argues thrived on the perverse "neuropolitics" and "necropolitics" of Japanese imperialism and its exploitative economy of prostitution, pimping, drug dealing, and human trafficking. Mark Driscoll, *Absolute Erotic, Absolute Grotesque: The Living, Dead, and Undead in Japan's Imperialism, 1895–1945* (Durham, NC: Duke University Press, 2010).

25. This image continued to be well-known into the twentieth century. The left portion of this double-page spread was reproduced soon after the 1923 quake in Miyatake Gaikotsu, *Shinsai gahō* (Earthquake Disaster Pictorial), 5 November 1923 (Tokyo: Hankyōdō), 3:51. Reproduced in *Miyatake Gaikotsu chosakushū* (Collected Works of Miyatake Gaikotsu), ed. Tanizawa Eiichi and Yoshino Takao (Tokyo: Kawade Shobō Shinsha, 1986), 1:515.

26. The site was formerly occupied by a factory that manufactured uniforms for the armed forces, but the building had been demolished and the large open space was slated to be converted into a public park and playground.

27. Nihon Shinbun Hakubutsukan, ed., *Daishinsai to hōdō-ten,* 7.

28. Some color postcard versions are labeled "Marunouchi district refugees," not "Honjo Cloth-

ing Depot refugees," but it is clear that they are the same image. In several instances, an image, usually one without any distinctive landmarks, was labeled as different places, indicating the interchangeability of certain kinds of generic disaster scenes throughout the city as well as the tremendous effect that the identifying caption could have on the reading of the image.

29. Richard H. Mitchell, *Janus-Faced Justice: Political Criminals in Imperial Japan* (Honolulu: University of Hawaii Press, 1992). Newspaper coverage of the Ōsugi murders and the Kameido Incident did not appear until 25 September and 10 October, respectively. The leftist magazine *Kaizō* devoted a full special issue to Ōsugi in November 1923 in which writers also called for justice in the Kameido Incident.

30. Other papers, such as *Hōchi shinbun* (25 September 1923), also covered the murder with prominent photographs, including images of the boxes with their cremated remains (*Hōchi shinbun,* 26 September 1923). The details of the incident and a photograph of the well where the victims' bodies had been dumped were featured on the first two pages of the 8 October evening edition of the *Tōkyō nichinichi shinbun*. This paper also published images of the trial of the assailant Amakasu and other officers on 9 October and ran individual photographs of the victims of the Kameido Incident on 11 October.

31. For a sociological study of the rumors that arose in the absence or "information blank space" after the Great Kantō Earthquake, see Satō Kenji, "Kantō Daishinsaigo ni okeru shakai no henyō" (Social Change after the Great Kantō Earthquake), in *Ritsumeikan Daigaku Kanagawa Daigaku 21 Seiki COE Puroguramu Jointo Wākushoppu Hōhokusho: Rekishi saigai to toshi, Kyōto Tōkyō o chūshin ni* (Kanagawa: Kanagawa Daigaku 21 Seiki COE Puroguramu, 2007), 81–85.

32. J. Michael Allen, "The Price of Identity: The 1923 Kanto Earthquake and Its Aftermath," *Korean Studies* 20 (1996).

33. The role of the authorities is discussed extensively in Sonia Ryang, "The Great Kanto Earthquake and the Massacre of Koreans in 1923: Notes on Japan's Modern National Sovereignty," *Anthropological Quarterly* 76, no. 4 (2003).

34. Mark Caprio, *Japanese Assimilation Policies in Colonial Korea, 1910–1945,* Korean Studies of the Henry M. Jackson School of International Studies (Seattle: University of Washington Press, 2009).

35. Ryang, "The Great Kanto Earthquake and the Massacre of Koreans in 1923," 734.

36. This film was produced by Tōkyō Shinema Shōkai and the Ministry of Education in 1923. Shirai, *Kamera to jinsei,* 48–49.

37. Marita Sturken, *Tourists of History: Memory, Kitsch, and Consumerism from Oklahoma City to Ground Zero* (Durham, NC: Duke University Press, 2007), 7–9, 15–18, 31–32.

38. Bates, "Fractured Communities," 34–36, 37–38. For an extensive discussion of this publication, see Edward Thomas Mack, *Manufacturing Modern Japanese Literature: Publishing, Prizes, and the Ascription of Literary Value,* Asia-Pacific: Culture, Politics, and Society (Durham, NC: Duke University Press, 2010), 51–89.

39. Kimura Matsuo and Ishii Toshio, *Ehagaki ga kataru Kantō Daishinsai: Ishii Toshio korekushon* (The Story Postcards Tell of the Great Kantō Earthquake: The Ishii Toshio Collection) (Tokyo: Tsuge Shobō, 1990).

40. Sales of individual postcards of the Russo-Japanese War regularly amounted to hundreds

of thousands of copies. People were known to line up for hours to buy new issues. *Art of the Japanese Postcard: The Leonard A. Lauder Collection at the Museum of Fine Arts, Boston,* exhibition catalogue, ed. Anne Nishimura Morse, J. Thomas Rimer, and Kendall H. Brown (Boston, MA: MFA Publications, 2004). Bates notes that postcards were issued after the Shiga earthquake of 1910, the volcanic eruption of Sakurajima in 1914, and the Tokyo floods of 1917 ("Fractured Communities," 51). For a discussion of the postcard imaging of the Tokyo floods, particularly those in the Asakusa area, which was devastated again in the earthquake, see Hosoma Hiromichi, *Ehagaki no jidai* (The Modernity of Postcards) (Tokyo: Seidosha, 2006), 244–60.

41. From the website for the Institute for Dark Tourism Research, www.dark-tourism.org.uk/ (accessed 25 July 2007).

42. Newspapers had special terms for some of the prevalent kinds of disaster sightseeing/spectatorship (*kenbutsu*), from "fire ruin sightseeing" (*yakeato kenbutsu*) to "barrack sightseeing" (*barakku kenbutsu*). Renowned author Kawabata Yasunari published an often-read essay describing his experiences viewing horrific scenes of death in the capital titled "Taika kenbutsu" (Great Fire Sightseeing), *Bungei shunjū* 1, no. 11 (1923): 66–68; Bates, "Fractured Communities," 58. The term *yakeato kenbutsu* seems to have been in use in the press as far back as 1891, when it appears in relation to a fire in the Yoshiwara prostitution quarter. "Chinkago yakeato kenbutsu" (Ruin Sightseeing after the Fires Are Extinguished), *Yomiuri shinbun,* 24 January 1891, A.M. ed., 3.

43. By 15 September, the *Ōsaka asahi shinbun* was reporting that sightseers/spectators (*kenbutsunin*) were becoming obstacles to recovery and reconstruction and that entry to the city would again be restricted. "Nyūkyōsha sarani seigen, kenbutsunin no kōtsu sogai" (People Entering [the City] Are Restricted Again, Traffic Obstructions by Sightseers), *Ōsaka asahi shinbun,* 15 September 1923, P.M. ed., 2.

44. Bates, "Fractured Communities," 57–59, 110.

45. Japanese newspapers carried extensive lists of donors and the amounts they had given to relief operations. This served as an inducement for fellow citizens to contribute.

46. See Bates, "Fractured Communities"; Janet Borland, "Stories of Ideal Japanese Subjects from the Great Kantō Earthquake of 1923," *Japanese Studies* 25, no. 1 (2005); Borland, "Capitalising on Catastrophe."

47. *San Francisco Chronicle,* 4 September 1923, 22. Gendaishi no Kai, *Dokyumento Kantō Daishinsai* (Documents of the Great Kantō Earthquake) (Tokyo: Sōfūkan, 1983), n.p. Pictured in *Taiyō* 30, no. 1 (January 1924): 37. Foreign reports of the nature of the disaster were often wildly inaccurate. There were even reports that all of Japan had been destroyed. In the early hours and days of the disaster, reporters could not send out reliable information because of damage to telecommunications and bureaucratic delays in relaying wireless messages. Calls for relief were given priority.

48. Relief ships from Britain, France, and Italy arrived soon after the American fleet did. For a discussion of Japanese-American diplomacy and foreign relations in the wake of the Great Kantō Earthquake, see Hatano Masaru and Iimori Akiko, *Kantō Daishinsai to Nichi-Bei gaikō* (The Great Kantō Earthquake and Japanese-American Diplomacy) (Tokyo: Sōshisha, 1999).

49. "Coolidge Cables Sympathy to Emperor of Japan: Navy Orders Vessels to Yokohama for Relief," *New York Times,* 2 September 1923, 1.

50. Nihon Shinbun Hakubutsukan, *Daishinsai to hōdō-ten,* 37.

51. From the "San Francisco Relief Survey: The Organization and Methods of Relief Used After the Earthquake and Fire of April 18, 1906," conducted by the Russell Sage Foundation in 1913 and posted on the website of the Virtual Museum of the City of San Francisco, www.sfmuseum.org/conflag/relief1.html (accessed 2 July 2009). Daniel T. Oliver, "Charity Done Right," *World and I,* vol. 16 (January 2001), www.worldandi.com (accessed 2 July 2009). Racist concern about Japanese earthquake refugees immigrating to the United States also spurred American fund-raising for earthquake relief.

52. There are reliable recorded accounts of local authorities refusing foreign assistance and of embassy personnel being bewildered by the resentment directed at them by the Japanese, who seemed to view them as high-handed interlopers. In this atmosphere, the American embassy staff decided to funnel all U.S. aid through official Japanese channels—in effect, to put a Japanese face on the relief effort.

53. "Red Cross Asks $5,000,000," *New York Times,* 5 September 1923, 1.

54. Bates, "Fractured Communities," 28. For an account of the relief efforts of the Japanese community in Hawaii, see Hilo Times and Hawai-tō Nihonjin Imin Shiryōkan, *Kantō Daishinsai to Hawai / The Tokyo Earthquake, 1923* (The Great Kantō Earthquake and Hawaii) (Hawai-shō Hiroshi: Hawai-tō Nihonjin Imin Shiryōkan / Hiro Taimusu Shinbunsha, 1980).

55. In addition to widely distributing English-language publications on the postquake restoration efforts, the Japanese government expressed its appreciation for aid during and after the disaster on the tenth anniversary of the earthquake, 1 September 1933, by distributing to select organizations a limited edition of 584 commemorative bas-relief bronze plaques designed by well-known sculptor Saitō Sogan (1889–1974). The plaques were commissioned by the Home Ministry and presented by the prime minister and the cabinet. The side panels of the plaque triptych depict forlorn refugees fleeing and being assisted by a female allegorical figure (perhaps Worldwide Sympathy), and the central panel features three stalwart male workers gazing hopefully at the bright rising sun in the background, which heralds both the dawn of a new day and the resilience of Japan, the land of the rising sun. The plaque is titled "Memorial Plaque for the Reconstruction of the Imperial Capital" (*Teito fukkō kinen pai*), and the details of the dedication are inscribed on the back. A rare example of it survives in the United States in the collection of Helen Kulas, who kindly shared photographs of it with me. I am grateful to Mizusawa Tsutomu, Maemura Fumihiro, and Hamazaki Reiji for helping me identify Saitō as the artist and confirming the circumstances of the commission.

56. Kevin Rozario, "'Delicious Horrors': Mass Culture, the Red Cross, and the Appeal of Modern American Humanitarianism," *American Quarterly* 55, no. 3 (2003): 418–19.

57. Jeffrey Hughes, "Destroy & Reclaim: Artists and Disaster Sites," *New Art Examiner* 29, no. 5 (2002): 70.

3. DISASTER AS SPECTACLE

1. *Fire Seen in Okubo, 3 P.M.* is a provisional title rendered in an English-language publication. The same image appears in numerous publications with various descriptive titles in Japanese and English.

2. Paul Virilio, *The Vision Machine* (Bloomington: Indiana University Press, 1994).

3. Elizabeth Cowie, "The Spectacle of Actuality," in *Collecting Visible Evidence,* ed. Jane Gaines and Michael Renov (Minneapolis: University of Minnesota Press, 1999), 19.

4. A number of print collections (woodblocks and lithographs) were sold after the quake. Some of the major ones are: *Teito Daishinsai gahō* (Pictorial Account of the Great Earthquake in the Imperial Capital) (Tokyo: Tenshōdō, Shūgadō, and Urashimadō, 1923); *Tōkyō Daishinsai gahō* (Pictorial Account of the Great Tokyo Earthquake) (Tokyo: Shōbidō, Ishikawa Shōten, 1923); Hiratsuka Unichi, *Tōkyō Shinsai ato fūkei* (Sights after the Tokyo Earthquake Disaster) (Kobe: Hanga no Ie, 1924–25); Nishizawa Tekiho et al., *Taishō Shinkasai mokuhangashū zen sanjū rokkei* (Collection of Woodblock Prints of the Taishō Earthquake—Thirty-Six Scenes) (Tokyo: Gahōsha, 1924); *Taishō Shinsai gashū* (Collected Prints of the Taishō Earthquake) (Tokyo: Zue Kenkyūkai, 1926). Some of the works in these publications were first publicly exhibited as original drawings; selected pieces from some of the collections were also republished, usually in black and white, in books, which guaranteed them wider, sustained distribution.

5. Economist Joseph Schumpeter coined the term "creative destruction" in 1942 to describe the entrepreneurial and innovative nature of progress in a free-market, capitalist economy that constantly needs to destroy to create. Kevin Rozario, *The Culture of Calamity: Disaster and the Making of Modern America* (Chicago: University of Chicago Press, 2007), 110.

6. Christopher Lasch and Kevin Rozario have both argued that governments capitalized on disaster images to expand executive power, government secrecy, and the national security state. Lasch quoted in Charles R. Garoian and Yvonne Gaudelius, *Spectacle Pedagogy: Art, Politics, and Visual Culture* (Albany: State University of New York Press, 2008), 27; Rozario, *The Culture of Calamity,* 135–73.

7. Quoted in Garoian and Gaudelius, *Spectacle Pedagogy,* 24.

8. Fusae Kanda, "Behind the Sensationalism: Images of a Decaying Corpse in Japanese Buddhist Art," *Art Bulletin* 87, no. 1 (2005).

9. Spectacular depictions of bloody battle scenes with fiery death and destruction date back to at least the Kamakura period. Kitahara Itoko has explored the war and fire representations of the 1615 Summer Siege of Osaka (Ōsaka Udōshizu) as a form of disaster journalism in *Saigai jānarizumu: Mukashi-hen* (Disaster Journalism: Ancient Times), Rekihaku bukkuretto 21 (Sakura-shi: Rekishi Minzoku Hakubutsukan Shinkōkai, 2001), 16–20.

10. See the vivid picture galleries and explanations on the Massachusetts Institute of Technology Visualizing Cultures website, particularly the units authored by John Dower on "Throwing Off Asia," http://ocw.mit.edu/ans7870/21f/21f.027/home/vis_menu.html (accessed 10 April 2010). The website also includes many photographs of the Russo-Japanese War. Several excellent English-language publications have been issued on modern Japanese war prints. See, for example, Julia Meech-Pekarik and Metropolitan Museum of Art, *The World of the Meiji Print: Impressions of a New Civilization* (New York: Weatherhill, 1986); Elizabeth de Sabato Swinton, *In Battle's Light: Woodblock Prints of Japan's Early Modern Wars* (Worcester, MA: Worcester Art Museum, 1991); Frederic A. Sharf, Anne Nishimura Morse, and Sebastian Dobson, *A Much Recorded War: The Russo-Japanese War in History and Imagery* (Boston: MFA Publications, 2005).

11. Marco Gerbig-Fabel, "Photographic Artefacts of War 1904–1905: The Russo-Japanese War as Transnational Media Event," *European Review of History* 15, no. 6 (2008); Jon D. Carlson, "Postcards

and Propaganda: Cartographic Postcards as Soft News Images of the Russo-Japanese War," *Political Communication* 26, no. 2 (2009); Markus Nornes, *Japanese Documentary Film: The Meiji Era through Hiroshima* (Minneapolis: University of Minnesota Press, 2003), 4–6.

12. For a complete discussion of the news reportage on the eruption of Mount Bandai, see Kitahara Itoko, *Bandaisan funka: Saii kara saigai no kagaku e* (Volcanic Eruption of Mount Bandai: From Uncommon Disasters to Disaster Science) (Tokyo: Yoshikawa Kōbunkan, 1998); and the National Museum of Nature and Science's Earthquake Archive (Kokuritsu Kagaku Hakubutsukan Jishin Shiryō Shitsu) website, http://research.kahaku.go.jp/rikou/namazu/ (accessed 20 July 2011).

13. Rozario, *The Culture of Calamity,* 110.

14. Robert Hariman and John Louis Lucaites, *No Caption Needed: Iconic Photographs, Public Culture, and Liberal Democracy* (Chicago: University of Chicago Press, 2007), 244.

15. Gregory Clancey notes the existence of a fanciful print in the Nagoya City Museum Collection in "The Meiji Earthquake: Nature, Nation, and the Ambiguities of Catastrophe," *Modern Asian Studies* 40, no. 4 (2006): 924.

16. The overturned train, the destruction of trains more generally, and their repair for vital relief operations were all topics of tremendous interest, prominently featured in Dai Nihon Yūbenkai Kōdansha, *Taishō Daishinsai daikasai* (The Great Taishō Earthquake and Conflagration) (Tokyo: Dai Nihon Yūbenkai Kōdansha, 1923), 75–82.

17. On the tenth anniversary of the quake, in 1933, a series of spectacular, haunting photographs of the Nebukawa commuter train, number 109, at the bottom of the sea, taken by the Yomiuri newspaper, were exhibited at the Earthquake Memorial Hall in Honjo. The exhibit overlapped with the annual Buddhist ritual funerary services at the memorial hall to ensure the peaceful eternal rest of the dead. Following the exhibition, Yomiuri donated the photographs to the memorial for its permanent archive. "Kinendō ni kazaru Nebukawa kaitei shashin" (Underwater Photographs of Nebukawa That Decorate the Memorial Hall), *Yomiuri shinbun,* 1 September 1933, A.M. ed., 7.

18. Igawa trained in *ukiyo-e* printmaking under Inano Toshitsune in Osaka and Tomioka Eisen in Tokyo. In addition to providing regular illustrations for *Miyako shinbun,* he illustrated stories in Kōdansha's popular magazine *Kingu* (King).

19. Takehisa Yumeji specifically uses the Buddhist term for suffering from the *Lotus Sutra kugen* here to imply a kind of repeated karmic suffering. Yumeji did daily sketches for the *Yomiuri shinbun* beginning in 1907. In 1909, these images were collected into an illustrated volume that was a runaway bestseller, going through seven reprintings. His popularity and celebrity prompted him to open a storefront called the Minato-ya (Harbor Store) in the Nihonbashi neighborhood to sell goods with his signature designs. Takehisa Yumeji, Kawamura Karyō, and Yamamura Kōka, *Yumeji to Karyō, Kōka no Kantō Daishinsai rupo* (Yumeji, Karyō, and Kōka's Reportage of the Great Kantō Earthquake) (Tokyo: Kuresu Shuppan, 2003), 16.

20. Dora Apel, "Review: On Looking: Lynching Photographs and Legacies of Lynching after 9/11," *American Quarterly* 55, no. 3 (2003): 468.

21. Ibid., 469. See also Dora Apel, *Imagery of Lynching: Black Men, White Women, and the Mob* (New Brunswick, NJ: Rutgers University Press, 2004); Dora Apel and Shawn Michelle Smith, *Lynching Photographs,* Defining Moments in American Photography (Berkeley: University of California Press, 2007).

22. Kanda, "Behind the Sensationalism," 24.

23. Together with another pioneering watercolorist, Ōshita Tōjirō, Maruyama Banka was a founding member of the Nihon Suisaigakai Kenkyūjo (Research Center for the Association of Japanese Watercolor Painting) in 1907.

24. See the Massachusetts Institute of Technology Visualizing Cultures website, http://ocw.mit.edu/ans7870/21f/21f.027/throwing_off_asia_03/toa_vis_02.html (accessed 10 April 2010). James H. Hare, ed., *A Photographic Record of the Russo-Japanese War* (New York: P. F. Collier & Son, 1905). Both sides suffered enormous casualties at the lengthy siege of Port Arthur, which resulted in fifty-eight thousand Japanese and thirty-one thousand Russians dead, and the immense battle of Mukden, which is estimated to have caused eighty-five thousand Russian casualties and seventy thousand Japanese casualties. Nish quoted in Carlson, "Postcards and Propaganda," 217.

25. Bill Nichols, *Introduction to Documentary* (Bloomington: Indiana University Press, 2001), 39.

26. Don Slater, "Photography and Modern Vision: The Spectacle of 'Natural Magic,'" in *Visual Culture,* ed. Chris Jenks (London: Routledge, 1995), 218–38.

27. The prints are all signed "SK," but the identity of the print designer is not known. Only two printing companies capable of lithography were operational in September. By 20 October, from print number 9 until the end, the series shifted its production and printing to Urashimadō, also in Asakusa Kōen. Tōkyō-to Edo Tōkyō Hakubutsukan Toshi Rekishi Kenkyūshitsu, *Kantō Daishinsai to Ansei Edo jishin,* Tōkyō-to Edo Tōkyō Hakubutsukan Chōsa Hōkokusho, vol. 10 (Tokyo: Tōkyō-to Edo Tōkyō Hakubutsukan: Tōkyō-to Rekishi Bunka Zaidan, 2000), 200.

28. There are several known illustrated handscrolls of the Great Kantō Earthquake: Kiriya Senrin's (1877–1932) scroll, *Taishō Shinsai emaki* (Taishō Earthquake Disaster), is dated 6 October 1923. Nishimura Goun (1877–1938), a Kyoto-based *nihonga* painter and disciple of well-known master Takeuchi Seihō, created *Kantō Daishinsai emaki* (Scenes of the 1923 Earthquake), a series of monoscenic vignettes of the quake in ink and color mounted on a handscroll. The piece is very similar to works in the ink painting collections. Katayama Nanpū's (1887–1980) *Daishinsai jissha zukan* (Illustrated Handscroll Depicting Actual Scenes of the Great Earthquake, 1925), like Senrin's, more effectively uses the continuous compositional format of the unfolding handscroll to animate the pictorial narrative so that the viewer experiences the moment-by-moment sensations of the disaster experience. Hosono Masanobu, ed., *Katayama Nanpū,* Gendai nihonga zenshū, vol. 1 (Tokyo: Shūeisha, 1983), 108–10. Kayahara Hakudō's (1896–1951) handscroll is titled *Tōto Daishinsai kaganroku* (Visual Record of the Great Earthquake in the Great Eastern Capital). Kitahara Itoko discusses a range of media that were used to represent the earthquake, including Kayahara's handscroll, in "Egakareta Kantō Daishinsai: Emaki, hanga, sobyō" (The Great Kantō Earthquake Pictured: Illustrated Handscrolls, Prints, and Drawings), *Himoji shiryō kenkyū nenpō,* no. 6 (2010): 19. For an excellent reproduction of Kayahara's handscroll, see Kayahara Hakudō, *Kantō Daishinsai o egaku: Tōto daishinsai kaganroku emaki zen san-kan: Emaki manga kodomo no e tenji zuroku* (Picturing the Great Kantō Earthquake: All Three Scrolls of the Illustrated Handscroll of the Visual Record of the Great Earthquake in the Great Eastern Capital: Exhibition Catalogue of Handscrolls, Cartoons, and Children's Drawings), ed. Kanagawa Daigaku, Himoji Shiryō Kenkyū Sentā (Yokohama-shi: Kanagawa Daigaku Himoji Shiryō Kenkyū Sentā, 2010).

29. For an encyclopedic history of precinematic media, see Hermann Hecht, Ann Hecht, and

British Film Institute, *Pre-Cinema History: An Encyclopaedia and Annotated Bibliography of the Moving Image before 1896* (London: Bowker Saur, 1993).

30. S. K., *Actual Situation of the Large Whirlwind of Fire on Nakanomachi Street in Shin Yoshiwara (Shin Yoshiwara Nakanomachidōri takuhi daisenpū no jikkyō)*, lithograph, *Teito Daishinsai gahō* (Pictorial Account of the Great Earthquake in the Imperial Capital), Asakusa, Tenshōdō, Shūgadō, 30 September 1923.

31. This blend of "fact" and "fiction" parallels the intersplicing of the "actualities" (*jikkyō*) of the Russo-Japanese War represented in newsreels with dramatic reenactments. Nornes, *Japanese Documentary Film,* 6–10.

32. The prints in the series clearly display the censor's seal of approval on the lower left-hand corner, indicating official sanctioning of this narrative of the event.

33. Nagata Mikihiko, *Daichi wa furū* (The Earth Shakes) (Tokyo, Shunyōdō, 1923). Alex Bates analyzes this novel in detail and compares it with earthquake literary works in other genres in Peter Alexander Bates, "Fractured Communities: Class and Ethnicity in Representations of the Great Kanto Earthquake (Ph.D. diss., University of Michigan, 2006), 196.

34. This translation of the song appears in Soeda Azenbō, *A Life Adrift: Soeda Azembō, Popular Song and Modern Mass Culture in Japan,* trans. Michael Lawrence Lewis, Routledge Contemporary Japan Series (New York: Routledge, 2008), 224–26. The cover and some of the music booklet titled "Songs of the Great Earthquake," which includes four songs, is reproduced on the National Museum of Japanese History's website, www.rekihaku.ac.jp/e-rekihaku/109/index.html (accessed 1 March 2010).

35. This subgenre of disaster story was known as *ibun,* or strange stories.

36. This is one of many cross-references between Buddhist subject matter and biblical imagery, which were already prevalent among painters from the Meiji period.

37. The complete series is loosely dated to after 1923. Maruyama Banka (1867–1942), Ogawa Senyō (1886–1971), Nakazawa Hiromitsu (1874–1964), Nakamura Fusetsu (1866–1943), Okamoto Ippei (1886–1948), Watanabe Shinya (1875–1950), Nagano Sōfu (1885–1949), Hattori Ryōei (1877–1955), Izawa Sosui (1870–?), and Ikeda Eiji (1889–1950) each produced a series of paintings.

38. Nihon Mangakai, *Daishinsai gashū* (Pictorial Collection of the Great Earthquake Disaster) (Tokyo: Kanao Bunendō, 1923).

39. Tōkyō-to Edo Tōkyō Hakubutsukan Toshi Rekishi Kenkyūshitsu, *Kantō Daishinsai to Ansei Edo jishin,* 48.

40. The emerging nation-state and its newly minted imperial universities inaugurated a host of professional art academies, including the esteemed Tokyo School of Fine Arts (Tōkyō Bijutsu Gakkō, founded in 1887), as a critical part of the country's cultural institutional infrastructure, which already included a number of art ateliers. In 1907, the Ministry of Education began to sponsor a prestigious annual juried art salon (Monbushō Bijutsu Tenrankai, or Bunten, for short, renamed the Imperial Art Academy Exhibition [Teikoku Bijutsuin Tenrankai, or Teiten] in 1919), where artists from across the nation competed for highly coveted prizes."Japanese-style painting" (*nihonga*) is a neologism of the Meiji period that represents a modern amalgamation of a range of traditional painting practices, from ink painting to polychrome painting in mineral pigments. *Nihonga* sometimes incorporated elements of Western-style painting such as perspectival rendering and shading.

41. Kōka's painting *Matsuchiyama* was also pictured in the *Tokyo nichinichi shinbun,* 30 October 1923, 7, along with the painting *The Feeling of Reality at Atagoyama* (*Atagoyama ue no jikkan*) by artist Itō Shinsui. Both works were included in the Earthquake Disaster Exhibition (Shinsai-ten) held in Tokyo that month.

42. The entire series is reproduced in Takehisa, Kawamura, and Yamamura, *Yumeji to Karyō, Kōka no Kantō Daishinsai rupo.*

43. Asai Ryōi's account was subsequently republished throughout the Edo period and became well known to later generations as the definitive account of this tragic calamity that destroyed three-quarters of the shogunal capital. Excerpt translated in James L. McClain, "Edobashi: Power, Space, and Popular Culture in Edo," in *Edo and Paris: Urban Life and the State in the Early Modern Era,* ed. James L. McClain, John M. Merriman, and Kaoru Ugawa (Ithaca, NY: Cornell University Press, 1994), 105–6.

44. Takehisa, Kawamura, and Yamamura, *Yumeji to Karyō, Kōka no Kantō Daishinsai rupo,* 66–67.

45. Nihon Mangakai, *Daishinsai gashū.* This publication was a catalogue of works exhibited by artists in mid-November 1923 at the Osaka Mitsukoshi department store. Ninety of the works were assembled within ten days after the quake. Many of the cartoonists who exhibited in the show regularly worked for major news organizations. A graduate in *nihonga* painting from the illustrious Tokyo School of Fine Arts (Tōkyō Bijutsu Gakkō, now Tōkyō Geijutsu Daigaku or the Tokyo School of Fine Arts and Music), Mizushima did illustrations for the *Ōsaka asahi shinbun,* and Miyao Shigeo worked for the *Chūō shinbun.* Kitahara Itoko, "Egakareta Kantō Daishinsai," 29–30.

46. Takehisa, Kawamura, and Yamamura, *Yumeji to Karyō, Kōka no Kantō Daishinsai rupo,* 12.

47. *Teito fukkō* (Reconstruction of the Imperial Capital), dir. Oda Hamatarō (Tokyo: Fukkō Kyoku [Reconstruction Agency], 1930), silent, 107 minutes.

48. *Hōchi shinbun,* 5 September 1923, 1. On the front page of the 7 September 1923 issue of the *Tōkyō nichinichi shinbun,* the tower is pictured with the caption "The Twelve Stories That Broke with a Snap" (*Bokkiri oreta jūnikai*), Tano Yasunori Collection.

49. The elevators were taken out of commission within six months of the building's opening because of safety concerns.

50. Hosoma Hiromichi, *Asakusa Jūnikai: Tō no nagame to "kindai" no manazashi* (Asakusa Twelves Stories: The View from the Tower and the Gaze of "Modernity") (Tokyo: Seidosha, 2001). The tower was designed by Scottish engineer W.K. Burton in the late 1880s. The structure had already been damaged by a strong earthquake in 1894 and was reinforced with steel girders.

51. Born in Saga prefecture on the southern island of Kyūshū, Okada Saburōsuke was also a graduate of the Tokyo School of Fine Arts who studied Western-style painting under renowned master Kuroda Seiki, later becoming a professor at his alma mater in 1896. In 1891, he became a member of the conservative Western-style painting society the Meiji Art Association (Meiji Bijutsu Gakkai), but he later joined Kuroda in founding the acclaimed progressive academic group the White Horse Society (Hakubakai). From 1897 to 1902, Okada studied painting in France under the sponsorship of the Ministry of Education; there, he learned from the late academic painter Raphael Collin, who excelled at painting *en plein air.* Collin also taught Okada the Western technique of engraving. In 1931, Okada founded the Japan Print Association (Nihon Hanga Kyōkai) and became its first chairman. In 1937, he received the Order of Cultural Merit. Saga Kenritsu Bijutsukan, *Nihon kindai*

yōga no eiga: Okada Saburōsuke (The Splendor of Japanese Modern Western-Style Painting: Okada Saburōsuke) (Saga: Saga Kenritsu Bijutsukan, 1993), 60.

52. A dramatic "strange and real story" in the Kōdansha volume told of three visitors who were sightseeing in Asakusa and survived the collapse of the Twelve Stories when they suddenly fell from the upper stories of the falling skyscraper into the pond below. Dai Nihon Yūbenkai Kōdansha, *Taishō Daishinsai daikasai,* 181.

53. After studying with leading Western-style academic painters Koyama Shōtarō and Asai Chū, Nakamura went to Paris from 1901 to 1905, where he worked under academician Jean-Paul Laurens at the Académie Julian. When he returned to Japan, Nakamura joined the conservative artists' group the Pacific Painting Association (Taiheiyō Gakkai), successor to the group founded by Koyama and Asai, the Meiji Art Association, which was a declared opponent of the White Horse Society (Hakubakai), Kuroda and Okada's more colorful offshoot group. A frequent exhibitor at the annual salon and later a member of the imperial academy committee, Nakamura was well known as a history painter and a calligrapher. He also served as school principal for the Pacific Painting Academy (Taiheiyō Bijutsu Gakkō).

54. Kawasaki Shōko studied under Kobori Tomoto, a mentor to many acclaimed *nihonga* painters in the early twentieth century. He then graduated from the *nihonga* division of the Tokyo School of Fine Arts in 1910 and had several works accepted into the national annual salon early in his career. He also regularly served as a salon juror. This print collection was based on a series of original sketches by six artists exhibited in Tokyo in October 1923. Records indicate that there were two versions of the print portfolio, one with thirty-six images and the other with twenty-five. The set was published sometime after January 1924. Nishizawa Tekiho et al., *Taishō Shinkasai mokuhangashū zen sanjū rokkei* (Collection of Woodblock Prints of the Taishō Earthquake—Thirty-Six Scenes) (Tokyo: Gahōsha, 1924). The choice of the number thirty-six undoubtedly referred to Katsushika Hokusai's famous *meisho-e* series of the Edo period depicting thirty-six views of Mount Fuji.

55. David Campany, *Photography and Cinema* (London: Reaktion, 2008), 27–28.

56. Jane Gaines discusses director Eisenstein's term "emotive vibration" in her analysis of sensual documentaries that evoke bodily responses to political mimesis. Jane Gaines, "Political Mimesis," in *Collecting Visible Evidence,* ed. Jane Gaines and Michael Renov (Minneapolis: University of Minnesota Press, 1999), 92.

57. Cowie, "The Spectacle of Actuality," 19.

58. *Kantō taishin taika jikkyō* (Actual Conditions of the Great Kantō Earthquake and Fire) (Tokyo: Monbushō Shakai Kyoikuka, Tōkyō Shinema Shōkai, 1923), silent, 64 minutes. Tōkyō Daigaku Daigakuin Jinbun Shakai Kakari Kenkyūka, ed., *Kantō Daishinsai to kiroku eiga: Toshi no shi to saisei* (The Great Kantō Earthquake an Documentary Film: The Death and Regeneration of the City) (Tokyo: Tōkyō Daigaku Daigakuin Jinbun Shakai Kakari Kenkyūka 21 Seiki COE Puroguramu, 2004), 11–12. Tochigi Akira, "Daishinsai ga eiga hyōgen ni motarashita mono" (Forms of Expression That the Great Kantō Earthquake Brought to Film), in *Kantō Daishinsai to kiroku eiga: Toshi no shi to saisei,* ed. Tōkyō Daigaku Daigakuin Jinbun Shakai Kakari Kenkyūka (Tokyo: Tōkyō Daigaku Daigakuin Jinbun Shakai Kakari Kenkyūka 21 Seiki COE Puroguramu, 2004), 39.

59. Shirai Shigeru, *Kamera to jinsei: Shirai Shigeru kaikoroku* (Camera and Life: The Memoirs of Shirai Shigeru) (Tokyo: Kikuchi Tokuji: Yuni Tsūshinsha, 1983), 46–49.

60. Translated in Soeda Azenbō, *A Life Adrift,* 218–20.

61. Quoted in Jeffrey T. Schnapp and Matthew Tiews, "Introduction: A Book of Crowds," in *Crowds*, ed. Jeffrey T. Schnapp and Matthew Tiews (Stanford, CA: Stanford University Press, 2006), x. See also Jeffrey T. Schnapp, "Mob Porn," in *Crowds*, 1–12.

4. THE SUBLIME NATURE OF RUINS

Epigraph: The poem, translated by Janet Borland, was written by a student at the Kiyoshima Primary School in Asakusa ward. It was originally published in a collection of essays written by children who survived the earthquake and fires: Tōkyō-shi, *Shinsai kinen bunshū: Tōkyō shiritsu shōgakkō jidō* (Earthquake Commemorative Essay Collection: Children of the Tokyo Municipal Elementary Schools), ed. Tōkyō-shi Gakumuka (Tokyo: Baifūkan, 1924), 3:353. Janet Borland translates and discusses the poem in "Makeshift Schools and Education in the Ruins of Tokyo, 1923," *Japanese Studies* 29, no. 1 (May 2009): 131.

1. Born in Okayama prefecture, Kanokogi went to the United States in 1900 and then to Paris until 1904 to study oil painting. Upon his return to Japan, he moved back to the Kansai region and opened the academic Western-style painting school Kansai Bijutsuin (Kansai Art Academy) in Kyoto. For a selection of Kanokogi's sketches, see Tōkyō-to Gendai Bijutsukan, *Tōkyō toshi to geijutsu: Kindai toshi to geijutsu-ten, 1870–1996 (Tokyo, La ville moderne: Visions urbaines d'artistes et d'architectes)* (Tokyo: Tōkyō-to Gendai Bijutsukan, 1996), 25–27. Kanokogi exhibited his sketches in April 1924, and postcards of the works were produced for the show as publicity and souvenirs. Artworks from exhibitions were often reproduced as postcards for viewers to take home. Ikeda Yōson, Kanokogi's younger colleague who accompanied him on sketching trips, recorded in his memoirs that during their sketching outings they were met with extreme anger and hostility by people on the street and felt constantly threatened by violence until they were able to secure press armbands from a friend at the *Asahi shinbun.* The Tokyo city government invited Yōson to exhibit some of his sketches at its 1924 Earthquake Reconstruction Exhibition (Shinsai Fukkō tenrankai) in Ueno. Kitahara Itoko, "Egakareta Kantō Daishinsai: Emaki, hanga, sobyō" (The Great Kantō Earthquake Pictured: Illustrated Handscrolls, Prints, and Drawings), *Himoji shiryō kenkyū nenpō,* no. 6 (2010): 26, 28, 30.

2. The first public showing of original drawings and sketches by the Risōsha group of *nihonga* artists and printmakers took place just a month after the quake at the Earthquake Sketch Exhibition (Shinsai Suketchi-ten) in October 1923. The show included works by Nishizawa Tekiho, Kiriya Senrin, Kawasaki Shōko, Tamura Saiten, and Isoda Chōshū. In December, a selection of these works was published in a book by Gahōsha: Risōsha, ed., *Shinsai suketsuchi gashū* (Collection of Sketches of the Earthquake) (Tokyo: Gahōsha, December 1923). Then followed a commercial portfolio collection of original woodblock prints titled *Collection of Woodblock Prints of the Taishō Earthquake—Thirty-Six Scenes.* The practice and extent of sketching after the earthquake are discussed in Ozaki Masato, "Kiroku ka? Sozai ka? Bijutsuka ni totte no Daishinsai" (Archive? Material? The Great Kantō Earthquake for Artists), *Gekkan bijutsu* 21, no. 5 (1995). Ozaki tries to correlate extant sketches with written accounts in artists' memoirs to ascertain whether the drawings were actual records (reportage) of their visual experiences or imaginative re-creations using the earthquake as inspiration. He concludes that many of the representations were produced from direct observation, although later image compilations tended to adopt the themes in earlier ones, indicating that some tropes became codified.

3. Nihon Shinbun Hakubutsukan, ed., *Daishinsai to hōdō-ten: Kantō daishinsai 80-shūnen kikaku* (Exhibition of the Great Earthquake and the News: Plans for the 80th Anniversary of the Great Kantō Earthquake) (Yokohama-shi: Nihon Shinbun Hakubutsukan, 2003), 28. *True Pictures of the Tokyo Calamity* is reproduced in Takehisa Yumeji, Kawamura Karyō, and Yamamura Kōka, *Yumeji to Karyō, Kōka no Kantō Daishinsai rupo* (Yumeji, Karyo, and Kōka's Reportage of the Great Kantō Earthquake) (Tokyo: Kuresu Shuppan, 2003), 3–43. Yumeji also published his personal account of the experience as *Hensai zakki* (Journal of the Unusual Disaster). Discussed in Gendaishi no Kai, *Dokyumento Kantō Daishinsai* (Documents of the Great Kantō Earthquake) (Tokyo: Sōfūkan 1983), 226.

4. Takehisa, Kawamura, and Yamamura, *Yumeji to Karyō, Kōka no Kantō Daishinsai rupo,* 42–43. "The design favored by the actor Kikugorō" refers to *Kikugorō-gōshi,* which the famous Kabuki actor Onoe Kikugorō III had created himself.

5. I borrow the term "magnificent destruction" from French writer Bernardin de Saint-Pierre, who used it in the chapter titled "The Pleasure of Ruin" in his 1796 book *Studies of Nature.* Discussed in Donald J. McNutt, *Urban Revelations: Images of Ruin in the American City, 1790–1860,* Literary Criticism and Cultural Theory (New York: Routledge, 2006), 1–2.

6. Charles Morris, ed., *The San Francisco Calamity by Earthquake and Fire . . . Told by Eye Witnesses, Including Graphic and Reliable Accounts of All Great Earthquakes and Volcanic Eruptions in the World's History, and Scientific Explanations of Their Causes* (Philadelphia: W. E. Scull, 1906), 30.

7. In *Taishō Shinsai gashū* (Collected Prints of the Taishō Earthquake) (Tokyo: Zue Kenkyūkai, 1926). Like many of his colleagues who produced works for this print portfolio collection, Kiriya was a *nihonga* division graduate of the Tokyo School of Fine Arts and actively worked as a painter of Buddhist subjects.

8. Iain Hampsher-Monk (1998), "Edmund Burke," in E. Craig, ed., *Routledge Encyclopedia of Philosophy,* www.rep.routledge.com/article/DB011SECT1 (accessed 9 December 2009). Edmund Burke, *A Philosophical Enquiry into the Origin of Our Ideas of the Sublime and Beautiful* (London: J. Dodsley, 1777), 58–59.

9. Paul Crowther (1998), "The Sublime," in E. Craig, ed., *Routledge Encyclopedia of Philosophy,* www.rep.routledge.com/article/M040 (accessed 10 December 2011).

10. Some prominent examples of such images of ruin are: Reimeisha Henshūbu, *Shinsai gafu;* Kojima Chūshū, *Shinsai suketchi: Kinen gachō* (Earthquake Sketches: Commemorative Sketchbook), vol. 2 (Tokyo: Banchō Shoin, November and December 1923); Kojima Chūshū, *Shinsai kinen gachō* (Earthquake Disaster Commemorative Sketchbook) (Tokyo: Banchō Shoin, March 1924).

11. A graduate of the sculpture division of the Tokyo School of Fine Arts, Shimizu was a prominent sculptor and prolific illustrator.

12. Reimeisha Henshūbu, *Shinsai gafu,* 88–89.

13. Ibid., 86. For the term "mournful skeletons," see Comte de Volney, *Ruins, or Meditations on the Revolution of Empires* (1791), quoted in McNutt, *Urban Revelations,* 1.

14. Christopher Woodward, *In Ruins* (London: Chatto & Windus, 2001), 108–59.

15. Odawarajo, *Hanga ni miru Kantō Daishinsai: Taishō Shinkasai mokuhangashū zen sanjū rokkei* (The Great Kantō Earthquake Seen in Prints: Collection of Woodblock Prints of the Taishō Earthquake—Thirty-Six Scenes) (Odawara: Odawarajo, 1998), 12.

16. The House of Prints (Hanga no Ie) in Kobe, owned by Yamaguchi Hisayoshi (publisher of

the premier quarterly print portfolio for the creative prints movement, *Hanga* [Print], from 1924 to 1930), published and sold Unichi's print series by subscription. The series had approximately three hundred subscribers. A picture of the promotional pamphlet for the set, including an order form, appears in Chiba-shi Bijutsukan, *Nihon no hanga,* vol. 2, *1911–1920: Kizamareta "ko" no kyōen* (Japanese Prints, 1911–1920: A Banquet of Carved "Individuality"), exhibition catalogue (Tokyo: Tōkyō Shinbun, 1999), 135.

17. Born in Tokyo, Tekiho was a well-established *nihonga* painter who studied with Araki Kanpo and regularly exhibited at the annual salon. He was best known for his detailed paintings of dolls, which he extensively researched and collected.

18. Mark Dery, "A Terrible Beauty," *Print* 58, no. 1 (2004): 54, 59.

19. Reimeisha Henshūbu, *Shinsai gafu,* 82–83.

20. The owner of the residence in Shōko's image, Yasuda Zenjirō (1838–1921), was the founder of the powerful Yasuda financial group (*zaibatsu*), the fourth largest of the financial cartels that dominated the Japanese economy until the end of the Asia-Pacific War. Yasuda had already been associated with sensationalized tragedy in the news when he was assassinated almost exactly two years before the quake, in September 1921, by a right-wing zealot who saw him as a symbol of the corrupting influence of business on government.

21. Saijō Yaso, *Aa Tōkyō: Sanbun, shishū* (Ah Tokyo! Tales of Tragedy Poetry Collection) (Tokyo: Kōransha, 1923). Reproduced in Wada Hirofumi, ed., *Kantō Daishinsai* (The Great Kantō Earthquake), vol. 26 of *Korekushon modan toshi bunka* (Tokyo: Yumani Shobō, 2004), 4–201.

22. Kawabata was associated with the White Birch Group (Shirakabaha). A famous 1918 portrait of him by renowned Western-style painter Kishida Ryūsei resides in the collection of the National Museum of Modern Art, Tokyo.

23. *Abtei im Eichenwald* is in the collection of the Berlin Schloss Charlottenburg. On Friedrich's juxtaposition of life and death in his work, see Linda Siegel, "Synaesthesia and the Paintings of Caspar David Friedrich," *Art Journal* 33, no. 3 (1974): 200–201.

24. Stereoscopic photographs of the earthquake are in the Collection of Prints and Photographs, Library of Congress, Washington, DC. The caption quotations are taken from objects 14918 and 14920, respectively.

25. Morris, ed., *The San Francisco Calamity by Earthquake and Fire,* xi.

26. Kevin Rozario, *The Culture of Calamity: Disaster and the Making of Modern America* (Chicago: University of Chicago Press, 2007), 73.

27. The dynamiting of the Twelve Stories was captured on film in Tōkyō Shinema Shōkai, *Kantō taishin taika jikkyō* (Actual Conditions of the Great Kantō Earthquake and Fire) (Tokyo: Monbushō, 1923), silent, 64 minutes.

28. Frank Lloyd Wright and Bruce Brooks Pfeiffer, *Letters to Architects* (Fresno: Press at California State University, Fresno, 1984), 39.

29. Interestingly, Tekiho seems to have drawn his image of the tangled power lines from a photograph of the Ginza that ran in the *Asahi shinbun* and then combined it with the distant image of Asakusa.

30. Gregory Clancey, *Earthquake Nation: The Cultural Politics of Japanese Seismicity, 1868–1930* (Berkeley: University of California Press, 2006), 15, 116.

31. Ibid., 17.

32. Ibid., 16.

33. Reimeisha Henshūbu, *Shinsai gafu,* 48.

34. Odawarajo, *Hanga ni miru Kantō Daishinsai,* 11.

35. Yoneyama notes that ill-informed government statements by the prime minister and the minister of agriculture and commerce urging insurers to pay victims without regard for the legality of their exclusion clauses created undue pressure on insurance companies. Yoneyama Takau, *The Great Kanto Earthquake and the Response of Insurance Companies: A Historical Lesson on the Impact of a Major Disaster* (Tokyo: Hitotsubashi University, 2008). Kaizōsha, *Taishō Daishin kasaishi* (Chronicle of the Great Taishō Earthquake and Fire) (Tokyo: Kaizōsha, 1924), 104–21. In her study of Japanese media after the earthquake, Satō Miya notes that the problem of disaster insurance was one of the most heavily discussed topics in newspapers. Satō Miya, "Media no naka no fukkō: Kantō Daishinsaigo no shakai ishiki to tenrankai" (Reconstruction in the Media: Social Consciousness and Exhibitions of the Great Kantō Earthquake), *Jinmin no rekishigaku,* no. 178 (2008): 25.

36. One *sun* equals 3.03 cm.

37. The Dōjunkai Aoyama Apartments in Tokyo's Shibuya ward were built after the earthquake in 1927 as Japan's first ferro-concrete multifamily housing. Tōkyō Shisei Chōsakai, *Nihon no kindai o dezain shita senkusha: Seitan 150-shōnen kinen Gotō Shinpei-ten zuroku = Goto Shinpei, Designer of Modern Japan* (Tokyo: Tōkyō Shisei Chōsakai, 2007), 120. Kaizōsha, *Taishō Daishin kasaishi,* 45–61.

38. *Shinsai isshūnen kinen* (One-Year Anniversary of the Earthquake), special edition, *Jiji shinpō,* 1 September 1924, 1.

39. Back cover of Ōsaka Asahi Shinbunsha, *Daishinsai shashin gahō* (Photographic Pictorial of the Great Earthquake), vol. 3 (Osaka: Ōsaka Asahi Shinbunsha, 7 October 1923).

5. RECLAIMING DISASTER

1. Dai Nihon Yūbenkai Kōdansha, *Taishō Daishinsai daikasai* (The Great Taishō Earthquake and Conflagration) (Tokyo: Dai Nihon Yūbenkai Kōdansha, 1923), 6.

2. This can also be translated as "circumstances being what they are."

3. Timothy P. Brown, "Trauma, Museums and the Future of Pedagogy," *Third Text* 18, no. 4 (2004): 254.

4. Allen H. Barton, *Communities in Disaster: A Sociological Analysis of Collective Stress Situations* (Garden City, NY: Doubleday, 1969); Kai Erikson, *A New Species of Trouble: Explorations in Disaster, Trauma, and Community* (New York: W. W. Norton, 1994), 236; Erikson, "Notes on Trauma and Community," in *Trauma: Explorations in Memory,* ed. Cathy Caruth (Baltimore: Johns Hopkins University Press, 1995), 189.

5. Kitahara Itoko, "Egakareta Kantō Daishinsai: Emaki, hanga, sobyō" (The Great Kantō Earthquake Pictured: Illustrated Handscrolls, Prints, and Drawings), *Himoji shiryō kenkyū nenpō,* no. 6 (2010): 26.

6. The statue was designed by the celebrated Western-style sculptor Takamura Kōun.

7. Nihon Shinbun Hakubutsukan, ed., *Daishinsai to hōdō-ten: Kantō daishinsai 80-shūnen kikaku* (Exhibition of the Great Earthquake and the News: Plans for the 80th Anniversary of the Great Kantō Earthquake) (Yokohama-shi: Nihon Shinbun Hakubutsukan, 2003), 27. Ōsaka Asahi Shin-

bunsha, *Daishinsai shashin gahō* (Photographic Pictorial of the Great Earthquake), vol. 1, cover illustration (Osaka: Ōsaka Asahi Shinbunsha, 15 September 1923). A similar message center was created on the equestrian statue of imperial hero Kusunoki Masashige outside the Imperial Palace.

8. The image appeared in Miyatake Gaikotsu, *Shinsai gahō* (Earthquake Disaster Pictorial) (Tokyo: Hankyōdō, 25 December 1923),1:5; *Miyatake Gaikotsu chosakushū* (Collected Works of Miyatake Gaikotsu), ed. Tanizawa Eiichi and Yoshino Takao (Tokyo: Kawade Shobō Shinsha, 1986), 3:469; Kawasaki-shi Shimin Myūjiamu, *Nihon no manga 300-nen* (300 Years of Japanese Cartoons) (Kawasaki-shi: Kawasaki-shi Shimin Myūjiamu, 1996), 99, 173.

9. Image 14914, Prints and Photographs Collection, Library of Congress, Washington, DC. The photograph also hangs in Japan's Reconstruction Memorial Museum (Fukkō Kinenkan) in the Ryōgoku area of Tokyo.

10. Gregory Clancey, *Earthquake Nation: The Cultural Politics of Japanese Seismicity, 1868–1930* (Berkeley: University of California Press, 2006), 131–33.

11. Japan Naimushō Shakaikyoku and Fujisawa Morihiko, *The Great Earthquake of 1923 in Japan* (Tokyo: Naimushō Shakaikyoku, 1926), n.p.

12. Erikson, *A New Species of Trouble,* 236.

13. Tokyo Municipal Office, *The Reconstruction of Tokyo* (Tokyo: Tokyo Municipal Office, 1933), 85.

14. See, for example, Dai Nihon Yūbenkai Kōdansha, *Taishō Daishinsai daikasai,* 104–69. Narita Ryūichi, "Kantō Daishinsai no meta hisutorii no tame ni: Hōdō, aiwa, bidan" (A Metahistory of the Great Kantō Earthquake: Reportage, Tragic Tales, and Heroic Tales), *Shisō,* no. 866 (August 1996).

15. "Food Riots in Japanese Capital; Millions Homeless and Starving—Police Are Compelled to Charge Hungry Mobs with Swords—New Reports Increase the Total Havoc Wrought by Earthquake, Tidal Waves and Fires," *New York Times,* 4 September 1923.

16. Ikeda is considered one of the premier Japanese cartoonists of the early twentieth century and is particularly celebrated for his regular contributions to the Sunday supplement of the *Yomiuri shinbun.* He was also an accomplished painter who exhibited at the annual salon.

17. A few other paintings that depict the murderous rampage against Koreans have come to light in recent years, although these works were produced privately and do not seem to have been publicly exhibited at the time. See Kitahara Itoko, "Egakareta Kantō Daishinsai," 19; Kayahara Hakudō, *Kantō Daishinsai o egaku: Tōto daishinsai kaganroku emaki zen san-kan: Emaki manga kodomo no e tenji zuroku* (Picturing the Great Kantō Earthquake: All Three Scrolls of the Illustrated Handscroll of the Visual Record of the Great Earthquake in the Great Eastern Capital: Exhibition Catalogue of Handscrolls, Cartoons, and Children's Drawings), ed. Kanagawa Daigaku, Himoji Shiryō Kenkyū Sentā (Yokohama-shi: Kanagawa Daigaku Himoji Shiryō Kenkyū Sentā, 2010).

18. *Taishō Shinsai gashū* (Collected Prints of the Taishō Earthquake) (Tokyo: Zue Kenkyūkai, 1926).

19. Jin-hee Lee argues that while the government took swift action to prosecute select individuals for illegal vigilante activities, primarily to forestall international criticism of the massacres, it never took responsibility for instigating the insidious rumors that spurred the violence. Jin-hee Lee, "The Enemy Within: Earthquake, Rumours and Massacre in the Japanese Empire," in *Violence: "Mercurial Gestalt,"* ed. Tobe Levin (Amsterdam: Rodopi, 2008). Yamada Shōji has argued that by placing the blame for the massacre on unnamed, and ultimately unpunished, rogue vigilantes, the

Japanese government effectively whitewashed the event and never admitted the central role played by the officials who issued decrees about the potential violence of Koreans. Yamada Shōji, *Kantō Daishinsaiji no Chōsenjin gyakusatsu: Sono kokka sekinin to minshū sekinin* (The Great Kantō Earthquake and the Massacre of Koreans: National Responsibility and the Responsibility of the People) (Tokyo: Sōshisha, 2003), 8–12.

20. Yumeji's image was reproduced again in the book *Shinsai suketchi gashū* (Collection of Sketches of the Earthquake) published in December 1923 with the new title *Hibiya Observation* (*Hibiya shoken*). Risōsha, ed., *Shinsai suketsuchi gashū* (Collection of Sketches of the Earthquake) (Tokyo: Gahōsha, December 1923). Reproduced in Takehisa Yumeji, Kawamura Karyō, and Yamamura Kōka, *Yumeji to Karyō, Kōka no Kantō Daishinsai rupo* (Yumeji, Karyō, and Kōka's Reportage of the Great Kantō Earthquake) (Tokyo: Kuresu Shuppan, 2003), 14–15.

21. Jin-hee Lee notes that there were numerous surveys of children's quake experiences published in the years following the disaster. Lee, "The Enemy Within," 190, 201nn9–10.

22. *Daijishin, Ansei no Daijishin to Kantō Daishinsai* (Great Earthquakes, the Great Ansei Earthquake and the Great Kantō Earthquake), special issue, *Taiyō* 2 (1977): 66; Tōkyō-shi and Tōkyō-shi Gakumuka, ed., *Shinsai kinen bunshū: Tōkyō shiritsu shōgakkō jidō* (Earthquake Commemorative Essay Collection: Children of the Tokyo Municipal Elementary Schools), vol. 6 (Tokyo: Baifūkan, 1924). The children's drawings were exhibited at the Jichi Kaikan in Ueno. "Itamashii inshō kara jidō ga egakitsukutta shinsai kinen sakuhin tenrankai" (Exhibition of Earthquake Commemorative Works Drawn by Children from Their Sad Impressions), *Tōkyō asahi shinbun,* 28 February 1924, 10.

23. Yanase Masamu, "Jijoden" (Autobiography), *Kirkos: Musashino Bijutsu Daigaku Shiryō Toshokan nyūsu*, no. 2 (October 1990): 2, 7.

24. Reimeisha Henshūbu, *Shinsai gafu: Gaka no me* (Earthquake Sketchbook: The Eye of the Artist) (Tokyo: Reimeisha, December 1923), 76–77; Wada Hirofumi, ed., *Kantō Daishinsai* (The Great Kantō Earthquake), Korekushon modan toshi bunka, vol. 26 (Tokyo: Yumani Shobō, 2004), 560–61.

25. Miyatake Gaikotsu, *Shinsai gahō* (Earthquake Disaster Pictorial) (Tokyo: Hankyōdō, 10 October 1923), 2:37; *Miyatake Gaikotsu chosakushū,* 3:499.

26. Nihon Mangakai, *Daishinsai gashū* (Pictorial Collection of the Great Earthquake Disaster) (Tokyo: Kanao Bunendō, 1923), n.p.

27. Before becoming a full-time professional cartoonist, Kitazawa Rakuten studied Western-style painting. He was broadly conversant in international trends in cartooning, following popular comic strips like *The Yellow Kid* by Richard Felton Outcault and *The Katzenjammer Kids* by Rudolph Dirks. He is considered to be one of the first professional modern cartoonists who used the terms *manga* and *mangaka* in Japan. While working for Jiji Shinpōsha, in 1905, he also founded the important satirical *manga* journal *Tokyo Puck,* which had an international readership. He opened one of the first cartoon schools in Japan and taught some of the best-known cartoonists of the era, including Shimokawa Hekoten and Okamoto Ippei. Shimizu Isao, *Jiji manga: Taishōki* (*Jiji Manga:* Taishō Period), vol. 6 of *Manga zasshi hakubutsukan* (Tokyo: Kokusho Kankōkai, 1986).

28. *Jiji manga* 140, 9 December 1923, 1.

29. Japan Naimushō Shakaikyoku and Fujisawa Morihiko, *The Great Earthquake of 1923 in Japan,* 11–12.

30. *Jiji manga* 133, 14 October 1923, 3.

31. Janet Borland, "Capitalising on Catastrophe: Reinvigorating the Japanese State with Moral Values through Education following the 1923 Great Kantō Earthquake," *Modern Asian Studies* 40, no. 4 (2006): 887; Haruno Ogasawara, "Living with Natural Disasters: Narratives of the Great Kantō and the Great Hanshin Earthquakes (Ph.D. diss., Northwestern University, 1999), 88–99.

32. Jelle Zeilinga de Boer and Donald Theodore Sanders, *Earthquakes in Human History: The Far-Reaching Effects of Seismic Disruptions* (Princeton, NJ: Princeton University Press, 2005), 99–101.

33. *Jiji shinpō,* 17 October 1923, P.M. ed., 3.

34. Translated in Ogasawara, "Living with Natural Disasters," 89–90. This statement refers to the highly publicized double suicide of novelist Arishima Takeo and his lover in June 1923, just a few months before the quake.

35. Dated autumn 1923 and signed with Shibusawa's pen name "Seien." This text was based on a passage in *The Great Learning* (*Daigaku*), one of the four main books of Confucianism. Dai Nihon Yūbenkai Kōdansha, *Taishō Daishinsai daikasai,* 4. I am grateful to Koide Izumi, Inoue Jun, and Kimura Masato for their assistance in confirming this quote by Shibusawa. For a discussion of Shibusawa's central role in the postquake reconstruction effort, see Zaidan Hōjin Shibusawa Eiichi Kinen Zaidan and Shibusawa Shiryōkan, eds., *Shibusawa Eiichi to Kantō Daishinsai: Fukkō e no manazashi* (Shibusawa Eiichi and the Great Kantō Earthquake: Looking toward Reconstruction) (Tokyo: Shibusawa Shiryōkan, 2010).

36. Quoted in Borland, "Capitalising on Catastrophe," 888.

37. Quoted in ibid., 889.

38. Gladys C. Hansen and Emmet Condon, *Denial of Disaster* (San Francisco: Cameron, 1989), 48.

39. Janet Borland, "Stories of Ideal Japanese Subjects from the Great Kantō Earthquake of 1923," *Japanese Studies* 25, no. 1 (2005): 25.

40. Dai Nihon Yūbenkai Kōdansha, *Taishō Daishinsai daikasai,* 3. I am extremely grateful to John Carpenter for his assistance in translating this text.

41. Japan Naimushō Shakaikyoku and Fujisawa Morihiko, *The Great Earthquake of 1923 in Japan.*

42. The first Peace Preservation Law was enacted in 1925. It criminalized any activity that threatened to alter the Japanese national polity (*kokutai no henkaku*). In 1928, the law was revised, and the crime was elevated to a capital offense.

43. Kusakabe Shirōta, "Jishin to namazu to kanameishi ni tsuite" (Concerning Earthquakes and Catfish and the Protective Keystone), in *Taishō Daishinsai daikasai,* ed. Dai Nihon Yūbenkai Kōdansha (Tokyo: Dai Nihon Yūbenkai Kōdansha, 1923).

44. Ibid.

45. "Heeded Myth on Disaster," *New York Times,* 5 September 1923.

46. Peter Alexander Bates, "Fractured Communities: Class and Ethnicity in Representations of the Great Kanto Earthquake (Ph.D. diss., University of Michigan, 2006), 26–27.

47. Borland, "Stories of Ideal Japanese Subjects from the Great Kantō Earthquake of 1923," 22.

48. Charles Morris, ed. *The San Francisco Calamity by Earthquake and Fire . . . Told by Eye Witnesses, Including Graphic and Reliable Accounts of All Great Earthquakes and Volcanic Eruptions in the World's History, and Scientific Explanations of Their Causes* (Philadelphia: W. E. Scull, 1906), xi.

49. Ultimately, the Kantō quake bore out Imamura's predictions of an imminent earthquake rather than Ōmori's belief in a longer cycle of recurrence. Imamura and his calls for immediate

earthquake preparedness were vindicated after years of professional difficulties and obstacles created by Ōmori, who died soon after the quake. Imamura replaced him as chair of seismology at Tokyo Imperial University. Imamura's "Stories of Earthquakes" (*Jishin no hanashi*) is featured in Dai Nihon Yūbenkai Kōdansha, *Taishō Daishinsai daikasai,* 246–54.

50. Kevin Rozario, *The Culture of Calamity: Disaster and the Making of Modern America* (Chicago: University of Chicago Press, 2007), 15–17.

51. Clancey, *Earthquake Nation,* 216.

52. Fujimori Terunobu, "Kon Wajirō to Barakku Sōshokusha" (Kon Wajirō and the Barrack Decoration Company), *Quarterly Column,* no. 88 (1983): 59.

53. There were also large barrack communities in Ueno Park, Hibiya Park, and on the grounds of the Hama Detached Imperial Palace (Hama Rikyū) near Shinbashi.

54. Born in Okayama in 1871, Tokunaga studied Western-style painting with a student of early master Goseda Horyū I named Matsubara Sangorō. He moved to Tokyo in 1888 to study Western-style painting under Goseda Horyū II, after which he opened his own studio to teach drawing and watercolor painting in 1893. In 1901, he was hired by *Yorozu chōhō* to supervise the newspaper's illustrated journal and worked there until 1911, when he left for France with fellow artist Mizutani Kunishirō to study with prominent French academic master Jean-Paul Laurens at the Académie Julian, where he stayed until 1914. After returning to Japan, he exhibited his work at the annual national salon and with the Kōfukai group. Since he was active in fund-raising for earthquake relief aid after the Kantō disaster, he was commissioned to produce twenty-five large-scale oil paintings for the Earthquake Memorial Hall (Shinsai Kinendō). Takashina Shūji, "Akademii Jurian to Tokunaga Hitoomi," *Kindai gasetsu,* no. 5 (March 1997).

55. Murayama Tomoyoshi and Hon no Izumisha, *Murayama Tomoyoshi gurafikku no shigoto* (The Graphic Art of Murayama Tomoyoshi) (Tokyo: Hon no Izumisha, 2001), 41.

56. Borland, "Stories of Ideal Japanese Subjects," 24.

57. According to David Ambaras, this pervasive public discourse, drawing on sociological data from Europe and the United States, decisively shaped the Japanese public imagination of slums and other perceived marginal spaces in the city. David Ambaras, "Topographies of Distress," in *Noir Urbanisms,* ed. Gyan Prakash (Princeton, NJ: Princeton University Press, 2010), 190, 201. Government bureaucrats were keenly aware of the housing problems and shortages in the capital and tried to enact legislation for slum reform (starting in 1909) and housing creation (in 1919). Despite this concern, however, they were not able to make much progress in alleviating the problems. See Jordan Sand, *House and Home in Modern Japan: Architecture, Domestic Space, and Bourgeois Culture, 1880–1930* (Cambridge, MA: Harvard University Asia Center, 2003), 170–74.

58. Dai Nihon Yūbenkai Kōdansha, *Taishō Daishinsai daikasai,* 198.

59. Alice Tseng, "Kuroda Seiki's Morning Toilette on Exhibition in Modern Kyoto," *Art Bulletin* 90, no. 3 (2008): 431.

60. Miyatake Gaikotsu, "Saisho ni kōkan sareta hakurai no rataiga" (The First Published Nude Paintings Imported from Abroad), *Meiji kibun* (Tokyo: Hankyōdō, March 1925), 2:25; *Miyatake Gaikotsu chosakushū,* 1:85.

61. Miyatake Gaikotsu, *Shinsai gahō* (Earthquake Disaster Pictorial) (Tokyo: Hankyōdō, 15 January 1924), 5:90–91; *Miyatake Gaikotsu chosakushū,* 3:554–55.

62. Takehisa, Kawamura, and Yamamura, *Yumeji to Karyō, Kōka no Kantō Daishinsai rupo*, 100–101.

63. Edgar Degas and Richard Kendall, *Degas by Himself: Drawings, Prints, Paintings, Writings* (Boston: Little, Brown, 1987), 311.

64. Miyatake Gaikotsu, *Shinsai gahō*, 4:1; *Miyatake Gaikotsu chosakushū*, 3:535.

65. Academically trained at the Tokyo School of Fine Arts, Okamoto Ippei was an innovative force in the cartoon world, developing and refining the form of cartoons and comic strips (*manga manbun*) in his work for the *Asahi shinbun* and other widely circulated publications.

66. Okamoto Ippei, "Fukkō ni arawaretaru yūmoa jūdai" (Ten Humorous Topics Appearing during Reconstruction), *Taiyō* 30, no. 1 (1924): 103.

67. *Jiji manga* 133, 14 October 1923, 2.

68. Kitazawa Rakuten and Nagasaki Batten, *Rakuten manga shutaisei* (Survey of Manga Works by Rakuten) (Omiya-shi: Kitazawa Rakuten Kenshōkai, 1973), 2:19.

69. Reimeisha Henshūbu, *Shinsai gafu*, 54–55. Trained in both *nihonga* and Western-style painting at the well-known private art atelier Kawabata Gakkō under the tutelage of academic master Fujishima Takeji, Shimizu had a long, illustrious career as a cartoonist and illustrator, first for the *Yorozu chōhō* newspaper until 1922 and then for a range of magazines published by Kōdansha, such as *King* (*Kingu*) and *Boy's Club* (*Shōnen kurabu*).

70. See Gennifer Weisenfeld, "Designing after Disaster: Barrack Decoration and the Great Kantō Earthquake," *Japanese Studies* 18, no. 3 (1998).

71. *Tōkyō nichinichi shinbun*, 29 October 1923, P.M. ed., 2.

72. "Shinsaigo no shinshokugyō: Ude o furū zekkō no kikai" (New Occupations after the Earthquake: They Skillfully Display Their Abilities, the Best Machine), *Chūō shinbun*, 6 March 1924, 3.

73. Kon was a graduate of the Tokyo School of Fine Arts, where he concentrated his studies on design while also pursuing a course in architecture under the prominent architect Okada Shinichirō. After graduating in 1912, Kon was hired to teach at Waseda University's newly founded architecture department, run by Satō Kōichi, where he stayed for most of his academic career. The company listed its initial members as Nakagawa Kigen (1892–1972), Kambara Tai (1898–1997), Asano Mōfu (1900–84), Yokoyama Junnosuke (1903–71), Yoshida Kenkichi (1897–1982), Yoshimura Jirō (1899–1942), Ōtsubo Shigechika (1899–1998), Asuka Tetsuo (1895–1997), Tōyama Shizuo (1895–1986), and Kon. Yoshida, Kon's primary collaborator, was a multitalented artist, graphic designer, and stage designer who had graduated from the design section of the Tokyo School of Fine Arts. Very little is known about the members of the Forefront Company: Asuka, Ōtsubo, and Tōyama. Their collegial relationship with Kon through art school was the basis for their association with the company. Records describe Tōyama as a specialist in lighting design working for the Imperial Theater (Teikoku Gekijō). "Barakku no bijutsuka 3" (Barrack Artists, part 3), *Miyako shinbun*, 29 October 1923, 1.

74. "Gaitō ni deru gakkatachi: Akushonsha to Dōjinsha ga" (Artists Out on the Street: Action and Dōjinsha), *Tōkyō asahi shinbun*, 2 October 1923, 3; "Atorie kara gairo e" (From the Atelier to the Streets), *Miyako shinbun*, 9 October 1923, 5.

75. Ogura Uichirō (1881–1962), who joined the company soon after its formation, used the expression. "Botsubotsu arawareru geijutsuteki no barakku" (The Artistic Barracks That Are Appearing

Little by Little), *Miyako shinbun,* 14 October 1923, 7. Fujimori Terunobu, "Ginza no toshi ishō to kenchikukatachi" (The Urban Design of Tokyo and Architects), in *Ginza modan to toshi inshō,* ed. Shiseido Gyararii (Tokyo: Shiseido, 1993), 19–26.

76. "Botsubotsu arawareru geijutsuteki no barakku," 7. The company's other projects are discussed in Weisenfeld, "Designing after Disaster."

77. Yukihiko Shigenobu, "Through the Eyes of a Good Deed Investigator," in *Frontiers of Social Research: Japan and Beyond,* ed. Akira Furukawa, Advanced Social Research Series (Melbourne: Trans Pacific Press, 2007).

78. Photograph reproduced in *Asahi Graph,* 14 November 1923, 4.

79. Sōga Takaaki, "Taishō makki ni okeru shinkō geijutsu undō no kōsatsu: Zōkei bijutsu to kenchiku no kakawari o megutte" (Thoughts on the New Art Movement of the Late Taishō Period: On the Relationship between the Plastic Arts and Architecture) (master's thesis, Waseda University, 1990), 53, 64.

80. Fujimori, "Kon Wajirō to Barakku Sōshokusha," 60.

81. Asuka Tetsuo, "Shigoto no ato de kanjita koto" (What I Felt after the [Barrack] Job), *Mizue,* no. 226 (1923): 22.

82. Kambara Tai, "Barakku Sōshokusha no shimei to watashi no kibō" (The Mission of the Barrack Decoration Company and My Hopes), *Mizue,* no. 225 (1923): 25.

83. Kon Wajirō, "Soboku to iroiro no bi" (Simplicity and Various Aesthetics), *Mizue,* no. 225 (1923).

84. Kon Wajirō, "Poketto no naka no ni en yon jū hachi sen" (The Two Yen and Forty-Eight Sen in My Pocket), *Mizue,* no. 226 (1923): 25.

85. Naitō Sadako, "Teidai Setorumento o miru" (Looking at the Tokyo Imperial University Settlement), *Fujin no tomo* 18, no. 8 (1924): 96–99. The ground plan and elevation for the project are reproduced in Fujimori, "Ginza no Toshi Ishō to Kenchikukatachi," 24–25.

86. Naitō, "Teidai Setorumento o miru," 96.

87. Mary Beard, "The Supreme Test of the Japanese Nation," *Shūkan asahi* 4, no. 25 (1923): 20.

88. Ōtsubo Shigechika, "Shigoto no nikki" (Diary of a Job), *Mizue,* no. 226 (1923).

89. Shūkan Asahi, ed., *Nedanshi nenpyō: Meiji, Taishō, Shōwa* (A Chronological History of Prices: Meiji, Taishō, Shōwa) (Tokyo: Asahi Shinbunsha, 1988), 161, 98.

90. Ōtsubo, "Shigoto no nikki," 24; Yoshida Kenkichi, "Kyatatsu no ue de" (On the Top of the Ladder), *Mizue,* no. 226 (1923): 22; "Barakku no sōshoku ga mirairashii moyō" (Barrack Decoration Has Futuristic Design), *Jiji shinpō,* 25 October 1923.

91. Saitō Sogan and Hinako Jitsuzō, "Barakku kenbutsu: Kasō no Ginza to Asakusa (6)" (Barrack Sightseeing: The Disguise of Ginza and Asakusa, part 6), *Yomiuri shinbun,* 11 March 1924, 7. *Barakku kenchiku* (Barrack Architecture [vol. 1]), Kenchiku shashin ruijū, vol. 4, no. 12 (Tokyo: Kōyōsha, 1923), ills. 35, 36.

92. Unfortunately, the morning after Café Kirin was completed, half of it was destroyed in a fire ignited by sparks from an adjacent building.

93. Omuka Toshiharu, *Taishōki shinkō bijutsu undō no kenkyū* (Research on the Taishō-Era Avant-Garde Art Movements) (Tokyo: Sukaidoa, 1995), 295, 98; Saitō and Hinako, "Barakku kenbutsu: Kasō no Ginza to Asakusa (6)," 7.

94. Omuka, *Taishōki shinkō bijutsu undō no kenkyū,* 292.

95. Ibid., 301.

96. Kon Wajirō, "Sōshoku geijutsu no kaimei" (Clarification of Decoration Art), *Kenchiku shinchō* 5, no. 2 (1924).

97. Kishida Hideto, "Sōanten shokan (kenchiku)" (Impressions of the Exhibition of Plans [for the Reconstruction of the Imperial Capital] [Architecture]), *Kenchiku shinchō* 5, no. 6 (1924): 2.

98. For a full discussion of Mavo's work and the group's other earthquake-related projects, see Gennifer Weisenfeld, *Mavo: Japanese Artists and the Avant-Garde, 1905–1931,* Twentieth-Century Japan (Berkeley: University of California Press, 2002), 77–91, 143–58.

6. RECONSTRUCTION'S VISUAL RHETORIC

Epigraph: The "Edokko" mentioned in the lyrics are native Edo-Tokyoites, generally from the low city, whose families had lived in the city for at least three generations. Although some sources cite the composer as "anonymous," this song has been attributed to Soeda Azenbō, who also claimed credit for composing it in his memoir. It can be heard on a CD recording of works by Soeda Tomomichi, Saeki Toshio, and Orimoto Yoshikazu titled *Taishō no hayariuta minshū no kidoairaku to tomo ni* (Popular Songs of the Taishō Period) (Tokyo: King Record Co., 2008). The song has also been attributed to Azenbō's son Tomomichi (commonly known as Soeda Satsuki) and Shibuya Hakurui. In the lyrics, the refrain "*ara-ma, oya-ma*" means literally "Oh, what's that?" but it is used more for its rhythmical cadence than for meaning, so I have left it in the original Japanese. The same is true of "*eezo eezo,*" which translates as "Hey, it's all right, it's all right." Imagawa cakes, named after the Kanda Imagawabashi Bridge in Edo near which they were first sold in the eighteenth century, are a popular Japanese dessert made with sweet batter on the outside and sweet bean paste on the inside. They are often served at festivals. The first character in the name Shintarō means "shake" (*furueru*) and is the same as the character *shin* in the word for "earthquake" (*jishin*). The song suggests that parents gave their children the name to play on this double meaning of the character. Similarly, the names Shinji, Shinsaku, and Shinko play on the use of the character "shake" in regular names. The girl's name Fukuko uses the character meaning "restore" or "return" (*fuku*), which is the first character in "reconstruction" (*fukkō*).

1. Carola Hein, "Resilient Tokyo: Disaster and Transformation in the Japanese City," in *The Resilient City,* ed. Lawrence Vale and Thomas Campanella (New York: Oxford University Press, 2005), 214.

2. Lawrence Vale and Thomas Campanella, eds., *The Resilient City* (New York: Oxford University Press, 2005), 9.

3. For general studies of the architectural and urban outcomes of the reconstruction process, see Edward Seidensticker, *Tokyo Rising: The City Since the Great Earthquake* (New York: Knopf, 1990); Koshizawa Akira, *Tōkyō no toshi keikaku* (Urban Planning of Tokyo) (Tokyo: Iwanami Shoten, 1991); Narita Ryūichi, "Kantō Daishinsai no meta hisutorii no tame ni: Hōdō, aiwa, bidan" (A Metahistory of the Great Kantō Earthquake: Reportage, Tragic Tales, and Heroic Tales), *Shisō,* no. 866 (August 1996); Jeffrey Hanes, "Urban Planning as an Urban Problem: The Reconstruction of Tokyo after the Great Kanto Earthquake," *Seisaku kagaku* 7, no. 3 (2000); Yoshimi Shunya, "Taishō bunka kenkyū to modaniti no bunka seiji" (Research on Taishō Culture and the Political Culture of Modernity), in *Kakudaisuru modaniti: 1920–30-nendai,* vol. 6 of *Kindai Nihon no*

bunkashi (Tokyo: Iwanami Shoten, 2002); Andre Sorensen, *The Making of Urban Japan: Cities and Planning from Edo to the Twenty-First Century* (London: Routledge, 2002); Hein, "Resilient Tokyo"; J. Charles Schencking, "Catastrophe, Opportunism, Contestation: The Fractured Politics of Reconstructing Tokyo Following the Great Kantō Earthquake of 1923," *Modern Asian Studies* 40, no. 4 (2006); Tōkyō Shisei Chōsakai, *Nihon no kindai o dezain shita senkusha: Seitan 150-shōnen kinen Gotō Shinpei-ten zuroku / Goto Shinpei, Designer of Modern Japan* (Tokyo: Tōkyō Shisei Chōsakai, 2007).

Only in the past few years have scholars in Japan begun to broach this topic from the standpoint of visual media. See Satō Miya, "Media no naka no fukkō: Kantō Daishinsaigo no shakai ishiki to tenrankai" (Reconstruction in the Media: Social Consciousness and Exhibitions of the Great Kantō Earthquake), *Jinmin no rekishigaku,* nos. 175 (2007) and 178 (2008); Takano Hiroyasu, "'Shinsai no kioku' no hensen to tenji" (Exhibition and the Vicissitudes of "Earthquake Memory"), *Himoji shiryō kenkyū nenpō,* no. 6 (2010).

4. This is one of the three types of master narratives of remembrance identified by Edward Linenthal. Vale and Campanella, eds., *The Resilient City,* 14, 338.

5. I use the term *remembrance* to emphasize broadly the commemorative mechanisms of collective and individual memory production in visual, spatial, and material forms as well as in discursive and social practices. Daniel Sherman's superb study of memory in interwar France lays out the general historiography of memory studies and the theoretical issues regarding the formation of historical consciousness in a society's relationship to the past as constituted in the construction of "sites of memory." His definition of commemoration is interchangeable with my use of remembrance: "Commemoration mobilizes a variety of discourses and practices into a representation of an event or epoch; this representation contains within it a social and cultural vision it casts as inherent in the 'memory' of the commemorated event." Sherman, *The Construction of Memory in Interwar France* (Chicago: University of Chicago Press, 1999), 6. My understanding of the stakes of memory production and remembrance is indebted to Andreas Huyssen's study of the postmodern obsession with memory and musealization in the context of the contemporary topography of memory politics. He sees the intense contemporary inclination toward preservation escalating with the global, technological world's compression of time and space and its steady erosion of stable geographical cultural boundaries. These trends produce a fear of forgetting that drives the impulse to remember, although Huyssen questions whether anything is actually ever remembered. Andreas Huyssen, *Present Pasts: Urban Palimpsests and the Politics of Memory* (Stanford, CA: Stanford University Press, 2003), 1–10.

6. Tōkyō Shisei Chōsakai, *Nihon no kindai o dezain shita senkusha,* 94.

7. James C. Scott, *Seeing Like a State: How Certain Schemes to Improve the Human Condition Have Failed,* Yale Agrarian Studies (New Haven, CT: Yale University Press, 1998).

8. See in particular Yoshimi Shunya's section "Teito fukkō no naka no modan toshi" (The Modern City in the Reconstruction of the Imperial Capital) in his "Taishō bunka kenkyū to modaniti no bunka seiji," 12–47.

9. Tokyo Municipal Office, *The Reconstruction of Tokyo* (Tokyo: Tokyo Municipal Office, 1933), 271.

10. Japanese bureaucrats and urban planners expressed concern about improving and regulating the development of Tokyo as early as the 1880s, particularly to facilitate the capital's economic development and make it less susceptible to fire, but real urban planning reform did not come until

enactment of the City Planning Law and the Urban Building Law in 1919. While these laws provided future plans for urban infrastructure development, they were less concerned with problems of housing or urban amenities. At the same time, a range of social reformers, including novelist Mori Ōgai, wrote compellingly about the need for urban planning to improve the welfare and hygiene of the working classes. Inspired by reforms in Germany, Home Ministry bureaucrats took up the cause of social welfare, although a number of historians have interpreted their interest as less about altruistic social concern and more about the realpolitik need to reduce the potential for class conflict and maintain social stability. Sorensen, *The Making of Urban Japan,* 87, 92–96.

11. Gotō had also established a third-party institution dedicated to the research of Tokyo's government called the Tokyo Institute of Municipal Research (Tōkyō Shisei Chōsakai).

12. Hanes, "Urban Planning as an Urban Problem," 129.

13. A budget diagram of Gotō's original plan appears in Uchiyama Zenzaburō and Tōkyō Shisei Chōsakai, eds., *Teito fukkō jigyō taikan* (General Survey of the Imperial Capital Reconstruction Project) (Tokyo: Nihon Tōkei Fukyūkai,1930), 1:ch9, 14. The two volumes of this survey went into multiple printings.

14. Tōkyō Shisei Chōsakai, *Nihon no kindai o dezain shita senkusha,* 80, 85.

15. Gotō is quoted in Schencking, "Catastrophe, Opportunism, Contestation," 834.

16. Toyo Nakamura and John Dolan, "Early Pop Song Writers and Their Backgrounds," *Popular Music* 10, no. 3 (1991): 266. I am grateful to Hosokawa Shuhei for bringing this parody to my attention.

17. *Taishō Chic: Japanese Modernity, Nostalgia, and Deco,* ed. Kendall H. Brown and Sharon Minichiello, exhibition catalogue (Honolulu: Honolulu Academy of Arts, 2001), 166.

18. The council's members included statesmen, business leaders, and leaders of the two main political parties in the lower house. For a detailed discussion of the various government entities that were formed to plan for Tokyo's reconstruction and of the internal politics of reconstruction, see Schencking's excellent account, "Catastrophe, Opportunism, Contestation," 854–57. Schencking also discusses the debates that took place in the council for and against Gotō's bureau's plans.

19. Among those on the Imperial Capital Reconstruction Consultative Committee, Tokyo Imperial University Professor Sano Toshikata was head of the architectural division, Ikeda Hiroshi was in charge of planning, and Naoki Rintarō was supervisor of civil engineering. Carola Hein, "Hiroshima: The Atomic Bomb and Kenzo Tange's Hiroshima Peace Center," in *Out of Ground Zero: Case Studies in Urban Reinvention,* ed. Joan Ockman (Munich: Prestel / New York: Temple Hoyne Buell Center for the Study of American Architecture, Columbia University, 2002), 4.

20. Gotō was known in Japan and around the world as the "Japanese Roosevelt." The *Los Angeles Times* noted, "Because Goto has a strong, resourceful manner, loving power and influence, the Japanese have come to name him 'the Japanese Roosevelt.'" "Jap Cabinet Is Balanced: Two Strong Men of Empire in Terauchi's Confidence; Policy toward United States Is Well Received; Declaration of Premier Has Pleased the People," *Los Angeles Times,* 26 November 1916. This reputation also related to his general position outside party politics and his strong-handed approach in colonial rule and urban planning in Formosa (Taiwan) when he was governor-general. I am very grateful to Ed Balleisen for helping me elucidate this reference. The expression "*uma no ashi*" ("Horse legs, inexperienced actor") in the cartoon refers to a bad and inexperienced actor, drawing on the fact that a

less skilled actor plays the role of a horse's hind legs in kabuki theater. The reference to Prime Minister Yamamoto as "Kagoshima-ya" alludes to the leader's home town of Kagoshima in Kyūshū. It is probably being used like a *yago* from kabuki, which was a shop name associated with an actor that audience members often shouted out in appreciation during performances. "Don't drop your glasses" refers to Gotō's signature pince-nez. *Kakaa denka* ("petticoat government") is a disparaging expression about strong-willed women running a household; it is also a synonym for gynarchy. The text, literally, "White radish [*daikon*]! Sweet potato [*satsumaimo*]!" invokes a common kabuki epithet called out by the audience to bad actors [*daikon*], and "sweet potato" is probably a reference to Yamamoto, since Kagoshima Prefecture was the old province of Satsuma, and the word for sweet potato in Japanese is *satsumaimo*.

21. Schencking, "Catastrophe, Opportunism, Contestation," 836.

22. *Sojō no teito fukkōan* plays on the expression *sojō no sakana*, which meant to be helpless (as a fish on a cutting board).

23. Sorensen, *The Making of Urban Japan*, 110.

24. *Jiji manga* 140, 9 December 1923, 2. The acronym *naishō* for home minister is substituted for the near homophone *naisho*, meaning confidence or secret, giving the term *naishō banashi* a double meaning.

25. Tōkyō Shisei Chōsakai, *Nihon no kindai o dezain shita senkusha*, 92.

26. *Asahi Graph*, 5 December 1923. Reproduced in Asahi Shinbunsha, *Asahi gurafu ni miru Shōwa zenshi (Taishō 12nen)* (Early Shōwa History Seen in *Asahi Graph*: Taishō 12) (Tokyo: Asahi Shinbunsha, 1975), 1:239.

27. Hein, "Resilient Tokyo," 213, 30. Andre Sorensen disagrees with Hein on this point and affirms the successful modernizing impact of the reconstruction plans, saying that they "provided much of the infrastructure that has supported Tokyo's growth until the present." He also notes that the reconstruction program established the use of land readjustment for prospective urban planning projects (*The Making of Urban Japan*, 129, 31). Sorensen's position concurs with Koshizawa Akira's influential historical account designating the postquake reconstruction as a major success. See, for example, Koshizawa's *Tōkyō no toshi keikaku*.

28. Only Hanes questions the "rational" logic of urban planning, in "Urban Planning as an Urban Problem"

29. According to the municipal government, land adjustment was "the work of promoting and increasing, in a rational way, the utilizability of the residential ground as contained in a given plot of land; by effecting the exchange and combinations, or making alterations in the division and shape, or again by carrying out a transformation or change with respect to the existing roads and ways, rivers and canals, or public parks, etc." Tokyo Municipal Office, *The Reconstruction of Tokyo*, 237. The term meant both relocation of property owners to other sites and, more commonly, slight adjustments in plot boundaries to regularize plot sizes and even out the general disposition of the layout of the city to accommodate the expansion of roads and the creation of public spaces. For the most part, the process involved more fine-tuning than wholesale appropriation of land. Land adjustment generally reduced plots by 9 percent or less, a strategic calculation arising from a Japanese law requiring the government to compensate landowners if it commandeered more than 10 percent of their property.

30. Hein, "Resilient Tokyo," 221.

31. For a detailed discussion of the final reconstruction scheme and annual expenditures, see Tokyo Municipal Office, *The Reconstruction of Tokyo,* 237.

32. Kitazawa Rakuten, *Jiji manga* 144, 1 January 1924, 1.

33. Kevin Rozario, "Making Progress: Disaster Narratives and the Art of Optimism in Modern America," in Vale and Campanella, eds., *The Resilient City,* 41.

34. Nagata was mayor of Tokyo from 29 May 1923 to 8 September 1924 and again from 30 May 1930 to 25 January 1933. In the interim, Ichiki Otohiko (1872–1956) and Horikiri Zenjirō (1884–1979) served.

35. Tōkyō Shisei Chōsakai, *Nihon no kindai o dezain shita senkusha,* 96, 80. Among the surviving pamphlets are *Kukaku seiri ni tsuite shimin shokun ni tsugu* (To the Citizens about Land Boundary Readjustment, 1923) and *Teito shimin shokun ni tsugu* (To the Citizens of the Imperial Capital, 1930). The latter was published by the Tōkyō Jichi Kaikan, the Self-Government Hall, which Gotō had helped establish in his push to strengthen local autonomy and the consciousness of self-government among Tokyo's citizenry. Uchiyama and Tōkyō Shisei Chōsakai, eds., *Teito fukkō jigyō taikan,* vol. 2.

36. The Reconstruction Agency (Fukkō Kyoku) was dissolved on 1 April 1930 and succeeded by the Reconstruction Secretariat (Fukkō Jimukyoku), which was closed on 1 April 1932, marking the end of official reconstruction work. Tōkyō Shisei Chōsakai, *Nihon no kindai o dezain shita senkusha,* 111. Publication of this pamphlet was announced in "Kukaku seiri hayawakari," *Tōkyō asahi shinbun,* 16 June 1924, A.M. ed., 4.

37. Scott, *Seeing Like a State.*

38. Satō Kenji, "Shinsai to shikaku media" (The Earthquake and Visual Media), in *Kantō Daishinsai to kiroku eiga: Toshi no shi to saisei,* ed. Tōkyō Daigaku Daigakuin Jinbun Shakai Kakari Kenkyūka (Tokyo: Tōkyō Daigaku Daigakuin Jinbun Shakai Kakari Kenkyūka 21 Seiki COE Puroguramu, 2004), 15; Tanaka Junichirō, *Nihon kyōiku eiga hattatsushi* (History of the Development of Japanese Education Films) (Tokyo: Kagyūsha, 1979), 49–52; Tochigi Akira, "Daishinsai ga eiga hyōgen ni motarashita mono" (Forms of Expression That the Great Kantō Earthquake Brought to Film), in *Kantō Daishinsai to kiroku eiga,* 35–44.

39. Japanese artists, architects, and designers were well acquainted with the work of the Dutch De Stijl movement in the 1920s through a range of popular contemporary publications.

40. Scott, *Seeing Like a State,* 108–12.

41. Tōkyō Shiyakusho, ed., *Kukaku seiri hayawakari: Teito fukkō no kiso* (Quick Guide to the Land Boundary Readjustment: Basics of the Imperial Capital Reconstruction) (Tokyo: Fukkō Kyoku, 1924), 1, 6, 9, 13. Gotō had realized this type of large-scale rationalized urban planning in Japan's colonies while governor-general of Taiwan, where he oversaw the development of the capital, Taipei; see Joseph R. Allen, "Mapping Taipei: Representation and Ideology," *Studies on Asia* Series III, vol. 2, no. 2 (2005): 71–75. The same planners who proposed the reconstruction plan for Tokyo eventually took their ideas back to the colonies even after Gotō's death, where they envisioned the landscape as a "blank slate" or "white paper" (*hakushi*) on which to build a new "utopian" city. They tried to realize their plans in the redesigned capital of the Japanese puppet state of Manchukuo, Hsinking (Shinkyō in Japanese), formerly Changchun, in Manchuria. See David Tucker, "City Planning without Cities: Order and Chaos in Utopian Manchukuo," in *Crossed Histories: Manchuria in the Age of Empire,* ed. Mariko Tamanoi (Honolulu: Association for Asian Studies / University of Hawaii Press, 2005).

42. Uchiyama and Tōkyō Shisei Chōsakai, eds., *Teito fukkō jigyō taikan,* 1:sec 10, 2–14.

43. Ibid., 1:sec 4, 3. The chart also indicates that by 1925, Tokyo had close to 2 million residents, still half a million short of its prequake population.

44. Ibid., 1:sec 7.

45. Ibid., 1:sec 8, 14.

46. Ibid., 1:sec 4, 27.

47. Vale and Campanella, eds., *The Resilient City,* 10.

48. Yoshimi Shunya, "Taishō bunka kenkyū to modaniti no bunka seiji"; Yamamoto Tadahito, "Kantō Daishinsai kinenbutsu shiryō hōzon katsudō to 'Fukkō Kinenkan'" (Conserving Commemorative Materials from the Great Kantō Earthquake and the "Reconstruction Memorial Museum"), *Shakaigaku zasshi,* no. 23 (2006); Satō Miya, "Media no naka no fukkō" (2007 and 2008); Takano Hiroyasu, "'Shinsai no kioku' no hensen to tenji."

49. Satō Miya, "Media no naka no fukkō" (2008): 31–33.

50. "Shinsai kinenbutsu shūshū" (Gathered Collection of Earthquake Commemorative Items), *Tōkyō asahi shinbun,* 22 February 1924, A.M. ed., 11; "Atsumaru atsumaru sugoi kinen no chinbutsu" (Gathering, Gathering Extraordinary Commemorative Items), *Tōkyō asahi shinbun,* 23 February 1924, A.M. ed., 7.

51. "Kaijō ga wareru yōna daininki no shinsaiten" (The Enormously Popular Earthquake Exhibition Where the Display Space Seemed about to Fracture), *Yomiuri shinbun,* 2 September 1924, A.M. ed., 5; "Kaikai suru Shinsaiten" (Mysterious Earthquake Exhibition), *Tōkyō asahi shinbun,* 1 September 1924, A.M. ed., 3

52. "Kaijō ga wareru yōna daininki no shinsaiten," 5.

53. Collaborating on this exhibition with the Tokyo municipal government and the Tokyo Association for the Earthquake Memorial Project (Tōkyō Shinsai Kinen Jigyō Kyōkai), the Tokyo Institute for Municipal Research dedicated a special issue of its monthly research journal to this exhibition, providing expert interpretations of the show's vision of reconstruction: *Toshi mondai* 10, no. 1 (January 1930). This issue documents the objects exhibited and records visitors' reactions. Responses indicate that viewers had an overwhelmingly positive view of the exhibition and took away strong social and moral messages about the benefits of reconstruction and the imperative to avoid the frivolousness (*keichō fuhaku*) of modern culture. Yamamoto, "Kantō Daishinsai kinenbutsu shiryō hōzon katsudō to 'Fukkō Kinenkan,'" 9–11; Takano, "'Shinsai no kioku' no hensen to tenji," 48–51.

54. "Iyoyo kyō kaikan no teito fukkōten" (Finally Today the Exhibition of the Reconstruction of the Imperial Capital Opens), *Yomiuri shinbun,* 19 October 1929, A.M. ed., 7; "Fukkōten hinobe" (Extension of the Reconstruction Exhibition), *Yomiuri shinbun,* 8 November 1929, A.M. ed., 7; "Kōhyō no fukkōten Ōsaka-shi de kaisai" (Highly Praised Reconstruction Exhibition Opens in Osaka), *Yomiuri shinbun,* 3 November 1929, A.M. ed., 7.

55. The comprehensive catalogue for the exhibition includes floor plans designating the spaces for particular sponsors. These are in the beginning of the catalogue and are unpaginated. Tōkyō Shisei Chōsakai, *Teito Fukkō Tenrankai shuppin mokuroku* (Catalogue of Exhibited Works in the Exhibition of the Reconstruction of the Imperial Capital) (Tokyo: Tōkyō Shisei Chōsakai, 1929), 1.

56. "Bijutsukai: Teito Fukkō no Sōanten" (Art World: Exhibition of Reconstruction of the Imperial Capital), *Yomiuri shinbun,* 22 February 1924, A.M. ed., 6. The eighteen-member jury was an illustrious group that included well-known architects and artists such as Satō Kōichi, Okada

Shinichirō, Chūjō Seiichirō, Asakura Fumio, Okada Saburōsuke, and Ishii Hakutei. The Citizens' Art Association formed in 1913.

57. Both artists are quoted in Satō, "Media no naka no fukkō" (2008): 27–28.

58. The association distributed a catalogue of the exhibited work, but unfortunately no copies appear to have survived. The catalogue's stipulations about customer payment procedures and assignment of a commission of 10 percent of the artist's selling price to the sponsor indicate that all the works were for sale, although there is no documentation showing what items sold.

59. See, for example, Mori Hitoshi, ed., *Taishō-hen* (Taishō Volume), vol. 25 of *Sōsho kindai Nihon no dezain* (Tokyo: Yumani Shobō, 2008).

60. Hijiribashi was completed in 1927. Ken Tadashi Oshima, "Hijiribashi: Spanning Time and Crossing Place," *Review of Japanese Culture and Society,* no. 13 (December 2001): 2, 10–12.

61. Satō, "Media no naka no fukkō" (2008), 28–30. Some of the groups that exhibited were Bunriha Kenchikukai, Sōusha Kenchikukai, Mokuzai Kōgei Gakkai (Wood Craft Society), Teien Kyōkai (Garden Association), Yōfūkai (Rising Wind Association), Sōgō Bijutsu Kyōkai (Synthetic Art Association), Meteorsha (Meteor Group), Ratō (Ratō), and the Barakku Sōshokusha.

62. Chūjō Seiichirō, "Teito Fukkō Sōan Tenrankai ni tsuki" (Concerning the Exhibition of Plans for Reconstruction of the Imperial Capital), *Kokumin bijutsu* 1, no. 5 (1924): 71.

63. Gennifer Weisenfeld, *Mavo: Japanese Artists and the Avant-Garde, 1905–1931* (Berkeley: University of California Press, 2002), 87–91.

64. Kishida Hideto, "Sōanten shokan (kenchiku)" (Impressions of the Exhibition of Plans [for the Reconstruction of the Imperial Capital] [Architecture]), *Kenchiku shinchō* 5, no. 6 (1924): 2.

65. "Yomiuri shō" (Yomiuri Extract), *Yomiuri shinbun,* 23 April 1924, A.M. ed., 5. Hinako (also sometimes read Hinago) was a student of famed sculptor and competition judge Asakura Fumio, who taught at the Tokyo School of Fine Arts. He also participated in the projects of the Barrack Decoration Company. The winning entry is illustrated in "Fukkō sōan tenrankai (Kokumin Bijutsu Kyōkai)" (Exhibition of Plans for the Reconstruction [Citizens' Art Association]), *Kenchiku shinchō* 5, no. 6 (1924): 2.

66. *Kokumin bijutsu* 1, no. 5 (May 1924): 92.

7. REMEMBRANCE

1. Andrew M. Shanken, "Planning Memory: Living Memorials in the United States during World War II," *Art Bulletin* 84, no. 1 (2002): 141.

2. By the tenth day after cremation, the bones were transferred to seventy large earthenware pots and some wooden boxes, which were also often pictured on commemorative postcards. Posthumous Buddhist names are bestowed on the dead as another means of preventing them from inadvertently being called back when their regular names are spoken by the living.

3. Jacqueline Ilyse Stone and Mariko Namba Walter, *Death and the Afterlife in Japanese Buddhism* (Honolulu: University of Hawaii Press, 2008), 9–10.

4. Nagano was a student of the distinguished *nihonga* master Kawakami Gyokudō, and together with the next generation of *nihonga* innovators—Yasuda Yukihiko, Isoda Chōshū, Koyama Eitatsu, and Imamura Shikō—he was a founding member of the celebrated art group the Shikōkai, later renamed the Kōjikai.

5. The hall was renamed the Tokyo Metropolitan Memorial Hall, or Tōkyō-to Ireidō, in 1951 to accommodate commemoration of the Great Fire Bombing of Tokyo on 9–10 March 1945.

6. Both Shinto and Buddhist religious associations brought groups to mourn at the memorial. Tōkyō Shinsai Kinen Jigyō Kyōkai Seisan Jimusho, *Hifukushōato* (The Site of the Former Army Clothing Depot) (Tokyo: Tokyo Shinsai Kinen Jigyō Kyōkai Seisan Jimusho, 1932), 10–11, 147–52, 313–20.

7. Kitahara Itoko, ed., *Nihon saigaishi* (The History of Disasters in Japan) (Tokyo: Yoshikawa Kōbunkan, 2006), 329–32.

8. "Shinsai kinen kakureta kōrōsha" (Earthquake Commemoration Hidden Heroes), *Tōkyō asahi shinbun,* 4 August 1924, A.M. ed., 7.

9. Tōkyōfu, *Taishō Shinsai biseki* (Excellent Deeds of the Taishō Earthquake) (Tokyo: Tōkyōfu, 1924).

10. Mai Denawa, "Behind the Accounts of the Great Kanto Earthquake of 1923" (2005), Brown University Library Center for Digital Scholarship, http://dl.lib.brown.edu/kanto/denewa.html (accessed 19 June 2007).

11. Yamada Shōji, *Kantō Daishinsaiji no Chōsenjin gyakusatsu: Sono kokka sekinin to minshū sekinin* (The Great Kantō Earthquake and the Massacre of Koreans: National Responsibility and the Responsibility of the People) (Tokyo: Sōshisha, 2003), 217–27.

12. Edward Linenthal, "'The Predicament of Aftermath': Oklahoma City and September 11," in *The Resilient City,* ed. Lawrence Vale and Thomas Campanella (New York: Oxford University Press, 2005), 57.

13. *Asahi Graph* 5, no. 10, 2 September 1925.

14. "Yatto mokei ga dekita" (Finally the Model Is Completed), *Tōkyō asahi shinbun,* 19 August 1926, P.M. ed., 2. Tōkyō Shinsai Kinen Jigyō Kyōkai Seisan Jimusho, *Hifukushōato.* The four primary members of the committee and their various advisors are pictured in the unpaginated opening pages of the final report. Gotō, by that time deceased, was pictured as number one. The others were Nagata, Shibusawa (deceased), Sakatani, Horikiri Zenjirō (politician and former mayor of Tokyo), Nishikubō Hiromichi (bureaucrat and former mayor of Tokyo), Ichiki Otohiko (former minister of finance, head of Nippon Ginkō, and former mayor of Tokyo), Izawa Takio (politician and former mayor of Tokyo), Nakamura Yoshikoto (deceased, politician, businessman, president of the South Manchurian Railway Company, and former mayor of Tokyo).

15. Tōkyō Shinsai Kinen Jigyō Kyōkai Seisan Jimusho, *Hifukushōato,* 92–93.

16. "Shinsai Kinendō no sekkeizu" (Design Plans of the Earthquake Memorial Hall), *Tōkyō asahi shinbun,* 17 March 1925, P.M. ed., 2. Tōkyō Shinsai Kinen Jigyō Kyōkai Seisan Jimusho, *Hifukushōato,* 95–97.

17. Hinako traveled to Europe in 1927 and visited many European war memorials, collecting commemorative postcards. It is clear that the 1913 *Monument to the Battle of the Nations* (*Völkerschlachtdenkmal*) in Leipzig, Germany, that commemorated the Battle of Leipzig of 1813 had a strong stylistic influence on him. His design is pictured in *Kenchiku shinchō* 5, no. 5, 1924, n.p. Hirota Toshikazu, *Shōwa shoki chōkoku no kisai: Hinago Jitsuzō no sekai* (The World of Hinago Jitsuzō: The Remarkable Talent of an Early Shōwa-Period Sculptor) (Kyoto: Shibunkaku Shuppan, 2008), 34, 58, 128–31.

18. Maeda Kenjirō's nickname was "*ken Ken*" (literally, competition Ken). Shiseido Arusu Māre Kikakushitsu, Shiseido Gyararī, and Ginza Āto Supēsu, *Ginza modan to toshi ishō: Dai 3-kai Shiseido Gyararī to sono ātisutotachi: Kon Wajirō, Maeda Kenjirō, Yamawaki Iwao Michiko, Yamaguchi Bunzō* (Modern Ginza and Urban Design: The Third Shiseido Gallery and Its Artists: Kon Wajirō, Maeda Kenjirō, Yamawaki Iwao Michiko, Yamaguchi Bunzō) (Tokyo: Shiseido Gallery, 1993). Second prize was awarded to Ozawa Hiroshi, and three others shared third prize: Katō Kanichi, Ōkura Saburō, and Aiga Kensuke. "Shinsai Kinendō no sekkeizu."

19. Tōkyō Shinsai Kinen Jigyō Kyōkai Seisan Jimusho, *Hifukushōato,* 98–99. Pictures of all the competition design submissions appear in Tōkyō Daishinsai Kinen Jigyō Kyōkai, ed., *Taishō Daishinsai kinen kenzōbutsu: Kyōgi sekkei zushū* (Memorial Structures of the Great Taishō Earthquake: Collection of Design Plans from the Competition) (Tokyo: Kōyōsha, 1925).

20. "Yatto mokei ga dekita" (Finally the Model Is Completed), *Tōkyō asahi shinbun,* 19 August 1926, A.M. ed., 2; Tōkyō Shinsai Kinen Jigyō Kyōkai Seisan Jimusho, *Hifukushōato,* 98–102; Tōkyō Daishinsai Kinen Jigyō Kyōkai, ed., *Taishō Daishinsai kinen kenzōbutsu.*

21. "Shinsai Kinendō sekkei tenkan" (Design Change for the Earthquake Memorial Hall), *Tōkyō asahi shinbun,* 19 March 1925, A.M. ed., 10.

22. For a general discussion of Sano's architectural philosophy, see Gregory K. Clancey, *Earthquake Nation: The Cultural Politics of Japanese Seismicity, 1868–1930* (Berkeley: University of California Press, 2006), 212–23.

23. Like many of his colleagues who were unable to fully realize their urban planning visions in postquake Tokyo, Sano was able to implement rationalized city planning and widespread earthquake-resistant construction in Japan's colonial cities, such as Hsinking (Xinjing in Chinese pinyin transliteration; formerly, Changchun), the capital of Manchukuo, Japan's colonial puppet state in North China. Qinghua Guo, "Changchun: Unfinished Capital Planning of Manzhouguo, 1932–42," *Urban History* 31, no. 1 (2004): 103–5.

24. "Bukkyōto kara sanjūman en" (300,000 Yen from the Buddhist Groups), *Tōkyō asahi shinbun,* 4 February 1926, A.M. ed., 6.

25. "Sen sanbyakunin no bōsan ga kyō kara shichū o takuhatsu" (One Thousand Three Hundred Monks Today Walk through the City Asking for Alms), *Tōkyō asahi shinbun,* 17 March 1926, A.M. ed., 10.

26. "'Shinsai kyōdō kikinkai' no seiritsu" (Founding of the "Earthquake Collaborative Fund-Raising Association"), *Tōkyō asahi shinbun,* 30 June 1926, A.M. ed., 3; "Yatto mokei ga dekita"; "Shinsai kyōdō kikin" (Earthquake Collaborative Fund-Raising Association), *Tōkyō asahi shinbun,* 26 August 1926, P.M. ed., 6.

27. "Shinsai Kinendō shinsekkei naru" (A New Design for the Earthquake Memorial Hall), *Tōkyō asahi shinbun,* 4 December 1926, A.M. ed., 6.

28. Tōkyō Shinsai Kinen Jigyō Kyōkai Seisan Jimusho, *Hifukushōato,* 102–5.

29. Toshio Watanabe, "Japanese Imperial Architecture: From Thomas Ranger Smith to Itō Chūta," in *Challenging Past and Present: The Metamorphosis of Nineteenth-Century Japanese Art,* ed. Ellen P. Conant (Honolulu: University of Hawaii Press, 2006), 243–50. Cherie Wendelken, "Pan-Asianism and the Pure Japanese Thing: Japanese Identity and Architecture in the Late 1930s," *positions: east asia cultures critique* 8, no. 3 (2000). For an informative overview of Itō's career, see Suzuki Hiroyuki, *Itō Chūta o shitte imasu ka* (Do You Know Itō Chūta?) (Matsudo-shi: Ōkokusha, 2003).

30. Mayor Nagata used the term *eirei* in his preface to Tōkyō Shinsai Kinen Jigyō Kyōkai Seisan Jimusho, *Hifukushōato,* 2. The bones are contained in over two hundred boxes stacked up in the tower, which has a separate entrance from the main memorial hall to facilitate access to the relics.

31. Linenthal, "'The Predicament of Aftermath' Oklahoma City and September 11," 60.

32. The Allied Buddhist Association (Bukkyō Rengōkai) was put in charge of the religious services, and the Tokyo municipal government took responsibility for guiding the general public through the memorial and for its administration. Tōkyō Shinsai Kinen Jigyō Kyōkai Seisan Jimusho, *Hifukushōato,* 204.

33. Hirochika Nakamaki, "Memorial Monuments of Interrupted Lives in Modern Japan: From *Ex Post Facto* Treatment to Intensification Devices," in *Perspectives on Social Memory in Japan,* ed. Tsu Yun Hui, Jan van Bremen, and Eyal Ben-Ari (Kent, England: Global Oriental, 2005), 44.

34. "Shinsai Kinendō iyoiyo raigetsu kikō su" (Finally Next Month Construction Starts on the Earthquake Memorial Hall), *Tōkyō asahi shinbun,* 5 October 1927, A.M. ed., 11.

35. Koizumi's entire series is illustrated in Koizumi Kishio et al., *Tokyo: The Imperial Capital* (Miami Beach, FL: Wolfsonian, Florida International University, 2003), 61.

36. "Shinsai Kinendō to kazaritsukerareru mayoke" (Decorative Talismans on the Earthquake Memorial Hall), *Tōkyō asahi shinbun,* 5 October 1927, A.M. ed., 11.

37. Japanese gardens had often been safe sanctuaries from disaster in the past. The complex also included small administrative structures for the parks officials, various religious groups, rest areas, and public toilets. Tōkyō Shinsai Kinen Jigyō Kyōkai Seisan Jimusho, *Hifukushōato,* 177–80.

38. Ibid., 202.

39. The official published record of the memorial complex project reproduces the original donation request announcements, forms, and receipts. Since many people were still living in barracks and did not have permanent residences, representatives of the committee went around to each ward to publicize their collection efforts. For a comprehensive, categorized list of the more than two thousand donated objects and their donors, see ibid., 212–303.

40. Ibid., 163–76. However, this monument made no mention of the Japanese massacre of Chinese along with Koreans after the quake, effectively silencing the rage that the Chinese government expressed over the atrocities by folding their deaths into the general memorialization of tragic victims. Yamamoto Tadahito, "Kantō Daishinsai kinenbutsu shiryō hōzon katsudō to 'Fukkō Kinenkan'" (Conserving Commemorative Materials from the Great Kantō Earthquake and the "Reconstruction Memorial Museum"), *Shakaigaku zasshi,* no. 23 (2006): 14.

41. Tōkyō Shinsai Kinen Jigyō Kyōkai Seisan Jimusho, *Hifukushōato,* 305–8. After this, Ogura became well known for his memorials to fallen soldiers (*chūkonhi*), the most famous of which was installed at the national Yasukuni Shrine in the capital.

42. The construction was actually completed in August, and the formal completion ceremony was held on the quake anniversary 1 September 1930.

43. For texts of some of the speeches made at the commemoration, see Tokyo Municipal Office, *The Reconstruction of Tokyo* (Tokyo: Tokyo Municipal Office, 1933), 369–85.

44. "Go-junkō iyoiyo asu" (Imperial Procession Finally Tomorrow), *Yomiuri shinbun,* 23 March 1930, A.M. ed., 7; Tōkyō-shi, ed., *Teito fukkōsai shi* (History of the Reconstruction Festival) (Tokyo: Tōkyō-shi, 1932).

45. Amid the general fanfare for the occasion, a small but vocal group of about three hundred

people registered their opposition to the reconstruction festival in a three-day protest. *Yomiuri shinbun* described the protesters as "left wingers," though they were from a range of groups. Their slogan was "Dead set against the Reconstruction Festival" (*Fukkōsai zettai hantai*). Without explaining the protesters' cause, the press merely indicated that they were being investigated by the police. "'Fukkōsai zettai hantai' mikkakan daidemo keikaku" ("Dead Set against the Reconstruction Festival" Plans for a Three-Day Large Demonstration), *Yomiuri shinbun,* 23 March 1930, A.M. ed., 7.

46. The exhibits were opened to the public after the imperial visit. Takano Hiroyasu, "'Shinsai no kioku' no hensen to tenji" (Exhibition and the Vicissitudes of "Earthquake Memory"), *Himoji shiryō kenkyū nenpō,* no. 6 (2010): 52–54.

47. The construction of the separate structure, referred to as the Earthquake Reconstruction Museum (Shinsai Fukkō Hakubutsukan), was announced in "Shinsai fukkō hakubutsukan no kenzō" (Building the Earthquake Reconstruction Museum), *Tōkyō asahi shinbun,* 22 June 1930, P.M. ed., 1. Its completion was announced in "Fukkō kinenkan shunkō" (Construction Completed on the Reconstruction Memorial Museum), *Yomiuri shinbun,* 15 August 1931, A.M. ed., 7.

48. Nagata Hidejirō, preface to Tōkyō Shinsai Kinen Jigyō Kyōkai Seisan Jimusho, *Hifukushōato,* 1.

49. Ibid., 12. Yamamoto Tadahito discusses the process of constructing a "public discourse" or shared understanding (*kōron*) of the earthquake in the public sphere after the disaster and emphasizes that diverse voices collectively produced a "memory legacy" (*kioku no keishō*). Yamamoto was one of the first scholars to highlight the critical role of exhibitions and the memorial complex in publicly solidifying the official narrative and historical legacy of the quake through the purposeful collection and organization of "historical materials" (*rekishi shiryō*). He argues that as a public institution for "social education" (*shakai kyōka*), the memorial hall, particularly the exhibition rooms, emphasized moral and civic lessons for the next generation over mourning for the dead, as officials feared that overemphasizing the tragedy might have a negative impact on the public psyche, jeopardizing reconstruction and economic recovery. He charts the genesis of the reconstruction exhibitions, and later the displays in the Reconstruction Memorial Museum, out of this concern, finding a strong official emphasis on the project of reconstruction. These efforts were also related to upcoming elections, as authorities sought to counteract dips in public confidence in the government and the economy after high-profile bribery scandals in 1928. Yamamoto also notes that this process continues to evolve in response to new historical events and issues, as I will discuss in the epilogue. Yamamoto Tadahito, "Kantō Daishinsai kinenbutsu shiryō hōzon katsudō to 'Fukkō Kinenkan.'"

50. Timothy P. Brown, "Trauma, Museums and the Future of Pedagogy," *Third Text* 18, no. 4 (2004): 255.

51. Linenthal, "'The Predicament of Aftermath' Oklahoma City and September 11," 55.

52. Nakamaki, "Memorial Monuments of Interrupted Lives in Modern Japan," 56.

53. For the original floor plans, see Tōkyō Shinsai Kinen Jigyō Kyōkai Seisan Jimusho, *Hifukushōato,* 207–12. For a discussion of the postwar reorganization of the displays, see Yamamoto Tadahito, "Tōkyō-to Ireidō no genzai" (The Tokyo Metropolitan Hall of Repose Now), *Rekishi hyōron* 616 (2001).

54. This is an excellent source for excavating the memorial's originary holdings. Tōkyō Shinsai Kinen Jigyō Kyōkai Seisan Jimusho, *Hifukushōato.* The committee, which eventually had close to twenty-two thousand members around the country, formally disbanded on 31 August 1931.

55. Yamamoto Tadahito, "Kantō Daishinsai kinenbutsu shiryō hōzon katsudō to 'Fukkō Kinenkan,'" 13; Takano Hiroyasu, "'Shinsai no kioku' no hensen to tenji," 54–56.

56. The Muramatsu Clock Store (Muramatsu Tokeiten) advertised its clock medal as an earthquake commemorative medal (*shinsai kinen metaru*) in the *Tōkyō asahi shinbun,* 21 April 1924, P.M. ed., 2.

57. The literal translation of *yurumu kokoro no neji o make* is "Wind the clock of your slackened spirit."

58. Many artworks like this one were donated to Tokyo city for display at the Reconstruction Memorial Museum after its inauguration.

59. Arishima's painting was originally exhibited across the country in the eighteenth Nikakai art exhibition in 1931, which was in Tokyo from 3 September to 4 October, in Nagoya from 10 to 18 October, and in Osaka from 23 October to 3 November. The records do not indicate when the painting entered the museum collection.

60. A chart identifying the figures in Arishima's painting is reproduced in Arishima Ikuma, *Hitotsu no yogen* (One Prediction), ed. Kōno Toshiro, Takumi Hideo, and Arishima Akiko (Tokyo: Keishōsha, 1979).

61. "Atorie nite: Konshū no sakuhin to seisakuka" (In the Studio: This Week's Works and Artists), *Sandē mainichi* 10, no. 39 (1931); Arishima Ikuma, *Hitotsu no yogen.* Gotō was the founder of the Boy Scouts organization in Japan and was often pictured in his scout-leader uniform.

62. Both the crown prince and the empress purchased works from this show. Willy vande Walle, *Japan & Belgium: Four Centuries of Exchange* (Aichi, Japan: Commissioners-General of the Belgian Government at the Universal Exposition, 2005), 238–39.

63. The figures to the left of Bassompierre include the prominent oil painters Yamashita Shintarō, Yasui Sōtarō, and Fujishima Takeji, and the novelist Shimazaki Tōson. Arishima Ikuma, *Hitotsu no yogen,* n.p.

64. In her memoirs, Arishima's daughter Akiko recounts that her father told her that the equestrian imagery referred to Satomi Ton's short story about Ikuma called "Gaki daishō Ikuma" (Ikuma Head of the Neighborhood Gang), which depicted him as one of the invincible heroes of popular children's stories. Arishima Akiko, "Hasuike no aze" (Banks of the Lotus Pond), in *Arishima Ikumaten: Kindai yōga no senkusha no zenbō,* ed. Kanagawa Kenritsu Kindai Bijutsukan et al. (Tokyo: Kanagawa Kenritsu Kindai Bijutsukan, 1977), n.p.

65. Arishima's experience of the quake and harrowing escape from the city are recounted in "Bibōki" (Notes) in Kaizōsha, *Taishō Daishin kasaishi* (Chronicle of the Great Taishō Earthquake and Fire) (Tokyo: Kaizōsha, 1924), 42–52. His personal reminiscences about Ōsugi appeared in Arishima Ikuma, "Ōsugi Sakae tsuisō kaiko" (Ōsugi Sakae Remembrance and Reminiscences), *Kaizō* 5, no. 11 (1923): 157.

66. Tokyo Municipal Office, *The Reconstruction of Tokyo,* 361.

8. EPILOGUE

Epigraph: Isozaki Arata quoted in Ken Tadashi Oshima, "Paradoxical Processes," in *Arata Isozaki,* ed. Arata Isozaki and Ken Tadashi Oshima (London: Phaidon, 2009), 12.

1. Michael Lucken, "Tōkyō Daikūshū no imēji: Gunyō shashin kara hakai no ikon e" (Images of

the Great Firebombing of Tokyo: From Military Photographs to Icons of Destruction), in *Kioku to rekishi: Nihon ni okeru kako no shikakuka o megutte,* ed. Tano Yasunori (Tokyo: Waseda Daigakuku Aizu Yaichi Kinen Hakubutsukan, 2006).

2. Cary Karacas, "Tokyo from the Fire: War, Occupation, and the Remaking of a Metropolis" (Ph.D. diss., University of California, Berkeley, 2006), 40.

3. National Archives of Japan, poster collection, Fire Bombing Prevention, http://jpimg.digital.archives.go.jp/kouseisai/category/poster/ippanbouku.html (accessed 28 May, 2010).

4. Karacas, "Tokyo from the Fire," 272.

5. Yamamoto Tadahito, "Tōkyō-to Ireidō no genzai" (The Tokyo Metropolitan Hall of Repose Now), *Rekishi hyōron* 616 (2001); Karacas, "Tokyo from the Fire," 264. Lisa Yoneyama, *Hiroshima Traces: Time, Space, and the Dialectics of Memory* (Berkeley: University of California Press, 1999).

6. Yamamoto Tadahito, "Tōkyō-to Ireidō no genzai," 40–41; Karacas, "Tokyo from the Fire," 265. A sculptural memorial to the air raid victims with an underground space housing all of their names designed by Tsuchiya Kimio called *Dwelling of Remembrance* (Kioku no Basho) was erected within the earthquake memorial complex in 2001. The inscription begins, "During the Second World War, repeated air attacks on Tokyo by the American forces starting on April 18, 1942, and continuing till the end of the war on August 15, 1945, inflicted untold damage on the city and cost the lives of a vast number of its people, who were mostly civilians. This monument was erected so that the memory of these air raids and their victims will not fade but live on to remind succeeding generations that today's peace and prosperity was built on the sacrifice of many precious lives. It embodies the profound hope that this peace will be everlasting." For an insightful discussion of the larger debates surrounding the establishment of a peace museum, see Ellen Hammond, "Commemoration Controversies: The War, the Peace, and Democracy in Japan," in *Living with the Bomb: American and Japanese Cultural Conflicts in the Nuclear Age,* ed. Laura Elizabeth Hein and Mark Selden (Armonk, NY: M.E. Sharpe, 1997). See also the online historical archive of air raid information created by Cary Karacas, Bret Fisk, and Eri Tsuji at www.japanairraids.org/ (accessed 20 June 2011).

7. Yoneyama, *Hiroshima Traces,* 13.

8. Carola Hein, "Hiroshima: The Atomic Bomb and Kenzo Tange's Hiroshima Peace Center," in *Out of Ground Zero: Case Studies in Urban Reinvention,* ed. Joan Ockman (Munich: Prestel, 2002). See the pocket watch in the virtual tour at the Hiroshima Peace Memorial website, www.pcf.city.hiroshima.jp/virtual/cgi-bin/museum.cgi?no=0002a&l=e (accessed 15 April 2010).

9. Domon Ken, *Hiroshima* (Tokyo: Kenkosha, 1958).

10. Domon Ken and Tōmatsu Shōmei, *Hiroshima-Nagasaki Document 1961* (Tokyo: Japan Council against A and H Bombs, 1961).

11. See the lunch box in the virtual tour at the Hiroshima Peace Memorial website, www.pcf.city.hiroshima.jp/virtual/VirtualMuseum_e/visit_e/vit_ex_e/vit_ex4_e.html (accessed 15 April 2010).

12. There seems to be little standardization of information about the quake and the massacre in textbooks. One current Japanese high school textbook notes that amid the great chaos after the quake, rumors spread about Koreans causing violence, the government declared martial law and mobilized troops, and vigilante groups formed. Civilians who believed the rumors murdered six thousand Koreans and eight hundred Chinese. The book also says that some members of the special higher police and the regular police got caught up in the chaos and murdered prominent leftists,

in particular during the Amakasu and Kameido Incidents. Yamamoto Hirofumi et al., *Nihon shi: B* (Japanese History: B) (Tokyo: Tōkyō Shoseki, 2004), 318–19. One standard middle-school Japanese history textbook devotes a single paragraph to the earthquake and adds a side note that reads, "In the chaos of the earthquake, rumors circulated that Koreans and socialists were causing disturbances. Many Koreans, Chinese, and socialists were murdered." Tanabe Hiroshi et al., *Atarashii shakai: Rekishi* (New Society: History) (Tokyo: Tokyo Shoseki, 2002), 162. Another standard 2004 middle-school textbook is more direct, explaining that army officers, police, and residents of the city formed self-defense groups (vigilante squads) in response to rumors circulating about Koreans poisoning well water and then murdered thousands of Koreans and hundreds of Chinese. It also presents Igawa Sengai's powerful print image *Nightwatch Vigilante Groups after the Earthquake (near Negishi)* (*Shinsaigo jikeidan [Negishi hōmen]*), 1926. Kodama Kōta et al., *Watashitachi no chūgaku shakai: Rekishiteki bunya* (Our Middle School Society: Historical Field) (Tokyo: Nihon Shoseki Kabushiki Shinsha, 2004), 165, fig. 35.

13. Inscriptions on the Memorial Monument to the Korean Victims of the Great Kantō Earthquake (Kantō Daishinsai Chōsenjin Giseisha Tsuitōhi) commissioned by the Association for Enacting Rites for the Korean Victims of the Great Kantō Earthquake (Kantō Daishinsai Chōsenjin Giseisha Gyōji Jikkō Iinkai), 1973.

14. Haruno Ogasawara, "Living with Natural Disasters: Narratives of the Great Kantō and the Great Hanshin Earthquakes (Ph.D. diss., Northwestern University, 1999), 54–55.

15. Ichinose offers this memory on the National Museum of Japanese History website, www.rekihaku.ac.jp/e-rekihaku/109/index.html (accessed 28 May 2010).

SELECTED BIBLIOGRAPHY

WORKS OF THE PERIOD: FILM

Fukkō no Tōkyō (The Reconstruction of Tokyo). Tokyo, n.d. Silent, 7 minutes.

Fukkō teito shinfonii (Reconstruction of the Imperial Capital Symphony). Tokyo: Tōkyō Shisei Chōsakai, 1929. Silent, 32 minutes.

Kantō Daishinsai Ina Seiichi ban (Great Kantō Earthquake Ina Seiichi Version). Filmed by Ina Seiichi. Tokyo, 1923. Silent, 16 minutes.

Kantō Daishinsai to fukkō (The Great Kantō Earthquake and Reconstruction). Tokyo, n.d. Talkie, 10 minutes. (Tentative title, as identified by film archivists.)

Kantō taishin taika jikkyō (Actual Conditions of the Great Kantō Earthquake and Fire). Filmed by Shirai Shigeru. Sponsored by Monbushō Shakai Kyoikuka. Tokyo: Tōkyō Shinema Shōkai, 1923. Silent, 64 minutes.

Kokusen nite fukkō no teito e (Reconstruction of the Imperial Capital Seen from an Airplane). Filmed by Shirai Shigeru. Tokyo: Monbushō, 1926. Silent, 15 minutes.

Mōka to shikabane no Tōkyō o fumite (Walking amidst the Raging Fires and the Dead of Tokyo). Osaka: Hayakawa Geijutsu Eiga Seisakujo, 1923. Silent, 10 minutes.

Teito Daishinsai Taishō 12-nen 9-gatsu tsuitachi (Great Earthquake of the Imperial Capital, 1 September 1923). Tokyo: [attributed] Monbushō and Nikkatsu, produced 26 September 1923. Silent, 31 minutes.

Teito fukkō (Reconstruction of the Imperial Capital). Directed by Oda Hamatarō. Tokyo: Fukkō Kyoku, 1930. Silent, 107 minutes.

Tōkyō Kantō chihō Daishinsai sangai jikkyō (The Actual Conditions of the Ravages of the Great Tokyo Kantō Region Earthquake). Sasayama: Ōsaka Shinbunsha Katsudō Shashin Sendenbu, 1923. Silent, 26 minutes.

WORKS OF THE PERIOD: PRINT

Arishima Ikuma. *Hitotsu no yogen* (One Prediction). Edited by Kōno Toshiro, Takumi Hideo, and Arishima Akiko. Tokyo: Keishōsha, 1979.

———. "Ōsugi Sakae tsuisō kaiko" (Ōsugi Sakae Remembrance and Reminiscences). *Kaizō* 5, no. 11 (1923): 157.

Asuka Tetsuo. "Shigoto no ato de kanjita koto" (What I Felt After the [Barrack] Job). *Mizue,* no. 226 (1923): 22.

"Atorie kara gairo e" (From the Atelier to the Streets). *Miyako shinbun,* 9 October 1923, A.M. ed., 5.

"Atorie nite: Konshū no sakuhin to seisakuka" (In the Studio: This Week's Works and Artists). *Sandē mainichi* 10, no. 39 (1931).

"Atsumaru atsumaru sugoi kinen no chinbutsu" (Gathering, Gathering Extraordinary Commemorative Items). *Tōkyō asahi shinbun,* 23 February 1924, A.M. ed., 7.

Barakku kenchiku (Barrack Architecture). Kenchiku shashin ruijū, vol. 4. Tokyo: Kōyōsha, 1923.

"Barakku no bijutsuka 3" (Barrack Artists, part 3). *Miyako shinbun,* 29 October 1923, 1.

"Barakku no sōshoku ga mirairashii moyō" (Barrack Decoration Has Futuristic Design). *Jiji shinpō,* 25 October 1923, A.M. ed., 7.

Beard, Mary. "The Supreme Test of the Japanese Nation. *Shūkan asahi* 4, no. 25 (1923): 20.

"Bijutsukai: Teito fukkō no sōanten" (Art World: Exhibition of Reconstruction of the Imperial Capital). *Yomiuri shinbun,* 22 February 1924, A.M. ed., 6.

"Botsubotsu arawareru geijutsuteki no barakku" (The Artistic Barracks That Are Appearing Little by Little). *Miyako shinbun,* 14 October 1923, A.M., 7.

"Bukkyōto kara sanjūman en" (300,000 Yen from the Buddhist Groups). *Tōkyō asahi shinbun,* 4 February 1926, A.M. ed., 6.

"Chinkago yakeato kenbutsu" (Ruin Sightseeing after the Fires Are Extinguished). *Yomiuri shinbun,* 24 January 1891, A.M. ed., 3.

Chūjō Seiichirō. "Teito fukkō sōan tenrankai ni tsuki" (Concerning the Exhibition of Plans for Reconstruction of the Imperial Capital). *Kokumin bijutsu* 1, no. 5 (1924): 71.

"Coolidge Cables Sympathy to Emperor of Japan: Navy Orders Vessels to Yokohama for Relief." *New York Times,* 2 September 1923, 1.

Dahlmann, Joseph, and Victor Felix Gettelman. *The Great Tokyo Earthquake, September 1, 1923, Experiences and Impressions of an Eye-Witness.* New York: America Press, 1924.

Dai Nihon Yūbenkai Kōdansha. *Taishō Daishinsai daikasai* (The Great Taishō Earthquake and Conflagration). Tokyo: Dai Nihon Yūbenkai Kōdansha, 1923.

"Daikonran no Tōkyō" (The Great Chaos of Tokyo). Special edition, *Ōsaka asahi shinbun,* 4 September 1923.

"Daishinsai zenki" (Full Record of the Great Earthquake). Special edition, *Asahi Graph,* October 1923.

Davison, Charles. *The Japanese Earthquake of 1923.* London: T. Murby, 1931.

"Dead Are in Heaps in Tokio Streets: Path of Destruction Extends for 50 Miles Around the Capital—More Earthshocks Sunday." *New York Times,* 3 September 1923, 1.

"Food Riots in Japanese Capital; Millions Homeless and Starving—Police Are Compelled to Charge Hungry Mobs with Swords—New Reports Increase the Total Havoc Wrought by Earthquake, Tidal Waves and Fires." *New York Times,* 4 September 1923, 1.

"Fukkō kinenkan shunkō" (Construction Completed on the Reconstruction Memorial Hall). *Yomiuri shinbun,* 15 August 1931, A.M. ed., 7.

"'Fukkōsai zettai hantai' mikkakan daidemo keikaku" ("Dead Set against the Reconstruction

Festival" Plans for a Three-Day Large Demonstration). *Yomiuri shinbun,* 23 March 1930, A.M. ed., 7.

"Fukkō sōanten no jushōkō nayamu" (Worrying about Awarding the Prize for the Exhibition of Plans for Reconstruction). *Chūō shinbun,* April 22 1924, A.M. ed., 2.

"Fukkō sōan tenrankai (Kokumin Bijutsu Kyōkai)" (Exhibition of Plans for the Reconstruction [Citizens' Art Association]). *Kenchiku shinchō* 5, no. 6 (1924): 2–5.

"Fukkōten hinobe" (Extension of the Reconstruction Exhibition). *Yomiuri shinbun,* 8 November 1929, A.M. ed., 7.

Fūzoku gahō (Manners and Customs in Pictures) 35, 30 November 1891.

"Gaitō ni deru gakkatachi: Akushonsha to Dōjinsha ga" (Artists Out on the Street: Action and Dōjinsha). *Tōkyō asahi shinbun,* 2 October 1923, P.M. ed., 3.

Gendaishi no Kai. *Dokyumento Kantō Daishinsai* (Documents of the Great Kantō Earthquake). Tokyo: Sōfūkan, 1983.

"Go-junkō iyoiyo asu" (Imperial Procession Finally Tomorrow). *Yomiuri shinbun,* 23 March 1930, A.M. ed., 7.

"Heeded Myth on Disaster." *New York Times,* 5 September 1923.

Hiratsuka Unichi. *Tōkyō Shinsai ato fūkei* (Sights after the Tokyo Earthquake). Kobe: Hanga no Ie, 1924–25.

"Iyoyo kyō kaikan no teito fukkōten" (Finally Today the Exhibition of the Reconstruction of the Imperial Capital Opens). *Yomiuri shinbun,* 19 October 1929, A.M. ed., 7.

"Jap Cabinet Is Balanced: Two Strong Men of Empire in Terauchi's Confidence; Policy toward United States Is Well Received; Declaration of Premier Has Pleased the People." *Los Angeles Times,* 26 November 1916.

Japan Naimushō Shakaikyoku and Fujisawa Morihiko. *The Great Earthquake of 1923 in Japan.* Tokyo: Naimushō Shakaikyoku, 1926.

"Kaijō ga wareru yōna daininki no shinsaiten" (The Enormously Popular Earthquake Exhibition Where the Display Space Seemed about to Fracture). *Yomiuri shinbun,* 2 September 1924, A.M. ed., 5.

"Kaikai suru Shinsaiten" (Mysterious Earthquake Exhibition). *Tōkyō asahi shinbun,* 1 September 1924, A.M. ed., 3.

Kaizōsha. *Taishō Daishin kasaishi* (Chronicle of the Great Taishō Earthquake and Fire). Tokyo: Kaizōsha, 1924.

Kambara Tai. "Barakku Sōshokusha no shimei to watashi no kibō" (The Mission of the Barrack Decoration Company and My Hopes). *Mizue* 225 (1923): 25.

Kantō Daishinsai risai hangashū (Collection of Prints of the Suffering of the Great Kantō Earthquake). Tokyo: Gahōsha, 1923–24.

Kawabata Yasunari. "Taika kenbutsu" (Great Fire Sightseeing). *Bungei shunjū* 1, no. 11 (1923): 66–68.

Kiitsu Kyōkai, ed. *Shinsai ni kansuru shūkyō dōtokuteki kansatsu* (Religious and Moral Observations Related to the Earthquake). Tokyo: Kiitsu Kyōkai, 1925.

"Kinendō ni kazaru Nebukawa kaitei shashin" (Underwater Photographs of Nebukawa That Decorate the Memorial Hall). *Yomiuri shinbun,* 1 September 1933, A.M. ed., 7.

Kishida Hideto. "Sōanten shokan (kenchiku)" (Impressions of the Exhibition of Plans [Architecture]). *Kenchiku shinchō* 5, no. 6 (1924): 2.

"Kōhyō no fukkōten Ōsaka-shi de kaisai" (Highly Praised Reconstruction Exhibition Opens in Osaka). *Yomiuri shinbun,* 3 November 1929, A.M. ed., 7.

Kojima Chūshū. *Shinsai kinen gachō* (Earthquake Disaster Commemorative Sketchbook). Tokyo: Banchō Shoin, March 1924.

———. *Shinsai suketchi: Kinen gachō* (Earthquake Sketches: Commemorative Sketchbook). 2 vols. Tokyo: Banchō Shoin, November 1923 (vol. 1) and December 1923 (vol. 2).

Kon Wajirō. "Poketto no naka no ni en yon jū hachi sen" (The Two Yen and Forty-Eight Sen in My Pocket). *Mizue,* no. 226 (1923): 25–26.

———. "Soboku to iroiro no bi" (Simplicity and Various Aesthetics). *Mizue,* no. 225 (1923): 4–7.

———. "Sōshoku geijutsu no kaimei" (Clarification of Decoration Art). *Kenchiku shinchō* 5, no. 2 (1924): 33–35.

"Kukaku seiri hayawakari" (Quick Guide to Understanding Land Readjustment). *Tōkyō asahi shinbun,* 16 June 1924, A.M. ed., 4.

Kusakabe Shirōta. "Jishin to namazu to kanameishi ni tsuite" (Concerning Earthquakes and Catfish and the Protective Keystone). In *Taishō Daishinsai daikasai,* edited by Dai Nihon Yūbenkai Kōdansha, n.p. Tokyo: Dai Nihon Yūbenkai Kōdansha, 1923.

Miyatake Gaikotsu. *Hentai chishiki* (Abnormal Knowledge), no. 9 (September 1924).

———. *Meiji kibun* (Meiji Strange News). 6 vols. Tokyo: Hankyōdō, January 1925–June 1926.

———. *Miyatake Gaikotsu chosakushū* (Collected Works of Miyatake Gaikotsu). Edited by Tanizawa Eiichi and Yoshino Takao. 8 vols. Tokyo: Kawade Shobō Shinsha, 1986.

———. "Saisho ni kōkan sareta hakurai no rataiga" (The First Published Nude Paintings Imported from Abroad). *Meiji kibun* (Tokyo: Hankyōdō, March 1925), 2:25.

———. *Shinsai gahō* (Earthquake Disaster Pictorial). 6 vols. Tokyo: Hankyōdō, September 1923–January 1924.

Morris, Charles, ed. *The San Francisco Calamity by Earthquake and Fire . . . Told by Eye Witnesses, Including Graphic and Reliable Accounts of All Great Earthquakes and Volcanic Eruptions in the World's History, and Scientific Explanations of Their Causes.* Philadelphia: W. E. Scull, 1906.

Naimushō Shakaikyoku. *Taishō Shinsaishi shashinchō* (Photograph Volume of the History of the Taishō Earthquake). Edited by Fujisawa Morihiko. Tokyo: Naimushō Shakaikyoku, 1926.

Naitō Sadako. "Teidai Setorumento o miru" (Looking at the Tokyo Imperial University Settlement). *Fujin no tomo* 18, no. 8 (1924): 96–99.

Nihon Mangakai. *Daishinsai gashū* (Pictorial Collection of the Great Earthquake Disaster). Tokyo: Kanao Bunendō, 1923.

Nishizawa Tekiho, Kawasaki Shōko, Isoda Chōshū, Oda Kanchō, Tamura Seitan, and Kiriya Senrin. *Taishō shinkasai mokuhangashū zen sanjū rokkei* (Collection of Woodblock Prints of the Taishō Earthquake—Thirty-Six Scenes). Tokyo: Gahōsha, 1924.

"Nyūkyōsha sarani seigen, kenbutsunin no kōtsu sogai" (People Entering [the City] Are Restricted Again, Traffic Obstructions by Sightseers). *Ōsaka asahi shinbun,* 15 September 1923, P.M. ed., 2.

Okamoto Ippei. "Fukkō ni arawaretaru yūmoa jūdai" (Ten Humorous Topics Appearing during Reconstruction). *Taiyō* 30, no. 1 (1924): 97–109.

Ōsaka Asahi Shinbunsha. *Daishinsai shashin gahō* (Photographic Pictorial of the Great Earthquake). 3 vols. Osaka: Ōsaka Asahi Shinbunsha, 15 September, 25 September, and 7 October 1923.

Ōsaka Mainichi Shinbunsha. *Kantō shinsai gahō* (Kantō Earthquake Pictorial). Vol. 3. Osaka: Ōsaka Mainichi Shinbunsha; Tokyo: Hatsubaijo Tōkyō Nichinichi Shinbunsha, 28 October 1923.

Ōtsubo Shigechika. "Shigoto no nikki" (Diary of a Job). *Mizue,* no. 226 (December 1923): 24.

"Red Cross Asks $5,000,000." *New York Times,* 5 September 1923, 1.

Reimeisha Henshūbu. *Shinsai gafu: Gaka no me* (Earthquake Sketchbook: The Eye of the Artist). Tokyo: Reimeisha, December 1923.

Rekishi Shashin Kai. *Kantō Taishin taika kinengō* (Great Kantō Earthquake Commemorative Issue). Edited by Tada Tetsuo. Vols. 123 and 124 of *Rekishi shashin.* Tokyo, February 1923 and June 1923.

Risōsha, ed. *Shinsai suketsuchi gashū* (Collection of Sketches of the Earthquake). Tokyo: Gahōsha, December 1923.

Saijō Yaso. *Aa Tōkyō: Sanbun, shishū* (Ah Tokyo! Tales of Tragedy Poetry Collection). Tokyo: Kōransha, 1923.

Saitō Sogan and Hinako Jitsuzō. "Barakku kenbutsu: Kasō no Ginza to Asakusa (6)" (Barrack Sightseeing: The Disguise of Ginza and Asakusa). *Yomiuri shinbun,* 11 March 1924, A.M. ed., 7.

Seki Rokō and Hatori Kozan (illustrator). *Shinsai gojū hakkei* (Fifty-Eight Views of the Earthquake Disaster). Tokyo: Banchō Shoin, July 1924.

"Sen sanbyakunin no bōsan ga kyō kara shichū o takuhatsu" (One Thousand Three Hundred Monks Ask for Alms throughout the City Today). *Tōkyō asahi shinbun,* 17 March 1926, A.M. ed., 10.

"Shinsai fukkō hakubutsukan no kenzō" (Building the Earthquake Reconstruction Museum). *Tōkyō asahi shinbun,* 22 June 1930, P.M. ed., 1.

"Shinsai isshūnen kinen" (One-Year Anniversary of the Great Earthquake). Special issue, *Jiji shinpō,* 1 September 1924.

"Shinsaigo no shinshokugyō: Ude o furū zekkō no kikai" (New Occupations after the Earthquake: They Skillfully Display Their Abilities, the Best Machine). *Chūō shinbun,* 6 March 1924, A.M. ed. "Shinsai kinenbutsu shūshū" (Gathered Collection of Earthquake Commemorative Items). *Tōkyō asahi shinbun,* 22 February 1924, A.M. ed., 11.

"Shinsai Kinendō iyoiyo raigetsu kikō su" (Finally Next Month Construction Starts on the Earthquake Memorial Hall). *Tōkyō asahi shinbun,* 5 October 1927, A.M. ed., 11.

"Shinsai Kinendō no sekkeizu" (Design Plans of the Earthquake Memorial Hall). *Tōkyō asahi shinbun,* 17 March 1925, P.M. ed., 2.

"Shinsai Kinendō sekkei tenkan" (Design Change for the Earthquake Memorial Hall). *Tōkyō asahi shinbun,* 19 March 1925, A.M. ed., 10.

"Shinsai Kinendō shinsekkei naru" (The Earthquake Memorial Hall Gets a New Design). *Tōkyō asahi shinbun,* 4 December 1926, A.M. ed., 6.

"Shinsai Kinendō to kazaritsukerareru mayoke" (Decorative Talismans on the Earthquake Memorial Hall). *Tōkyō asahi shinbun,* 5 October 1927, A.M. ed., 11.

"Shinsai kinen kakureta kōrōsha" (Earthquake Commemoration Hidden Heroes). *Tōkyō asahi shinbun,* 4 August 1924, A.M. ed., 7.

"Shinsai kyōdō kikin" (Earthquake Collaborative Fund-Raising Association). *Tōkyō asahi shinbun,* 26 August 1926, P.M. ed., 6.

"'Shinsai kyōdō kikinkai' no seiritsu" (Founding of the "Earthquake Collaborative Fund-Raising Association"). *Tōkyō asahi shinbun,* 30 June 1926, A.M. ed., 3.

Taishō Shinsai gashū (Collected Prints of the Taishō Earthquake). Tokyo: Zue Kenkyūkai, 1926.

The Tale of the Heike/Heike monogatari. Translated by Hiroshi Kitagawa and Bruce T. Tsuchida. Vol. 1 (Books 1–6). Tokyo: University of Tokyo Press, 1975. Reprint, 1989.

Teito daishinsai gahō (Pictorial Account of the Great Earthquake in the Imperial Capital). Tokyo: Tenshōdō, Shūgadō, Urashimadō, 1923.

Tōkyō Daishinsai Kinen Jigyō Kyōkai, ed. *Taishō Daishinsai kinen kenzōbutsu: Kyōgi sekkei zushū* (Memorial Structures of the Great Taishō Earthquake: Collection of Design Plans from the Competition). Tokyo: Kōyōsha, 1925.

Tōkyōfu. *Taishō Shinsai biseki* (Excellent Deeds of the Taishō Earthquake). Tokyo: Tōkyōfu, 1924.

Tokyo Municipal Office. *The Reconstruction of Tokyo.* Tokyo: Tokyo Municipal Office, 1933.

Tōkyō-shi. *Shinsai kinen bunshū: Tōkyō shiritsu shōgakkō jidō* (Earthquake Commemorative Essay Collection: Children of the Tokyo Municipal Elementary Schools). 6 vols. Edited by Tōkyō-shi Gakumuka. Tokyo: Baifūkan, 1924.

———, ed. *Teito fukkōsai shi* (History of the Reconstruction Festival). Tokyo: Tōkyō-shi, 1932.

Tōkyō Shinsai Kinen Jigyō Kyōkai Seisan Jimusho. *Hifukushōato* (The Site of the Former Army Clothing Depot). Tokyo: Tokyo Shinsai Kinen Jigyō Kyōkai Seisan Jimusho, 1932.

———. *Nihon no kindai o dezain shita senkusha: Seitan 150-shūnen kinen Gotō Shinpei-ten zuroku/Goto Shinpei, Designer of Modern Japan.* Tokyo: Tōkyō Shisei Chōsakai, 2007.

———. *Teito Fukkō Tenrankai shuppin mokuroku* (Catalogue of Exhibited Works in the Exhibition of the Reconstruction of the Imperial Capital). Tokyo: Tōkyō Shisei Chōsakai, 1929.

Tōkyō Shiyakusho, ed. *Kukaku seiri hayawakari: Teito fukkō no kiso* (Quick Guide to the Land Boundary Readjustment: Basics of the Imperial Capital Reconstruction). Tokyo: Fukkō Kyoku, 1924.

———. *Tokyo, Capital of Japan: Reconstruction Work, 1930.* Tokyo: Municipal Office, 1930.

Toshi mondai 10, no. 1 (January 1930).

Tōyama Shizuo. "Oboegaki" (A Note). *Mizue* 226 (1923): 26.

Uchiyama Zenzaburō, ed. (Tōkyō Shisei Chōsakai, dir.). *Teito fukkō jigyō taikan* (General Survey of the Imperial Capital Reconstruction Project). 2 vols. Tokyo: Nihon Tōkei Fukyūkai, 1930.

"Yatto mokei ga dekita" (Finally the Model Is Completed). *Tōkyō asahi shinbun,* 19 August 1926, P.M. ed., 2.

"Yomiuri shō" (Yomiuri Extract). *Yomiuri shinbun,* 23 April 1924, A.M. ed., 5.

Yoshida Kenkichi. "Baraku Tōkyō no kanbanbi" (The Beauty of Signs in the Barracks of Tokyo). *Kenchiku shinchō* 5, no. 1 (1924): 21–25.

———. "Kyatatsu no ue de" (On the Top of the Ladder). *Mizue,* no. 226 (1923): 22.

CONTEMPORARY SOURCES

Allen, J. Michael. "The Price of Identity: The 1923 Kanto Earthquake and Its Aftermath." *Korean Studies* 20 (1996): 64–93.

Allen, Joseph R. "Mapping Taipei: Representation and Ideology." *Studies on Asia,* Series III, vol. 2, no. 2 (Fall 2005): 5–80.

Ambaras, David. "Topographies of Distress." In *Noir Urbanisms,* edited by Gyan Prakash, 187–217. Princeton, NJ: Princeton University Press, 2010.

Apel, Dora. *Imagery of Lynching: Black Men, White Women, and the Mob.* New Brunswick, NJ: Rutgers University Press, 2004.

———. "Review: On Looking: Lynching Photographs and Legacies of Lynching after 9/11. *American Quarterly* 55, no. 3 (2003): 457–78.

Apel, Dora, and Shawn Michelle Smith. *Lynching Photographs.* Defining Moments in American Photography. Berkeley: University of California Press, 2007.

Arishima Akiko. "Hasuike no aze" (Banks of the Lotus Pond). In *Arishima Ikuma-ten: Kindai yōga no senkusha no zenbō,* edited by Kanagawa Kenritsu Kindai Bijutsukan, Kokukakai, Nihon Bijutsukan Kikaku Kyōgikai, and Asahi Shinbunsha, n.p. Tokyo: Kanagawa Kenritsu Kindai Bijutsukan, 1977.

Arishima Ikuma-ten: Kindai yōga no senkusha no zenbō (Exhibition of Works by Arishima Ikuma: The Whole Story of the Pioneer of Western-Style Painting). Edited by Kanagawa Kenritsu Kindai Bijutsukan, Kokukakai, Nihon Bijutsukan Kikaku Kyōgikai, and Asahi Shinbunsha. Tokyo: Kanagawa Kenritsu Kindai Bijutsukan, 1977. Exhibition catalogue.

Art of the Japanese Postcard: The Leonard A. Lauder Collection at the Museum of Fine Arts, Boston. Edited by Anne Nishimura Morse, J. Thomas Rimer, and Kendall H. Brown. Boston: MFA Publications, 2004. Exhibition catalogue.

Asahi Shinbunsha. *Asahi gurafu ni miru Shōwa zenshi (Taishō 12nen)* (Early Shōwa History Seen in *Asahi Graph:* Taishō 12). Vol. 1. Tokyo: Asahi Shinbunsha, 1975.

Azoulay, Ariella. *The Civil Contract of Photography.* New York: Zone Books, 2008.

Barton, Allen H. *Communities in Disaster: A Sociological Analysis of Collective Stress Situations.* Garden City, NY: Doubleday, 1969.

Bates, Peter Alexander. "Fractured Communities: Class and Ethnicity in Representations of the Great Kanto Earthquake." Ph.D. diss., University of Michigan, 2006.

Borland, Janet. "Capitalising on Catastrophe: Reinvigorating the Japanese State with Moral Values through Education Following the 1923 Great Kantō Earthquake." *Modern Asian Studies* 40, no. 4 (2006): 875–907.

———. "Makeshift Schools and Education in the Ruins of Tokyo, 1923." *Japanese Studies* 29, no. 1 (May 2009): 131–43.

———. "Stories of Ideal Japanese Subjects from the Great Kantō Earthquake of 1923." *Japanese Studies* 25, no. 1 (2005): 21–34.

Brown, Timothy P. "Trauma, Museums and the Future of Pedagogy." *Third Text* 18, no. 4 (2004): 247–59.

Busch, Noel Fairchild. *Two Minutes to Noon.* New York: Simon and Schuster, 1962.

Cadava, Eduardo. "Lapsus Imaginis: The Image in Ruins." *October,* no. 96 (2001): 35–60.

———. *Words of Light: Theses on the Photography of History.* Princeton, NJ: Princeton University Press, 1997.

Campany, David. *Photography and Cinema.* London: Reaktion, 2008.

Caprio, Mark. *Japanese Assimilation Policies in Colonial Korea, 1910–1945.* Korean Studies of the Henry M. Jackson School of International Studies. Seattle: University of Washington Press, 2009.

Carlson, Jon D. "Postcards and Propaganda: Cartographic Postcards as Soft News Images of the Russo-Japanese War." *Political Communication* 26, no. 2 (2009): 212–37.

Clancey, Gregory. *Earthquake Nation: The Cultural Politics of Japanese Seismicity, 1868–1930.* Berkeley: University of California Press, 2006.

———. "The Meiji Earthquake: Nature, Nation, and the Ambiguities of Catastrophe." *Modern Asian Studies* 40, no. 4 (2006): 909–51.

Cowie, Elizabeth. "The Spectacle of Actuality." In *Collecting Visible Evidence,* edited by Jane Gaines and Michael Renov, 19–45. Minneapolis: University of Minnesota Press, 1999.

"Daijishin, Ansei no Daijishin to Kantō Daishinsai" (Great Earthquakes, the Great Ansei Earthquake and the Great Kantō Earthquake). Special issue, *Taiyō* 2, 1977.

Degas, Edgar, and Richard Kendall. *Degas by Himself: Drawings, Prints, Paintings, Writings.* Boston: Little, Brown, 1987.

Dery, Mark. "A Terrible Beauty." *Print* 58, no. 1 (2004): 54–59.

Doi Tadao, Morita Takeshi, and Chōnan Minoru, eds. *Hōyaku nippo jisho* (Japanese Translation Daily Dictionary). Tokyo: Iwanami Shoten, 1980.

Domon Ken. *Hiroshima.* Tokyo: Kenkosha, 1958.

Domon Ken and Tōmatsu Shōmei. *Hiroshima-Nagasaki Document 1961.* Tokyo: Japan Council against A and H Bombs, 1961.

Driscoll, Mark. *Absolute Erotic, Absolute Grotesque: The Living, Dead, and Undead in Japan's Imperialism, 1895–1945.* Durham, NC: Duke University Press, 2010.

80-nenme no kioku: Kantō Daishinsai to ima (Memories in the 80th Year: The Great Kantō Earthquake and Now). Edited by Kanagawa Kenritsu Rekishi Hakubutsukan. Yokohama: Kanagawa Kenritsu Rekishi Hakubutsukan, 2003. Exhibition catalogue.

Erikson, Kai. *A New Species of Trouble: Explorations in Disaster, Trauma, and Community.* New York: W. W. Norton, 1994.

———. "Notes on Trauma and Community." In *Trauma: Explorations in Memory,* edited by Cathy Caruth, 183–99. Baltimore: Johns Hopkins University Press, 1995.

Fradkin, Philip L. *The Great Earthquake and Firestorms of 1906: How San Francisco Nearly Destroyed Itself.* Berkeley: University of California Press, 2005.

Fujimori Terunobu. "Ginza no toshi ishō to kenchikukatachi" (The Urban Design of Tokyo and Architects). In *Ginza modan to toshi inshō,* edited by Shiseido Gyararī, 19–26. Tokyo: Shiseido, 1993.

———. "Kon Wajirō to Barakku Sōshokusha" (Kon Wajirō and the Barrack Decoration Company). *Quarterly Column,* no. 88 (1983): 59–64.

Gaines, Jane. "Political Mimesis." In *Collecting Visible Evidence,* edited by Jane Gaines and Michael Renov, 84–102. Minneapolis: University of Minnesota Press, 1999.

Gaines, Jane, and Michael Renov. *Collecting Visible Evidence.* Minneapolis: University of Minnesota Press, 1999.

Garoian, Charles R., and Yvonne Gaudelius. *Spectacle Pedagogy: Art, Politics, and Visual Culture.* Albany: State University of New York Press, 2008.

Gerbig-Fabel, Marco. "Photographic Artefacts of War 1904–1905: The Russo-Japanese War as Transnational Media Event." *European Review of History* 15, no. 6 (2008): 629–42.

Ginza modan to toshi ishō: Dai 3-kai Shiseido Gyararī to sono ātisutotachi: Kon Wajirō, Maeda Kenjirō, Yamawaki Iwao Michiko, Yamaguchi Bunzō (Modern Ginza and Urban Design: The Third Shiseido Gallery and Its Artists: Kon Wajirō, Maeda Kenjirō, Yamawaki Iwao Michiko, Yamaguchi Bunzō). Edited by Shiseido, Arusu Māre Kikakushitsu, Shiseido Gyararī, and Ginza Āto Supēsu. Tokyo: Shiseido Gallery, 1993. Exhibition catalogue.

Guo, Qinghua. "Changchun: Unfinished Capital Planning of Manzhouguo, 1932–42." *Urban History* 31, no. 1 (2004): 100–17.

Hammer, Joshua. *Yokohama Burning: The Deadly 1923 Earthquake and Fire That Helped Forge the Path to World War II.* New York: Free Press, 2006.

Hammond, Ellen. "Commemoration Controversies: The War, the Peace, and Democracy in Japan." In *Living with the Bomb: American and Japanese Cultural Conflicts in the Nuclear Age,* edited by Laura Elizabeth Hein and Mark Selden, 100–21. Armonk, NY: M.E. Sharpe, 1997.

Hanes, Jeffrey. "Urban Planning as an Urban Problem: The Reconstruction of Tokyo after the Great Kanto Earthquake." *Seisaku kagaku* 7, no. 3 (2000): 123–37.

Hanga ni miru Tōkyō fūkei: Kantō Daishinsai kara senzen made (The Tokyo Landscape in Prints: From the Great Kantō Earthquake until before the War). Edited by Ōta Kuritsu Kyōdo Hakubutsukan. Tokyo: Ōta Kuritsu Kyōdo Hakubutsukan, 2002. Exhibition catalogue.

Hanga: Taishō kara Shōwa e: Hikari to kage no modanizumu (Prints: From Taishō to Shōwa: The Light and Shadow of Modernism). Edited by Machida Shiritsu Kokusai Hanga Bijutsukan. Machida: Machida Shiritsu Kokusai Hanga Bijutsukan, 2003. Exhibition catalogue.

Hansen, Gladys C., and Emmet Condon. *Denial of Disaster.* San Francisco: Cameron and Co., 1989.

Hariman, Robert, and John Louis Lucaites. *No Caption Needed: Iconic Photographs, Public Culture, and Liberal Democracy.* Chicago: University of Chicago Press, 2007.

Hatano Masaru and Iimori Akiko. *Kantō Daishinsai to Nichi-Bei gaikō* (The Great Kantō Earthquake and Japanese-American Diplomacy). Tokyo: Sōshisha, 1999.

Hecht, Hermann, Ann Hecht, and British Film Institute. *Pre-Cinema History: An Encyclopaedia and Annotated Bibliography of the Moving Image Before 1896.* London: Bowker Saur, 1993.

Hein, Carola. "Hiroshima: The Atomic Bomb and Kenzo Tange's Hiroshima Peace Center." In *Out of Ground Zero: Case Studies in Urban Reinvention,* edited by Joan Ockman, 63–83. Munich: Prestel / New York: Temple Hoyne Buell Center for the Study of American Architecture, Columbia University, 2002.

———. "Resilient Tokyo: Disaster and Transformation in the Japanese City." In *The Resilient City,* edited by Lawrence Vale and Thomas Campanella, 213–34. New York: Oxford University Press, 2005.

———. "Visionary Plans and Planners." In *Japanese Capitals in Historical Perspective: Place, Power and Memory in Kyoto, Edo and Tokyo,* edited by Nicolas Fiévé and Paul Waley, 309–46. London: Routledge Curzon, 2003.

Hein, Laura Elizabeth, and Mark Selden, eds. *Living with the Bomb: American and Japanese Cultural Conflicts in the Nuclear Age.* Armonk, NY: M.E. Sharpe, 1997.

Hilo Times and Hawai-tō Nihonjin Imin Shiryōkan. *Kantō Daishinsai to Hawai / The Tokyo Earthquake, 1923* (The Great Kantō Earthquake and Hawaii). Hawai-shō Hiro-shi: Hawai-tō Nihonjin Imin Shiryōkan, Hiro Taimusu Shinbunsha, 1980.

Hirota Toshikazu. *Shōwa shoki chōkoku no kisai: Hinago Jitsuzō no sekai* (The World of Hinago Jitsuzō: The Remarkable Talent of an Early Shōwa-Period Sculptor). Kyoto: Shibunkaku Shuppan, 2008.

Hosoma Hiromichi. *Ehagaki no jidai* (The Modernity of Postcards). Tokyo: Seidosha, 2006.

Hughes, Jeffrey. "Destroy & Reclaim: Artists and Disaster Sites." *New Art Examiner* 29, no. 5 (2002): 66–73.

Huyssen, Andreas. *Present Pasts: Urban Palimpsests and the Politics of Memory.* Stanford, CA: Stanford University Press, 2003.

Isozaki, Arata, and Ken Tadashi Oshima, eds. *Arata Isozaki.* London: Phaidon, 2009.

Kanda, Fusae. "Behind the Sensationalism: Images of a Decaying Corpse in Japanese Buddhist Art." *Art Bulletin* 87, no. 1 (2005): 24–49.

Karacas, Cary. "Tokyo from the Fire: War, Occupation, and the Remaking of a Metropolis." Ph.D. diss., University of California, Berkeley, 2006.

Kasza, Gregory James. *The State and the Mass Media in Japan, 1918–1945.* Berkeley: University of California Press, 1988.

Kawazoe Noboru. *Kon Wajirō: Sono kōgengaku* (Kon Wajirō: His Modernology). Vol. 9 of *Minkan nihon gakusha.* Tokyo: Liburopōto, 1987.

Kayahara Hakudō. *Kantō Daishinsai o egaku: Tōto daishinsai kaganroku emaki zen san-kan: Emaki manga kodomo no e tenji zuroku* (Picturing the Great Kantō Earthquake: All Three Scrolls of the Illustrated Handscroll of the Visual Record of the Great Earthquake in the Great Eastern Capital: Exhibition Catalogue of Handscrolls, Cartoons, and Children's Drawings). Edited by Kanagawa Daigaku, Himoji Shiryō Kenkyū Sentā. Yokohama-shi: Kanagawa Daigaku Himoji Shiryō Kenkyū Sentā, 2010.

Kimura Matsuo and Ishii Toshio. *Ehagaki ga kataru Kantō Daishinsai: Ishii Toshio korekushon* (The Story Postcards Tell of the Great Kantō Earthquake: The Ishii Toshio Collection). Tokyo: Tsuge Shobō, 1990.

Kindai hanga ni miru Tōkyō: Utsuriyuku fūkei (The Changing City as Depicted in Modern Woodblock Prints: Tokyo in Transition). Edited by Tōkyō-to Edo Tōkyō Hakubutsukan. Tokyo: Tōkyō-to Edo Tōkyō Hakubutsukan, 1996. Exhibition catalogue.

Kinoshita Naoyuki, Yoshimi Shunya, Ono Hideo, Tōkyō Daigaku Sōgō Kenkyū Hakubutsukan, and Tōkyō Daigaku Shakai Jōhō Kenkyūjo. *Nyūsu no tanjō: Kawaraban to shinbun nishikie no jōhō sekai/The Birth of the News, Visual Media in Nineteenth-Century Japan,* Tōkyō Daigaku korekushon. Tokyo: Tōkyō Daigaku Sōgō Kenkyū Hakubutsukan, 1999.

Kitahara Itoko. *Ansei Daijishin to minshū: Jishin no shakaishi* (The Great Ansei Earthquake and the People: A Social History of Earthquakes). Tokyo: Sanichi Shobō, 1983.

———. *Bandaisan funka: Saii kara saigai no kagaku e* (Volcanic Eruption of Mount Bandai: From Uncommon Disasters to Disaster Science). Tokyo: Yoshikawa Kōbunkan, 1998.

———. "Egakareta Kantō Daishinsai: Emaki, hanga, sobyō" (The Great Kantō Earthquake Pictured: Illustrated Handscrolls, Prints, and Drawings). *Himoji shiryō kenkyū nenpō,* no. 6 (2010): 15–36.

———. *Kantō Daishinsai no shashin (Tōkyō-to Ireidō hokan) ni tsuite* (Concerning Photographs of the Great Kantō Earthquake [The Collection of the Tokyo Ireidō]). Japan: n.p., 2007.

———. *Kinsei saigai jōhōron* (Information Theory of Early Modern Disasters). Tokyo: Hanawa Shobō, 2003.

———, ed. *Nihon saigaishi* (The History of Disasters in Japan). Tokyo: Yoshikawa Kōbunkan, 2006.

———. "Saigai to shashin media: 1894 Shōnai Jishin no kēsu sutadī" (Disaster and Photography: A Case Study of the 1894 Shōnai Earthquake). *Chōsa kenkyū shiryō,* no. 4 (March 2007): 77–126.

Kitazawa Rakuten and Nagasaki Batten. *Rakuten manga shūtaisei* (Survey of Manga Works by Rakuten). 3 vols. Omiya-shi: Kitazawa Rakuten Kenshōkai, 1973.

Kodama Kōta et al. *Watashitachi no chūgaku shakai: Rekishiteki bunya* (Our Middle School Society: Historical Field). Tokyo: Nihon Shoseki Kabushiki Shinsha, 2004.

Köhn, Stephan. "Between Fiction and Non-Fiction: Documentary Literature in the Late Edo Period." In *Written Texts—Visual Texts: Woodblock-Printed Media in Early Modern Japan,* edited by Susanne Formanek and Sepp Linhart, 283–310. Hotei Academic European Studies on Japan. Amsterdam: Hotei, 2005.

Koizumi Kishio, Marianne Lamonaca, James T. Ulak, Frederic Alan Sharf, and Wolfsonian-Florida International University. *Tokyo: The Imperial Capital.* Miami Beach, FL: Wolfsonian, Florida International University, 2003. Exhibition catalogue.

Koshizawa Akira. *Tōkyō no toshi keikaku* (Urban Planning of Tokyo). Tokyo: Iwanami Shoten, 1991.

Lee, Jin-hee. "The Enemy Within: Earthquake, Rumours and Massacre in the Japanese Empire." In *Violence: "Mercurial Gestalt,"* edited by Tobe Levin, 187–211. Amsterdam: Rodopi, 2008.

Linenthal, Edward. "'The Predicament of Aftermath': Oklahoma City and September 11." In *The Resilient City,* edited by Lawrence Vale and Thomas Campanella, 55–74. New York: Oxford University Press, 2005.

———. *The Unfinished Bombing: Oklahoma City in American Memory.* New York: Oxford University Press, 2001.

Lucken, Michael. "Tōkyō Daikūshū no imēji: Gunyō shashin kara hakai no ikon e" (Images of the Great Firebombing of Tokyo: From Military Photographs to Icons of Destruction). In *Kioku to rekishi: Nihon ni okeru kako no shikakuka o megutte,* edited by Tano Yasunori, 54–61. Tokyo: Waseda Daigakuku Aizu Yaichi Kinen Hakubutsukan, 2006.

Mabuchi Miho. "Maruyama Ōkyo pitsu *Nanpuku Zukan*—'Nan' no zu o chūshin ni" (A Study of the "Misfortune" Pictures of the *Nanpuku Zukan* Painted by Maruyama Ōkyo). *Bijutsushi* 47, no. 1 (1997): 65–81.

———. "Maruyama Ōkyo pitsu *Nanpuku Zukan* no fuku no zu ni tsuite" (A Study of the "Fortune" Pictures of the *Nanpuku Zukan* Painted by Maruyama Ōkyo). *Bijutsushi ronsō* 14, no. 3 (1998): 109–31.

———. "Maruyama Ōkyo pitsu *Nanpuku Zukan* to *Kannon Kyō Kannon Kyō-e*" (A Study of the *Nanpuku Zukan* Painted by Maruyama Ōkyo and the *Kannon Sutra* and *Kannon Sutra Paintings*). *Bijutsushi ronsō* 19 (2003): 85–111.

Mack, Edward Thomas. *Manufacturing Modern Japanese Literature: Publishing, Prizes, and the Ascription of Literary Value.* Asia-Pacific: Culture, Politics, and Society. Durham, NC: Duke University Press, 2010.

———. "The Value of Literature: Cultural Authority in Interwar Japan." Ph.D. diss., Harvard University, 2002.

Mainichi Shinbunsha, ed. *Kantō Daishinsai 69-nen* (69th Anniversary of the Great Kantō Earthquake). Mainichi Graph Bessatsu. Osaka: Mainichi Shinbunsha, 1992.

Markus, Andrew. "*Gesaku* Authors and the Ansei Earthquake of 1855." In *Studies in Modern Japanese Literature: Essays and Translations in Honor of Edwin McClellan,* edited by Edwin

McClellan, Dennis C. Washburn, and Alan Tansman, 53–87. Ann Arbor: Center for Japanese Studies, University of Michigan, 1997.

McNutt, Donald J. *Urban Revelations: Images of Ruin in the American City, 1790–1860.* Literary Criticism and Cultural Theory. New York: Routledge, 2006.

Mitchell, Richard H. *Janus-Faced Justice: Political Criminals in Imperial Japan.* Honolulu: University of Hawaii Press, 1992.

Mitchell, W. J. T. "Responses to Mieke Bal's 'Visual Essentialism and the Object of Visual Culture' (2003): The Obscure Object of Visual Culture." *Journal of Visual Culture* 2, no. 2 (2003): 249–52.

———. "Showing Seeing: A Critique of Visual Culture." *Journal of Visual Culture* 1, no. 2 (August 2002): 165–81.

———. *What Do Pictures Want? The Lives and Loves of Images.* Chicago: University of Chicago Press, 2005.

Miyata Noboru and Takada Mamoru. *Namazu-e: Shinsai to Nihon bunka* (Catfish Pictures: Earthquakes and Japanese Culture). Tokyo: Ribun Shuppan, 1995.

Mori Hitoshi, ed. *Taishō-hen* (Taishō Volume). Vol. 25 of *Sōsho kindai Nihon no dezain.* Tokyo: Yumani Shobō, 2008.

Murayama Tomoyoshi and Hon no Izumisha. *Murayama Tomoyoshi gurafikku no shigoto* (The Graphic Art of Murayama Tomoyoshi). Tokyo: Hon no Izumisha, 2001.

Nakamaki, Hirochika. "Memorial Monuments of Interrupted Lives in Modern Japan: From *Ex Post Facto* Treatment to Intensification Devices." In *Perspectives on Social Memory in Japan,* edited by Tsu Yun Hui, Jan van Bremen, and Eyal Ben-Ari, 44–57. Kent: Global Oriental, 2005.

Narita Ryūichi. "Kantō Daishinsai no meta hisutorii no tame ni: Hōdō, aiwa, bidan" (A Meta-history of the Great Kantō Earthquake: Reportage, Tragic Tales, and Heroic Tales). *Shisō* 866 (August 1996).

Nichiren. *Selected Writings of Nichiren.* Translated by Philip Yampolsky and Burton Watson. Translations from the Oriental Classics. New York: Columbia University Press, 1990.

Nichols, Bill. *Introduction to Documentary.* Bloomington: Indiana University Press, 2001.

Nihon chōki tōkei sōran (Survey of Japanese Long-Term Statistics). Vol. 3. Tokyo: Nihon Tōkei Kyōkai, 1987–88.

Nihon kindai yōga no eiga: Okada Saburōsuke (The Splendor of Japanese Modern Western-Style Painting: Okada Saburōsuke). Edited by Saga Kenritsu Bijutsukan. Saga: Saga Kenritsu Bijutsukan, 1993. Exhibition catalogue.

Nihon no hanga (Japanese Prints). Vol. 2, *1911–1920: Kizamareta "ko" no kyōen* (1911–1920: A Banquet of Carved "Individuality"). Edited by Chiba-shi Bijutsukan. Tokyo: Tōkyō Shinbun, 1999. Exhibition catalogue.

Nihon no hanga (Japanese Prints). Vol. 3, *1921–1930: Toshi to onna to hikari to kage to* (1921–1930: The City, Women, Shadow, and Light). Edited by Chiba-shi Bijutsukan. Tokyo: Tōkyō Shinbun, 2001. Exhibition catalogue.

Nihon no manga 300-nen (300 Years of Japanese Cartoons). Edited by Kawasaki-shi Shimin Myūjiamu. Kawasaki-shi: Kawasaki-shi Shimin Myūjiamu, 1996. Exhibition catalogue.

Nihon Shinbun Hakubutsukan, ed. *Daishinsai to hōdō-ten: Kantō daishinsai 80-shūnen kikaku* (Exhibition of the Great Earthquake and the News: Plans for the 80th Anniversary of the

Great Kantō Earthquake). Yokohama-shi: Nihon Shinbun Hakubutsukan, 2003. Exhibition catalogue.

Nolte, Sharon H. *Liberalism in Modern Japan: Ishibashi Tanzan and His Teachers, 1905–1960.* Berkeley: University of California Press, 1987.

Nornes, Markus. *Japanese Documentary Film: The Meiji Era through Hiroshima.* Minneapolis: University of Minnesota Press, 2003.

Nourbakhsh, Illah, Randy Sargent, Anne Wright, Kathryn Cramer, Brian McClendon, and Michael Jones. "Mapping Disaster Zones." *Nature* 439, no. 7078 (2006): 787–88.

Odawarajo. *Hanga ni miru Kantō Daishinsai: Taishō Shinkasai mokuhangashū zen sanjū rokkei* (The Great Kantō Earthquake Seen in Prints: Collection of Woodblock Prints of the Taishō Earthquake Thirty-Six Scenes). Odawara: Odawarajo, 1998. Exhibition catalogue.

Ogasawara, Haruno. "Living with Natural Disasters: Narratives of the Great Kantō and the Great Hanshin Earthquakes." Ph.D. diss., Northwestern University, 1999.

Oguma Eiji. *A Genealogy of "Japanese" Self-Images.* Japanese Society Series. Melbourne: Trans Pacific Press, 2002.

Omuka Toshiharu. *Taishōki shinkō bijutsu undō no kenkyū* (Research on the Taishō-Era Avant-Garde Art Movements). Tokyo: Sukaidoa, 1995.

Oshima, Ken Tadashi. "Hijiribashi: Spanning Time and Crossing Place." *Review of Japanese Culture and Society* 13 (December 2001): 1–21.

———. "Paradoxical Processes." In *Arata Isozaki,* edited by Arata Isozaki and Ken Tadashi Oshima, 10–18. London: Phaidon, 2009.

Ouwehand, Cornells. *Namazu-e and Their Themes: An Interpretative Approach to Some Aspects of Japanese Folk Religion.* Leiden: E. J. Brill, 1964.

Ozaki Masato. "Kiroku ka? Sozai ka? Bijutsuka ni totte no Daishinsai" (Archive? Material? The Great Kantō Earthquake for Artists). *Gekkan bijutsu* 21, no. 5 (1995): 45–48.

Ozawa Takeshi, ed. *Shashin de miru Kantō Daishinsai* (The Great Kantō Earthquake Seen through Photographs). Tokyo: Chikuma Shobō, 2003.

Rozario, Kevin. *The Culture of Calamity: Disaster and the Making of Modern America.* Chicago: University of Chicago Press, 2007.

———. "'Delicious Horrors': Mass Culture, the Red Cross, and the Appeal of Modern American Humanitarianism." *American Quarterly* 55, no. 3 (2003): 417–55.

———. "Making Progress: Disaster Narratives and the Art of Optimism in Modern America." In *The Resilient City,* edited by Lawrence Vale and Thomas Campanella, 27–54. New York: Oxford University Press, 2005.

Ryang, Sonia. "The Great Kanto Earthquake and the Massacre of Koreans in 1923: Notes on Japan's Modern National Sovereignty." *Anthropological Quarterly* 76, no. 4 (2003): 731–48.

———. "The Tongue That Divided Life and Death: The 1923 Tokyo Earthquake and the Massacre of Koreans." *Japan Focus* (3 September 2007), www.japanfocus.org/-Sonia-Ryang/2513.

Saigai jānarizumu: Mukashi-hen (Disaster Journalism: Ancient Times). Rekihaku bukkuretto 21. Sakura-shi: Rekishi Minzoku Hakubutsukan Shinkōkai, 2001.

Sand, Jordan. *House and Home in Modern Japan: Architecture, Domestic Space, and Bourgeois Culture, 1880–1930.* Cambridge, MA: Harvard University Asia Center, 2003.

Sato, Barbara Hamill. *The New Japanese Woman: Modernity, Media, and Women in Interwar Japan.* Durham, NC: Duke University Press, 2003.

Satō Kenji. "Kantō Daishinsaigo ni okeru shakai no henyō" (Social Change after the Great Kantō Earthquake). In *Rekishi saigai to toshi: Kyōto Tōkyō o chūshin ni (Ritsumeikan Daigaku Kanagawa Daigaku 21 Seiki COE Puroguramu Jointo Wākushoppu Hōhokusho),* 81–89. Kanagawa: Kanagawa Daigaku 21 Seiki COE Puroguramu, 2007.

———. "Shinsai to shikaku media" (The Earthquake and Visual Media). In *Kantō Daishinsai to kiroku eiga: Toshi no shi to saisei,* edited by Tōkyō Daigaku Daigakuin Jinbun Shakai Kakari Kenkyūka, 45–72. Tokyo: Tōkyō Daigaku Daigakuin Jinbun Shakai Kakari Kenkyūka 21 Seiki COE Puroguramu, 2004.

Satō Miya. "Media no naka no fukkō: Kantō Daishinsaigo no shakai ishiki to tenrankai" (Reconstruction in the Media: Social Consciousness and Exhibitions of the Great Kantō Earthquake). *Jinmin no rekishigaku,* no. 175 (2007): 36–37 and no. 178 (2008): 24–38.

Schencking, J. Charles. "Catastrophe, Opportunism, Contestation: The Fractured Politics of Reconstructing Tokyo Following the Great Kantō Earthquake of 1923." *Modern Asian Studies* 40, no. 4 (2006): 833–73.

———. "The Great Kanto Earthquake and the Culture of Catastrophe and Reconstruction in 1920s Japan." *Journal of Japanese Studies* 34, no. 2 (2008): 295–331.

———. "1923 Tokyo as a Devastated War and Occupation Zone: The Catastrophe One Confronted in Post Earthquake Japan." *Japanese Studies* 29, no. 1 (2009): 111–29.

Schnapp, Jeffrey T. "Mob Porn." In *Crowds,* edited by Jeffrey T. Schnapp and Matthew Tiews, 1–47. Stanford, CA: Stanford University Press, 2006.

Schnapp, Jeffrey T., and Matthew Tiews. "Introduction: A Book of Crowds." In *Crowds,* edited by Jeffrey T. Schnapp and Matthew Tiews, ix–xvi. Stanford, CA: Stanford University Press, 2006.

Scott, James C. *Seeing Like a State: How Certain Schemes to Improve the Human Condition Have Failed.* Yale Agrarian Studies. New Haven, CT: Yale University Press, 1998.

Seidensticker, Edward. *Low City, High City: Tokyo from Edo to the Earthquake.* Cambridge, MA: Harvard University Press, 1991.

———. *Tokyo Rising: The City Since the Great Earthquake.* New York: Knopf, 1990.

Shanken, Andrew M. "Planning Memory: Living Memorials in the United States during World War II." *Art Bulletin* 84, no. 1 (2002): 130–47.

Sherman, Daniel J. *The Construction of Memory in Interwar France.* Chicago: University of Chicago Press, 1999.

Shibusawa Eiichi to Kantō Daishinsai: Fukkō e no manazashi (Shibusawa Eiichi and the Great Kantō Earthquake: Looking toward Reconstruction). Edited by Zaidan Hōjin Shibusawa Eiichi Kinen Zaidan and Shibusawa Shiryōkan. Tokyo: Shibusawa Shiryōkan, 2010. Exhibition catalogue.

Shiga Hidemi. "The Catfish Underground: Japan's Earthquake Folklore and Popular Responses to Disaster." *Orientations* 37, no. 3 (2006): 77–80.

Shigenobu, Yukihiko. "Through the Eyes of a Good Deed Investigator." In *Frontiers of Social Research: Japan and Beyond,* edited by Akira Furukawa, 158–82. Melbourne: Trans Pacific Press, 2007.

Shimizu Isao. *Jiji manga: Taishōki* (Jiji Manga: Taishō Period). Vol. 6 of *Manga zasshi hakubutsukan*. Tokyo: Kokusho Kankōkai, 1986.

Shinbun Shiryō Raiburarii. *Kantō Daishinsai* (Great Kantō Earthquake). 2 vols. Tokyo: Ōzorasha, 1992.

Shirai Shigeru. *Kamera to jinsei: Shirai Shigeru kaikoroku* (Camera and Life: The Memoirs of Shirai Shigeru). Tokyo: Yuni Tsūshinsha, 1983.

Siegel, Linda. "Synaesthesia and the Paintings of Caspar David Friedrich." *Art Journal* 33, no. 3 (1974): 196–204.

Slater, Don. "Photography and Modern Vision: The Spectacle of 'Natural Magic.'" In *Visual Culture,* edited by Chris Jenks, 218–37. London: Routledge, 1995.

Smits, Gregory. "Shaking Up Japan: Edo Society and the 1855 Catfish Picture Prints." *Journal of Social History* 39, no. 4 (Summer 2006): 1045–78.

Soeda Azenbō. *A Life Adrift: Soeda Azembō, Popular Song and Modern Mass Culture in Japan.* Translated by Michael Lawrence Lewis. Routledge Contemporary Japan Series. New York: Routledge, 2008.

Soeda Tomomichi, Saeki Toshio, and Orimoto Yoshikazu. *Taishō no hayariuta minshū no kidoairaku to tomo ni* (Popular Songs of the Taishō Period) Tokyo: King Record Co., 2008. Sound recording (CD).

Sōga Takaaki. "Taishō makki ni okeru shinkō geijutsu undō no kōsatsu: Zōkei bijutsu to kenchiku no kakawari o megutte" (Thoughts on the New Art Movement of the Late Taishō Period: On the Relationship between the Plastic Arts and Architecture). Master's thesis, Waseda University, 1990.

Sontag, Susan. *Regarding the Pain of Others.* New York: Farrar, Straus and Giroux, 2003.

Sorensen, Andre. *The Making of Urban Japan: Cities and Planning from Edo to the Twenty-First Century.* London: Routledge, 2002.

Stone, Jacqueline Ilyse, and Mariko Namba Walter. *Death and the Afterlife in Japanese Buddhism.* Honolulu: University of Hawaii Press, 2008.

Sturken, Marita. *Tourists of History: Memory, Kitsch, and Consumerism from Oklahoma City to Ground Zero.* Durham, NC: Duke University Press, 2007.

Suzuki Hiroyuki. *Itō Chūta o shitte imasu ka* (Do You Know Itō Chūta?). Matsudo-shi: Ōkokusha, 2003.

Taishō Chic: Japanese Modernity, Nostalgia, and Deco. Edited by Kendall H. Brown and Sharon Minichiello. Honolulu: Honolulu Academy of Arts, 2001. Exhibition catalogue.

"Taishō jidai" (The Taishō Period). Special issue, *Taiyō* 5, 1974.

Takano Hiroyasu. "'Shinsai no kioku' no hensen to tenji" (Exhibition and the Vicissitudes of "Earthquake Memory"). *Himoji shiryō kenkyū nenpō,* no. 6 (2010): 37–75.

Takashina Shūji. "Akademii Jurian to Tokunaga Hitoomi." *Kindai gasetsu* 5 (March 1997): 132–36.

Takehisa Yumeji, Kawamura Karyō, and Yamamura Kōka. *Yumeji to Karyō, Kōka no Kantō Daishinsai rupo* (Yumeji, Karyō, and Kōka's Reportage of the Great Kantō Earthquake). Tokyo: Kuresu Shuppan, 2003.

Tanabe Hiroshi et al. *Atarashii shakai: Rekishi* (New Society: History). Tokyo: Tōkyō Shoseki, 2002.

Taylor, John. *Body Horror: Photojournalism, Catastrophe, and War.* Critical Image. New York: New York University Press, 1998.

Tochigi Akira. "Daishinsai ga eiga hyōgen ni motarashita mono" (Forms of Expression That the Great Kantō Earthquake Brought to Film). In *Kantō Daishinsai to kiroku eiga: Toshi no shi to saisei,* edited by Tōkyō Daigaku Daigakuin Jinbun Shakai Kakari Kenkyūka, 35–44. Tokyo: Tōkyō Daigaku Daigakuin Jinbun Shakai Kakari Kenkyūka 21 Seiki COE Puroguramu, 2004.

Tōkyō Daigaku Daigakuin Jinbun Shakai Kakari Kenkyūka, ed. *Kantō Daishinsai to kiroku eiga: Toshi no shi to saisei* (The Great Kantō Earthquake and Documentary Film: The Death and Regeneration of the City). Tokyo: Tōkyō Daigaku Daigakuin Jinbun Shakai Kakari Kenkyūka 21 Seiki COE Puroguramu, 2004.

Tōkyō-to Edo Tōkyō Hakubutsukan Toshi Rekishi Kenkyūshitsu. *Kantō Daishinsai to Ansei Edo jishin* (The Great Kantō Earthquake and the Ansei Edo Earthquake). Tōkyō-to Edo Tōkyō Hakubutsukan Chōsa Hōkokusho, vol. 10. Tokyo: Tōkyō-to Edo Tōkyō Hakubutsukan: Tōkyō-to Rekishi Bunka Zaidan, 2000.

Tōkyō toshi to geijutsu: 1870–1996 (Tokyo, La ville moderne: visions urbaines d'artistes et d'architectes). Edited by Tōkyō-to Gendai Bijutsukan. Tokyo: Tōkyō-to Gendai Bijutsukan, 1996. Exhibition catalogue.

Tseng, Alice. "Kuroda Seiki's Morning Toilette on Exhibition in Modern Kyoto." *Art Bulletin* 90, no. 3 (2008): 418–41.

Tsuchida Mitsufumi. *Tōkyō kiroku bungaku jiten* (Dictionary of Tokyo Documentary Literature). Tokyo: Kashiwagi Shobō, 1994.

Tucker, David. "City Planning without Cities: Order and Chaos in Utopian Manchukuo." In *Crossed Histories: Manchuria in the Age of Empire,* edited by Mariko Tamanoi, 51–81. Honolulu: Association for Asian Studies / University of Hawaii Press, 2005.

Tufte, Edward R. *Visual Explanations: Images and Quantities, Evidence and Narrative.* Cheshire, CT: Graphics Press, 1997.

Unno Kazutaka. "Maps of Japan Used in Prayer Rites or as Charms." *Imago Mundi,* no. 46 (1994): 65–83.

Vale, Lawrence, and Thomas Campanella, eds. *The Resilient City.* New York: Oxford University Press, 2005.

Vande Walle, Willy. *Japan & Belgium: Four Centuries of Exchange.* Aichi, Japan: Commissioners-General of the Belgian Government at the Universal Exposition, 2005.

Virilio, Paul. *Unknown Quantity.* London: Thames & Hudson; Paris: Fondation Cartier pour l'art contemporain, 2003.

———. *The Vision Machine.* Bloomington: Indiana University Press, 1994.

———. *War and Cinema: The Logistics of Perception.* London: Verso, 1989.

Wada Hirofumi, ed. *Kantō Daishinsai* (The Great Kantō Earthquake). Vol. 26 of *Korekushon modan toshi bunka.* Tokyo: Yumani Shobō, 2004.

Watanabe, Toshio. "Japanese Imperial Architecture: From Thomas Ranger Smith to Itō Chūta." In *Challenging Past and Present: The Metamorphosis of Nineteenth-Century Japanese Art,* edited by Ellen P. Conant, 240–53. Honolulu: University of Hawaii Press, 2006.

Weisenfeld, Gennifer. "Designing after Disaster: Barrack Decoration and the Great Kantō Earthquake." *Japanese Studies* 18, no. 3 (1998): 229–46.

———. "The Expanding Arts of the Interwar Period." In *Since Meiji: Perspectives on the Japanese*

Visual Arts, 1868–2000, edited by J. Thomas Rimer, 66–98. Honolulu: University of Hawaii Press, 2011.

———. "Imaging Calamity: Artists in the Capital after the Great Kantō Earthquake." In *MOBO MOGA/Modern Boy, Modern Girl: Japanese Modern Art 1910–1935,* 125–29. Sydney: Art Gallery of New South Wales, 1998. Published in Japanese as "Omoiegakareru sanka: Kantō Daishinsaigo no teito o meguru gakatachi" (Imaging Calamity: Artists in the Capital after the Great Kantō Earthquake). In *Mobo moga-ten 1910–1935,* edited by Kamakura Museum of Modern Art, 241–45. Kamakura: Museum of Modern Art, 1998. Exhibition catalogue.

———. *Mavo: Japanese Artists and the Avant-Garde, 1905–1931.* Berkeley: University of California Press, 2002.

———. "Saigai to shikaku hyōshō: Kantō Daishinsai no imējigun o megutte" (Disaster and Vision: On the Visual Representations of the Great Kantō Earthquake). In *Kioku to rekishi: Nihon ni okeru kako no shikakuka o megutte,* edited by Tano Yasunori, 42–53. Tokyo: Waseda Daigakuku Aizu Yaichi Kinen Hakubutsukan, 2006.

Wells, Marguerite. "Satire and Constraint in Japanese Culture." In *Understanding Humor in Japan,* edited by Jessica Milner Davis, 194–217. Detroit: Wayne State University Press, 2006.

Wendelken, Cherie. "Pan-Asianism and the Pure Japanese Thing: Japanese Identity and Architecture in the Late 1930s." *positions: east asia cultures critique* 8, no. 3 (2000): 819–28.

Wood, Denis, John Fels, and John Krygier. *Rethinking the Power of Maps.* New York: Guilford Press, 2010.

Woodward, Christopher. *In Ruins.* London: Chatto & Windus, 2001.

Wright, Frank Lloyd, and Bruce Brooks Pfeiffer. *Letters to Architects.* Fresno, CA: Press at California State University, Fresno, 1984.

Yamada Shōji. *Kantō Daishinsaiji no Chōsenjin gyakusatsu: Sono kokka sekinin to minshū sekinin* (The Great Kantō Earthquake and the Massacre of Koreans: National Responsibility and the Responsibility of the People). Tokyo: Sōshisha, 2003.

Yamamoto Hirofumi et al. *Nihon shi: B* (Japanese History: B). Tokyo: Tōkyō Shoseki, 2004.

Yamamoto Tadahito. "Kantō Daishinsai kinenbutsu shiryō hōzon katsudō to 'Fukkō Kinenkan'" (Conserving Commemorative Materials from the Great Kantō Earthquake and the "Reconstruction Memorial Museum"). *Shakaigaku zasshi,* no. 23 (2006).

———. "Tōkyō-to Ireidō no genzai" (The Tokyo Metropolitan Memorial Hall Now). *Rekishi hyōron* 616 (2001): 40–52.

Yoneyama, Lisa. *Hiroshima Traces: Time, Space, and the Dialectics of Memory.* Berkeley: University of California Press, 1999.

Yoneyama, Takau. "The Great Kanto Earthquake and the Response of Insurance Companies: A Historical Lesson on the Impact of a Major Disaster." Tokyo: Hitotsubashi University, 2008.

Yoshimi Shunya. "Taishō bunka kenkyū to modaniti no bunka seiji" (Research on Taishō Culture and the Political Culture of Modernity). In *Kakudaisuru modaniti: 1920–30-nendai,* vol. 6 of *Kindai Nihon no Bunkashi,* 3–61. Tokyo: Iwanami Shoten, 2002.

Zeilinga de Boer, Jelle, and Donald Theodore Sanders. *Earthquakes in Human History: The Far-Reaching Effects of Seismic Disruptions.* Princeton, NJ: Princeton University Press, 2005.

ILLUSTRATIONS

INDEX

Italicized page references indicate illustrations.

ASIA: LOCAL STUDIES/GLOBAL THEMES

Jeffrey N. Wasserstrom, Kären Wigen, and Hue-Tam Ho Tai, Editors

1. *Bicycle Citizens: The Political World of the Japanese Housewife,* by Robin M. LeBlanc
2. *The Nanjing Massacre in History and Historiography,* edited by Joshua A. Fogel
3. *The Country of Memory: Remaking the Past in Late Socialist Vietnam,* by Hue-Tam Ho Tai
4. *Chinese Femininities/Chinese Masculinities: A Reader,* edited by Susan Brownell and Jeffrey N. Wasserstrom
5. *Chinese Visions of Family and State, 1915–1953,* by Susan L. Glosser
6. *An Artistic Exile: A Life of Feng Zikai (1898–1975),* by Geremie R. Barmé
7. *Mapping Early Modern Japan: Space, Place, and Culture in the Tokugawa Period, 1603–1868,* by Marcia Yonemoto
8. *Republican Beijing: The City and Its Histories,* by Madeleine Yue Dong
9. *Hygienic Modernity: Meanings of Health and Disease in Treaty-Port China,* by Ruth Rogaski
10. *Marrow of the Nation: A History of Sport and Physical Culture in Republican China,* by Andrew D. Morris
11. *Vicarious Language: Gender and Linguistic Modernity in Japan,* by Miyako Inoue
12. *Japan in Print: Information and Nation in the Early Modern Period,* by Mary Elizabeth Berry
13. *Millennial Monsters: Japanese Toys and the Global Imagination,* by Anne Allison
14. *After the Massacre: Commemoration and Consolation in Ha My and My Lai,* by Heonik Kwon
15. *Tears from Iron: Cultural Responses to Famine in Nineteenth-Century China,* by Kathryn Edgerton-Tarpley
16. *Speaking to History: The Story of King Goujian in Twentieth-Century China,* by Paul A. Cohen
17. *A Malleable Map: Geographies of Restoration in Central Japan, 1600–1912,* by Kären Wigen
18. *Coming to Terms with the Nation: Ethnic Classification in Modern China,* by Thomas S. Mullaney
19. *Fabricating Consumers: The Sewing Machine in Modern Japan,* by Andrew Gordon
20. *Recreating Japanese Men,* edited by Sabine Frühstück and Anne Walthall
21. *Selling Women: Prostitution, Markets, and the Household in Early Modern Japan,* by Amy Stanley
22. *Imaging Disaster: Tokyo and the Visual Culture of Japan's Great Earthquake of 1923,* by Gennifer Weisenfeld
23. *Taiko Boom: Japanese Drumming in Place and Motion,* by Shawn Bender
24. *Anyuan: Mining China's Revolutionary Tradition,* by Elizabeth J. Perry

DESIGNER: SANDY DROOKER
TEXT: 10.5/15 ADOBE GARAMOND
DISPLAY: UNIVERS
COMPOSITOR: BOOKMATTERS, BERKELEY
PREPRESS: IOCOLOR
PRINTER AND BINDER: QUALIBRE